FIRST INTERNATIONAL CONFERENCE
ON
VISUALIZATION AND INTELLIGENT DESIGN
IN ENGINEERING AND ARCHITECTURE

owledgement is made to S. Menoni and G. Vignati for the use of Photograph 7 on
485, which appears on the front cover of this book.

VISUALIZATION AND INTELLIGENT DESIGN IN ENGINEERING AND ARCHITECTURE

Ack
page

VISUALIZATION AND INTELLIGENT DESIGN IN ENGINEERING AND ARCHITECTURE

Editors:
J.J. Connor
Massachusetts Institute of Technology, Cambridge, USA
S. Hernandez
Universidad de Zaragoza, Zaragoza, Spain
T.K.S. Murthy
Wessex Institute of Technology, Southampton, UK
H. Power
Wessex Institute of Technology, Southampton, UK

Computational Mechanics Publications
Southampton Boston

Co-published with

Elsevier Science Publishers
London New York

CMP

ELSEVIER

J.J. Connor
Department of Civil Engineering
Massachusetts Institute of Technology
Cambridge, MA 02139
USA

T.K.S. Murthy
Associate Director of Extension Programmes
Wessex Institute of Technology
University of Portsmouth
Ashurst Lodge, Ashurst
Southampton SO4 2AA
UK

S. Hernandez
Departamento de Ingenieria Mecanica
Centro Politecnico Superior de Ingenieros
Universidad de Zaragoza
50015 Zaragoza
Spain

H. Power
Wessex Institute of Technology
University of Portsmouth
Ashurst Lodge, Ashurst
Southampton SO4 2AA
UK

Co-published by

Computational Mechanics Publications
Ashurst Lodge, Ashurst, Southampton, UK

Computational Mechanics Publications Ltd
Sole Distributor in the USA and Canada:

Computational Mechanics Inc
25 Bridge Street, Billerica, MA 01821, USA

and

Elsevier Science Publishers Ltd
Crown House, Linton Road, Barking, Essex IG11 8JU, UK

British Library Cataloguing-in-Publication Data

A Catalogue record for this book is available
from the British Library

ISBN 1-85166-837-3 Elsevier Applied Science, London, New York
ISBN 1-85312-227-0 Computational Mechanics Publications, Southampton
ISBN 1-56252-151-9 Computational Mechanics Publications, Boston, USA

Library of Congress Catalog Card Number 92-75800

Printed and bound in Great Britain by Cromwell Press Ltd

PREFACE

In recent years, remarkable advances in computer visualization of objects and physical phenomena have been made. Computer images can now represent physical objects very accurately. By using these capabilities, virtual universes can be generated to represent whole environments for fields as varied as engineering, architecture, and others. These techniques can be enhanced by defining any desired path, creating animation, moving computer views and real world video models, as well as sound track, resulting in multimedia representations. The development of these techniques has been made possible because of the improvements in computer graphics devices, better algorithms and faster processors, which allow workstations and high speed PCs to be suitable platforms for visualization.

These developments have been essential for design techniques which rely on wider visualization as well as other advances such as knowledge based support systems and object bases. The emphasis now is to provide intelligent design systems, as a result of integration of these capabilities and considerable research has recently been carried out in all these topics and their relationship to design methodology.

This book contains the edited version of most of the papers presented at the First International Conference on Visualization and Intelligent Design in Engineering and Architecture (VIDEA) held in Southampton from 28-30 April 1993. The meeting was organized by the Wessex Institute of Technology and co-sponsored by the Flow Visualization and the Information Processing Societies of Japan, as well as the International Society for Boundary Elements (ISBE). The objectives of the meeting as evidenced by the papers published in this book were to discuss advances in visualization and intelligent design and their applications. The conference brought together a series of internationally renowned researchers in computational mathematics and in industrial software as well as scientists, engineers and architects developing practical applications in their various fields.

The papers contained in the book are divided into seven different sections and a number of them are written by many of the foremost scientists working on the topic of Visualization and Intelligent Design. The sections are as follows:

1. Rendering Techniques
2. Animation
3. Virtual Reality
4. Computational Geometry, Imaging and Visualization

5. Fluid Flow Simulation and Visualization
6. Architecture and Urban Planning
7. Engineering Applications

The Editors trust that this book will be useful to all engineers, architects and scientists who wish to be aware of the state-of-the-art in Visualization and Intelligent Design. They are grateful to all authors for their contributions and in particular to the members of the Scientific Advisory Committee who helped during the various stages of the conference organization and editing of the book.

The Editors
April 1993

CONTENTS

SECTION 5: FLUID FLOW SIMULATION AND VISUALIZATION

SECTION 6: ARCHITECTURE AND URBAN PLANNING

SECTION 7: ENGINEERING APPLICATIONS

SECTION 1: RENDERING TECHNIQUES

Invited Paper

The design of a second generation visualization environment

G.J. Edwards

Silicon Graphics Computer Systems, Forum 1, Station Road, Theale, Reading, Berks, RG7 4RA, UK

INTRODUCTION

A number of visualization systems are now in the market and have successfully established the concept of interactive visualization as a tool for gaining insights into large or complex data sets. This paper examines the design requirements for a second-generation visualization environment, which learns from the performance and flexibility limitations of first-generation products. We will then illustrate solutions to these issues as implemented in the IRIS Explorer™ visualization environment from Silicon Graphics.

WHAT IS VISUALIZATION ?

We use the term here in the simplest and most generic sense: Visualization is any process using computer graphics to gain insight and understanding into complex processes and data. This definition covers a wide range of endeavours: from using image processing techniques to analyze satellite photos, or planning radiation treatments using volumetric renderings of 3D CAT scan data, or studying turbulence with geometric renderings of an aircraft in a simulated wind tunnel, or designing drugs with wire-frame or solid molecules.

THE VISUALIZATION CYCLE

The process of visualizing a large data set can be broken down into several phases, characterized by either reducing the data into a more relevant set or manageable size, or transforming it into a visual representation for viewing.

The first step is feature extraction or analysis, usually reducing the volume of data to be processed in further operations. For example, a metereologist might extract the atmospheric pressure at ground level from a full 3D simulation of the atmosphere, or take the gradient of the 3D pressure field to identify storm systems.

The second step is geometric mapping, or the transformation of the reduced data set into geometry for subsequent viewing. For example, a 3D scalar field can be visualized using everything from point primitives, linear contours, polygons, to volumetric data cells. Each representation may reveal different information contained in the original data set.

The final step is rendering, or the conversion of the chosen geometry into images. Image attributes chosen at this stage can include viewpoint and orientation, colours, transparency, texture, surface material properties, and lighting/shading methods. In visualization systems to date, this has been the best understood and implemented part of the process.

An important aspect of the steps above is that they should be highly interactive. The user needs to be able to make different choices at each stage of the processing sequence and assess the results quickly. The best subset and rendition of the data may not be known a priori; the user is truly "exploring" the data set.

WHO ARE THE USERS ?

A visualization environment is primarily for computational scientists and engineers who have data sets, especially large ones, to evaluate and analyze. It is for researchers who want to explore their data and simulations in unique ways, not addressed by existing special–purpose packages. They need to tailor the software to their requirements, yet they do not have the expertise or inclination to write custom programs. Their job is to concentrate on the science, not become experts in graphics programming or user interface design.

These users' work habits and environment present some constraints on the design of a visualization system:

· Their jobs are not routine; they seldom ask the same question twice.

· They allocate very little time to learning a new system and demand some tangible results within minutes of working on that system.

· They may work at different locations throughout the year and collaborate with other researchers on the same project.

· They work in a heterogeneous machine environment.

· They have existing code and data which they need to use in their visualization.

These characteristics lead to a clear set of requirements for a visualization system.

WHAT ARE THE REQUIREMENTS ?

The first requirement is that the software must be easy to use. The means of combining elements to produce an application prototype must be clear and intuitive. This dictates the use of a visual, point–and–click interface with the behaviour, capabilities and options of each element clearly evident from visual cues.

The system must be readily extensible, so that users can add their own processing modules and data import/export modules, while continuing to use all existing facilities.

The system must offer software–compatibility on high–end workstations and compute servers. This means the software must be based on standards, be portable, and run efficiently in a distributed network environment.

There must be multiple levels of interaction with the standard system, from pre–packaged analysis suites for common disciplines, through application prototyping with a large repetoire of building blocks, to interactive modification of the behaviour, input/output conversions and data types of the standard modules to address ad–hoc requirements.

In addition to the above there must clearly be powerful graphics and computational performance on the primary workstation, since interaction is the key to successful prototyping.

FIRST GENERATION SHORTCOMINGS

The currently available visualization products have some limitations associated with pioneering in a new software area. These limitations can be summarized in three major categories: 1) bringing data of many different types into the system, 2) moving data around within the system, and 3) allowing the user to add extensions to customize the system to a specific application.

IMPORTING DATA

The data import problem is essentially the process of bringing data in a foreign format into the visualization software environment. This includes reading any arbitrary datafile on disk whether it is ascii, binary direct, binary sequential access, byte swapped, or IEEE floating point Many users have their own data file formats, specifying, for example, which numbers are geometry, which are temperature, which are pressure, and so forth. The imported data has to be understood as it is read in and converted to a common format or structure for use within the visualization system. The ability to import a broad range of data into the system is crucial for a successful visualization environment.

IRIS Explorer addresses this problem with the Datascribe utility, which allows data import and reformatting using purely visual programming. This is described in more detail later.

MOVING DATA WITHIN THE SYSTEM

Once data is in the system, runtime application building is limited by the existing suite of functional modules and the datatypes by which the modules communicate. An important tradeoff in datatyping is whether to strongly type the system (have a set of restricted datatypes) or to open up the typing for end users to add their own data structures.

The strongly typed system results in more modules that can communicate with each other and so are more reusable. On the other hand if a user–extensible datatyping approach is used, the end user could add a new type but would then be required to write all other modules which need to share information in this datatype. Here flexibility has been gained at the cost of both a higher programming effort and a lower level of module reusability and interchangeability. An object oriented approach can help to minimize these problems, although not eliminate them entirely.

USER EXTENSIBILITY

This problem concerns the set of functionality provided with the system in the form of modules. These modules fall into four broad classes: data import (already discussed), feature extraction/analysis, visual representation or geometric mapping, and rendering and display modules.

There are about ten basic types of rendering styles, and a hundred or more distinct visual representation classes for scientific data, including such types as contour maps, vector plots, and isosurfaces. A comprehensive set of these rendering styles must be provided in a complete system.

A little trickier are the feature extraction and analysis modules. For these there are as many individual modules as there are users of the system. This is because each user has a unique data set and style of data exploration. Since no system could possibly include all the modules necessary to cover each user's needs, the solution is to open the system up for enhancement by the user.

USER INTERFACE AND SCREEN LAYOUT

Other limitations of user extensibility in first generation systems include limited end user customization of the user interface and screen layout. Furthermore, the visual connectivity diagrams are restrictive and unintuitive and are targeted for computer scientists and programmers instead of the real end users – computational scientists and engineers.

EFFICIENT EXECUTION

The final limitation in current systems is due simply to the nature of creating first generation products: putting off the implementation of efficient execution until later revisions. During these subsequent editions, the tradeoffs made early in the architecture design inhibit the efficient operation of the system. This usually

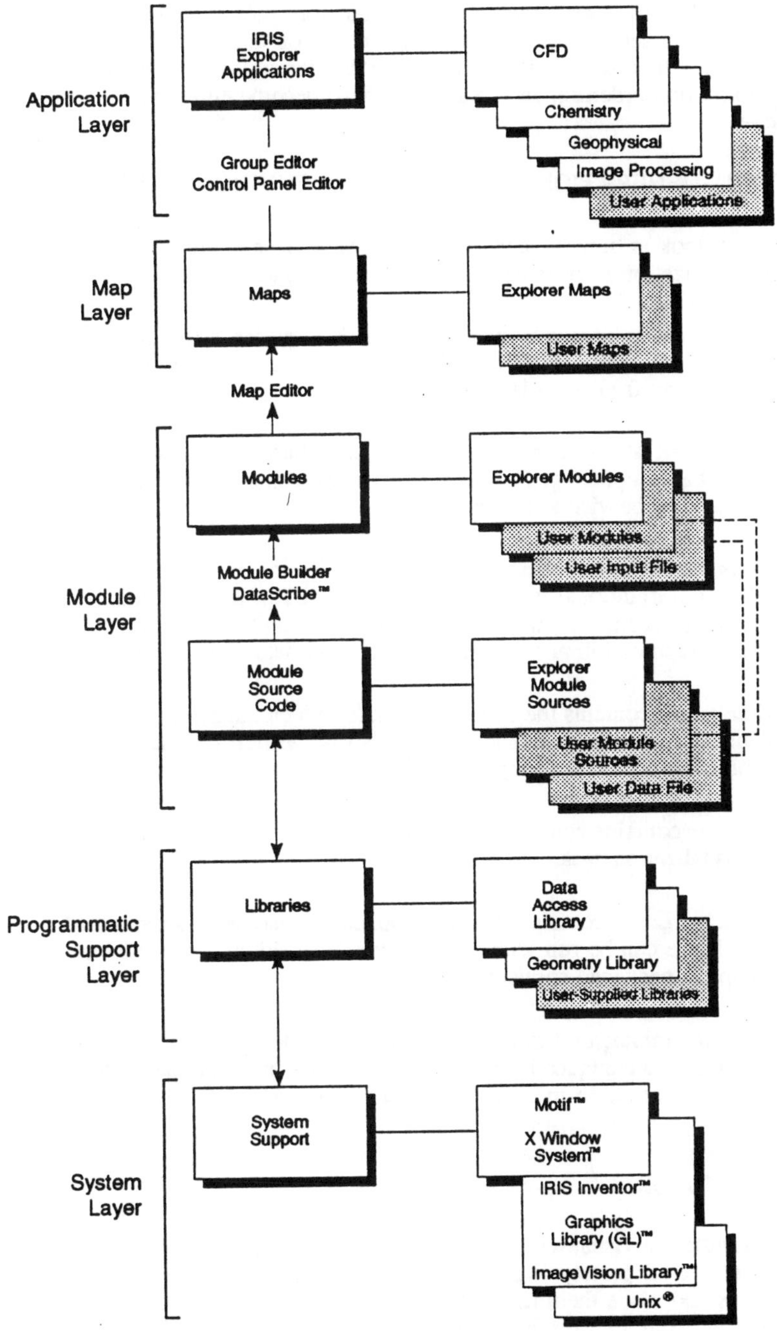

Figure 1. IRIS Explorer system architecture

shows itself when the user makes the transition from prototyping an application with a single data set or image to seeing the effect on tens to thousands of data sets. It is at this time that small inefficiencies in memory organization, data copying, execution ordering, and control overhead become enormous if not prohibitive.

A SECOND–GENERATION VISUALIZATION ENVIRONMENT

We will now look at how the above problems have been addressed in the IRIS Explorer visualization environment from Silicon Graphics.

Figure 1. gives an architectural overview of IRIS Explorer, showing the hierarchical, layered software foundation, and the tools which let users move up the hierarchy away from low–level design.

The *System Layer*, at the lowest level, contains technology and standards which the system is based upon, eg. X, Motif, Unix, networking and rendering libraries; it does not need to be visible to the user.

The *Programmatic Support Layer* contains resources for developing new functionality in a traditional language (eg. C, C++, F77). It includes libraries which shield the user from implementation details of IPC networking, Motif widgets, the Graphics Library, image handling, and data type conversions.

The *Module Layer* contains the suite of modules which handle the majority of standard visualization tasks. Modules may also be written by users to meet specific computation or data–import requirements.

The *Map Layer* contains connected networks of modules which perform a complete visualization task.

The *Application Layer* contains working application networks for standard disciplines. These can be run with the Explorer Map Editor, or standalone to form a complete packaged application.

With the above architectural model in mind, we will look at selected areas where second generation techniques have been incorporated into Explorer; specifically the Map Editor, execution model, system data types, Module Builder and DataScribe.

THE MAP EDITOR

The Map Editor is a visual tool for connecting and managing modules, and provides the overall prototyping environment for Explorer. Users select modules from palettes and place them in the Map Editor work space, where they are represented as icons in various styles, with interactor buttons, sliders and other visual indicators. Users connect these icons, or micro control panels, with mouse actions. Once a map has been connected, it can be saved for later use. In reality however, many simple maps are discarded after use – the process of prototyping can be so intuitive that the result can be regarded as a "disposable network".

Figure 2. shows a small application. The leftmost panel contains a browser palette of module types. The righthand panel contains a module at the top left reading in Plot3D data, other modules processing it, and a small render module at the right. All the elements are using a Motif interface and the windows are under the overall control of an X/Motif window manager. In normal use the render window would be enlarged for more detailed viewing.

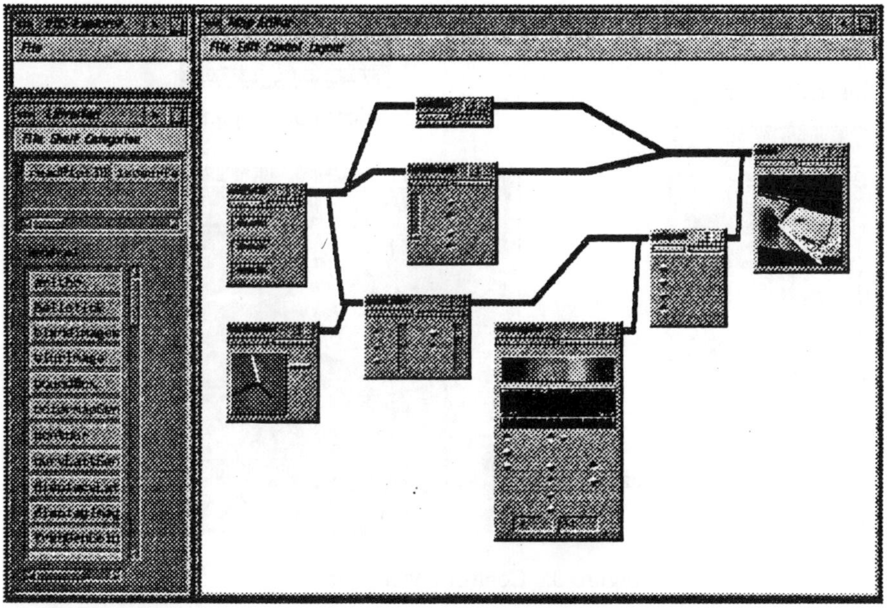

Figure 2. A simple application

The choice of best representation for the processing modules is a challenging one. If the full set of controls for each module is displayed with appropriate labelling, the size of each control panel can be quite significant. In an application with 50 or 100 modules, it will be nearly impossible to sensibly view and manage all the panels, and also impossible to clearly see the topological connections of the network. Explorer offers this module style but it is not the default.

On the other hand, a purely topological view with minimalised abstract icons and no controls occupies the smallest screen space, and keeps the network connections clear, but leaves the user needing an alternative technique to recall or check the module controls. Explorer offers this style as well, but again it is not the default.

Explorer approaches these problems with a unifying default view in which the topological view contains the control view. Each modules icon contains small,

Diminutif™ control widgets which present a fully functional yet concise interface to the module. The module is easily recognised by its unique configuration, without wasting screen real estate on an arbitrary pictorial icon of doubtful meaning to most users. All Diminutif controls are "live"; all sliders, dials, text slots, buttons and so on are fully operational.

Figure 3. shows the three styles for an example module, a contour generator. The user can flip between styles for each module as needed, by using the small buttons on the top right of each representation.

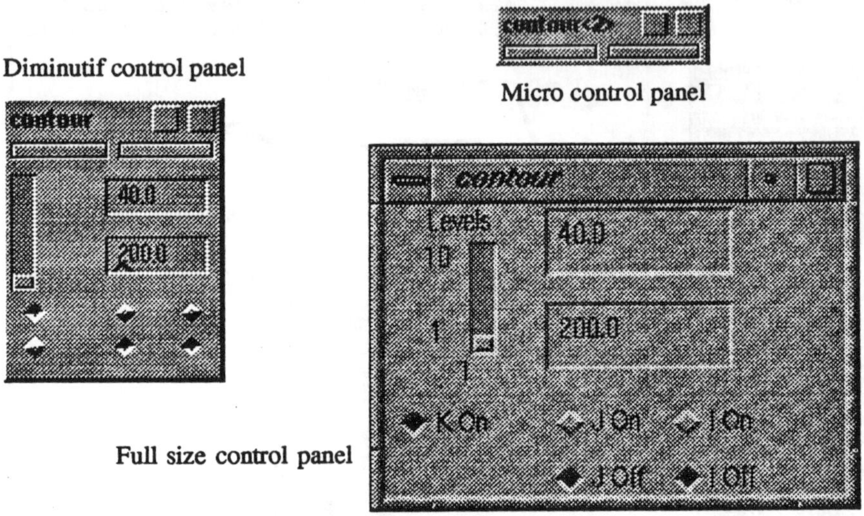

Figure 3. Control panel styles

When there are large numbers of modules in a network, application prototyping has been completed, and the application is expected to have a significant lifetime and/or be used by others, no combination of miniature vs. full views may be satisfactory. Future users (and even the original developer) may not need or have screen space for the full network connection map and all the available controls, most of which are probably not required. Explorer supplies two levels of abstraction for this situation: grouping and standalone applications, or "application–wrapping".

When a number of modules are gathered into a *group*, a new control panel can be created and populated using relevant widgets from any of the group members. Widgets with constant or irrelevant values can be omitted, building a streamlined control panel. A group will typically be a related set of modules whose overall input/output dataflow can be treated as a "black box" in traditional process modelling terms. In many cases all the modules of a map can be combined into one group, giving an application with a single coherent control panel.

A standalone application takes this process one stage further, by allowing a grouped map to run without the Map Editor and other Explorer control windows being present.

Finally, in the process of building an application it is often desirable to drive one control widget by another. Since all widgets are connected by default to parameter ports in the Map Editor, they behave like other data ports. This means that a module's widget output can be wired to another module's widget input so that when the upstream widget is moved, the downstream one also changes.

Often however a literal one–to–one connection will not be appropriate; some massaging needs to be applied to the control value on the way. The Map Editor contains a Parameter Function Interpreter for this purpose. This interpreter can be applied to any input parameter port of a module. Since connections in IRIS Explorer can be fan–in connections (that is, several outputs can be wired to the same input), the expression given to the interpreter can be a function of all fanned–in connections. Standard C single–expression syntax is supported, with all common arithmetic, relational, Boolean and mathematical operators.

This interpreter can also be used to make easy adjustments to basic data connections between modules, without defining new modules or datatypes.

EXECUTION MODEL

The execution of an IRIS Explorer computational map is based upon a distributed, decentralized data flow execution model.

The typical hardware environment for scientific computing consists of workstations from more than one manufacturer, and can include supercomputers and instrumentation as well. IRIS Explorer can execute on a heterogeneous machine network, within the basic constraints of the UNIX operating system, the X Window System, and Motif. Thus, while IRIS Explorer is provided on Silicon Graphics platforms, its designers have anticipated the need for it to run transparently across several machine architectures.

Figure 4. provides an overview of the communications and execution model for a simple network on two machines. Modules A, B, C and D are Unix processes, connected amongst themselves via shared memory locally, and network sockets remotely. A feeds C & D, which both feed B.

Figure 5. shows a larger heterogeneous network and includes the communications controll processes. There is a single copy of a Global Communications Server (GC), responsible for overall management of communication. Changes in control settings or network topology are forwarded to the local communications server (LC) on each machine. The LC is responsible for starting modules, establishing pipes and sockets between them, managing shared memory, and cleaning up after modules terminate.

When a user takes any action in the Graphical User Interface, the GUI sends a

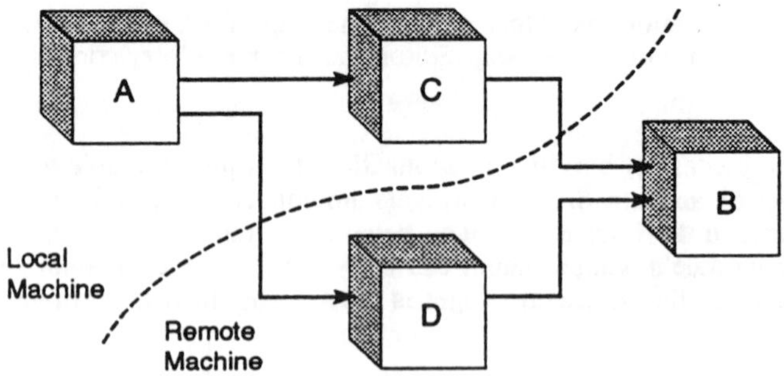

Figure 4. Communications and Execution model

message to the GC, which forwards it to the LC, which delivers it to the module. While this is not the most direct method of communication with the module (and indeed, a highly interactive module such as a renderer can accept input directly from the user), it decentralizes the communication process. IRIS Explorer is designed to optimize data communication bandwidths, and, as such, has several paths to convey information. In practice, it is impossible for a user to generate large amounts of data from the GUI to a module, so the packets from the GUI process through the GC and LC to the module are very small and occur a few times a second, at most.

The amount of data passed between modules, however, can be quite large. To accommodate this, all module–to–module communication is handled without any intermediary processing. When a module produces an IRIS Explorer data object, it sends it directly to those modules that are connected to it downstream, without waking the LC, GC, or GUI, thus avoiding bottlenecks where they can be the most disastrous. This makes it possible to build large networks of IRIS Explorer modules. Even when modules on different machines with different basic data representations, such as a Silicon Graphics IRIS and a Cray Y–MP, are communicating, no intermediary process is necessary; the modules convert their output into a network–neutral format.

Silicon Graphics machines and other machines with shared memory use named pipes and shared memory for data connections, conveying the reference of a data item through a named pipe. The actual data resides in shared memory, where the downstream module can access it without any data motion. This results in a minimum of data copies and memory transfers. Two modules communicating on a non–shared memory machine and modules on different machines transfer data through network sockets.

FIRING DETERMINATION AND DATA TAGGING

Modules fire when the right combination of mandatory and optional inputs is present on their ports. To prevent excessive module firings, data is tagged with information regarding who produced it. The GC constructs a dependency graph and sends it to all modules. In figure 4, when module A fires, it could result in two firings of module B. Data tagging reduces this to one firing. (A central controller which told modules when to execute could also avoid these extraneous firings, but only at the expense of a central bottleneck.) In IRIS Explorer, data tagging and upstream firing activity are piggybacked on the data objects that flow through the connections, to reduce messaging overhead. Because each module has knowledge of the topology of the network upstream, tags on the data allow the modules to avoid redundant firings.

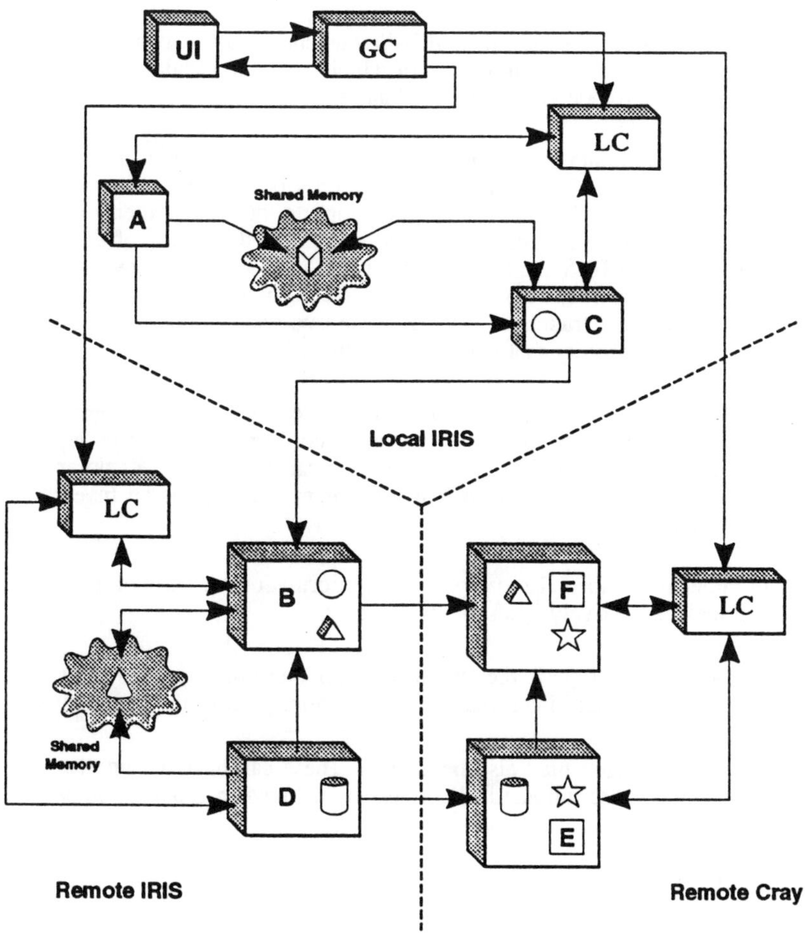

Figure 5. Communications and Execution model

DATA TYPES

Each of the six standard IRIS Explorer data types is highly abstracted, to permit maximum interconnectivity. Each type can represent an entire class of data on one hand and a very specific instantiation of data on the other. For example, the lattice, in its most general form, can represent any multidimensional array, whether it is byte structured, short, integer, floating point, or double precision. A specific instance may represent only a two–dimensional lattice in byte format. In general, the more abstract, or universal, the instance of the data type used, the more possible interconnections can be made with that module.

Parameter: The parameter type is used to communicate scalars between modules and between the Map Editor and a module. Parameters are available in long integer, double precision floating point, and character strings.

Lattice: The lattice is a generalized, multi–dimensional array. It consists of a regularly structured matrix of nodes, each of which contains a scalar or vector of data. In addition, each node can have coordinates.

Pyramid: The pyramid data type combines lattices with connectivity information in a hierarchical structure. The depth of this hierarchy is arbitrary, although it has two generally accepted semantic interpretations. A module written to use a pyramid with one of these interpretations can be connected to many of the other modules provided with IRIS Explorer. Without this interpretation, a pyramid can be built in any manner. These two interpretations are tailored for finite element and chemical molecular structures. Each of these two applications contains data with an inherent connectivity that can be hierarchical in nature.

Geometry: The geometry type is a general, hierarchical, geometrical scene description. The scene graph contains information concerning geometric objects and their attributes, cameras, lights, and viewing transformations. Explorer uses IRIS Inventor as a rendering layer, and the Geometry data type is an Inventor scene description.

Pick: This rather more specialized type carries information on parts of a displayed model interactively "picked" by the user.

Unknown: The last data type is the "unknown" type, an uninterpreted array of bytes. The organization and interpretation is left entirely up to the programmer.

User–Defined Data Types may also be created. These can have a complex internal structure and are automatically transported between modules as with standard system types.

MODULE BUILDER

The Module Builder is an interactive tool that lets a user transform C, C++, F77 or even pure executable code into an IRIS Explorer module. The whole process is entirely visual, using a series of forms and palettes. The steps in creating a module can be summarised as:

Input and Output Ports: A visual palette is used to describe all ports to the module, including data types and any breakdown of complex data types. This information is needed for type checking at module connection time.

Calling Sequence: The actual computational code's calling sequence is defined using another palette. The arguments to the code are often simple scalars, vectors and so on, if the code is a pre–existing computational unit, although they may also be any of the Explorer complex data types.

Connections: While the data types flowing between modules can be quite complicated C–like structures, the arguments to user–supplied functions can be quite simple. Each input/output port is decomposed if necessary into simpler pieces, which are then associated with function arguments. The interface resembles the map Editor in that data elements are visually "wired" from input/output ports to function arguments. Information can also be wired from input ports straight through to output ports.

Control Panel Creation: The Motif widget control panel is created using the Control Panel Editor. Each input port of type parameter (ie. control input) can have a widget associated with it. The type, location, and default values are interactively set with the Editor. The current widgets supported by IRIS Explorer are dials, sliders, text slots, text/file browsers, radio buttons, toggle buttons, and meters, as well as X and GL drawing areas. Figure 6 shows a session from the Control Panel Editor.

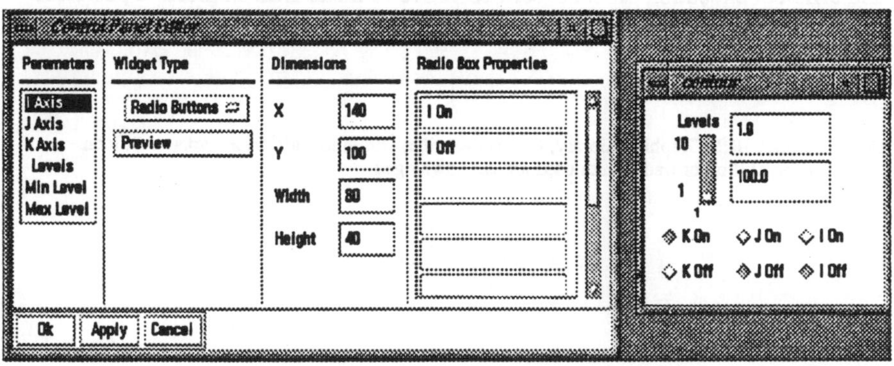

Figure 6. Control Panel Editor

After the module is entirely defined, a build is ordered which compiles and links all the necessary library code to produce an efficient, directly–executed module. Note that during the entire module building phase, the user did not program or specify any information concerning parallelism, distributed execution, machine independent data representation, the X Window System, Motif widgets, IRIS Explorer data structures, makefiles, compilers, or linkers.

DATASCRIBE

In a similar visual, point–and–click fashion to the Module Builder, DataScribe allows outside data formats from files or existing executable applications to be connected into IRIS Explorer.

The data format to be read in (or written out) is built up in a visual template. Not all of the data in a file need be sent through to Explorer, data may be selectively read and regrouped into new data types to match other module inputs. The "inputs" from file data are visually wired through to output ports and the templates created act as a convenient record of the file structure for future reference.

The result of the build is an IRIS Explorer module which can be plugged into the application network in the same way as analysis or render modules. This DataScribe module can read data files or data produced by another executing program, allowing loose integration of third–party applications without access to the source code.

CONCLUSION

We have omitted many features of IRIS Explorer in our focus on some of the second generation design aspects, not the least of which are the 130+ visualization modules provided as standard with Explorer and all their functionality. These however all rely on the comprehensive, intuitive visual tools and efficient execution model described above to make IRIS Explorer a powerful second–generation Visualization Environment.

Invited Paper

The use of visibility coherence for radiosity computation

X. Pueyo

Dept. Llenguatges i Sistemes Informátics, Universitat Politècnica de Catalunya, Av. Diagonal 647 pl.8, E-08028-Barcelona, Spain

ABSTRACT

Coherence has been widely used to improved the performance of visibility computation in traditional hidden line/surface algorithms. The appearance of radiosity algorithms has increased the importance of computing efficiently the visibility function given that it is performed a lot of times to obtain a single static image. We develop, in this paper, a characterization of the uses of coherence in radiosity algorithms and we give a brief and schematic description of proposed algorithms using them.

INTRODUCTION

As well known, radiosity was introduced as a new powerful method to model the behavior of light interacting between diffuse reflecting surfaces. This new technique dealt with a problem of illumination models not studied in depth up until that moment.

Derived from thermal engineering, its application to image synthesis was proposed by Goral et al. [1] with the following conclusions:

(1)

$$B_i = E_i + \rho_i \sum_{j=1}^{N} F_{ij} \, B_j$$

where

B_i= radiosity in surface i

E_i= emissivity of surface i

ρ_i= reflectivity of surface i

F_{ij} = form-factor between surface i and j

N = number of surfaces in the environment

Equation 1 yields a set of N linear equations with N B_i unknowns where E_i and ρ_i are application data, and F_{ij} must be derived from geometry of the environment:

(2)

$$\begin{bmatrix} 1-\rho_1 F_{11} & \cdots & -\rho_1 F_{1N} \\ \vdots & \ddots & \vdots \\ -\rho_1 F_{N1} & \cdots & 1-\rho_N F_{NN} \end{bmatrix} \begin{bmatrix} B_1 \\ \vdots \\ B_N \end{bmatrix} = \begin{bmatrix} E_1 \\ \vdots \\ E_N \end{bmatrix}$$

The computation of the form factor is obtained by,

(3)

$$F_{A_iA_j} = \frac{1}{A_i} \int_{A_i} \int_{A_j} \frac{\cos(\phi_i)\cos(\phi_j)}{\Pi r^2} dA_j dA_i$$

Where A is the area, r is the distance between patches and ϕ is the angle between r and the normal. Instead of solving this double area integral, Goral et al. used Stoke's theorem and solved a contour integral. In fact, the resolution of the integral in equation (3) is the key to the different implementation methods.

The computation of form factors and the solution of equation (2) for a discretization N of the environment, gives the radiosities (B_i) of the patches centre points, which are view independent. Finally, the rendering procedure computes the scene's visible points by z-buffering. The intensities are calculated by linear interpolation of B_i's in object space.

We have seen above that the form-factor is given by the geometric relationship of two surface entities (patch to patch), and represents the ratio of energy leaving one of these entities that lands on the other. In real environment, light leaving a surface and arriving on another depends not only on this relationship, but also on elements in the scene that intercept light and produce shadows. Then the form-factors need only to be computed between surfaces or pieces of surfaces that are visible to each another. So a hidden surface removal process takes place for the computation of each form-factor. Thus, equation (3) must be modified in the following way,

$$F_{A_iA_j} = \frac{1}{A_i} \int_{A_i} \int_{A_j} \frac{\cos(\phi_i)\cos(\phi_j)}{\Pi r^2} h \, dA_j dA_i$$

where h takes the value one or zero depending on whether dA_j is visible from dA_i.

In order to speed up the computation of the intervisibility (i.e. the h function of equation 4) a number of papers have appeared which try to reduce its computational cost by means of exploiting several types of coherence. We develop here a characterization of them, firstly pointed out in [6]. This characterization focuses on radiosity algorithms for static environments with static observers. Finally, we describe schematically some of the algorithms using the different types of coherence.

We do not consider in the present work those algorithms where speed up is obtained by means of minimising the number of patches into which the environment is divided. We deal with algorithms which reduce the cost of computing the form factors of a "previously fixed" number of patches.

COHERENCE IN RADIOSITY. CHARACTERIZATION

We develop, in this section, a characterization of the uses of coherence made in radiosity algorithms in order to efficiently compute form factors. Coherence is often used in these algorithms when computing the h function. So most of them are well known strategies widely used in hidden line/surface removal algorithms. Many of the techniques we are going to explain have already been introduced for form factor computation. Other strategies have not yet been used so we just introduce them as ideas to be more deeply studied and eventually implemented. Whether each of the types of coherence we are going to see is adequate or not depends on a number of parameters. Specially on:

- the "complexity" of the environment: number and location of the geometrical elements of the scene.

- the type of implementation: sequential or parallel, software or hardware.

General concepts

Previous to the description of different types of coherence we consider interesting to underline several concepts we are going to use widely below.

- Geometrical elements. When they evaluate the h function, classical radiosity algorithms focus on the patches. The problem to solve is whether patch i may or not see patch j. This is sometimes solved for elements

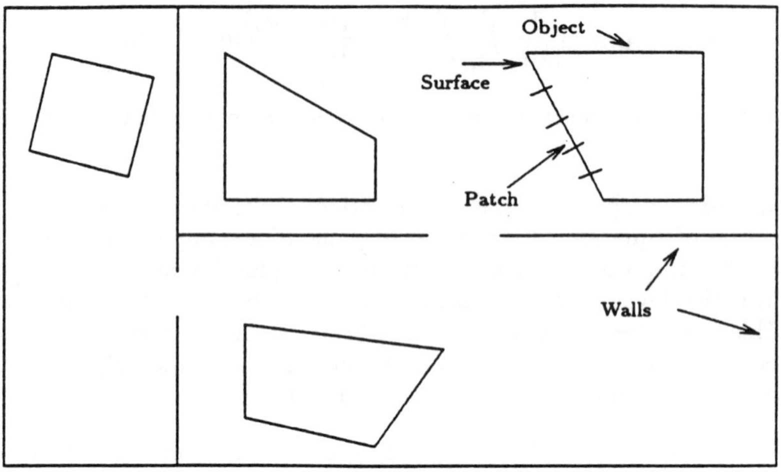

figure 1. Geometrical elements

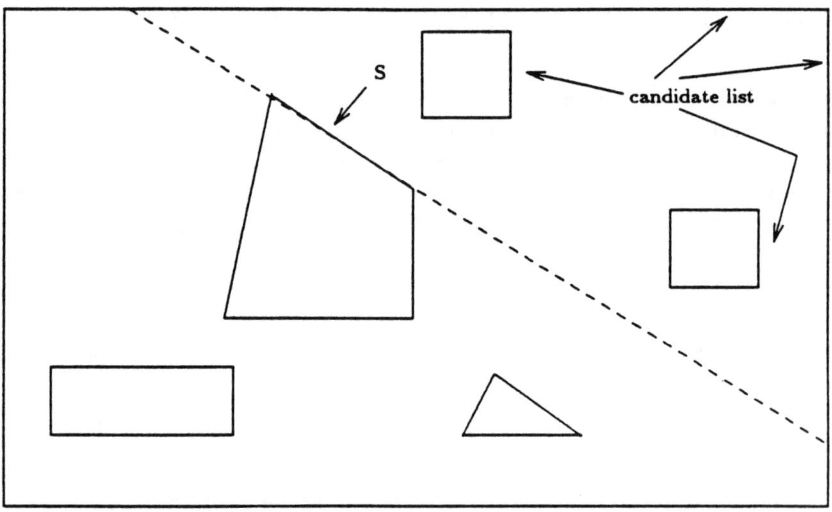

figure 2. Computing the candidate list

which result from the subdivision of patches. Algorithms using coherence to compute form factors often accounts on higher level geometrical elements: surfaces and objects. Finally, we find a special geometrical element, the wall. Sometimes the walls are thought as "big" surfaces, other they are considered as "thin" objects (figure 1).

- Attributes of the geometrical elements. Patches (surfaces,objects) are seen from different scopes in radiosity algorithms; as "observers" (visibility is computed from them) or as "observed". As a consequence of this, a same type of coherence will be usually exploited in a different way depending on the attribute "observer / observed".

- Goals of coherence. In form factor computation the targets of the use of coherence are two: to avoid computing form factors and to improve the performance of the computation of form factors taking advantage of the previously computed ones. Of course the former is the most interesting.

Characterization

Here we try to give a hierarchical view of the use of coherence based on the geometrical hierarchy of the scene elements. An environment is composed by a set of objects which are in turn composed by a set of surfaces (we assume here that surfaces are flat polygons). Finally, surfaces are split up into patches. We assume that lower level subdivision supply geometrical elements with the same behavior as patches; so we do not consider it. On the other hand, we consider that the scene may include a higher level of organization. A certain type of structure can be built in order to organize the elements (object and/or surfaces) of the environment. The success of these techniques in accelerating ray tracing makes us think that it is reasonable to use them.

In an observer-surface the visibility function must be computed for each of his patches. Given one of them, its visibility is presumably very similar to the visibility of his neighbours. Let's take a surface S of the environment. Global clipping, view transformation and depth sort can be, partially or fully, performed for all patches of S in a unique operation. In fact, the plane of S clips the environment providing a candidate set of objects and/or surfaces (figure 2). In this operation the number of form factors to be computed for each patch of S is reduced to the candidate list and this is done once for all the patches of S instead of repeating the operation for each patch. The elements of the candidate list (objects and/surfaces) may be presorted, independently of the specific "observer" (patch) of S, in order to store depth priority information valid for all the patches of the surface. These procedures concerning all the patches of S may be helped using a suitable organization of the environment. So structuring it using well known tools like hierarchies of bounding boxes, octtrees, BSP trees.

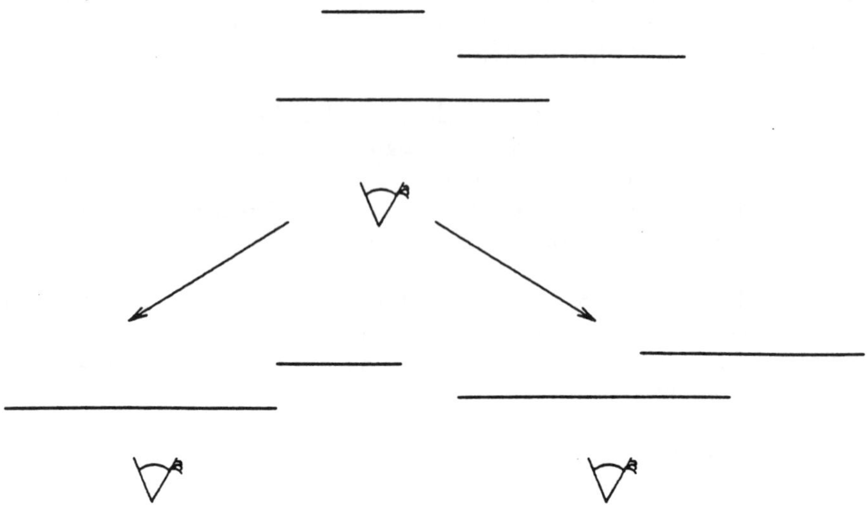

a) full computation

b) partial computation

figure 3. Visibility computation at surface level

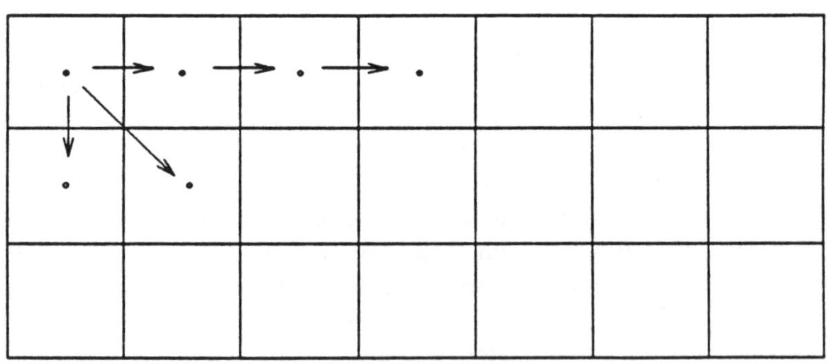

observer-surface

figure 4. Moving from patch to patch

If we focus on an observer-patch P of S, something similar happens. Now we must solve visibility for several frustrums (five if we use the hemicube technique). For each of them, visibility will probably be very similar to the visibility of his neighbours. So we may take advantage of this in order to reduce the computational cost. For example, to clip a surface S may be done once for two consecutive frustrums. Visibility relationships may also be transmitted from a frustum to its neighbours.

Once the list of observed-surfaces (or objects or patches) has been built, the algorithm must compute the visibility from each patch P of S; or, more accurately, from each frustum F of P. We recall that the list may or not contain depth information. Anyhow, the goal of this step is to determine which observed-patches are seen in the current frustrum. Doing this without account of the fact that patches are grouped on surfaces is a loss of efficiency. Actually, a good strategy may first compute visibility at surface level. This could imply an important reduction of the number of patches to deal with, so of the number of form factors to be computed. Two ways may be envisaged:

- Compute visible surfaces. This means to obtain a set of fully visible surfaces and a set of fully occluded surfaces. Then the surfaces of the visible set must be divided into patches which are fully visible. This technique has the drawback that the original surfaces are clipped by those which are in front of them. So the resulting surfaces are not regular and their discretization into patches will be more complex and must be performed for each view point. A more feasible approach is to compute patches only once; but, in this case, we must clip some patches in addition to the surfaces and these patches will not be regular (figure 3.a).

- Filter non visible surfaces. Now the idea is to remove all those original surfaces which are fully hidden. This means that surfaces partially hidden will be kept in the candidate list. So this list is bigger and must be reprocessed in order to compute the visibility of partially occluded surfaces, at patch level (figure 3.b).

Once the form factors have been computed for a given patch, the visibility computation from neighbouring patches may be accelerated accounting on the fact that this visibility will probably be very similar for neighbours belonging to the same surface (assumed flat here). In fact, this kind of coherence is very similar to frame-to-frame coherence used in animation. Actually, when the algorithm change from a patch to one of its neighbours, its behaviour is the same as the one of an observer moving on the surface with a linear uniform movement. Something similar happens when moving from a given frustum of a patch to a neighbour frustum of the same patch; but, now with a circular movement (figure 4).

As we have pointed out above, a way to help the use of coherence is

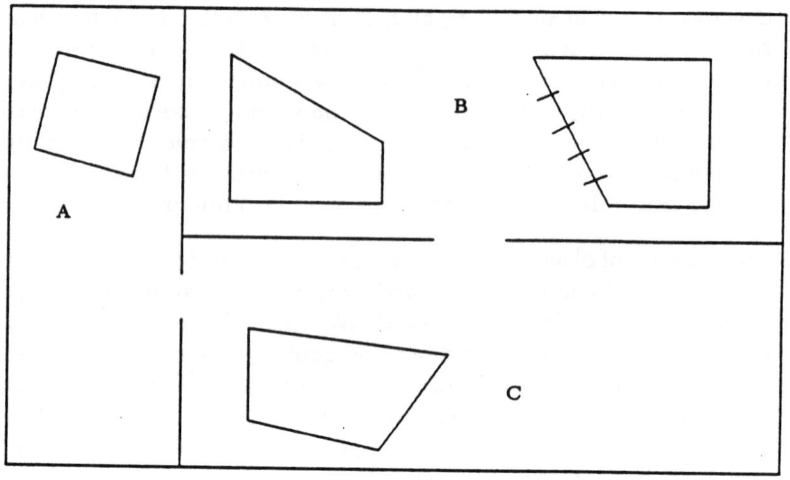

figure 5. Using walls

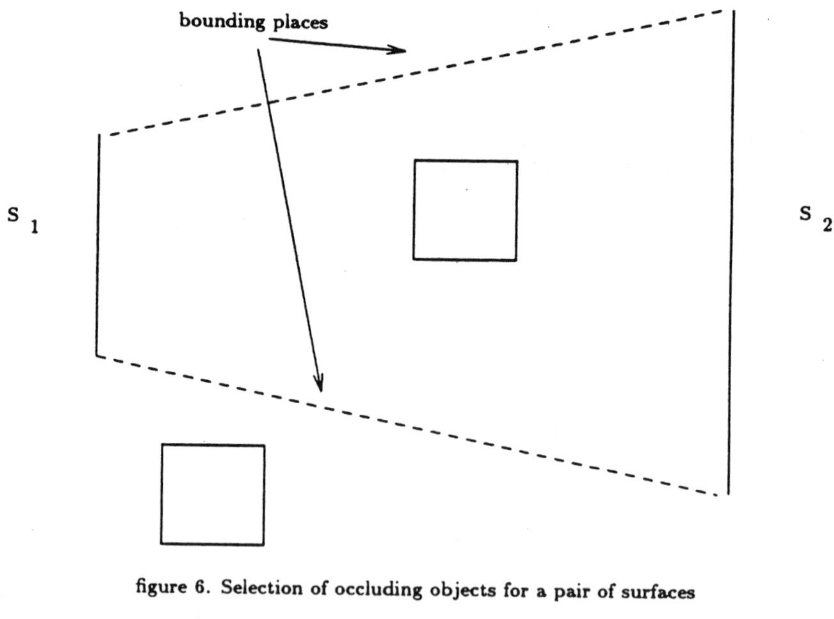

figure 6. Selection of occluding objects for a pair of surfaces

to introduce a suitable data structure of the scene. In some cases, these structures only contribute to quickly select candidate elements (objects, surfaces) by means of grouping them into bounding boxes like clusters. In other cases, the data structure also includes visibility information. It seems clear that the rear should be more suitable given the target.

Finally, we want to underline the fact that walls, very frequent in radiosity environments, may be very helpful to exploit coherence. Really, walls often divide the environment in such a way that they occlude a very big number of surfaces. So the visibility function would preprocess the environment (the candidate list) testing it against the walls of the current "subenvironment". In this way, the list of candidates to be submitted to the full process could be widely reduced. In figure 5, for example, objects in the subenvironment A have a low probability to see objects in the subenvironment B.

PROPOSED ALGORITHMS

In this section we give some references of papers where radiosity is computed using some kind of coherence and we match them with the characterization introduced above.

Marks et al. [2] and Haines et al. [3] have used a similar idea in order to account on the fact that patches of a same surface must have similar visibilities. More specifically, they give a tool to analyse whether there is or not any occlusion between two surfaces. In this way, they select surfaces or objects which occlude partially or fully one surface to the other. Afterwards, this reduced list of occluding elements is used to determine patches visibility. That selection is made by means of defining a box bounding both surfaces and using the faces of the box as clipping planes (figure 6). Actually, this idea was first exploited by Nishita and Nakamae in [4].

For a given observer-patch, visibility is computed at surface level in [5] and [6]. The former compute visible surfaces and is based on a BSP structure of the surfaces of each object. The rear discards the fully occluded surfaces using a z-buffer. Only the surfaces present at the buffer and the end of the process are taken into account for a second visibility process at patch level.

The technique exposed above, may be improved when the surfaces have been depth sorted like in the first algorithm of the previous paragraph. In this case, the candidate list may be processed in front to back order, so if we use a z-buffer, when a surface has been assigned to each of its pixels the process stops because all visible surfaces are already known. In additon to Campbell et al. this has been used by Wang et al. [7].

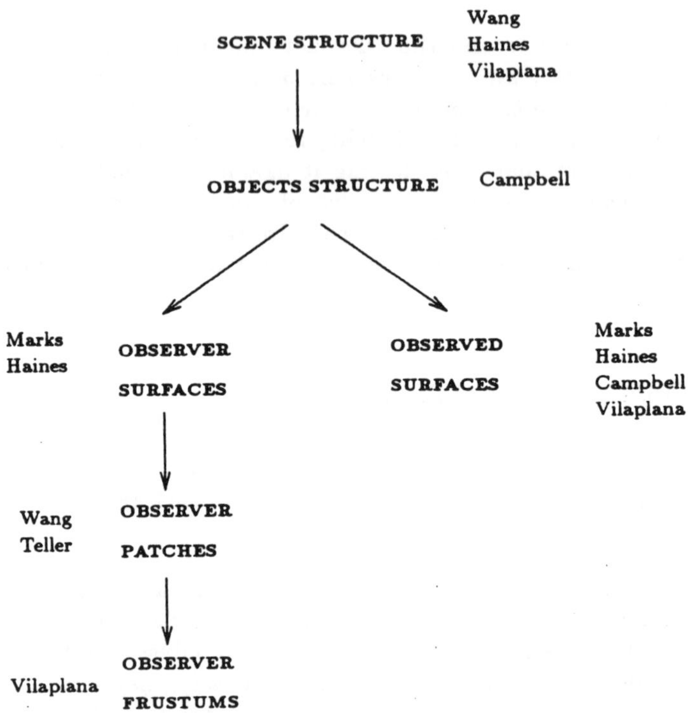

figure 7. Scheme of the levels of coherence used in proposed algorithms

The acceleration of form factor computation of a patch based on the previous computation of form factors of neighbouring patches have been used by Tellier et al. [8] in a ray tracing based algorithm. In this paper, the visible element for a patch in a given direction (so, the delta form factor) is derived from the results of the neighbour patch in the same direction. In [9] is proposed the use of coherence between neighbouring patches and frustums for clipping and view transformations.

In order to improve the efficiency of radiosity algorithms three types of data structures have been proposed which help the implementation of some coherence concepts.

- Wang et.al.[7] proposed the used of an octtree to organise the radiosity environments. As well known, this structure will give depth priority information which is very helpful; but must clip surfaces.

- Haines et al. [3] use a hierarchy of bounding boxes which offers an efficient pruning tool. Nevertheless, this structure does not include useful visibility information.

- Vilaplana et al.[6] propose the use of a BSP tree structure of the scene. The nodes of the structure are planes separating the objects of the scene and the leaves are the objects. The advantages of this solution are that the structure includes information of potential visibility (as well known for BSP trees) and that surfaces are not clipped.

Instead of or in addition to structuring the environment, these algorithms may use a suitable structure for the surfaces of each object. Campbell et al. [5] proposed to use a BSP tree of surfaces for each object.

In figure 7 we try to summarize the different types of coherence used by the algorithms briefly described above. This scheme clearly suggest the use of a top down analysis of the radiosity algorithm in order to optimize the use of coherence.

CONCLUSIONS AND FUTURE WORK

We have developed a characterization of the uses of coherence to compute visibility for form factor evaluation. This study shows that radiosity algorithms offers a hierarchical use of coherence like in traditional visibility algorithms; but adding some new types of coherence which are specific for the determination of form factors. We have also presented several algorithms which use coherence strategies that match with the previous characterization.

From this work we may derive two main tasks to be developed. First we could extent this work in order to introduce algorithms which work for dynamic environments and environments with moving observers. Secondly, we have observed that published algorithms very rarely present comparisons of the proposed methods with others. So it would be interesting to compare the efficiency of uses of coherence which belong to the same class.

ACKNOWLEDGEMENTS

The author wishes to thank J.Vilaplana for his helpfull observations and B.Garcia for her valuable aid during the editing of the paper.

REFERENCES

1 Goral,C.M., Torrance,K.E., Greenberg,D.P. and Battaile, B. 'Modeling the Interaction of Light Between Diffuse Surfaces' in *Proc. ACM SIGGRAPH* 1984, pp.213-222.

2 Marks,J., Walsh,R., Christense,J. and Friedell,M. 'Image and Intervisibility Coherence in Rendering' in *Proc. of Graphics Interface'90*, pp. 17-30, Canadian Information Processing Society, 1990.

3 Haines,E.A. and Wallace,J.R. 'Shaft Culling for Efficient Ray-Traced Radiosity' in *Proc. of the Second EUROGRAPHICS Workshop on Rendering*, May 1990.

4 Nishita,T. and Nakamae,E. 'Continuous Tone Representation of Three Dimensional Objects Taking Account of Shadows and Interreflection' in *Proc. ACM SIGGRAPH* 1985, pp. 23-30.

5 Campbell,A. and Fussell,D.S. 'Adaptative Mesh Generation for Global Diffuse Illumination' in *Proc. ACM SIGGRAPH* 1990, pp. 155-164.

6 Vilaplana,J. and Pueyo,X. 'Multilevel Use of Coherence for Complex Radiosity Environments' (submitted), *Research Report of LSI Dept., Universitat Politècnica de Catalunya*, 1993.

7 Wang,Y. and Davis,W.A. 'Octant Priority for Radiosity Image Rendering' in *Proc. Graphics Interface'90*, pp. 83-91, Canadian Information Processing Society, 1990.

8 Tellier,P., Maiselle,E., Bouatouch,K. and Languenou, E. 'Using Coherence to Accelerate Radiosity' to appear in *The Visual Computer*, IRISA Publication Interne 616, 1991.

9 Vilaplana,J. and Pueyo,X. 'Exploiting Coherence for Clipping and View Transformation in Radiosity Algorithms' *Photorealism in Computer Graphics*, pp. 137-149, Eurographics Seminars Series, Springer Verlag, 1991.

Invited Paper

Characterization of object shapes by singular points: with an application to the feature extraction of human facial expressions

H. Saji,[a] Y. Shinagawa,[a] T.L. Kunii,[a] H. Hioki,[a] K. Hara,[a] N. Asada,[b] M. Yasumoto[b]

[a] *Department of Information Science, Faculty of Science, The University of Tokyo, 7-3-1 Hongo, Bunkyo-Ku, 113 Tokyo, Japan*

[b] *Government Industrial Research Institute, Osaka, Japan*

ABSTRACT

The surface singular points of a function defined on an object surface represents the topological structure of the object. This paper proposes and reports the implementation and testing of a new method to characterize the facial expressions by extracting the singular points of the face. As the input, the surface normals of human faces obtained by the lighting switch photometry method are used. An averaging filter is used to reduce the noise in the raw data. The experimental results demonstrate the usefulness of the proposed method to characterize facial expressions.

INTRODUCTION

The importance of human face recognition as a new means of human-machine communications has been increasing rapidly. There are two main subjects in this field; One is the identification of persons [13, 6] and the other is the analysis of human facial expressions [11, 2].

Most of the researches on the analysis of facial shapes have been based on the two-dimensional (2D) data; characteristic regions such as a nose, eyes and a mouth are extracted from 2D facial images and their shapes, locations and lengths are analyzed [9, 15, 16]. Such 2D methods can analyze faces quickly. The results, however, are influenced by the orientation of the face and the illumination. For this reason, the 2D methods cannot extract stable information from the facial images.

Recently, research on the analysis of 3D data has begun. The methods extract and trace the characteristic points and regions such as the top of a nose [14], the markers attached to a face [1, 7, 8], the isodensity maps [10] and the surface curvature for face identification computed from the Moiré topography [4]. However, the number of characteristic points obtained by these methods is small and hence it has been impossible to analyze detailed facial shapes such as the wrinkles on the cheeks.

To analyze the detailed shapes, the Lighting Switch Photometry method [3] is used to obtain the data in this paper. This method provides the normal vectors at each point on the face. Using the normal vectors, we compute the singular points and use them for representing the characteristics of the facial expressions. This enables us to characterize subtle changes in the facial expressions.

FACIAL SHAPE CHARACTERIZATION BY THE SINGULAR POINTS

In the case of manufactured objects, their shapes can be represented by using combinatorial topology in terms of vertices, edges and faces. In the case of a human face, we cannot find such characteristics to describe the shape. For this reason, contemporary methods use characteristic points such as the top of the nose or marks attached to the face as is described previously. Recently, Shinagawa, Kergosien and Kunii [17] have proposed the characteristics that can be used instead of vertex-edge-face description: the singular points of a Morse function defined on the surface.

First of all, we define a height function on the object surface (see Figure 1). The singular points of this function are the peaks, saddle points and pits. The pits and peaks are to be understood as points of the surface at which the height function h has a local minimum and maximum. At a saddle point, h has a minimum when viewed from one direction and a maximum when viewed from another direction. At the singular points, the surface normals are parallel to the axis of the height function.

According to Morse theory, there is a one-to-one correspondence between a primitive called a cell and a non-degenerate singular point of a Morse function defined on the object surface that is a C^2 2-manifold. When we consider an object surface, there are 0D, 1D and 2D cells. Simply speak-

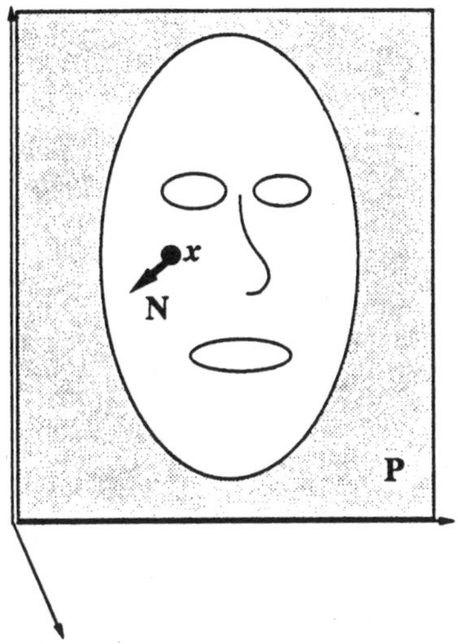

Figure 1: The height function

ing, a 1D cell is a point, a 2D is a chip of a string and a 2-cell is a patch of a surface. The object surface can be reconstructed by pasting the cells corresponding to the singular points of the function. A peak corresponds to a 2-cell, a saddle point to a 1-cell and a pit to a 0-cell. (A description of Morse theory can be found, for example, in [5].) Therefore, the singular points determine the topological structure of the object surface.

The singular points are important for object recognition. For example, we recognize the terrain by the peaks, pits and saddle points. When we consider the energy potential surface, the peaks, pits and the saddle points represent the equilibrium states. As can be seen from these examples, the singular points play significant roles when we characterize the object.

For these reasons, we characterize the shapes of the faces corresponding to various facial expressions by the singular points. The lighting switch photometry method [3], which is summarized in the next section, outputs the surface normals of an object surface. The singular points of the surface can be directly computed from the data.

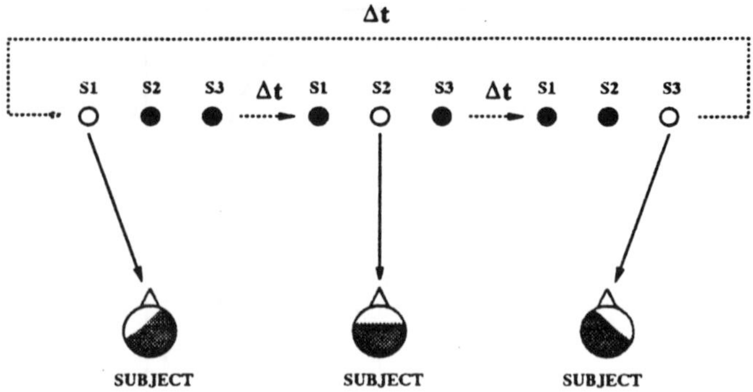

Figure 2: Light Switching

LIGHTING SWITCH PHOTOMETRY METHOD

The lighting switch photometry (LSP) method uses multiple light sources and one video camera as in the case of the photometric stereo method [12]. The three light sources are switched rapidly in turn and the images are obtained using a video camera when the subject is illuminated by one of the light sources (see Figure 2). For each light source, the images obtained when it illuminates are interpolated to obtain the intermediate images at a given instant. Three images with different light sources are computed for each instant. Assuming the Lambertian reflectance, the normal vector at each point on the face at a given instant is computed from the intensity of the corresponding pixels of the three images. The 3D shape of the face at each instant is reconstructed from these normal vectors using integration iteratively. Even if the human face moves, the detailed shapes such as the wrinkles on the face can be extracted by this method.

COMPUTATION OF SINGULAR POINTS FROM THE NORMAL VECTORS

The singular points on the surface of a face are detected using the normal vectors as follows.

Let us denote the surface of a face by A. First of all, we consider a plane P perpendicular to the axis of the height function. We define a map

Π that projects the normal vectors at each point to the plane P. The result of this mapping forms the "needle map".

For a point x on the face, we consider a closed curve that goes around x counterclockwisely. Let us denote the curve by $c : [0,1] \rightarrow A$ where $c(0) = c(1)$. Points on this curve are designated by $c(s)$, $s \in [0,1]$. Figure 3 shows the relationship between the singular points on the surface and the needle map.

Let $t(s)$ be the tangent vector of the curve c at the point $c(s)$, and $n(s)$ be the normal vector of the face at the point $c(s)$. Now $\alpha(s)$ is defined as the angle between $\Pi(t(s))$ and $\Pi(n(s))$ on the plane P (see Figure 4). If the projected normal vector $\Pi(n(s))$ is directed to the interior of the curve $\Pi(c)$, $\alpha(s)$ takes the negative value. Otherwise, $\alpha(s)$ takes the positive value.

The value of $\alpha(s)$ varies as we trace the curve c. We can classify x by the change in the value of $\alpha(s)$; The point x is a *peak* if $\alpha(s)$ is always positive, a *pit* if $\alpha(s)$ is always negative and a *saddle point* if the sign of $\alpha(s)$ changes four times. Figure 5 shows the variations of $\alpha(s)$ for the peak, pit, pass, and other cases. Here, we assume that the surface is smooth enough so that the classification of x does not depend on the choice of the curve c if c is small enough.

The LSP method provides one normal vector per pixel. The data is hence discretized. We consider a pixel as a square region with four vertices (Figure 6). $\Pi(x)$ is discretized and is located at a vertex. The curve $\Pi(c)$ is defined as the square passing through the centers of the four pixels containing $\Pi(x)$ as one of its vertices. $\alpha(s)$ is defined as the angle between $\Pi(t(s))$ and $\Pi(n(s))$ at the centers of the four pixels.

The original data of normal vectors is noisy. To reduce the noise, an averaging filter is used. A certain number of pixels are merged to form a region of equal size. The direction of the normal vector of each region is the average direction of the normal vectors in that area.
Applying the above procedure to each region, we can find the singular points on the surface.

In our experiment, the normal vectors are obtained for 640×480 pixels. By the averaging process, the original image is partitioned into 40×30 areas. As is stated previously, the surface is assumed to be smooth enough so that we can classify x correctly by the curve c of this size. When the person is not viewed in full face by the video camera, it is necessary to rotate the obtained normal vectors so that the facial expression is seen in full face.

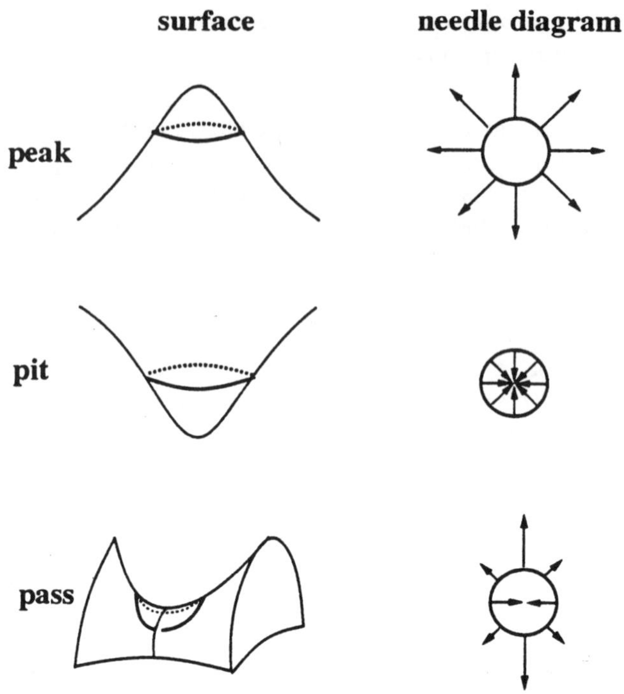

Figure 3: The singular points and the needle diagrams

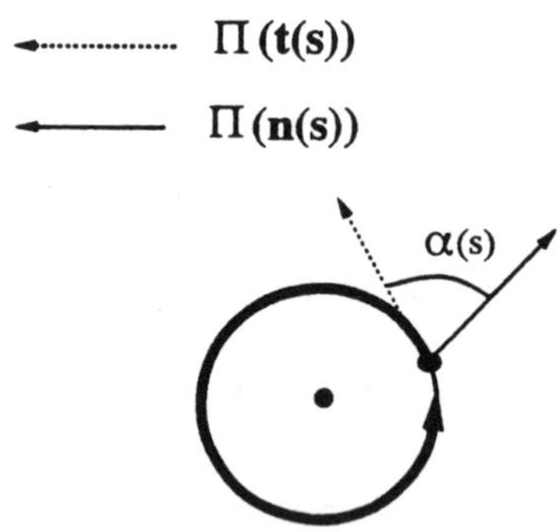

Figure 4: The angle α

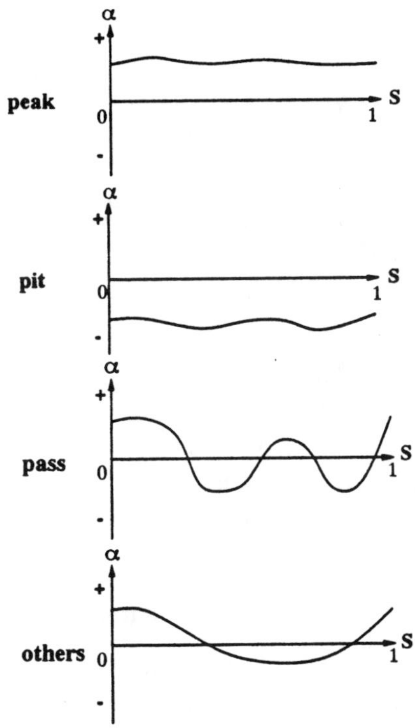

Figure 5: The change in α on the curve c

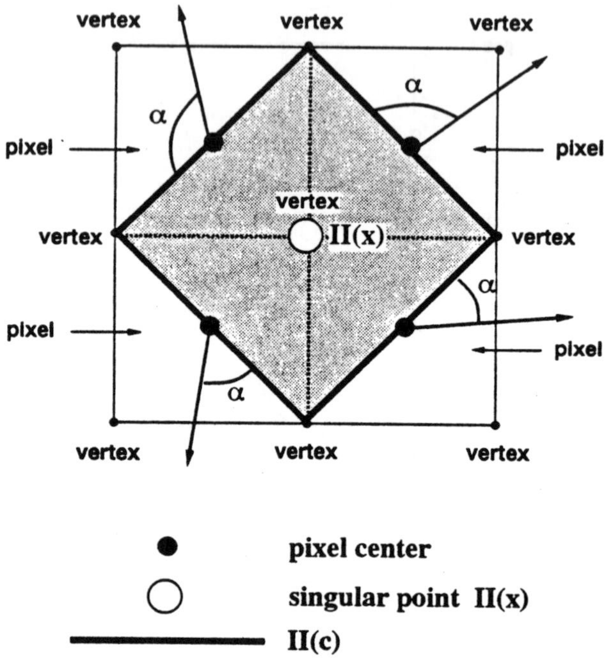

Figure 6: Discretization of the curve c and $\alpha(s)$

EXPERIMENTAL RESULTS

Figure 7 shows the change in the locations of the singular points on the needle diagram of a laughing face. A square stands for a pass, a black circle for the peak and a white circle for a pit. On the neutral face (Figure 7(a)), there are three peaks and two saddle points. When the face begins to laugh, the number of singular points increases (see Figure 7(a) and (b)). It can be also seen that the newly generated singular points are on the upper part of the face and the cheeks. Figure 8 shows the corresponding face images. Figure 9(a) shows the singular points of another person's laughing face.

Figure 9(b) shows the singular points of an angry face. Singular points are found between the eyes. There are, however, none on the cheeks. Figure 9(c) shows the singular points of a sad face. Figure 10 shows the corresponding face images.
It is hard to discriminate the sad face (Figure 9(c)) from the laughing face (Figure 9(a)) just looking at the singular points. This is because the both faces have similar wrinkles. To discriminate between the two, we need additional information such as the shape of the eyes and the mouth.

The distribution of the singular points characterizes the facial expressions as can be seen in these examples.

CONCLUSIONS AND FUTURE WORK

We have presented a method to characterize the facial expressions by the singular points. Automatic recognition of facial expressions is a future application of our method. Using additional information such as the shape of the eyes and the mouth to reinforce the information of the singular points is another extension we have in mind. Elimination of the noise in the data obtained by the LSP method is one of the future research. As another possible improvement of the system, we are considering the use of infra-red light sources to avoid the influence of the lights to the facial expressions.

ACKNOWLEDGEMENTS

We would like to express our sincere gratitude to Dr. Ichiro Hagiwara and Ms. Wakae Kozukue of Nissan Motor Co., Ltd. for their support to this research. Sincere thanks are also extended to Yokogawa Hewlet Packard Co., Ltd. for offering HP9000 and Silicon Graphics Co., Ltd. for personal IRIS.

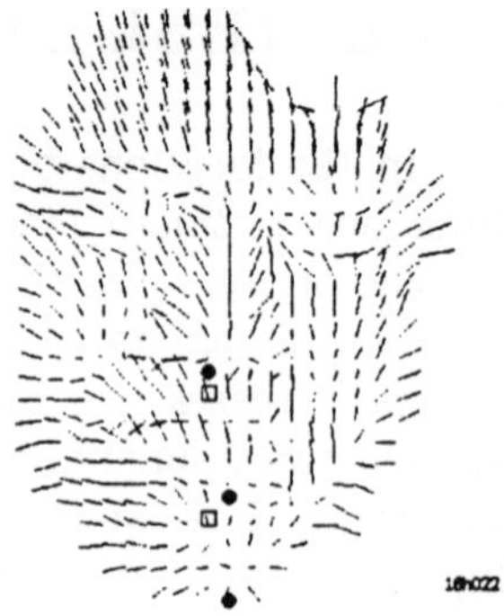

(a)

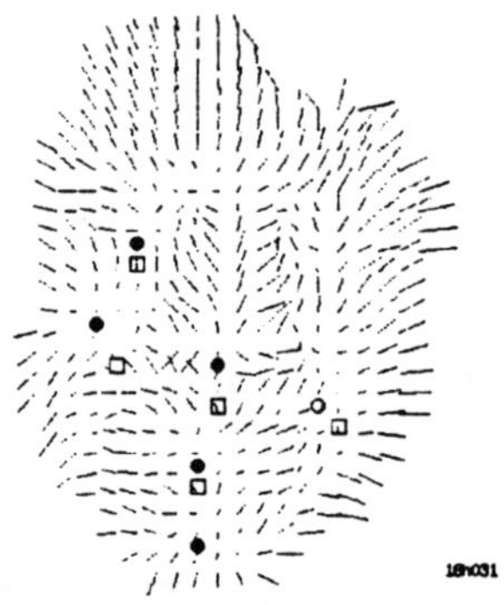

(b)

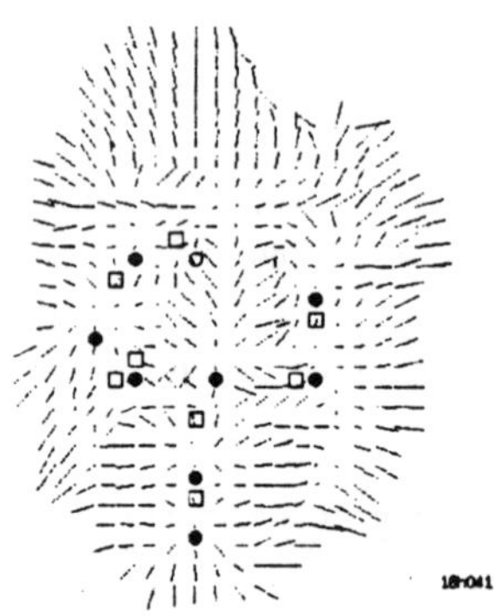

(c)

Figure 7: The distribution of the singular points when a person begins to laugh

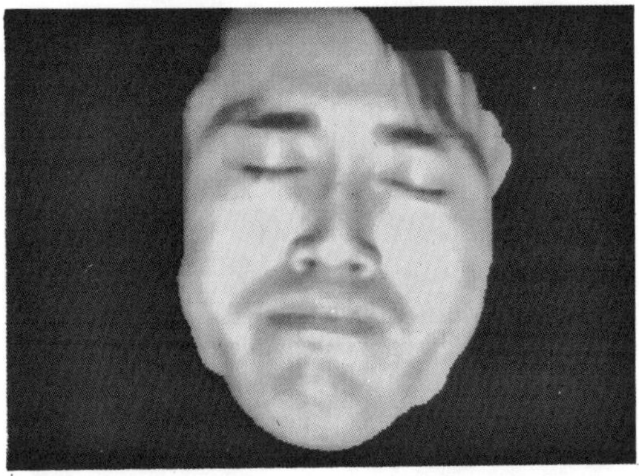

(a)

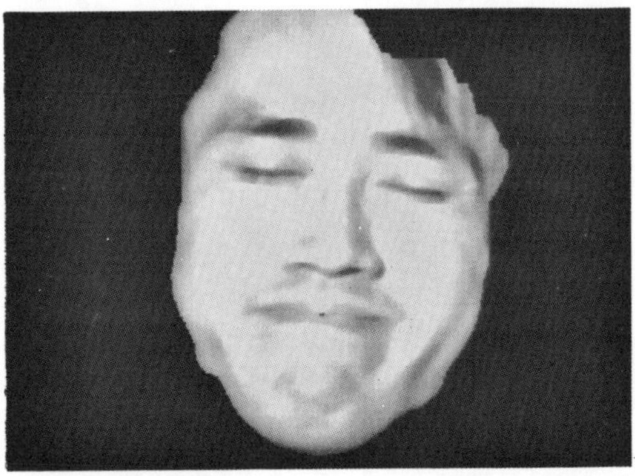

(b)

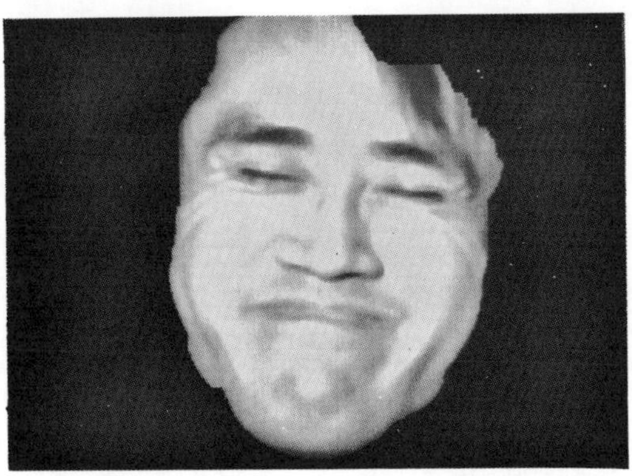

(c)

Figure 8: The images corresponding to Figure 7

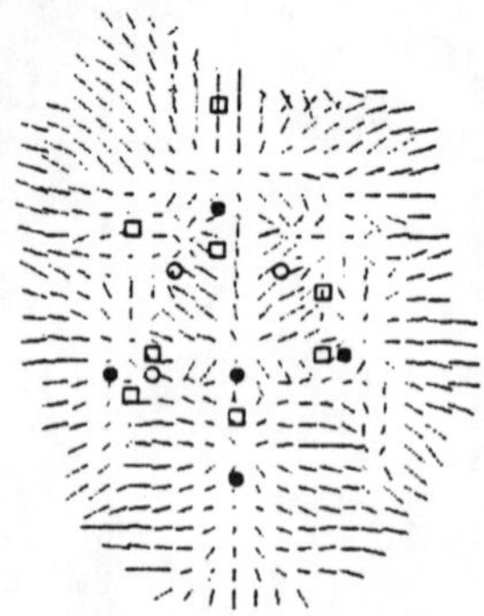

(a)

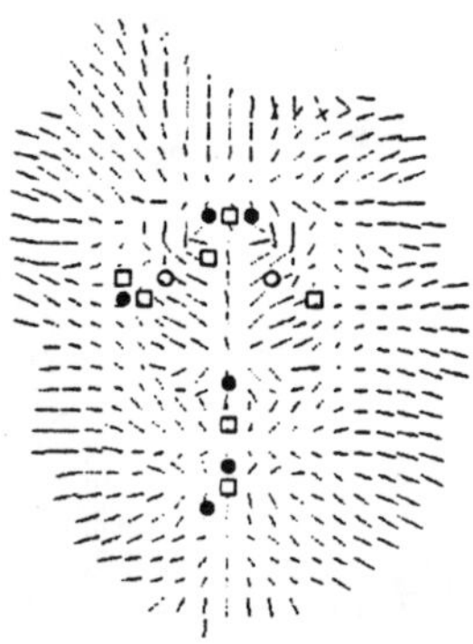

(b)

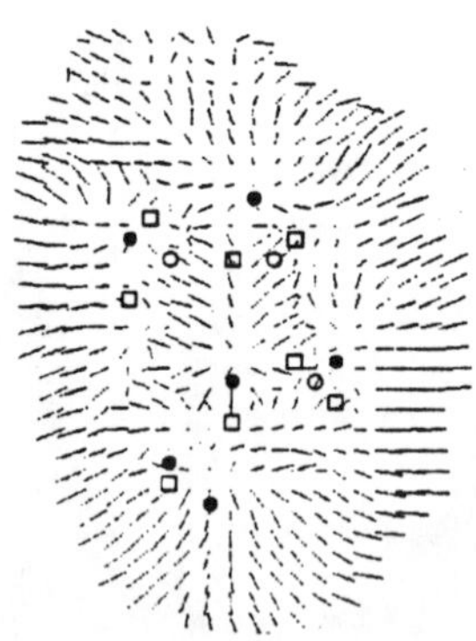

(c)

Figure 9: The distribution of the singular points on (a) a laughing face, (b) an angry face (c) and a sad face

(a)

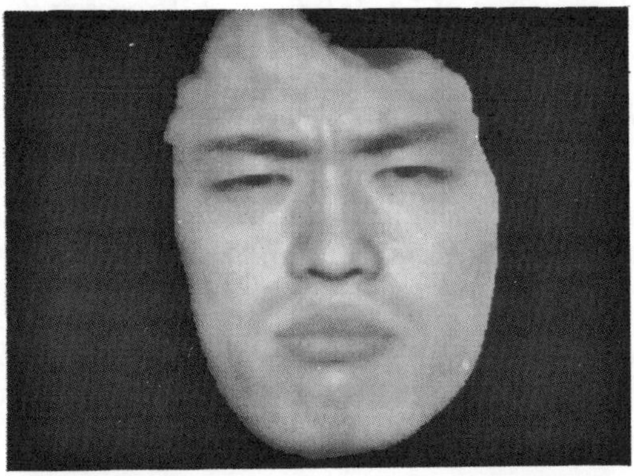

(b)

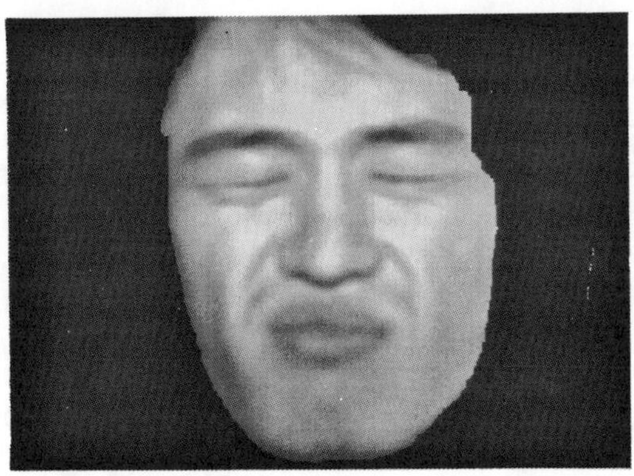

(c)

Figure 10: The images corresponding to Figure 7

REFERENCES

[1] Patterson E.C., Litwinowicz P.C., and Greene N. 'Facial Animation by Spatial Mapping'. In Thalmann N.M. and Thalmann D., editors, *Computer Animation '91*, pages 45–58. Springer-Verlag, 1991.

[2] Parke F.I. 'Parameterized Models for Facial Animation'. *IEEE Computer Graphics and Applications*, 2(9):61–68, November 1982.

[3] Saji H., Hioki H., Shinagawa Y., Yoshida K., and Kunii T.L. 'Extraction of 3D Shapes from the Moving Fuman Face Using Lighting Switch Photometry'. In Thalmann N.M. and Thalmann D., editors, *Computer Animation 92*, pages 69–86. Springer-Verlag, 1992.

[4] Tanaka H. and Kishino F. 'Surface Curvatures for Face Identification'. Technical Report 91-CV-73, Information Processing Society of Japan, 1991.

[5] Milnor J. *'Morse Theory'*. Princeton University Press, New Jersey, 1963.

[6] Ueno K., Kato M., Takahashi K., Nakamura O., and Minami T. 'Personal Identification through Facial Images'. Technical Report ISEC90-16, The Institute of Electronics, Information and Communication, 1990.

[7] Waters K. and Terzopoulos D. 'Modelling and Animating Faces using Scanned Data'. *The Journal of Visualization and Computer Animation*, 2(4):123–128, 1991.

[8] Williams L. 'Performance-Driven Facial Animation'. *Computer Graphics*, 24(4):235–242, 1990.

[9] Harmon L.D. 'Automatic Recognition of Human Face Profiles'. *IEEE Computer Graphics and Image Processing*, 6:135–156, 1977.

[10] Kato M., So I., Hishinuma Y., Nakamura O., and Minami T. 'Description and Synthesis of Facial Expression Based on Isodensity Maps'. In Kunii T.L., editor, *Visual Computing*, pages 39–56. Springer-Verlag, 1992.

[11] Ekman P. and Friesen W.V. *'Unmasking the Face'*. Prentice-Hall, Englewood Cliffs, New Jersey, 1975.

[12] Woodham R.J. 'Photometric Method for Determining Surface Orientation from Multiple Images'. *Optical Engineering*, 19(1):139–144, 1980.

[13] Abe T., Aso H., and Kimura M. 'Automatic Identification of Human Faces Using Profiles Extracted from 3-D Data'. *The Transactions of the Institute of Electronics and Communication Engineers of Japan*, J73-D-II(9):1468–1476, 1990.

[14] Kurihara T. and Arai K. 'A Transformation Method for Modeling and Animation of the Human Face from Photographs'. In Thalmann N.M. and Thalmann D., editors, *Computer Animation '91*, pages 31–44. Springer-Verlag, 1991.

[15] Sakai T., Nagao M., and Kanade T. 'Computer Analysis of Photographs of Human Faces'. *The Transactions of the Institute of Electronics and Communication Engineers of Japan*, 56-D(4):228–233, 1973.

[16] Sasaki T., Wallace R.S., Akamatsu S., and Suenaga Y. 'Discussion On A Face Recognition System Based On Image Segmentation And Labeling And Pattern Matching Method'. Technical Report PRU90-43, The Institute of Electronics, Information and Communication, 1990.

[17] Shinagawa Y., Kergosien Y.L., and Kunii T.L. 'Surface Coding Based on Morse Theory'. *IEEE Computer Graphics and Applications*, 11(5):66–78, September 1991.

Invited Paper

Automatic reconstruction of a solid from three orthographic views

T. Masuda, F. Yamaguchi

Department of Mechanical Engineering, School of Science and Engineering, Waseda University, 3-4-1 Ohkubo, Shinjuku, Tokyo, 169 Japan

Abstract

The present article proposes an efficient method of reconstructing a polyhedron solid from three orthographic views. First, a surface model (intermediate model) corresponding to these three views is created by examining the corresponding relations among the three views. The intermediate model is considered to include false faces, false edges and false vertices as well as true ones. The main problem is how to efficiently detect and remove false elements from the intermediate model against the exponential increase of the number of the combinations of candidate faces.

This article treats this false element removal process as a determination problem of status values(true, false) of all the geometric elements represented in a constraint relationship network. In this approach, many rules concerning consistency checking are systematically organized and the property of local value propagation is fully exploited, so that a systematic method has been developed and demonstrated to be very successful.

Introduction

Automatic Reconstruction of a solid from three orthographic views is an important input method in solid modelling. The first research probably dates back to Idesawa[1] and from then onwards many researches have been done on this theme (there is a good review paper(Nagendra & Gujar[2])). The earlier trend has been towards the generation of three dimensional objects containing only planar faces. Significant researches in the generation of three dimensional objects made up of both planar and curved faces are those by Aldefeld [3] , Sakurai [4] and Loquette [5] . Some algorithms are capable of generating multiple solutions, each satisfying the given input views(Sakurai [4]),(Markowsky [6]).

Let us confine our present concern to generating plane faced polyhedra. Although these algorithms are based on a variety of approaches, the major steps involved of a typical algorithm are the following:

1. Transformation of 2-D points to 3-D vertices,
2. Generation of 3-D edges from 3-D vertices,
3. Construction of faces from 3-D edges, and
4. Formation of a 3-D polyhedron from faces.

In step 3. above, a surface model(intermediate model) is created which is considered to include non-existent(false) vertices, edges and faces. The main problem is how to efficiently detect and remove such elements from the intermediate model against the exponential increase of the number of the combinations of candidate faces. Many papers (Idesawa [1], Senda [7] etc.) seem to suffer from this problem.

This article treats this false element removal process as a determination problem of status values(**true**, **false**) of all the geometric elements represented in a constraint relationship network. In this approach, many rules concerning consistency checking are systematically organized and the property of local value propagation is fully exploited, so that a systematic method has been developed and demonstrated to be very successful.

Problem Definition

Input and Output

Input data are three orthographic views represented in computer memory, where they are drawn in terms of line segments and endpoints. The line segments are flagged either as solid or dashed lines and the endpoints are represented in Cartesian coordinates. Output of the system is a manifold polyhedron solid in a boundary representation corresponding to given orthographic views.

Used Terms

In an orthographic view we will call a line element a line segment and an area surrounded by line segments a "range". We will stipulate the line segments such that they do not overlap or touch each other except at the endpoints. In a solution solid, a plane, line or point element is called a face, edge or vertex respectively. In the beginning of the process of determining a solution solid, an intermediate non-manifold model is created from the given orthographic views. The intermediate model consists of all possible plane, line and point elements corresponding to the views. It includes **false** elements which are not the elements of a solution solid. All the **false** elements are to be removed from the intermediate model before a final solution is output. Therefore, a plane, a line or a point element in the intermediate model is called a candidate face, a candidate edge or a candidate vertex respectively.

Main Problem

A solution solid must satisfy the next two conditions. That is,
(1) it must satisfy the constraints as a manifold polyhedron solid, and
(2) its projections must coincide with the given three orthographic views.

The objective of the present article is to present an efficient method of reconstructing a manifold polyhedron solid corresponding to three orthographic views. If there exists more than one solid, the reconstruction system is desired to generate all possible solution solids.

The key technique is how to efficiently remove **false** elements from an intermediate model against the exponential increase of the number of the combinations of candidate faces.

An Outline of the Whole Process

We have considered the **false** element removal process as a determination problem of the status values of all the geometric elements represented in a constraint relationship network. We will use three status values, i.e., **true**, **false** and **unfixed**. What we aim at is to find one or more combinations of **true** faces.

The whole process falls into three sub-processes, i.e.,
(1) intermediate model generation,
(2) **false** element removal, and
(3) solution solid generation.
Each time a solution solid is found, the reconstruction system outputs the solution and again tries to find another solution by returning to the second process (2).

In this section we will only explain the processes of the intermediate model generation and the solution solid generation. As for the **false** element removal process, we will explain it in detail later.

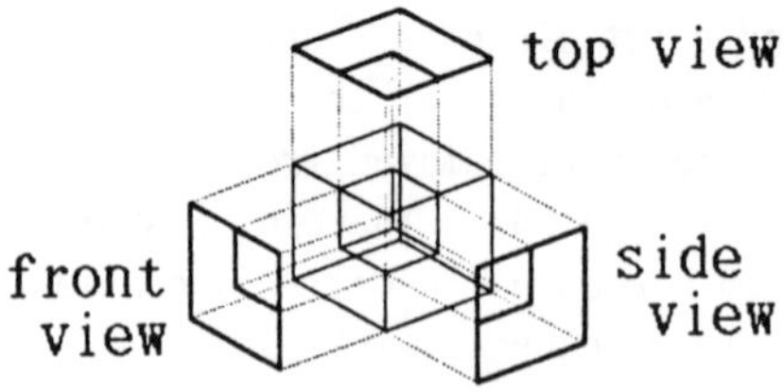

Figure 1 Candidate Vertices and Candidate Edges

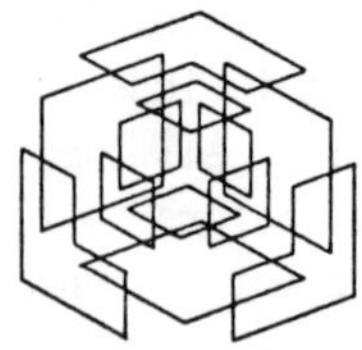

Figure 2 Candidate Faces

Intermediate Model Generation

All possible candidate vertices, edges and faces are generated from the given orthographic views in this order(see Figures 1 & 2), where the candidate edges and faces must satisfy the next conditions respectively. That is,

(a) candidate edges can not overlap or touch with each other except at each vertex, and

(b) candidate faces can not overlap with each other.

The condition (a) forces overlapping candidate edges to be split into collinearly connected smaller edges. The condition (b) generates coplanar candidate faces, shown in Figures 3 & 4. The unification of these edges or faces into an edge or a face is done in the solution solid generation process.

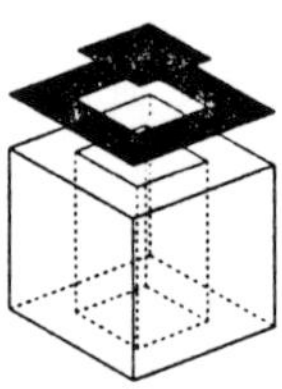

Figure 3 Candidate Faces

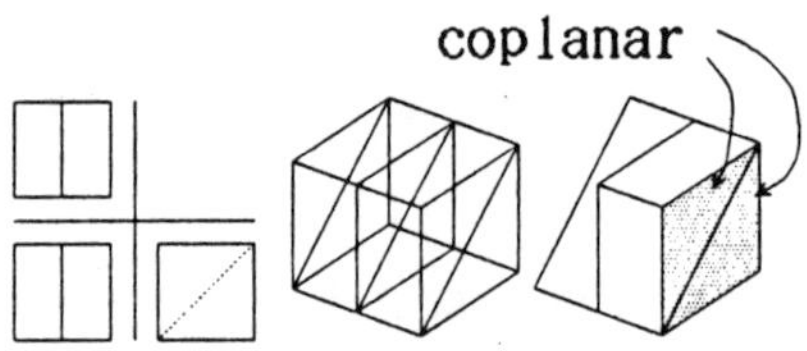

Figure 4 Coplanar Candidate Faces

Solution Solid Generation

It is the solution solid generation process that constructs a polyhedron solid by using the faces determined to be **true** in the **false** element removal process. Note that the solid is created separately from the intermediate model. Collinear edges and coplanar faces are united in the output solid as explained before.

Geometric Constraint Conditions

As mentioned before, a solution solid must satisfy the two conditions, i.e., that of a manifold polyhedron solid and that of coincidence with three orthographic views. These two conditions fall into seven practically usable ones in programs. Candidate faces, candidate edges, candidate vertices, ranges, line segments and "intersections" are checked with each other for consistency with respect to these geometric conditions.

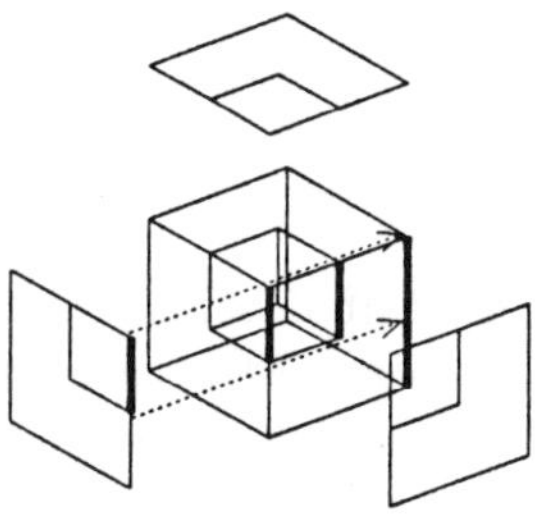

Figure 5 Candidate Edges Corresponding to a Line Segment

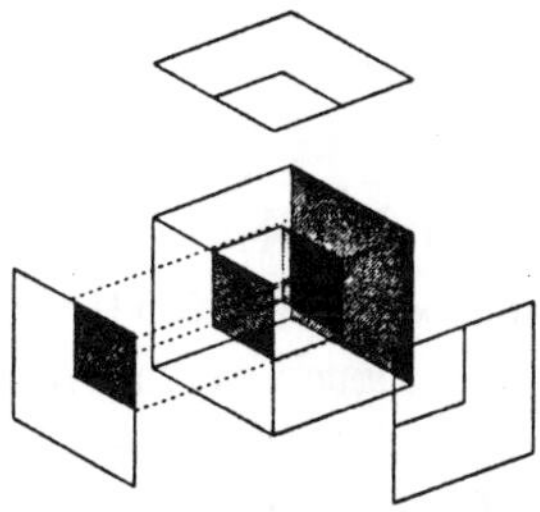

Figure 6 Candidate Faces Corresponding to a Range

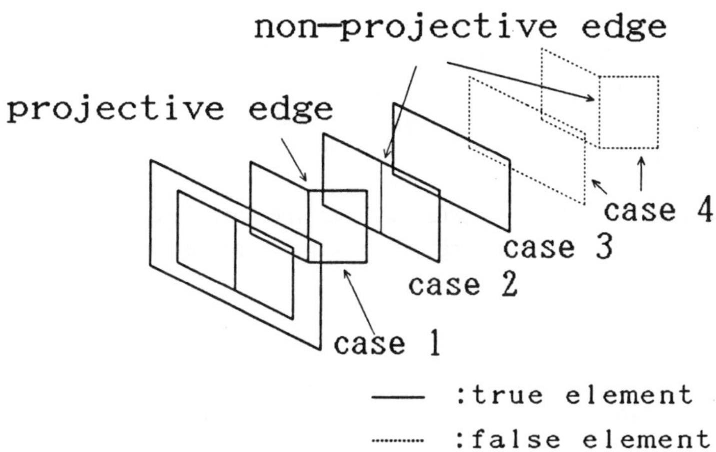

Figure 7 Candidate Faces and Candidate Edges Projected onto an Orthographic View

In the following descriptions, we will use the word "correspond" as shown below.

- Candidate edges "corresponding" to a line segment are those of which the projections overlap a line segment(see Figure 5).
- Candidate faces "corresponding" to a range are those of which the projections overlap a range(see Figure 6).

For the sake of explanation, we will name local geometries in the intermediate model in relation to projections(see Figure 7):

- case 1: a **true** edge and two non-coplanar **true** faces
- case 2: a **true** edge and two coplanar **true** faces
- case 3: a **true** face
- case 4: a **false** face, or a **false** edge and two **false** faces

Conditions as a Manifold Polyhedron Solid

① Relation between a Candidate Edge and its Incident Candidate Faces

Let n be the number of candidate faces incident to a candidate edge.

- When a candidate edge is **true**, the numbers of incident **true** faces and incident **false** faces are 2 and n-2 respectively.
- When a candidate edge is **false**, the numbers of incident **true** faces and incident **false** faces are 0 and n respectively.

② Relation between a Range and Corresponding Candidate Faces (see Figure 6)

Let n and m be the number of corresponding candidate faces and an even number between 2 and n.

- When a range represents a hole, the numbers of corresponding **true** and **false** faces are 0 and n respectively.
- When a range does not represent a hole, the numbers of corresponding **true** and **false** faces are m and n-m respectively.

③ Relation between a Candidate Vertex and its Incident Candidate Edges

When a candidate vertex is **false**, candidate edges incident to the vertex are all **false**.

④ Intersection(see Figure 8)

The intermediate model may have mutually intersecting candidate faces except at each vertex or along each edge. At most one face is **true** in a combination of intersecting faces.

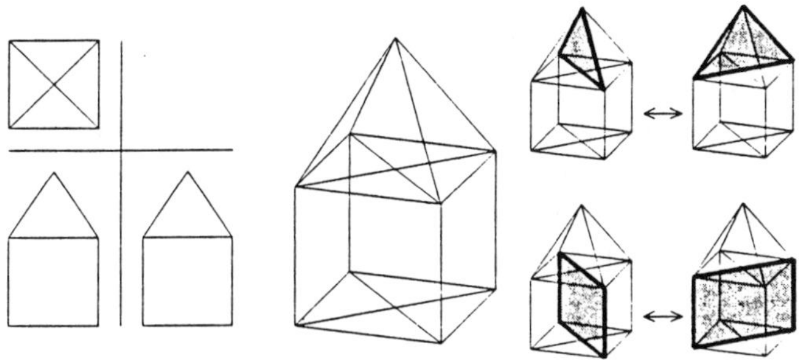

Figure 8 Examples of Intersecting Candidate Faces except at Each Vertex or along Each Edge

Conditions of Coincidence with Orthographic Views

⑤ Relation between a Line Segment and Corresponding Edges(see Figure 5)
Whether a line segment is solid or dased, it corresponds to at least one **true** edge of case 1.

⑥ Relation between a Line Segment and Incident Ranges
It is impossible that both ranges incident to a line segment correspond to holes.

⑦ Relation between Line Segment Attributes and Corresponding Geometries
- When a line segment is solid, the nearest corresponding local geometry in the intermediate model is the case 1.
- When a line segment is dashed, the nearest corresponding local geometry in the intermediate model is either case 2 or case 3.

False Element Removals

Status Values and Flags

Candidate faces, edges and vertices can take on three status values, i.e., **true**, **false** and **unfixed**. A candidate edge has a flag representing whether it is projected in an orthographic view, or not. In Figure 7, case 1 edge is projected in the view while case 2 and case 4 edges are not projected. A range takes on a flag representing whether it corresponds to a hole or not.

In the initialization of the false element removal process, all the status values and flags are initialized to be **unfixed**. In the course of the processing, the system continues to assume a status value of each candidate face systematically one by one, checks for consistency of the assumed status value of the face with respect to the seven geometric constraint conditions and gradually fixes the status values of other geometric elements. When all the status values are fixed and there are some **true** faces found, then the system is considered to have found a solution solid.

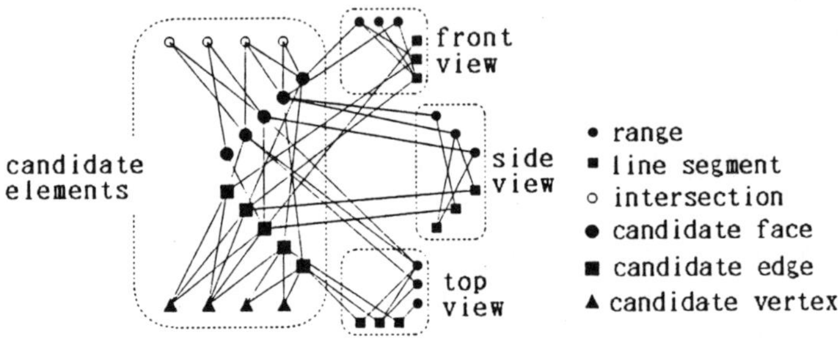

Figure 9 Constraint Relationship Network

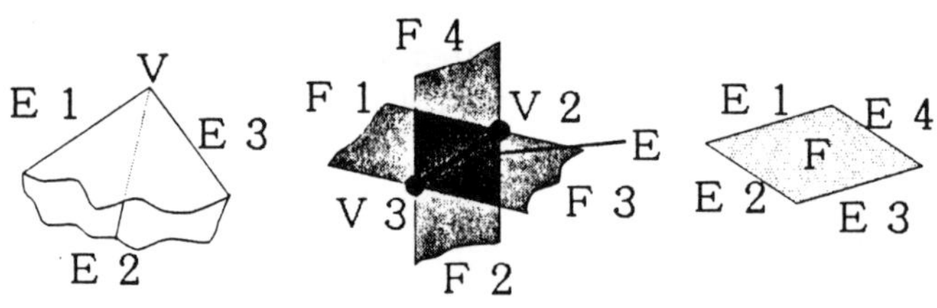

Figure 10 Local Topological Relations between Candidate Faces, Candidate Edges and Candidate Vertices

Constraint Relationship Network

The status values of geometric elements (candidate faces, candidate edges, candidate vertices, ranges, line segments, "intersections") are constrained by each other under the geometric conditions ①~⑦. This relation is illustrated in Figure 9. Prior to the processing of the false element removals, the reconstruction system creates a data structure representing the network topology shown in Figure 9. Let us call this data structure constraint relationship network. The constraint relationship network is composed of three types of information. That is,

- topological connection relations between candidate faces, edges and vertices(see Figure 10),
- correspondence relations between candidate edges and line segments (see Figure 5),
  correspondence relations between candidate faces and ranges (see Figure 6), and
- Relations between "intersections" and relevant candidate faces (see Figure 8).

Processing Pertaining to Constraint Relationship Network

Detections of Contradiction

Let us explain some examples of the cases where contradictions are detected in the course of the processing.

Suppose there exist more than three **true** candidate faces incident to a candidate edge in the constraint relationship network. Even if there are **unfixed** faces incident to the edge, we can conclude that there occurs a contradiction in the edge, because it is impossible to satisfy the constraint condition ① even if the **unfixed** faces take on any status values.

Another example is the case where a candidate edge has been fixed as **true** and all the incident candidate faces except one have been fixed as **false**. This case represents another contradiction.

The above two cases speak well that a contradiction can be detected even if all status values of all relevant geometric elements are not fixed. This is the reason why we were determined to introduce an **unfixed** status value in addition to **true** and **false**. By making use of the **unfixed** status value, all the cases of constraint conditions ①~⑦ can be expressed as the relations concerning the number of each status value in a unified manner.

The reconstruction system relies on the fact that with the progress of the processing, the number of **unfixed** elements decreases monotonously while the numbers of **true** and **false** increase monotonously.

Local Propagation of Status Values

Let us note an **unfixed** node Nu in a constraint relation with a node Nf whose status has just been fixed. In some cases the status value of Nu may be automatically fixed as **true** or **false** in order for the status of Nu not to produce a contradiction. Then the status value of another node in a constraint relation with Nu may also be fixed successively. We will call this automatic determination of status values of some nodes local propagation of status values.

Contradictions Detected through Local Propagation There is a case where a status value of the node reached by status value propagation has already been fixed. In this case it is necessary to check for consistency between the already fixed value and the propagated value. If a contradiction is detected, it means that the combination of the so far accumulated assumptions includes a wrong one which forces to assign two contradictory status values to a node.

The above explanation means that contradictions can be detected in some cases through local propagation.

Representation of Geometric Constraint Conditions The geometric constraint conditions ①~⑦ are expressed as "if-then" type rules and applied to each geometric element type for the processing of contradiction detection and status value propagation.

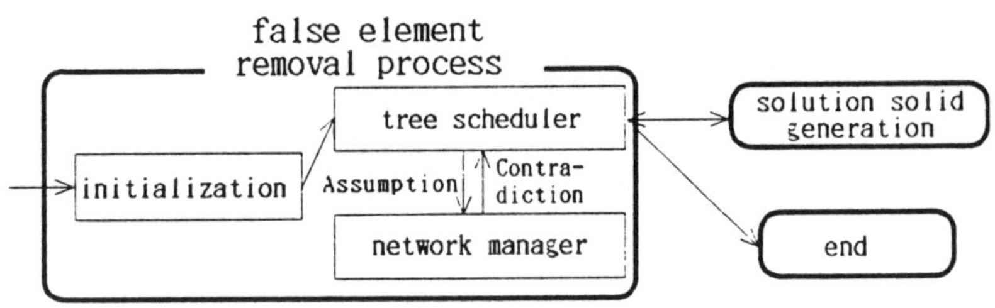

Figure 11 False Element Removal Process

Details of False Element Removal

False element removals are realized by performing two functions. One is to maintain consistency of the constraint relationship network, i.e.,

- to maintain consistency between status values of each node, and
- to fix all status values which can be determined.

The other is to control the assumptions of the status values to be placed on candidate faces systematically in order to avoid testing the same combination twice. This function is performed by the tree scheduler. A schematic diagram of the false element removal process is shown in Figure 11.

Network Manager The network manager consists of check procedures, check waiting queues and a consistency management procedure. These detailed explanations follow below.

Check Procedure

The check procedure tests only the local consistency between the status value of a node and those of its neighbouring nodes. There are six types of check procedures corresponding to the types of geometric elements, i.e., a candidate face, candidate edge, candidate vertex, line segment, range and intersection. Any constraint conditions are checked in either of the check procedures. The check procedure makes use of a check waiting queue.

Check Waiting Queue

The check waiting queue is a queue of the pointers which point to the nodes to be tested for consistency. The check waiting queue is controlled such that an existing node in the queue is not doubly entered into it. When a check procedure generates local propagation of status values, relevant nodes are entered into a check waiting queue. That is, when the status of a node is fixed, all the neighbouring nodes in constraint relations with it must be tested for consistency and therefore these nodes are entered into a check waiting queue for consistency.

Consistency Management Procedure

The consistency management procedure repeatedly takes out a node from a check waiting queue and calls its corresponding check procedure. The consistency management procedure terminates the processing either when a contradiction occurs, or when the check waiting queue becomes empty. In the latter case the constraint relationship network is considered to be in an equilibrium state. The consistency management procedure is called from the tree searching procedure which will be mentioned later.

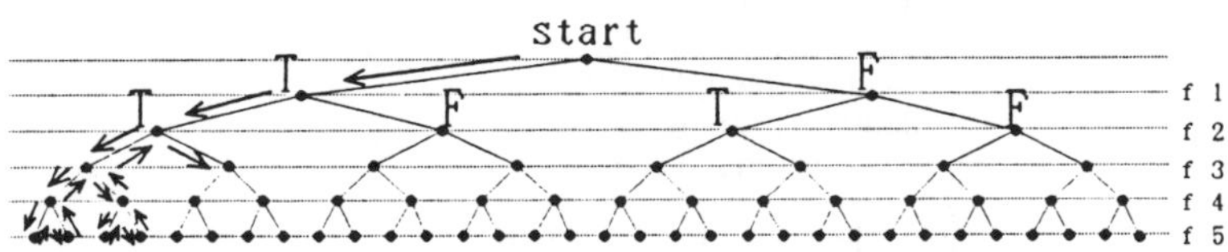

Figure 12 Depth-First Search

Tree Scheduler When the constraint relationship network is in an equilibrium state, there occurs neither contradiction detection nor status propagation. A trigger must be issued by assuming the status value of a candidate face node. The tree scheduler systematically performs this assumption controlon a depth-first search basis. There is no restriction to be placed on the order of candidate faces to be assumed. The tree scheduler scans a logical searching tree in the order of candidate faces determined quite arbitrarily(see Figure 12). A contradiction detection in the course of the processing means that any of the so far accumulated assumptions are erroneous. Then a backtracking is performed and therefore the tree pruning is attained effectively in this case.

Now, let us suppose a local propagation has occurred after the status value of a candidate face f_i was assumed. This means that status values of candidate faces f_{i+1}, f_{i+2}, •••• to be fixed have been fixed without assuming and processing these faces. This technique helps avoid testing unnecessary combinations to cope with the exponential increase of the combinations of candidate faces in addition to the pruning technique mentioned above.

Backtracking Procedure

A backtracking means to withdraw and reverse one of the assumptions placed on the candidate faces and to return to unfixed status all of the status values of geometric elements fixed under the wrong assumption of the status value of a candidate face.

```
search_tree( check waiting queue Q )
  {
    call manage_consistency( Q )
    if( contradiction ){
      call backtrack procedure
    }else{  /* equilibrium */
      if( All nodes are fixed ){
        call solution solid generator
      }else{
        let F be one of unfixed nodes
        add F to Q
        assume F to be False
        call search_tree(Q)
        assume F to be True
        call search_tree(Q)
      }
      call backtrack procedure
    }
  }
```

Figure 13 Tree Searching Procedure

Tree Searching Procedure

The tree scheduler controls the assumptions to be placed on candidate faces by calling the tree searching procedure recursively. The tree searching procedure receives a check waiting queue as input. Figure 13 describes the algorithm of the tree searching procedure.

Initialization The **false** element removal process is started by initializing all status values of all the geometric elements and calling the tree searching procedure "search_tree". As long as this first call of the tree searching procedure is concerned, no assumption is necessary to be placed on candidate faces. The first call of the procedure with the check waiting queue full of all the geometric elements can automatically fix the status values of some geometric elements in constraint relations with geometric elements of the given orthographic views.

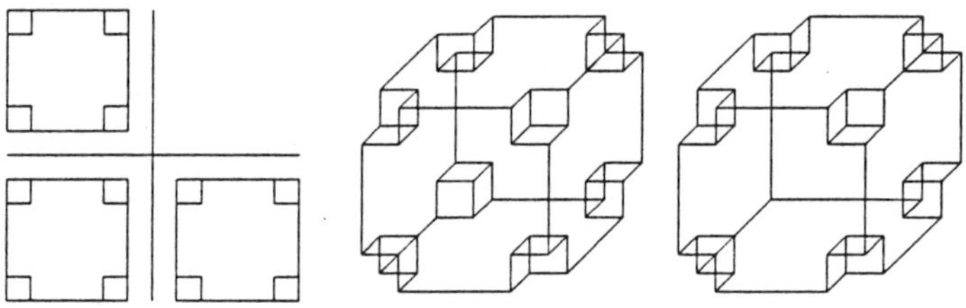

Figure 14 Example 1

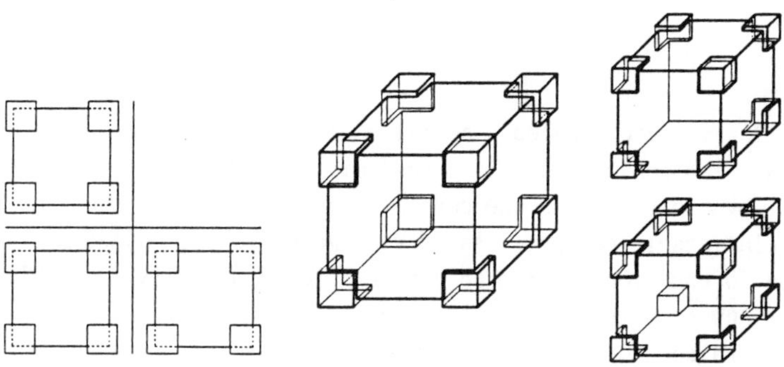

Figure 15 Example 2

Let us explain the reason in detail. Take a contour line segment of an orthographic view as an example. The outer range of the segment(background area) is considered to be a hole because there is no corresponding candidate face. Then the other range of the line segment is thought of being not a hole(see condition ⑥). If there are only two faces corresponding to the range, as is often the case, then these two faces are both **true** and there is a chance of local propagation of status values starting from the face nodes. In almost all cases this propagation fixes almost all status values of candidate faces without placing any assumptions on them.

Experiments

Some experimental examples are shown in Figs. 14 & 15. The reconstruction system is implemented in C-language and the experiments were performed on an EPSON personal computer(12 MHz). Execution times are shown in Table 1.

Table 1 Experimental Results

	No. of candidate faces	No. of solution solids	Execution time(sec)		
			intermediate model generation	false element removal	Total
Ex. 1	54	2	10.016	0.988	11.004
Ex. 2	102	3	23.663	3.651	27.314

Considering the fact that the number of combinations of candidate faces to be processed is 2^N(where N represents the number of candidate faces), the techniques proposed in this article can be considered to be very effective in coping with the increase of the number of candidate faces.

Conclusions

In this article we have treated the false element removal problem as a constraint satisfaction process with respect to status values of geometric elements by introducing the concepts of the status value and the constraint relationship network. And furthermore we introduced a status value **unfixed** in addition to **true** and **false**. This made it possible to represent various geometric constraint conditions in a unified mannner as "if-then" type rules applied to each node type. Eventually we have been successful in making the most use of the concepts of the contradiction detection and the local propagation of status values, and in solving the false element removal problem efficiently by largely eliminating the testings of unnecessary combinations.

We would like to mention some researches to be done in the future. The first problem is to extend the proposed method such that it can deal with simply curved faces as well. We are considering this extension as quite straightforward. The second is on adding a function of descriptive error detections of an orthographic view. If there are such errors, then they could be detected by using the geometric constraint conditions proposed in this article. The third is on further improvement of processing efficiency of the system. As already explained we separated the procedure(network manager) which performs each operation corresponding to each geometric constraint condition from the procedure (tree scheduler) which controls systematic assumptions of status values of face nodes. There is a possibility of finding an optimum order of the assumptions of face nodes by making use of the separation of these two functions.

References

1. Idesawa, M. 'A System to Generate a Solid Figure from Three Views' Transactions of the Japan Society of Mechanical Engineers, Vol. 38 No. 310, pp. 1267-1276, 1972

2. Nagendra, I.V. and Gujar, U.G.'3-D Objects from 2-D Orthographic Views -A Survey' Comput. and Graphics , Vol. 12 No. 1, pp. 111-114, 1988

3. Aldefeld, B. 'On automatic Recognition of 3-D Structures from 2-D representations' Comput. Aided Des. , Vol. 15 No. 2, pp. 59-64, 1983

4. Sakurai, H. and Gossard, D.C. 'Solid Model Input through Orthographic Views' ACM/SIGGRAPH, Vol. 17 No. 3, pp. 243-252, 1983

5. Lequette, R.'Automatic Construction of Curvilinear Solids from Wire-frame Views' Comput. Aided Des. , Vol. 20 No. 4, pp. 171-180, 1988

6. Markowsky, G. and Wesley, M.A. 'Fleshing Out Projections' IBM J. Res. and Dev. , Vol. 25 No. 6, pp. 934-954, 1981

7. Senda, T. 'Reconstruction of Solid from a Set of the Orthographical Three Views' Transactions of Information Processing Society of Japan, Vol. 31 No. 9, pp. 1312-1320, 1990

8. Markowsky, G. and Wesley, M.A. 'Fleshing Out Wire Frames' IBM J. Res. and Dev. , Vol. 24 No. 5, pp. 582-597, 1980

9. Preiss, K. 'Constructing the Solid Representation from Engineering Projections' Comput. and Graphics , Vol. 8 No. 4, pp. 381-389, 1984

10. Gujar, U.G. and Nagendra, I.V. 'Construction of 3D Solid Objects from Orthographic Views' Comput. and Graphics, Vol. 13 No. 4, pp. 505-521, 1989

Invited Paper

Coherence in computer graphics

E. Gröller, W. Purgathofer

Institute for Computer Graphics, Technical University Vienna, A-1040 Vienna, Austria

ABSTRACT

Coherence denotes similarities between items or entities. It describes the extent to which these items or entities are locally constant. An introduction to coherence and a survey of various types of coherence, that are used in computer graphics, are given. Techniques and data structures for exploiting coherence in computer graphics are described. Incremental techniques, bounding volume schemes, subdivision techniques and several geometric data structures are discussed in more detail. Applications of coherence principles to computer graphics are treated and a survey of previous research is done.

INTRODUCTION

General remarks

The widespread application of coherence principles allows only a vague definition of coherence. Without giving a formal definition coherence denotes, in the context of this paper, similarities between items or entities. It describes the extent to which these items or entities are locally constant. In many situations properties do not change drastically but rather in a smooth or continuous way ("natura saltum non facit"). In science coherence properties have already been used for quite a long time. Laser optics is one notable example, where coherent rays of light (similar frequencies and phases) are investigated. Coherence is based on the principle of locality, whereby "nearby" things have the same or similar characteristics. In the field of computer science and computer graphics coherence properties have been exploited in a variety of different methods and techniques. With coherence,processing of data can be accelerated considerably, e.g., hidden line/hidden surface removal. Data can be stored more efficiently by eliminating redundancies due to coherence, e.g., run length encoding of raster images. Some approximate or incremental techniques are only useful in coherent situations, e.g., Newton iteration to determine the roots of a polynomial. In a coherent situation incremental methods can be designed that process an item by reusing the results obtained for a similar (coherent) item. Exploiting coherence leads to algorithms based on an incremental application of very simple operations, that in many cases can use "integer logic" to replace time-consuming floating-point arithmetic. In the following sections coherence is classified according to three different criteria: types of coherence, techniques and data structures for exploiting coherence, applications of coherence principles. These sections contain certain redundancies as some methods or techniques are mentioned in more than one section. The advantage of considering coherence from different points of view was considered to outweigh these redundancies.

Types of coherence
In this section a survey of different types of coherence is given. Some of these types of coherence are not mutually exclusive and they are overlapping in the sense that in certain situations they can be used interchangeably.

Scan-line coherence Scan-line algorithms for visible-surface determination have been one of the earliest examples in computer graphics that make use of coherence properties. With scan-line algorithms an image is generated sequentially, one scan line at a time. The fact that there are relatively few changes from one scan line to the next and almost the same objects (e.g., polygons) are visible on consecutive scan lines is called scan-line coherence. Therefore a scan line is processed by only updating the information of the previous scan line. As a reference see Foley[16].

Span coherence Coherence does not only exist between consecutive scan lines but is usually present within a single scan line too. Spans are portions of a scan line with some constant property, e.g., the same object is visible over a given span. All scan line based algorithms (e.g., hidden surface removal, polygon filling) do exploit span coherence, which is given between adjacent spans and within a single span as well. For references see, for example, Crocker[11] or Foley[16].

Depth coherence Depth coherence expresses the fact that the depth or distance to the viewer at some surface point changes gradually. Adjacent parts of a surface are usually close in depth. Therefore the depth at some surface point can be efficiently calculated by incrementally updating the depth information of an already processed nearby surface point. Furthermore the depth ordering of surfaces at one pixel is likely to be similar to the ordering at adjacent pixels. An incremental update operation is again feasible (Bronsvoort[8], Crocker[11], Foley[16]).

Area coherence Area coherence follows from image coherence and is given if adjacent pixels of a raster image have the same or similar color or intensity values. This situation arises if, for example, a group of pixels is covered by the same object or the same visible surface. In area coherent portions of an image the calculation of one pixel value allows for the calculation of values of nearby pixels with significantly less computation (Joy[33]). Warnock's area subdivision algorithm for visible surface determination of a scene consisting of polygons is a prominent example of a method that makes use of area coherence (Foley[16]).

Object coherence Object coherence is based on some known relationships between objects or between parts of the same object. Objects may be disjoint, may be closely clustered or may consist of collections of low-level geometry, which are usually connected, smooth and bounded (e.g., polygons, surfaces). Local neighborhoods of space are likely to be occupied by the same objects. Depending on the given relations, manipulations like clipping, sorting, comparison, or intersection can first be performed at the object level to reduce the complexity and volume of calculations on low-level geometry. References to object coherence are further made in Crocker[12], Foley[16], Fujimoto[18], Glassner[22], Green[23], Haines[28], Joy[33] and Weghorst[60].

Spatial coherence Spatial coherence describes spatial homogeneities. These are a consequence of constant or slow varying relationships in the spatial arrangement of objects or data. Volume data like flow fields of a viscous fluid often exhibit a high degree of homogeneity. Considering the spatial arrangement of objects, spatial coherence is somewhat similar to object coherence. References to spatial coherence are in Chapman[10], Foley[16], Fujimoto[18], Gröller[24], Heckbert[29] and Watt[59].

Temporal coherence Temporal coherence occurs whenever a dynamic environment changes smoothly over time, e.g., small viewpoint and/or object movement (object space temporal coherence). Discretization of a temporal coherent situation allows the calculation of the relevant information by using results obtained for previous time steps. Temporal coherence has been used in a variety of methods for accelerating the calculation of animation sequences. Often only restricted situations with high temporal coherence are handled by these techniques. Restrictions can be one of the following: fixed viewpoint or restricted camera movement (Badt[3], Chapman[9,10], Huber[30], Murakami[40], Plantinga[45]), no object movement (Hubschman[31], Plantinga[45]), fixed light sources, restricted class of objects like planar or convex objects (Chapman[10], Hubschman[31]), etc. One attempt to exploit temporal coherence without putting any restrictions on the scene or object description is made in Gröller[24].

Frame coherence Frame coherence denotes the fact that successive frames of an animation sequence or video sequence are likely to be very similar if the difference in time is small, i.e., the projection of an environment tends to change continuously over time. Frame coherence is an immediate consequence of temporal coherence and object coherence and may be considered to be an image space temporal coherence. Frame coherence allows an efficient calculation and storage of video sequences. Frame coherence is sometimes called frame-to-frame coherence. References to frame coherence are in Chapman[9,10], Crocker[12], Foley[16], Glassner[22], Gröller[24], Green[23], Huber[30], Hubschman[31], Müller[41], Netravali[42], Stocker[55], Tost[56,57].

Image coherence Image coherence is the view-dependent analogue to object coherence. It results from the transfer of object coherence properties to the image plane by well-behaved projections, e.g., orthogonal or perspective projections. In the 2-D image plane there is at least the same degree of connectedness and smoothness as among the original 3-D objects (Glassner[22]). Local constancy of object space translates to local constancy in image space with only gradual changes. Some additional coherence is due to the projection technique itself. A given part of the image plane may be influenced by only a small subset of the items in object space. For references to image coherence see Amanatides[1], Arvo[2], Gervautz[19], Green[23], Haines[28], Marks[39], Netravali[42] and Weghorst[60].

Ray coherence, ray tree coherence Tracing rays through an object environment has been used extensively in computer graphics for image rendering purposes. The classical ray-tracing approach has been as follows: for each pixel of an image a ray is cast from the eye point into object space to determine the intersection point of the first visible object. To account for global illumination effects an approximation to the light distribution in the object scene is done by recursively casting further rays from this point of intersection. Depending on the surface properties of the intersected object, rays are cast in the direction of reflection and refraction. Shadow calculation is done by tracing rays from the point of intersection to the various light sources. Therefore, a ray tree is constructed to calculate the color information for each pixel. Due to the spatial coherence and object coherence of the scene, similar rays, i.e., rays with similar origin and direction, often have almost the same behavior. Similar rays often intersect the same object, the points of intersection are close together and so on. This property is called ray coherence. The similarity between ray trees is analogously called ray tree coherence. One of the major drawbacks of ray tracing has been its excessive computational cost. An extensive amount of research has been done to accelerate ray tracing by taking advantage of the various types of coherence that are inherent in the ray-tracing technique. Ray coherence and ray tree coherence are exploited in different ways (Amanatides[1], Arvo[2], Glassner[22], Gröller[24], Heckbert[29], Joy[32], Murakami[40], Ohta[43], Shinya[53], Speer[54]).

Other types of coherence The following less frequently used types of coherence are listed only for reasons of completeness: cell coherence (Semwal[51]), cluster coherence (Tost[56]), cube coherence (Haines[28]), data coherence (Arvo[2], Green[23], Pearce[44], Reghbati[49]), edge coherence (Foley[16]), external coherence (Ohta[43]), face coherence (Foley[16]), frustum coherence (Vilaplana[58]), geometric coherence (Baraff[5]), hierarchical coherence (Tost[56], Ballard[4]), implied edge coherence (Foley[16]), intervisibility coherence (Marks[39]), invisibility coherence (Crocker[11]), objective coherence (Tost[56,57]), patch-to-patch coherence (Vilaplana[58]), path coherence (Shelley[52]), predictive coherence (Tost[56,57]), screen-area coherence (Crocker[12]), shadow coherence (Haines[28], Pearce[44]), space coherence (Murakami[40]), surface coherence (Vilaplana[58], Ohta[43]), volume coherence (Max[38]).

TECHNIQUES AND DATA STRUCTURES FOR EXPLOITING COHERENCE

Incremental techniques

Incremental techniques are well suited to exploit coherence properties. Results from a previous, similar and therefore coherent situation are reused or modified to efficiently process the current situation. The Newton method for the incremental determination of the roots $f(x)=0$ of a continuous function is a typical example. Thereby the iteration formula $x_{i+1} = x_i - f(x_i)/f'(x_i)$ is used. Starting out from an approximation x_i to a root of $f(x)$ a better approximation x_{i+1} is determined. In certain situations there exist only incremental calculations to derive the desired result. In coherent situations incremental calculations are usually much faster than algorithms that do not reuse previous results. Incremental methods try to minimize costly operations during an iteration or replace them by easier operations, e.g., integer rather than floating-point arithmetic. Interpolation formulas that typically make sense only in coherent situations can be handled efficiently with incremental calculations. Some examples in computer graphics where incremental computations are used with great advantage are: raster conversion of lines, circle, ellipses, shading of polygons (Gouraud, Phong), intersection calculations (edge - scan line, ray - parametrically defined surface patches). One problem with incremental calculations is numerical stability. Care has to be taken that error propagation is kept under control. There are systems, namely chaotic deterministic dynamic systems, that are highly sensitive to calculation errors. In such cases incremental processing is not applicable.

Another often used application of incremental techniques is the incremental modification or traversal of data structures (Fujimoto[18], Glassner[20,21], Gröller[25], Murakami[40], Preparata[46], Speer[54]).

Data compression through incremental storage, i.e., storing only the hopefully small differences to previous data sets, is another example of coherence exploitation (Netravali[42], Stocker[55]).

Incremental techniques typically constitute a trade-off between computation cost and storage cost. Instead of doing a costly calculation of a result from scratch the previous result is stored and, usually, has to be modified only slightly. Incremental techniques introduce a sequential ordering into the processing which might interfere with parallelization efforts. Some further references that deal with incremental techniques are: Bennett[6], Bronsvoort[8], Crocker[11], Foley[16], Fujimoto[18], Glassner[20,22], Gröller[25,27], Hubschman[31], Joy[32,33], Kleij[36,37], Murakami[40], Plantinga[45], Preparata[46], Speer[54], Tost[56,57], Vilaplana[58].

Bounding volumes

The central idea of bounding volumes is the assumption that it is cheaper to process several simple objects (bounds) instead of a complex object. A complex (geometric) object is approximated by a simple bounding volume. This strategy pays off if most of the calculations can be done with the simple approximating bounding volume and only a few calculations must be done with the complex object itself. A typical example is the ray-object intersection test of ray tracing: A ray is first tested against the simple bounding volume. Only if a ray pierces the bounding volume, a more complicated ray-object intersection test is done. The additional overhead for those rays that intersect the bounding volume and hence have to be tested against the complex geometric object anyway is easily compensated by the fact that most rays are far away from the object. This can be determined efficiently by a simple ray - bounding volume intersection test. Bounding volumes are dealt with in more detail in Glassner[22] and Weghorst[60].

The usage of bounding volumes substitutes simple calculations for more costly ones. However, the number of calculations is not decreased. Computation expense may be thus reduced by a constant factor but no improvement on the linear time complexity (in the number of objects) is achieved. Bounding volume hierarchies have been developed to attain a logarithmic time behavior in the number of objects. Bounding volumes are grouped together and are enclosed within larger bounding volumes recursively so that a hierarchy (tree) of bounding volumes is generated. Processing a parent volume may make it superfluous to consider bounding volumes and objects contained within. For example, if a ray does not intersect a parent volume there is no need to check for an intersection with all the corresponding child volumes. With one intersection test it is possible to eliminate many objects from further consideration. Kay[35] lists some requirements a bounding volume hierarchy should meet to fully exploit coherence properties: Objects close together in the underlying space should be close together in the bounding volume hierarchy. The volume of each node in the hierarchy should be minimal. The sum of the volume of all bounding volumes should be minimal. The construction of the tree should concentrate on the nodes nearer the root of the tree. Pruning a branch of the tree allows a large subtree to be removed from further consideration. The preprocessing time spent on constructing the hierarchy tree must be compensated by faster hierarchy based object manipulation. A traversal of a bounding volume hierarchy in sorted order is facilitated if bounding volumes on the same level of hierarchy are disjoint. Therefore overlapping of bounding volumes of the same level should be avoided as far as possible.

Bounding volume hierarchies may be defined for objects in image space, object space, ray space and so on. Some references that treat bounding volumes and hierarchies thereof are Arvo[2], Ballard[4], Chapman[10], Gervautz[19], Glassner[21,22], Gröller[24,25], Heckbert[29], Kajiya[34], Kay[35], Ohta[43], Speer[54], Weghorst[60].

Subdivision techniques

Subdivision strategies are rather common in computer science. The divide-and-conquer approach is often used to subdivide a problem into subproblems that are easier to process. The solution of the problem is then generated from the solutions to the subproblems. In many situations subdivision increases coherence properties in the subproblems by, e.g., reducing the complexity or dimensionality of the original problem. The scan-line algorithm for visible surface determination is a typical example: A 2D image is subdivided into 1D scan lines, the 3D object scene is intersected with 2D scan planes and visibility is finally determined for 1D spans at 0D span boundaries. In computer graphics subdivision is done to make use of the principle of locality that exists in coherent situations. Thereby subdivision tries to obtain coherent homogeneous regions that can be manipulated efficiently. Subdivision is applied to a variety of different spaces, e.g., image space, object space, ray space, time space. With subdivision faster calculation, parallel processing and reduction in storage requirements are achievable. Subdivision may be uniform (Fujimoto[18]) or non-uniform (Glassner[20]). A subdivision technique is called overlapping if subspaces have non-empty intersections, otherwise it is called non-overlapping. With overlapping subdivisions portions of the underlying space, i.e., those parts that lie in more than one subspace, might be processed more than once. Sometimes it is convenient to process subspaces in sequential order, e.g., according to distance to an eye point. This can usually be done easier with disjoint subspaces. Therefore non-overlapping subdivisions are preferable whenever possible.

There is a fundamental difference between subdivision techniques and bounding volume hierarchies: subdivision techniques select sets of objects based on given subspaces (volumes), bounding volume hierarchies select subspaces (volumes) based on given sets of objects (Glassner[22]). Subdivision techniques proceed top-down, subdividing a given space into smaller subspaces, whereas bounding volume hierarchies are constructed bottom-up. Bounding volume hierarchies produce tight bounds but sometimes poor (overlapping) hierarchies, subdivision techniques produce good (non-overlapping) hierarchies but weak bounds. In Glassner[21] a combination of both strategies is presented.

Data structures

Some specific data structures have been developed that are well suited to exploit coherence properties. Only a few examples are shortly mentioned here. A grid is a subdivision of 3D space into regular cubical elements to exploit object and hierarchical coherence. A grid is used for example to accelerate ray - object intersection in the ray-tracing technique (Fujimoto[18]). An octree describes an adaptive hierarchical subdivision of 3D space (Glassner[20]). Traversal of an octree in sorted order, e.g., along a ray, is more complicated than with a grid structure. An octree can however adjust to varying levels of coherence by subdividing incoherent regions more often. The BSP-tree (Binary Space Partitioning) is a hierarchical data structure designed for storing polygonal objects (Fuchs[17]). The implicit ordering given by a BSP-tree is convenient for those applications that require a processing of the polygons in sorted order, e.g., displaying with hidden surface removal. CSG (Constructive Solid Geometry) defines an object through a binary tree (Roth[50]). A CSG tree usually contains an efficient bounding volume hierarchy (hierarchical coherence).

There are a large number of other data structures described in literature that make use of coherence. In computational geometry many data structures utilize subdivisions to reduce the manipulation cost from linear to logarithmic. Examples are interval trees, segment trees, range trees, etc. The following references deal with data structures that exploit coherence properties: Arvo[2], Ballard[4], Blinn[7], Foley[16], Fuchs[17], Fujimoto[18], Glassner[20,21,22], Gröller[25], Kajiya[34], Pearce[44], Preparata[46], Roth[50], Weghorst[60], Wyvill[62].

Other

Another strategy for exploiting coherence is to handle coherent sets of objects together instead of processing single objects by themselves. Bounding volume hierarchies do follow this principle for objects with a given hierarchical ordering. One example in the field of ray tracing is briefly mentioned. Several approaches have been tested where sets of coherent rays are processed instead of single rays. These ray sets are either called cones (Amanatides[1]), beams (Heckbert[29]), ray-bounds (Ohta[43]) or pencils (Shinya[53]). The higher cost of manipulating ray sets is amortized among the rays that make up such a ray set. Caching (Arvo[2], Baraff[5], Green[23], Pearce[44]) is a strategy to keep the small portion of a database that was relevant in the previous processing step in a local cache memory. In the current processing step testing the small amount of data in the cache memory might substitute a costly manipulation of the large database.

APPLICATIONS OF COHERENCE PRINCIPLES

Hidden line/ hidden surface removal

Visible-surface algorithms are an important application of coherence principles. Chapter 15 of Foley[16] gives an excellent survey of visible-surface algorithms, most of which take advantage of coherence properties. Joy[33] contains important research papers on visible-surface algorithms that are grouped together into z-buffer algorithms, painter's algorithms, scan-line algorithms, ray-tracing algorithms and area/volume subdivision algorithms.

Scan-line methods rely heavily on coherence, e.g., scan-line, span, depth, edge coherence. Crocker[11] exploits invisibility coherence to speed up visibility calculation of scan-line algorithms. Crocker[12] describes a scan-line algorithm for incrementally displaying a moderately modified object database. Warnock's area subdivision algorithm (Foley[16], Blinn[7]) hierarchically subdivides image space into subspaces until homogeneous areas are generated. Area coherence and hierarchical coherence are considered. Recursive adaptive subdivision ensures that only incoherent parts of image space are partitioned more often. Binary space-partitioning (Foley[16], Fuchs[17]) is a subdivision scheme for visibility calculation of static polygonal environments. In a preprocessing step a BSP-tree is generated. Hierarchical coherence and cluster coherence are then exploited by an inorder traversal of the BSP-tree.

Ray tracing

Ray tracing is a technique for high quality image generation. For each pixel of the final image a ray (primary ray) is cast through object space. From the intersection point, if there is any, additional rays (secondary rays) are recursively traced. Shadow rays determine if the intersection point is shadowed by an opaque object. Reflection and transparency rays account for global lighting effects. So for each pixel a ray tree is constructed. Thereby nodes correspond to an intersection point and edges correspond to rays. These ray trees are evaluated by a shading model to compute color values for the pixels. Glassner[22] gives an excellent survey of ray tracing acceleration techniques. Most of the acceleration techniques exploit various kinds of coherence properties (object coherence, hierarchical coherence, ray coherence). Bounding volumes (Gröller[25], Kajiya[34]), bounding volume hierarchies (Kay[35], Weghorst[60]) and subdivision strategies (Fujimoto[18], Glassner[20]) may be used to accelerate the time consuming ray-object intersection test.

Bounding volume hierarchies offer tight bounds but poor (overlapping) hierarchies. Space subdivisions offer bad bounds but good hierarchies. Both approaches are thus complementary in their strengths and weaknesses. Glassner[21] presents a combination of bounding volume hierarchies and subdivision strategies to take advantage of the benefits of both techniques.

With directional techniques ray directions are taken into consideration to allow the processing of whole sets of rays instead of single rays (ray coherence). Directional information is explicitly incorporated into data structures (Arvo[2], Gröller[24], Haines[28]). Directional techniques are, however, characterized by high storage requirements.

Radiosity

The radiosity technique calculates light transportation between diffuse reflecting surfaces in a polygonal environment. So called form factors have to be calculated, which make up most of the cost of the radiosity technique. The hemicube method has been the first approach to determine form factors. Vilaplana[58] investigates different types of coherence properties (frustum coherence, surface coherence, patch-to-patch coherence) to speed up hemicube processing. In Marks[39] intervisibility coherence is investigated to speed up form factor calculation. In Cohen[63] a sequence of images, that rapidly converges to the final image, is calculated incrementally by progressive refinement. Already after a few iterations a good approximation to the final image is achieved. Further references on incremental radiosity are George[64] and Chen[65].

Visualization

Max[38] presents a method for visualization of 3D scalar functions as a combination of density clouds and contour surfaces. Area and volume coherence are employed in the rendering process.

Wilhelms[61] describes a coherent projection approach to direct volume rendering. Projection approaches process volume data region by region as opposed to ray tracing methods which process volume data ray by ray. Projection approaches are typically much faster than ray tracing methods as they take advantage of coherence properties among adjacent regions of the volume data.

Antialiasing

Aliasing in ray tracing is due to point sampling finite sized pixel areas by an infinitesimally thin ray. Area sampling by processing sets of coherent rays instead of single rays has been investigated in several ways. Thereby generalized rays operate simultaneously on entire sets of rays that are bundled as beams (Heckbert[29]), cones (Amanatides[1]), pencils (Shinya[53]) or ray-bounds (Ohta[43]). Usually generalized rays are only applicable under certain restrictions, e.g., beam tracing of Heckbert[29] is limited to polygonal objects only.

Computational geometry

In computational geometry various subdivision schemes (e.g., Voronoi subdivision, Delaunay triangulation, subdivision into slabs), various data structures (e.g., interval trees, segment trees, range trees), and various algorithms (e.g., sweep techniques, geometric searching methods) are based on coherence properties. Applying the principle of locality many algorithms allow a fast focus on the small relevant portion of a large database. Preparata[46] gives a survey on computational geometry techniques. In Müller[41] an acceleration method for ray tracing sequences of frames with fixed viewpoint and only slightly changing objects is given that uses range trees. Bennett[6] applies different techniques of computational geometry to exploit temporal coherence in computer animation. Various incremental algorithms are investigated in detail.

Basic raster graphics algorithms

Basic raster graphics algorithms, e.g., raster conversion, filling, clipping, are executed frequently. Fast and easy algorithms are absolutely necessary to produce a reasonably quick performance of raster graphics packages. Foley[16] gives a survey on basic raster graphics algorithms. Coherence properties are widely used. Most raster graphics methods are based on incremental techniques to replace time consuming floating point operations by simple integer arithmetic. Bresenham developed a classic line conversion algorithm that proceeds incrementally and uses only integer arithmetic. The incremental technique may be applied to the integer computation of circles as well. Raster conversion and filling of polygons are typically done with scan-line type algorithms (Foley[16]).

Modeling

In Kajiya[34] hierarchical coherence is exploited to efficiently model and ray trace geometrically very complex fractal landscapes. To avoid excessive storage requirements the fractal landscape is not kept in memory explicitly. A bounding volume hierarchy ensures a fast ray - fractal landscape intersection test. Gröller[25] describes a somewhat similar approach for modeling fractal like objects with geometric grammars. One other example in modeling that takes advantage of coherence is, as already mentioned, CSG-modeling (application of bounding volumes).

Data compression

Raster images typically exhibit strong image coherence and raster-image sequences are characterized by frame coherence and temporal coherence. Coherence is the cause of redundancies which are the part of the input data that may be omitted without reducing the information content. Eliminating redundancies leads to compressed input data with, however, increased error sensitivity (Netravali[42], Stocker[55]).

Computer animation

Video and computer animation sequences consist of slowly varying single images that produce the impression of continuous changes. The inherent high frame and temporal coherence can be exploited in various ways to accelerate the calculation of computer generated animation sequences. Many techniques increase coherence further by imposing restrictions on the input data: only static objects or objects with limited movement, fixed camera position or only restrained camera movement, only certain types of objects, e.g., polygonal or convex objects. Some other techniques make a trade-off between calculation speed and image quality.

Bennett[6] investigates the use of temporal coherence in computer animation. Incremental and adaptive techniques are analyzed from the point of view of computational geometry. The asymptotical time complexities of various algorithms are studied. However, only polyhedral objects given through faces, edges and vertices are considered.

The method given in Glassner[21] is based on two central concepts: spacetime ray tracing and a hybrid adaptive space subdivision/bounding volume technique for generating efficient, non-overlapping bounding volume hierarchies. Spacetime is an extension of object space by a temporal dimension. Moving objects in 3D object space are static objects in 4D spacetime. Ray tracing dynamic 3D objects is reduced to ray tracing static 4D objects. A combination of 4D-spacetime subdivision and bounding-volume hierarchies significantly reduces the cost of ray-object intersection tests.

Hubschman[31] is one of the earliest works dealing with temporal coherence. The incremental hidden surface calculation for a sequence of images is described. Objects are restricted to closed, convex, non-intersecting polyhedra. Objects are static and the animation results from a continuous movement of the viewer.

In Plantinga[45] the real-time hidden-line elimination for a rotating polyhedral scene is presented. A static object scene is assumed and viewpoint movement is restricted to simple rotations.

In Shelley[52] a method is given that animates a static polyhedral environment by defining a smooth path for viewpoint movement (exploitation of path coherence).

Incremental ray tracing (Murakami[40]) accelerates the calculation of a sequence of images with fixed viewpoint and changing object scene. The ray tree calculated for a pixel is kept and incrementally updated to get the ray tree for the same pixel one time step later. The major drawback of incremental ray tracing is the excessive storage requirement.

In Gröller[24] a general method (no restrictions on camera and object movement) without loss of image quality is discussed that utilizes ray-space, object-space and temporal subdivision. This technique is again characterized by large storage requirements.

Further references on coherence in computer animation are Chapman[9,10].

Parallelization

Coherence properties are used for parallelizing algorithms as well. The principle of locality ("nearby" objects do have the same or similar characteristics, e.g., require the same small subset of a large database) is well transferable to multiple-processor systems with local memory resulting in a low communication overhead. As ray tracing is very time consuming but easy to implement in parallel some research has been going on to speed up ray tracing by parallelization. Partitioning of object space between processors results in a straightforward subdivision of the model database with efficient memory use, but load distribution between processors is difficult. Partitioning of image space between processors allows an easy load distribution, however, an efficient division of the database among the processors is difficult to determine (Dippe[13], Green[23]).

Vilaplana[58] shortly mentions the parallelization of form factor calculation for radiosity. Use of coherence properties is proposed to design efficient strategies for the distribution of the model database.

Transputer networks (parallel multiple-processor systems with local memory and non-centralized processor communication via so called "links") have also been employed for parallel implementations of ray tracing (Purgathofer[47]) and radiosity (Feda[14], Purgathofer[48]). Coherence principles are taken into account to allow efficient communication and load distribution.

Shading

Phong shading and Gouraud shading are two prominent examples that make use of coherence properties to speed up the intensity calculation of polygonal faces (Foley[16]). Gouraud shading (intensity interpolation shading) calculates intensities at polygon vertices. By scan converting the polygon, intensities at interior points are calculated through intensity interpolation along polygon edges and scan lines. Phong shading (normal-vector interpolation shading) interpolates normal-vectors instead of intensities. Intensity calculation at an interior point is then done by using the interpolated normal vector for the chosen illumination model. Interpolations across polygon edges and scan lines are performed by efficient incremental calculations.

FINAL REMARKS

A survey on coherence principles has been given and applications of coherence properties to computer graphics were investigated. Only a loose definition of coherence has been made and a classification of coherence according to types of coherence, techniques for exploiting coherence and applications of coherence principles was done. A more detailed description of coherence in computer graphics with in-depth treatment of special topics (scan-line algorithms for CSG, calculation and storage of animation sequences, modeling of fractal like objects) is done in Gröller[26].

It turned out that the term coherence is widely used in quite different situations. The definition of coherence in this paper ("similarities between items or entities") has been deliberately chosen to be rather vague to include vastly varying cases. However, the danger of overusing such a loosely defined term exists. Almost any non-chaotic and non-stochastic situation does contain some amount of continuity and could thus be called coherent. The inflationary usage of the term coherence should be avoided. Only situations, where the exploitation of strong similarities or continuities result in a significantly better, e.g., faster, manipulation as compared to approaches that do not use these similarities, should be called coherent.

Taking advantage of coherence offers a great potential for increasing efficiency. The additional effort to determine if in a certain situation the coherence properties are strong enough to allow an easy exploitation, is very often compensated many times by the resulting performance gains.

REFERENCES

1. Amanatides, J., "Ray Tracing with Cones", Computer Graphics 18(3), July 1984, pp. 129-135.

2. Arvo, J., Kirk, D., "Fast Ray Tracing by Ray Classification", Computer Graphics, 21(4), July 1987, pp. 196-205.

3. Badt, S., "Two algorithms for taking advantage of temporal coherence in ray tracing", The Visual Computer, No. 4, 1988, pp. 123-132.

4. Ballard, D.H., "Strip Trees: A Hierarchical Representation for Curves", Communications of the ACM, 24(5), May 1981, pp. 310-321.

5. Baraff, D., "Curved Surfaces and Coherence for Non-penetrating Rigid Body Simulation", Computer Graphics 24(4), August 1990, pp. 19-28.

6. Bennett, C.B., "The use of temporal coherence in computer animation", Dissertation, UMI Dissertation Information Service, 1986.

7. Blinn, J.F., "Triage Tables", IEEE Computer Graphics and Applications, January 1990, pp. 70-75.

8. Bronsvoort, W.F., "Direct Display Algorithms For Solid Modelling", PhD thesis, Delft University Press, 1990.

9. Chapman, J., Calvert, T.W., Dill, J., "Exploiting Temporal Coherence in Ray Tracing", Graphics Interface´90, Canadian Information Processing Society Editor, Canada 1990, pp. 196-204.

10. Chapman, J., Calvert, T.W., Dill, J., "Spatio-Temporal Coherence in Ray Tracing", Graphics Interface'91, Canadian Information Processing Society Editor, Canada 1991, pp. 101-108.

11. Crocker, G.A., "Invisibility coherence for faster scan-line hidden surface algorithms", Computer Graphics 18(3), July 1984, pp. 95-102.

12. Crocker, G.A., "Screen-Area Coherence for Interactive Scanline Display Algorithms", IEEE Computer Graphics and Applications, September 1987, pp. 10-17.

13. Dippe, M., Swensen, J., "An Adaptive Subdivision Algorithm and Parallel Architecture for Realistic Image Synthesis", Computer Graphics 18(3), July 1984, pp. 149-158.

14. Feda, M., Purgathofer, W., "Progressive Refinement Radiosity on a Transputer Network", Second Eurographics Workshop on Rendering, Barcelona, May 1991.

15. Feda, M., Purgathofer, W., "Accelerating Radiosity by Overshooting", Third Eurographics Workshop on Rendering, Bristol, May 1992, pp. 21-32.

16. Foley, J., van Dam, A., Feiner, St., Hughes, J., "Computer Graphics Principles and Practice", Addison-Wesley, 1990.

17. Fuchs, H., Abram, G., Grant, E.D., "Near Real-Time Shaded Display of Rigid Objects", Computer Graphics 17(3), July 1983, pp. 65-72.

18. Fujimoto, A., Tanaka, T., Iwata, K., "ARTS: Accelerated Ray-Tracing System", IEEE Computer Graphics and Applications, April 1986, pp. 16-26.

19. Gervautz, M., "Three improvements of the ray tracing algorithm for CSG trees", Computers&Graphics, 10(4), Pergamon Press, 1986, pp. 333-339.

20. Glassner, A., "Space Subdivision for Fast Ray Tracing", IEEE Computer Graphics and Applications, October 1984.

21. Glassner, A., "Spacetime Ray Tracing for Animation", IEEE Computer Graphics and Applications, March 1988, pp. 103-113.

22. Glassner, A. (ed.), "An Introduction to Ray Tracing", Academic Press Ltd., 1989.

23. Green, S.A., Paddon, D.J., "Exploiting coherence for multiprocessor ray tracing", IEEE Computer Graphics and Applications, November 1989, pp. 12-26.

24. Gröller, E., Purgathofer, W., "Using temporal and spatial coherence for accelerating the calculation of animation sequences", Proceedings of EUROGRAPHICS'91, Elsevier Science Publishers, Vienna 1991, pp. 103-113.

25. Gröller, E., "Fractals and Solid Modeling", Proceedings of EUROGRAPHICS'92, Cambridge, September 1992.

26. Gröller, E., "Coherence in Computer Graphics", PhD thesis, Technical University Vienna, September 1992.

27. Gröller, E., Stocker, W., "ACC-lossless data compression for animation sequences", submitted for publication.

28. Haines, E.A., Greenberg, D.P., "The Light Buffer: A Shadow-Testing Accelerator", IEEE Computer Graphics and Applications, September 1986, pp. 6-16.

29. Heckbert, P.S., Hanrahan, P., "Beam tracing polygonal objects", Computer Graphics 18(3), July 1984, pp. 119-127.

30. Huber, S., "Ausnützung zeitlicher Kohärenz beim Raytracing", master's thesis, Institute for practical computer science, Technical University Vienna, 1988.

31. Hubschman, H., Zucker, St., "Frame-To-Frame Coherence and the Hidden Surface Computation: Constraints For a Convex World", Computer Graphics, 15(3), August 1981, pp. 45-54.

32. Joy, K.I., Bhetanabhotla, M.N., "Ray Tracing Parametric Surface Patches Utilizing Numerical Techniques and Ray Coherence", Computer Graphics 20(4), July 1986, pp. 189-195.

33. Joy, K.I., Grant, Ch.W., Max, N.L., Hatfield, L., "Tutorial: Computer Graphics: Image Synthesis", IEEE Computer Society Press, 1988.

34. Kajiya, J.T., "New Techniques For Ray Tracing Procedurally Defined Objects", Computer Graphics 17(3), July 1983, pp. 91-102.

35. Kay, T.L., Kajiya, J.T., "Ray Tracing Complex Scenes", Computer Graphics 20(4), August 1986, pp. 269-278.

36. van Kleij, R., "Implementation of a solid modelling system with quadratic surfaces", Reports of the Faculty of Technical Mathematics and Informatics no. 90-41, Technical University Delft, 1990.

37. van Kleij, R., Roussou, R., "A CSG Scanline Algorithm for Quadrics", Reports of the Faculty of Technical Mathematics and Informatics no. 90-90, Technical University Delft, 1990.

38. Max, N., Hanrahan, P., Crawfis, R., "Area and Volume Coherence for Efficient Visualization of 3D Scalar Functions", Computer Graphics 24(5), November 1990, pp. 27-33.

39. Marks, J., Walsh, R., Christensen, J., Friedell, M., "Image and Intervisibility Coherence in Rendering", Graphics Interface´90, Canadian Information Processing Society Editor, Canada 1990, pp. 17-30.

40. Murakami, K., Hirota, K., "Incremental Ray Tracing", Eurographics Workshop on Photosimulation, Realism and Physics in Computer Graphics, Conference Proceedings, Rennes, June 1990, pp. 15-29.

41. Müller, H., "Time Coherence in Computer Animation by Raytracing", Lecture Notes in Computer Science No. 333, Springer, March 1988, pp. 187-201.

42. Netravali, A., Haskell, B., "Digital Pictures", Plenum Press, 1988.

43. Ohta, M., Maekawa, M., "Ray-bound tracing for perfect and efficient anti-aliasing", The Visual Computer, Vol. 6, 1990, pp. 125-133.

44. Pearce, A., Jevans, D., "Exploiting Shadow Coherence in Ray Tracing", Graphics Interface´90, Canadian Information Processing Society Editor, Canada 1991, pp. 109-116.

45. Plantinga, H., Dyer, Ch.R., Seales, B.W., "Real-Time Hidden-Line Elimination for a Rotating Polyhedral Scene Using the Aspect Representation", Graphics Interface´90, Canadian Information Processing Society Editor, Canada 1990, pp. 9-16.

46. Preparata F.P., Shamos, M.I., "Computational Geometry, An Introduction", Springer, 1985.

47. Purgathofer, W., Zeiller, M., "Configuring transputers for ray-tracing", Applications of Transputers, IOS Press, 1990, pp. 276-283.

48. Purgathofer, W., Zeiller, M., "Fast Radiosity by Parallelization", Eurographics Workshop on Photosimulation, Realism and Physics in Computer Graphics, Conference Proceedings, Rennes, June 1990, pp. 173-183.

49. Reghbati, H., "An Overview of Data Compression Techniques", IEEE Computer, April 1981, pp. 71-75.

50. Roth, S.D., "Ray Casting for Modeling Solids", Computer Graphics and Image Processing, Vol. 18, 1982, pp. 109-144.

51. Semwal, S.K., "The slicing extent technique for fast ray tracing", Dissertation, UMI Dissertation Information Service, 1987.

52. Shelley, K.L., Greenberg, D.P., "Path Specification and Path Coherence", Computer Graphics 16(3), July 1982, pp. 157-166.

53. Shinya, M., Takahashi, T., "Principles and Applications of Pencil Tracing", Computer Graphics 21(4), July 1987, pp. 45-54.

54. Speer, R., DeRose, T., Barsky, B., "A Theoretical and Empirical Analysis of Coherent Ray-Tracing", Proceedings of Graphics Interface ´85, 1985, pp. 11-25.

55. Stocker, W., "Kompression von Computer-Animationen", master's thesis, Institute for Computergraphics, Technical University Vienna, November 1991.

56. Tost, D., Brunet, P., "A Definition of Frame-to-Frame Coherence", Computer Animation´90, pp. 207-225, Editor: Magnenat-Thalmann, N., Thalmann, D., Springer 1990.

57. Tost, D., "An algorithm of hidden surface removal based on frame-to-frame coherence", Proceedings of EUROGRAPHICS´91, Elsevier Science Publishers, Vienna 1991, pp. 261-273.

58. Vilaplana, J., Pueyo, X., "Exploiting Coherence for Clipping and View Transformations in Radiosity Algorithms", Eurographics Workshop on Photosimulation, Realism and Physics in Computer Graphics, Conference Proceedings, Rennes, June 1990, pp. 137-150.

59. Watt, A., "Fundamentals of Three-Dimensional Computer Graphics", Addison Wesely, 1989.

60. Weghorst, H., Hooper, G., Greenberg, D., "Improved Computational Methods for Ray Tracing", ACM Transactions on Graphics, 3(1), January 1984, pp. 52-69.

61. Wilhelms, J., Van Gelder, A., "A Coherent Projection Approach for Direct Volume Rendering", Computer Graphics 25(4), July 1991, pp. 275-284.

62. Wyvill, G., Kunii, T., Shirai, Y., "Space Division for Ray Tracing in CSG", IEEE Computer Graphics and Applications, 6(4), April 1986, pp. 28-34.

63. Cohen, M., Chen, S., Wallace, J., Greenberg, D., "A Progressive Refinement Approach for Fast Radiosity Image Generation", Computer Graphics 22(4), August 1988, pp. 75-84.

64. George, D., Sillion, F., Greenberg, D., "Radiosity Redistribution for Dynamic Environments", IEEE Computer Graphics and Applications, July 1990, pp. 26-34.

65. Chen, S., "Incremental Radiosity: An Extension of Progressive Radiosity to an Interactive Image Synthesis System", Computer Graphics 24(4), August 1990, pp. 135-144.

ACKNOWLEDGMENT

The authors would like to thank Martin Feda for valuable suggestions during the preparation of this paper.

Invited Paper

Temporal coherence in progressive radiosity

D. Tost, X. Pueyo

Software Department, Polytechnical University of Catalonia, ETSEIB - Avda Diagonal 647, Pta 8, Barcelona 08028, Spain

ABSTRACT

In this paper we firstly present a study of the use of temporal coherence in radiosity algorithms. We underline the fact that this may be used even in static environments given that the behavior of a radiosity algorithm, when computing the form-factors of a surface is the same as the behavior of an algorithm computing visibility for a sequence of view points corresponding to a linear movement. Secondly, we study the proposed solutions for dynamic environments. Finally, we propose an algorithm for dynamic environments which target is to reach a wide use of temporal coherence.

INTRODUCTION

Radiosity based lighting models have proved to accurately simulate the interreflections of light between diffuse reflecting surfaces [1]. The balance of energy between N surface patches obtained from the division of a whole, energetically closed environment can be expressed as:

$$B_i = E_i + \rho_i \sum_{j=1}^{N} B_j F_{ij}$$

$$i = 1, ..., N$$

where B_i is the radiosity of surface i, E_i is the energy directly emitted from surface i, ρ_i is the reflectivity of surface i and F_{ij} is the form factor that represents the fraction of energy leaving surface j and landing onto surface j.

The expression of the form factor F_{ij} is:

$$F_{ij} = \frac{1}{A_i} \int_{A_i} \int_{A_j} \frac{cos\phi_i cos\phi_j}{\pi * r_{ij}^2} * V * dA_j * dA_i$$

where, A_i and A_j are the area of the surface patches, V is 1 or 0 depending on whether dA_j is visible to dA_i or not; r_{ij} is the distance between the two patches and ϕ_i and ϕ_j are the angles between the surface patches normals and the line passing by the centers of the two patches. Several methods have been proposed in order to compute them, including:

- Analytical solutions
- Hemispace projection
- Raytracing

A survey of these methods can be found in [2].

The computation of the form-factors is the bottleneck of the algorithms. However, being view independent, they only need to be computed once in order to obtain a sequence of images of an observer moving continuously around a static scene. In dynamic environments, form-factors may change so they must be recomputed at each frame. This is a serious drawback for using extensively the radiosity in computer animation. However, the scenes in which radiosity is applied are usually closed enviroments, for instance rooms. We believe that animation in such environments will often deal with a reduced number of *actors* in front of a large number of static *furniture*. Thus, from one frame to the next one, the number of changes in the geometry and the attributes of the scene will be small and they will provoke predictable modifications of the form-factors.

In a previous paper [3], we presented a frame-to-frame coherent algorithm for hidden surface removal that allows significant per-frame computations savings. We herein propose to extend this method to the radiosity determination in animated scenes where the objects motion is piecewise linear and is known a priori.

The paper is divided into 3 sections: we first analyse how temporal coherence may be used for radiosities computations (section 1), we next describe the existing solutions (section 2), then we propose an algorithm based on progressive radiosity and intended to make a wide use of temporal coherence (section 3). Finally, we give our conclusions and future work mainlines.

1. TEMPORAL COHERENCE IN RADIOSITY

Temporal coherence capitalizes on the environment continuity existing over time. It may be applied to visibility computations [4][5] as well as shading [6]. In radiosity it may have several meanings. First, the interest of radiosity itself is based on temporal coherence; i.e. it is based on the fact that diffuse environments don't change their aspect when the viewer changes his position, as stated in the introduction. If we think on temporal coherence as a way to improve the efficiency of radiosity algorithms, so to speed up the form-factor computation, then we may find different types of temporal coherence. In a first level, these may be divided into three classes depending on the features of the environment:

- static scene - view independence (classic case).
- static scene - view dependence
- dynamic scene

Static scene - view independence

Here the concept of time is virtual, i.e. we think on a virtual observer moving from one patch to another. We may include in this class the well known uses of objective coherence based on view independent presorted structures of objects or surfaces like BSP trees [7],[8],[9] and octrees [10].

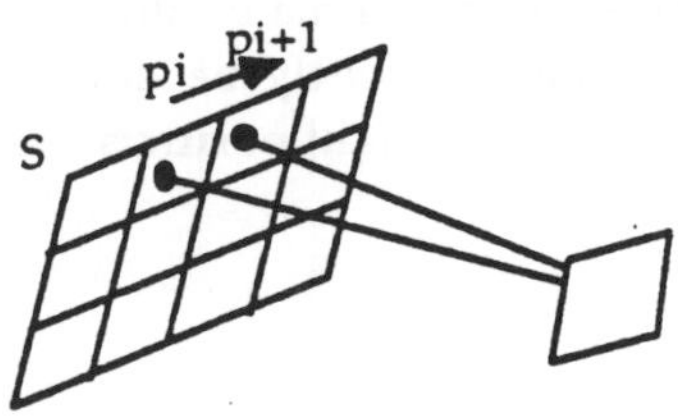

Figure 1. Virtual motions in form-factors computations: linear movement.

On the other hand, we find two kinds of virtual movements, when computing form factors, where predictable coherence may be used. First, when we proceed to compute the form factors of patches of a given surface, we usually "move" from a patch to its neighbour following a linear movement (figure 1). Secondly, the form factor of a patch is computed by means of performing the visibility function for several neighbouring directions. We may think, in this case, on several different circular movements around the center of the patch (figure 2). We may find examples of the use of these kinds of coherence in [11] and [9].

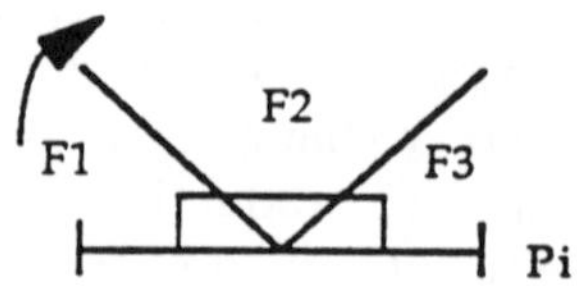

Figure 2. Virtual motions in form-factors computations: circular movement.

Static scene - view dependence

This class of algorithms includes the techniques proposed to speed up the radiosity process in complex architectural models [12]. In these papers the authors propose strategies to interactively update the radiosity results during walkthroughs. The main idea is to take advantage of the fact that the walls divide the environment into subenvironments with restricted visibility; i.e. that a viewer in a given subenvironment may only see a reduced number of subenvironments (figure 3). Then radiosity may be computed using only these subenvironments and some of their neighbours. Finally, the update of the radiosities may efficiently be done when the observer moves in a given direction towards a neighbouring subenvironment, using predictable coherence.

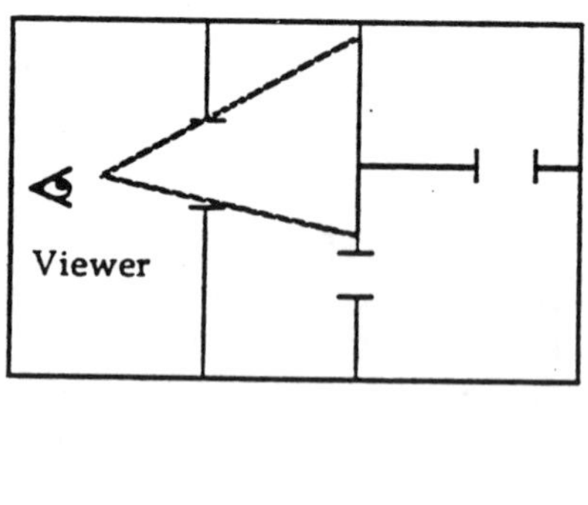

Figure 3. Visibility coherence during a walkthrough

Dynamic scenes

Now the goal is to efficiently compute radiosities at interactive rates in environments where the visibility relationships change. This happens when new objects appear in the scene or when the scene contains moving objects [13], [14]. Main concepts to be used in these cases are: precomputing static-to-static form-factors and considering the new occluding objects of a given frame, as surfaces shooting negative energy to the occluded patches. In the next section we describe these algorithms before introducing the proposed method which belongs to this class of temporal coherence uses.

2. RADIOSITY IN DYNAMIC SCENES

We describe in this section the two main approaches proposed to deal with an efficient computation of radiosity in dynamic scenes. Baum et. al. [14] used predictive coherence based on the knowledge of the paths of the dynamic objects, whereas Chen's algorithm [13] uses objective coherence and it is based on the progressive refinement approach.

The Back-buffer algorithm [14]

This technique is presented as a method for exploiting temporal coherence from the fixed "view point" of a static patch in a dynamic environment. In this case, the intervisibility only changes over time for a subset of patches. This subset is defined by the paths of the moving objects. More accurately, by their swept volumes. Three types of geometrical relationships are found in these environments:

- static patch to static patch.
- static patch to dynamic patch.
- dynamic patch to dynamic patch.

The temporal coherence is applied to the former relationship given its invariance over time.

A temporally occluded surface (patch) is defined to be a static surface (patch) that is partially or fully occluded from the "observer" (the reference patch) by a dynamic object at some point in time. These patches are recorded in a buffer (the back item buffer) so that a temporally occluded patch is known at each pixel (the hemicube technique is assumed). The back-buffer algorithm consists of two steps:

- Preprocessing: It computes all the static-to-static form factors and initializes the back-buffer. It is executed only once.
- Update: It computes visibility for the temporally occluded patches as well as for the dynamic patches. The former is obtained by comparing the current position of the dynamic object with the back buffer.

Incremental radiosity [13]

The progressive radiosity approach, differs from the classical one in that, at each cycle, a patch is chosen *(shooting patch)* and its unshot radiosity, B_i^u, is distributed to the other ones.

$$\begin{bmatrix} B_1 \\ B_2 \\ \cdot \\ \cdot \\ \cdot \\ B_i \end{bmatrix}^{new} = \begin{bmatrix} B_1 \\ B_2 \\ \cdot \\ \cdot \\ \cdot \\ B_i \end{bmatrix}^{old} + B_i^u * \begin{bmatrix} \rho_1 * F_{1i} \\ \rho_2 * F_{2i} \\ \cdot \\ \cdot \\ \cdot \\ \rho_n * F_{Ni} \end{bmatrix}$$

The process is performed iteratively until it converges. The shooting patches are chosen among those that contribute to the environment with the most energy, in order to enable the system to quickly converge.

The advantage of this method is that, as in other progressive refinement methods [15], in the first iterations, it allows to obtain, with a small cost, images of a low quality but sufficient in order to detect unwanted visual effects and to stop the process. This is particularly desirable in an animated sequence where many frames are to be produced and thus errors should be filtered early.

Incremental radiosity is an extension of progressive radiosity that allows to introduce changes in the environment at each solution cycle. The allowable changes are:

- surface attributes (emission or reflectance): solved by subtracting new and the old radiosity:

$$\Delta B_i = E_i' - E_i + \frac{(\rho_i' - \rho_i) * (B_i - E_i)}{\rho_i}$$

- scene geometry: solved by removing the contribution to the old geometry of the previously shooted patches and reshooting their radiosity to the new geometry:

$$B_j' = B_j + \rho_j * (B_i^s \Delta F_{ji} + B_i^u F_{ji}')$$

where $\Delta F_{ji} = F_{ji}' - F_{ji}$ is the incremental form factor and $B_i^s = B_i - B_i^u$ is the shot radiosity of patch i.

3. A FRAME-TO-FRAME COHERENT RADIOSITY ALGORITHM

Frame-to-frame Coherence in visibility computations

We first describe the main features of the frame-to-frame coherent algorithm of hidden surface removal proposed in [3]. The algorithm incrementally computes, for each frame, a depth sorted list of the scene faces which is used as input of the Painter's algorithm. It is based on a double structure: each object is represented by a temporal BSP-tree *(tBSP)* of its faces and the objects themselves are structured in a temporal graph of the depth priority relationships between them. In a pre-process, these structures are created and traversed in order to obtain the first depth sorted list of the scene faces. In successive frames, the list can be easily updated with partial traversals and with no geometrical computations.

We use BSP trees as representation schemes of the objects because they store information on the relative priorities of the objects faces, which, if no deformations are allowed, is independent on the geometry of the objects and therefore, valid throughout the whole animation sequence [7]. tBSP-trees are conventional BSP-trees in which, at each node, the sign of the dot product between the surface normal and the viewing direction is stored along with the time at which this dot product will change. Figure 4 illustrates this structure: *f* points at the faces list, *s* is the current sign, and *t* is the associated instant of change. These two fields are computed in the first traversal of the BSP tree and need only to be consulted at successive traversals. These traversals may even be avoided at some frames if the first instant of time in which a sign change will occur, is stored at the root level.

The temporal graph of the depth priority relationship between objects is computed based on their spacetime extents. It is represented by an adjacency list associated to each object, where, along with each adjacency the interval of time in which the occlusion exists, is stored (figure 4). It is showed in [17] that, a traversal of this graph can be performed, that maximizes the level (or number of ascendents along a tree path) of all graph nodes and thus structures the graph by levels. It is also demonstrated that the depth order of two nodes of same level is irrelevant. Therefore, the depth sorted list of objects is implicit in the levels of the structured graph. Two algorithms have been developed: the*maximum-level* topological sorting and the update of the graph.

A complete analysis of the complexity of the algorithms and implementations results may be found in [17], from which is it concluded that, if the rate of changes is moderate, they allow significant reductions in the computational time.

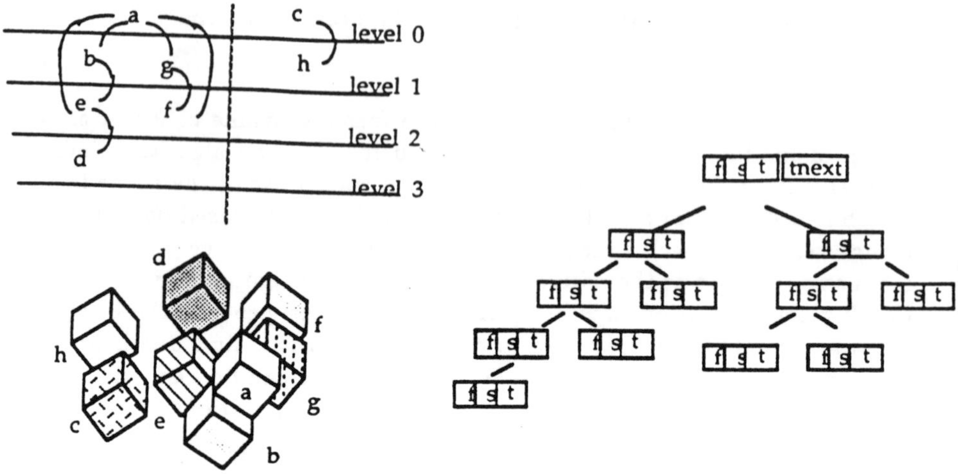

Figure 4. Data Structures of the Algorithm.

Extension of the algorithm to radiosities computation

As mentioned before, the computation of the form-factors involves the determination of intervisibilities between patches. Thus, an extension of the algorithm described above could be developed. We herein focus on the computations of the static-to-static and static-to-dynamic form-factors.

Given a static patch and a patch viewing direction, a conventional form-factor computation is first performed without taking into account the moving objects. The visible static patch in the patch viewing direction is the *background patch.* Its associated delta form-factor is computed and stored. Next, a temporal graph of occlusions between moving objects in the patch viewing direction is calculated. If none of the moving objects may occlude the background patches during the sequence, then the form-factors from the patch are definitely valid. Otherwise, the graph must be stored. At successive frames an update of the graph and a traversal of the t-BSP trees of the corresponding moving objects will give a depth sorted list of moving patches which last element is the background patch. A FTB traversal of the list will determine the visible patch. If it is not the background patch, its corresponding delta form-factor is computed but not stored. The temporal graphs may be computed at the surface level and at the object level. Research is currently being carried on in order to minimize the number of graphs.

Heuristic for choosing the shooting patch

As mentioned before, in the progressive approach, the shooting patches are chosen according to their potential contribution to the scene radiosities. A simple estimation of this contribution is: $B_i^u A_i$, where A_i is the patch area and B_i^u is the unshot radiosity of patch i [16].

For the incremental radiosity, Shenchang [13] suggests a similar test based on the following approximation of the form-factors:

$$F_{ji} = A_i / \sum_{j=1}^{N} A_j$$

We herein propose a different heuristic that, along with the patches contribution, takes into account two additional criteria:

- the visibility of the patches towards the observer
- the motion of the patches

As mentioned before, the computation of the radiosities is view-independent. The final picture is obtained in a separate process, generally a Z-buffer algorithm, in which only a small subset of the scene patches, the visible ones, is rendered. However, the radiosities of hidden patches must be computed because they contribute to the visible patches radiosities. We believe that, in a progressive approach, in order to choose the shooting patches, their visibility can be taken into account in two different ways by: first, at a same potential contribution, choosing visible patches in front of hidden ones, and second, avoiding to distribute the radiosities to low emittance hidden patches.

A main reason account to prefer visible patches rather than hidden ones: the probability that the energy shooted by a visible patch reaches, with one iteration, the other patches seen by the observer is higher. On the contrary, the energy leaving a hidden patch has a higher probability to be intercept when shooted to visible patches. This assertion is specially true when it is applied to culled patches.

In addition, if a hidden patch has a small potential contribution to the scene radiosities, then, not only it musn't be chosen as a shooting patch but even, the radiosities of the shooting patches may not need to be distributed to it. Hidden patches with a low potential radiosity are called *passive* patches, by opposite to the remainers, *active* ones. Passive patches are taken into account for visibility since the beginning of the process but, regarding to lighting, they are introduced progressively at given cycles. The algorithm is progressive in two different ways: by shooting a new patch at each iteration and by introducing new patches.

The difference between conventional and progressive radiosity has been expressed [18] in terms of *gathering in* light versus *shooting it out*. Our algorithm combines both approaches (figure 5). At each cycle where a patch is shooted, its contribution to the *nac* active patches is computed according to the formula:

$$for\ j = 1, .., nac,$$

$$B_j\ due\ to\ B_i = \rho_j * B_i^u * F_{ij} * \frac{A_i}{A_j}$$

At each cycle where a passive patch is activated, its radiosity is computed as:

$$B_i = E_i + \sum_{j=1}^{nac} \rho_i B_i * F_{ij} * \frac{A_i}{A_j}$$

$$nac = nac + 1$$

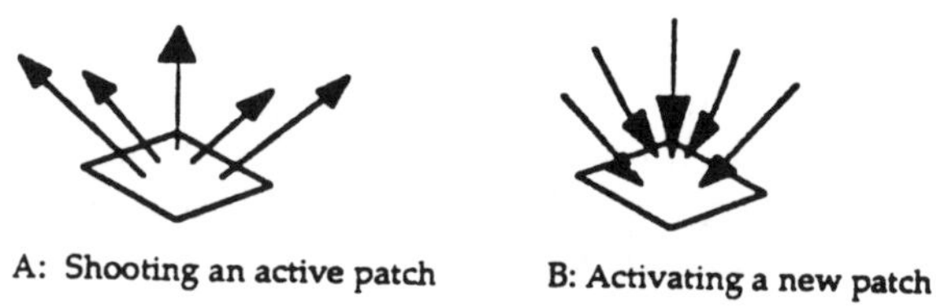

Figure 5. Active and Passive patches.

This approach is meaningless in static environments, where it is often desirable to change the viewpoint once all the radiosities have been computed, but, on the contrary, it is useful in a dynamic environment with a fixed viewpoint if a sequence of frames is to be obtained.

The scene structure allows to classify the patches at the object level and at the surface level. If an occluded object of low emittance is found, its faces, and consequently its surface patches, are classified as passive. Thus, it can be considered that low emittant back-faces are *culled* for lighting purposes although not for visibility.

Figure 6 illustrates our heuristic. In the perspective projection, objects A, B, C, H ang G are not visible because they are clipped off the viewing prism or they are occluded by other objects. The lamp C will be taken into account as an active object because, being a light source, it has a high potential radiosity. The same criterion can be applied at the surface level, to the computer's screen. Similarly, the high reflectivity of table H will determine that it is an active object. On the contrary, A, B and G as well as all their patches are to be considered as passive.

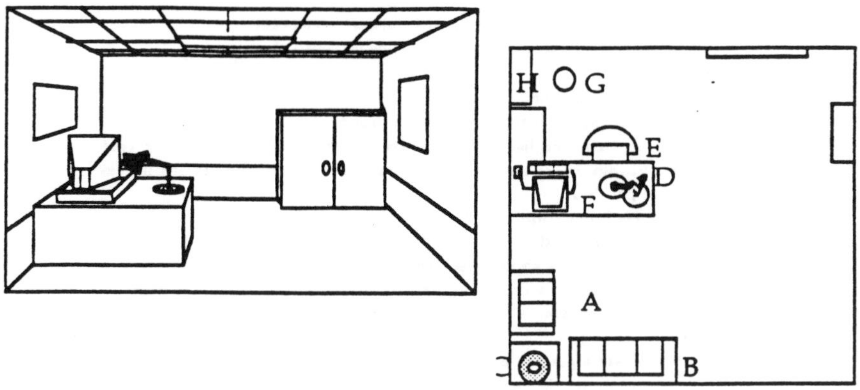

Figure 6. Heuristic for the shooting patch: a sample scene.

The application of this criterion supposes that the visibility status of each patch is known a priori. The three following steps must be realized in order to compute a single frame:

- determination of the patches visibility
- radiosities computations
- rendering and display of the visible patches

The visibility computation (step 1) can be performed identically as at the patch level: by traversing a temporal graph of the objects, constructed from the observer viewpoint, and the t-BSP trees. The resulting depth sorted list of the objects is the input of a FTB ray-casting that outputs a depth sorted list of the surfaces that really project onto the image plane. The third step of the process can here use this list as input for a FTB Z-Buffer. This allows to reduce the number of required radiosities interpolations.

In a dynamic environment, the three steps must be performed in turn at each frame. However, the visibilities determination is based on an update of the temporal graph. In addition, the radiosities can be computed according to the incremental approach developed in [13]. Motion will therefore be an additional criterion for choosing shooting patches, as visible, highly reflectant, moving patches will be selected in the first place, in order to subtract their previous contribution and add their new one.

4. THE ALGORITHM

We outline below the main steps of the proposed algorithm

```
Preprocessing
    Compute the tBSP trees
    Compute the temporal graph from the viewers viewpoint
    Compute static form-factors and visibilities intervals
First Frame
    Visibilities computations
    maximum-level topological sorting of the graph
    compute a depth sorted list of surfaces by traversing the objects tBSP-trees
    compute a visiblesurface list by ray-casting the depth sorted list of surfaces
    mark active patches
    Progressive radiosity computation
    while (not converged)do
        if ( non previously shooted active patches exist) then
            choose an active shooting patch distribute its radiosity to other
            active patches
        else
            activate a passive patch
            compute its radiosity
        endif
    endwhile
    Rendering
    Render visible surfaces
Next frames
    incremental visibilities computations
    incremental radiosities computations
    rendering of visible surfaces
```

CONCLUSIONS AND FUTURE WORK

Temporal coherence has been shown as a tool to improve radiosity computation not only in scenes with moving objects but also for walkthroughs and even in static environments with static observers.

We have analysed the solutions proposed up to now for an efficient computation of radiosity in dynamic environments confirming that only a few solutions have appeared.

We have proposed a framework for an extensive useof temporal coherence in dynamic environments. More accurately we propose to used it for computing the form-factors, by storing those are constant overtime and allowing to quickly compute the visibilities changes, and for speeding up the radiosities computation by reducing the number of patches involved in the environment's energy balance.

As future work we envisage to implement the different concepts developed in this paper in order to assess them empirically. We also foresee to apply temporal coherence to static environments and walkthroughs. Finally, we will try to combine temporal coherence with other types of coherences applied to form-factor computations.

REFERENCES

1 Goral, C, Torrance K., Greenberg D., Battaile B., 'Modeling the interaction of light between diffuse surfaces', *ACM Computer Graphics* , Vol. 18(3), pp. 213-222, 1984

2 Pueyo, X., 'Diffuse interreflections: techniques for form-factor computations: a survey',*The Visual Computer* , (7) pp. 200-209, 1991

3 Tost, D., 'An algorithm of hidden surface removal based on frame-to-frame coherence', *Eurographics'91*, Edited by Post,F. and Barth,W., North Holland, pp. 261-273, 1991

4 Hubshmann, H. Zucker, S., 'Frame-to-frame coherence and the hidden surface computation: Constraints for a convex world', *Computer Graphics* , Vol. 15 (3), pp 45-54, 1981

5 Shelley, K., Greenberg, D., 'Path specification and path coherence', *ACM Computer Graphics*, Vol. 16 (3), pp 157-166, 1982.

6 Groller, E., Purgathofer, W., 'Using temporal and spatial coherence for accelerating the calculation of animation sequences', *Eurographics'91*, Edited by Post,F. and Barth,W., North Holland, pp. 203-213, 1991

7 Fuchs, H.,Kedem, Z.,Naylor,B., 'On Visible Surface Generation by a priori tree structure', *ACM Computer Graphics*, Vol. 14(3), July 1980, pp 124-133.

8 Campbell, A., Fussell, D. 'Adaptative Mesh Generation for Global Diffuse Illumination',*ACM Computer Graphics*, Vol 24(4), August 1990

9 Vilaplana, J., Pueyo, X.,'Multi-level use of coherence for complex radiosity environment',*Research Report of LSI Dept.,Polytechnical University of Catalonia*, 1992

10 Wang, Y., Davis, W., 'Octant Priority for Radiosity Image Rendering'*Graphics Interface'90*, pp 17-30, 1990.

11 Tellier, P., Maiselle, E., Bouatouch, K., Languenou, E., 'Using coherence to accelerate radiosity',*Irisa, Publication Interne* , 616, November 1991

12 Teller, S., Sequin, C.'Visibility Preprocessing for Interactive Walkthroughs' *ACM Computer Graphics* Vol 24(4), pp 61-69, July 1991.

13 Chen, S.,'Incremental radiosity: an extension of progressive radiosity to an interactive image synthesis system', *ACM Computer Graphics*, Vol 24(4), August 1990

14 Baum, D., Wallace, J., Cohen M., Greenberg D., 'The Back-Buffer algorithm: an extention of the radiosity method to dynamic environments', The Visual Computer (2) pp.298-306, 1986.

15 Maillot, J., Carraro, L., Peroche, B.,'Progressive Ray Tracing', *Third Eurographics Workshop on Rendering*, Bristol, pp 9-19, 1992.

16 Cohen, M. F., Cheng S.,E., E.C.,Wallace, J.R., Greenberg, D.P., ' A Progressive Refinement Approach to Fast Radiosity Image Generation, '*ACM Computer Graphics*, Vol. 22(4), August 1984, pp 75-84.

17 Tost, D. 'Utilització de la coherència entre imatges successives per a la visualització d'escenes 3D en animació', PhD Thesis, Polytechnical University of Catalonia, March 1991.

18 Wallace, J.R., Elmquist, K.A., Haines, E.A. 'A Ray-Tracing Algorithm for Progressive Radiosity' *ACM Computer Graphics*, Vol. 23(3), July 1989, pp 315-324.

Invited Paper

Computer animation procedure with rendering techniques using low cost personal computers

A.F. Montagu

CAO Center, Faculty of Architecture, Design & Urbanism, University of Buenos Aires, R. Argentina

ABSTRACT

This paper discusses the advantages and limitations of existing software for video-animation procedures from the architects' and urban designers' point of view, considering how they conceive "morphological space".

One of the main advantages of using animation techniques is the simulation of the "space-time continuum" concept during the design process. For the first time in history, it has become possible to overcome the Renaissance concept of a view from a single point representing a static space.

Rendering and animation techniques also allow the simulation of textures, light and environmental conditions such as: natural lighting, shadows, transparency, fog, etc.

Most of the existing software is designed for multipurpose applications that respond to the demands of advertising, TV, video and films, particularly for creating special effects. The way that the "space" is built up in this type of applications can be considered as "hyper-realistic" and there is no connection with the ways in which the "urban-architectural space" was conceived in the past.

Therefore the main objective of this study is to develop a conceptual model that includes the constrains, the variables and parameters of the "design process" and introduce the power of "geometrical spatial generation" which is embedded in most of the video-animation and rendering techniques.

The paper includes some examples of architectural and urban visualizations, accompanied by a VHS video.

INTRODUCTION

The conception of urban-architectural space in different periods of history was related to the way of perceiving the outside world, and the methods by which this perception was represented through plans and drawings.
For the Renaissance master architects, architecture was thought of in terms of "three dimensional models" using drawing more as a conception mechanism than a technical description.

A.Palladio,1570 [1] and L.B.Alberti,1550 [2] systematized the architectural thinking process and geometric representation of space mainly using mathematical proportions as a design method [3] Fig. 1. The Renaissance urban space was also organized starting from a geometric conception based on focal points that derived from the laws of perspective, Fig. 2.

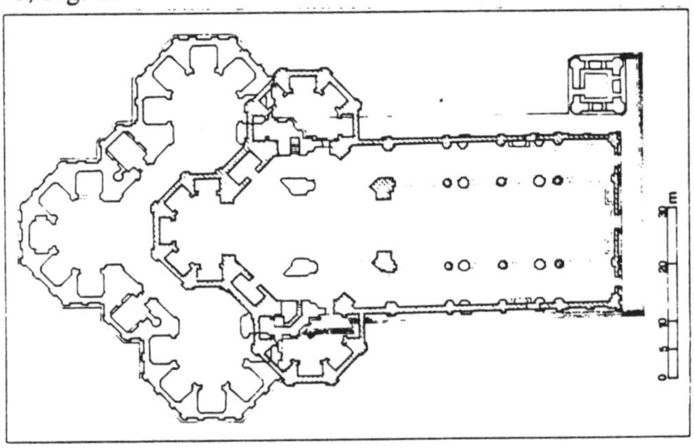

Fig. 1: St. Mary, Florence.

Fig. 2: St. Lorenz, Florence.

From this approach, it is possible to define a conceptual model as can be seen in Fig. 3 where the regularity of the relationship between components can be observed.

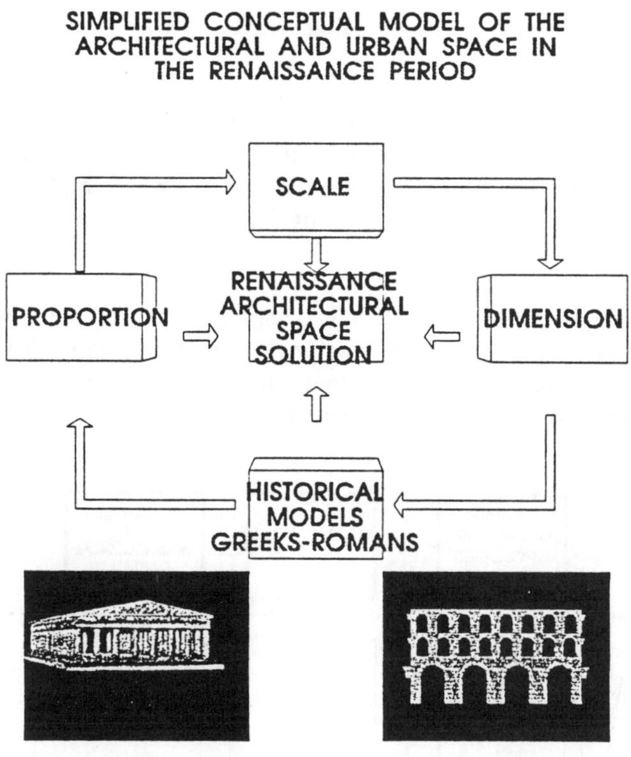

Fig. 3: Conceptual Model of the Space in the Renaissance Period.

The "Urban-Architecture Space" in the modern age has assumed a completely different character due to the influence of the "cubist movement" on the spatial concept of "modern architecture". The master works of Swiss-French architect Le Corbusier [6] Fig. 4, produced the turning point in terms of architectural evolution which also affected ideas about the urban environment.

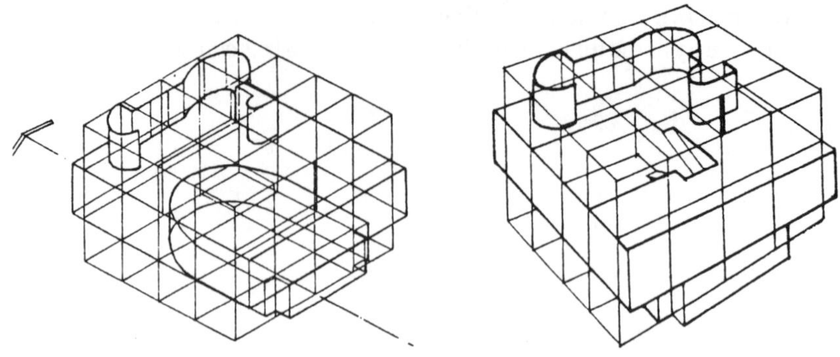

Fig. 4: Ville Savoie by Le Corbusier.

The "cubist movement" in painting and sculpture introduced the concept of "multiple perception" of reality which gave a totally different vision of the world.
This concept blends with the idea of space-time continuum which allows us to think of the "architectural space" as a "walk-through" experience rather than a view from a fixed point.

The "modern movement" in architecture completely changed the way to conceive "space" as a consequence of new functional requirements and a different design process. Fig. 5 shows a simplified conceptual model of the development of the "spatial organization" in modern architecture.

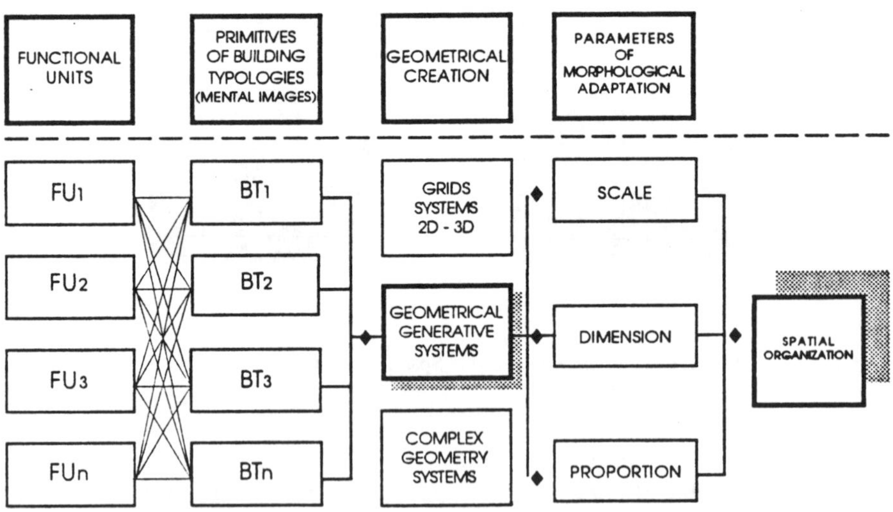

Fig. 5: Model of Spatial Organization in Modern Architecture.

DESIGN PROCESS AND MENTAL IMAGES

The design process movement lead by Alexander [4], Jones [5] and Archer [6] started in the sixties and has been highlighted again in the seventies and eighties with the advent of CAD languages using personal computers platforms, Moreyra, Ferrante, B. Videla & Montagu [7].

Design Process Methods were based in a kind of hierarchical decomposition of design problems in order to obtain partial solutions at each stage of the process. The process could be linear or non linear, but nevertheless the final results (architectural and urban forms or industrial design products) were not much different from those designed by traditional methods.
There is a gap between the functional requirements (the program) and the first sketches of the design idea (the solution) that, from the point of view of the designer, cannot be resolved by abstract methods as the designer is always thinking in terms of "mental images" like geometrical solutions for "building typologies" or "urban schemes".

These "mental images" can be numerical in a CAD environment showing the way to introduce this possibility at the beginning of the design process. As Porada and M. Porada [8] state: "Our hypothesis is that the numerical image can, under certain conditions, be used to express this core of information and to visualize the mental image of the future project as well as its representation, by means of intermediate representations which are some kinds of part shots".
Porada and Porada [op. cit.] also introduce the concept of "Infographical Work as the field of creation of numerical computer images born from three dimensional infographical models (mathematical and computer models). To summarize an infographical work is at the same time a sculpture, an image, a film and a written text".

The possibility to introduce render and animated images at the beginning of the design process, during the sketch design stage, gives a new tool for architects and designers not only in the sense of solid modelling with textures and varying lighting conditions but also as new mechanisms for geometrical creation.

DESIGN PROCESS WITH RENDERING AND ANIMATION PROCEDURES

The simulation of "space" in the urban and architectural environment was always the fundamental question from the point of view of "conception and creativity" during the design process.

The introduction of CAD procedures in the eighties and the extensive diffusion of these techniques during the decade has proved that without a proper design method embedded in the CAD environment, most of the designers are using these as drawing tools.

The three dimensional models, rendering and animation procedures of that time were to dull to compete with hand rendering techniques. At the end of the 80's, several programs for three dimensional modelling were developed for low cost PCs that completely changed the attitude of architects and designers.

For the first time, the simulation of textures (building materials) and physical environmental conditions is possible, to represent light (natural and artificial), shadows, reflectivity, transparency, fog and also all components of the outdoor environment such as: sky, green areas, waterpools, etc. It is also possible to simulate the indoor environment including: furniture, lighting conditions, etc.. Using "mapping textures", it is also feasible to simulate pictures.

This type of "reality" provides the architect with a very important simulation tool, that together with the "animation" procedure will allow the concept of "motion" to be incorporated for the first time in a three dimensional model.

This concept of motion gives the possibility to simulate the "walk-through, around or inside" the building or the city, given the concept of "space-time continuum" a real and "procedural-representational value" during the design process.

This procedural value has been determined after cheking two modelling systems: StrataVision 3d [9] and Topas [10] during the design process of a real scheme, obtaining very important feedback results from the designers and also selecting those attributes of these modelling systems that are most usefull during the design process.

Finally another advantage of these modelling systems is the relationship established with the standard format of most of the CAD files such as: DXF, IGES-DXF, TGA, GIF, etc., which reduce the problems of compatibility betwee PCs platforms.

DESIGN METHODS, CAD TECHNIQUES AND MODELLING SYSTEMS APPLIED

In the field of architecture and urban design, there should be a direct relation between design methods, CAD languages and modelling systems, since they are connected through the "creative strategy" developed by the project leader. This leader should have the ability to organize the coordinated participation of all specialists in the design team, Trum and Bax [11].

When this strategy is applied, it can be recognized as a "heuristic approach" in relation with the design methods used in the project. Fig.6 shows a simplified diagram of the relationship between design methods, CAD systems and 3D modelling systems which were applied in a rather complex urban project: the rehabilitation of the old Buenos Aires harbour.

The diagram shows the path that has been used through the design process and the principal nodes where feedback loops indicate the appearance of several design problems as a result of the sketches obtained using 3D modelling techniques.

The paper includes some prints obtained using the "Stratavision Modelling System" files; there is also a VHS Video with images produced with the "Topas System" as a complementary view of design information. In practice, architects and designers were reluctant to use video systems due to the lowresolutionofvideoimages comparedwith those of high resolution prints and on-line slides.

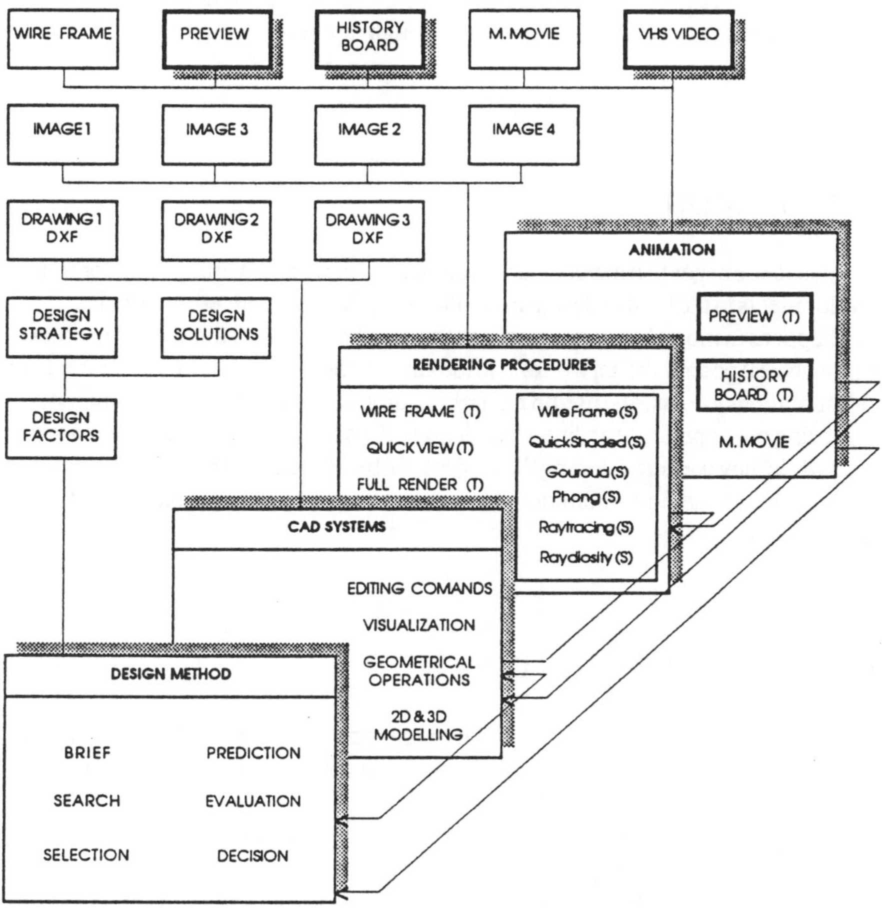

Fig. 6: Diagram of Design Methods, CAD Systems & 3D Modelling Software Applied.

The design process is not linear but in this case it can be summarized as follows:
a) At the beginning there is a design method approach, a"CAD group" and a "design by hand group" working together with somebody who acts as an "interface designer" between both groups.
b) Regarding the basic geometric planning information, the first CAD file was set up using the "Autocad System"[12]. The CAD file was translated into an interface format such as a DXF file which allows it to be processed with other 3D modelling systems.
c) This type of complex schemes always starts with several "mental image" solutions produced by the design group; these solutions are tested against a set of planning, architecture and environmental conditions, some of these conditions, such as: different lighting hours, high sun penetration, texture materials, pools of water, etc. can be simulated using certain rendering algorithms for: wire frame, quick shaded, flat shaded, gouraud, phong, raytracing and raydiosity [13] as can be observed in Fig. 7.
Figs. 8, 9, 10, 11, show several render examples of the same project. These examples are taking into consideration for the "walk through" approach as will be observed in the video.

CONCLUSION

The ideas and images that have been showed in this paper are the starting point for further developments in the field of design methods using CAD and rendering technics.
As it happens in most CAD systems about the possibilities of customization in order to exploit its real potentialities, we are also looking for the optimization of various render and animation systems to be used in the architectural and urban design environment.
Regarding our experience as "interface designer" and looking the architects reactions meanwhile they were developing the project, we believe that it is necessary to define a dynamic design method, open enough, in order to cover all the alternatives that occur during the process.

ACKNOWLEDGEMENT

I thank deeply to the master of argentinians architects Mr. Clorindo Testa who allows me to present in VIDEA 93 his design ideas about the renewall of the old harbour of Buenos Aires, competition award.
I would like to thanks Mr. Cristian Adamo, a young architectural student, who produced the printed images of this paper using the Stratavision system.
I also wish to thank Mr. Juan P. Cieri, Oscar Galdamez and Fabrizio Lazzazera, who prepared the VHS video using the Topas system.
I would like to mention Mrs Dora Castane and Mrs Laura Rivkind who worked actively during the design process producing the Autocad files.

BIBLIOGRAPHY

1. Palladio, A. "I Quattro Libri dell'Architettura" Venezia 1570, in Storia dell' Architettura del Rinascimento, Electra Editore, Milan 1978.

2. Alberti, L.B. "De re Aedificatoria" 1550, op. cit.

3. Baker, G.H. "Le Corbusier Analisis de la Forma", Gustavo Gili, Barcelona 1985.

4. Alexander, C. "Notes on the Synthesis of Form", Harvard University Press, Cambridge, Mass, 1964.

5. Jones, J.C. "A Method of Systematic Design" in Conference on Design Methods, Pergamon, Oxford 1963.

6. Archer, B. "Systematic Method for Designers" in Design, London 1963.

7. Ferrante, A.J., Moreira, L.F., B. Videla, J.M. and Montagu, A. "Computer Graphics for Engineers and Architects", Elsevier & Computational Mechanics Publications, Southampton-Boston, 1991.

8. Porada, S. and Porada, M. "Numerical Images, Mental Image and the Modern Architecture Triad", paper presented at the conference on Design & Decision Support Systems in Architecture & Urban Planning, Eindhoven University of Technology, Eindhoven, 1992.

9. Strata Inc. "Stratavision 3d" R 2.0 modelling system, users manual, 1991.

10. AT & T "Topas" modelling system, reference manual, 1992.

11. Bax T. and Trum, H. "The Concepts of Architecture", paper presented at the conference on Design & Decision Support System in Architecture & Urban Planning, Eindhoven University of Technology, Eindhoven, 1992.

12. Autodesk, "Autocad R 10", users manual, 1991.

13. Strata Inc. Rendering Algorithms of the Stratavision modelling system, 1991.

ADDITIONAL BIBLIOGRAPHY

14. Krier, R. "Urban Space", Academy Editions, London 1979.

15. Salmon, R. and Slater, M. "Computer Graphics, systems & concepts", Addison-Wesley, Wokingham, GB, 1987.

16. McCullogh, M., Mitchell, W.J. and Purcell, P. edit. "The Electronic Design Studio: Architectural Knowledge and Media in the Computer Era", MIT Press, Mass., 1990.

Wire Frame

Fig 7: View of the Harbour Using Rendering Algorithms

Quick Shaded

Flat Shaded

Gouraud

Phong

Raytracing

Raydiosity

Fig.8: Axonometric View of the Harbour Zone.

Fig.9: Perspective View with High Sun penetration (8a.m).

Fig.10: Perspective View with Sun penetration (6p.m).

Fig.11: Perspective View of the Harbour Zone with texture mapping at 6p.m

A distortion camera for ray tracing

P. Acquisto, E. Gröller

Institute for Computer Graphics, Technical University Vienna, Karlsplatz 13/186/2, A-1040 Vienna, Austria

ABSTRACT

Ray tracing is a powerful technique for realistic image generation. Typically a simple camera definition is used, whereby a 3D environment is mapped onto an image plane either by an orthographic or perspective projection. The concept of the usual simple camera definition is extended in several ways to achieve distorted views or projections of the object scene. The origins of primary rays are not required to lie on a plane anymore. Thus projections onto curved image surfaces are possible. The directions of primary rays may be chosen according to various nonlinear functions that allow nonstandard projections of the environment. The concept of centers of interest (coins) is introduced which enables to concentrate on especially interesting portions of object space. A center of interest is a 3D position that either distorts (attracts) a portion of the image surface or locally influences the directions of primary rays. The results of a test implementation are presented to show the feasibility of the methods presented. Several applications of distorted images are: creating elemental holographic images and raster omnimax images, virtual reality and arts.

INTRODUCTION

Ray tracing is a powerful technique in computer graphics for realistic image generation. Typically a simple camera definition is used to map the objects of a 3D environment onto a 2D image plane. The pinhole-camera model is a very simple approximation to the way the human eye perceives the 3D world (Figure 1).

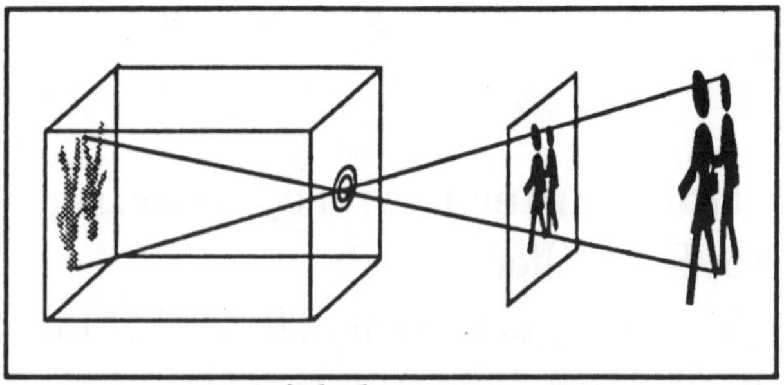

pinhole camera

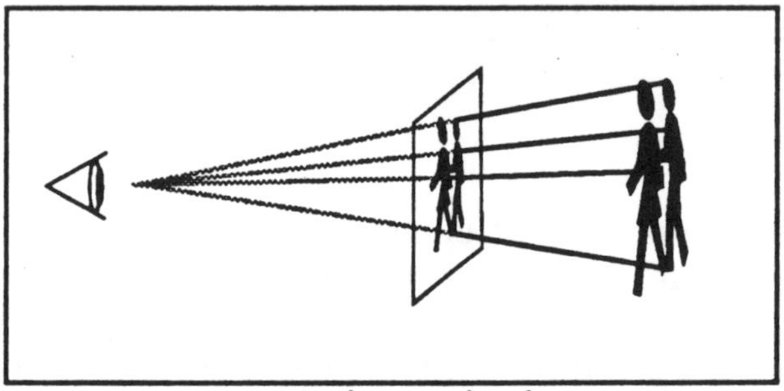

perspective projection

Figure 1: pinhole-camera model and perspective projection

The perspective projection (Foley[5]) is closely related to the pinhole-camera model and works as follows: given an eye point and an image plane a 3D point p of the environment is projected onto the image plane by taking the intersection of the image plane with the straight line segment between eye point and p (Figure 1). In the past a lot of projections differing from the perspective projection have been used that do not try to model human perception but are more appropriate for certain applications.

With Parallel projection the 3D environment is mapped onto the image plane using parallel rays (Figure 2). This kind of projection preserves length, angle and area on planes parallel to the image plane. Therefore certain properties of a 3D model can be easily deduced from its 2D image.

Cartographic projections (Paeth[14]) are another class of important projections that map the surface of the earth onto 2D maps (e.g., stereoscopic projection, cylinder projection, Mercator projection). Some of these projections are either length, angle or area preserving, although due to the nature of the problem (mapping a sphere onto a plane) no projection can fulfill all three properties at the same time. The angle preserving characteristic of the Mercator projection, for example, has been extensively used for navigational purposes.

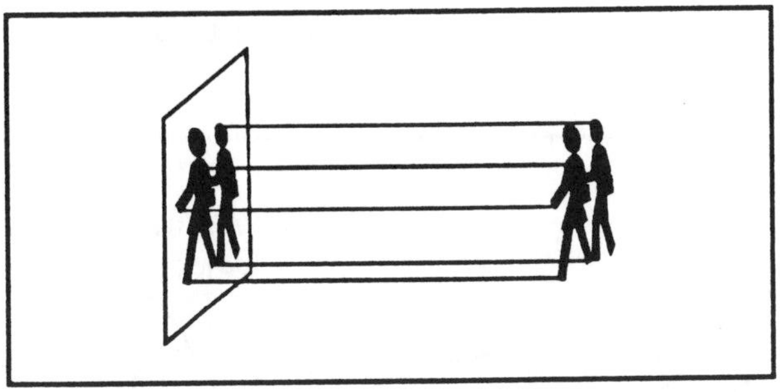

Figure 2: parallel projection

Nonstandard projections have been used by artists as well. In some of the paintings of Pablo Picasso more than one point of view is integrated within a single image thereby producing strange effects. For example, one might see a person from two differing views simultaneously. M.C. Escher is another prominent example of an artist that worked, among other things, with unusual projections, distorted views and distorted images (Ernst[2], Escher[3]). Multiple views within a single image are also used for generating Omnimax images (Greene[9], Max[12]) and holographic images (Haines[11]). Glassner[8] shortly mentions possible modifications to the camera model used in ray tracing. In the following section an extension of the pinhole-camera model (perspective projection) is defined and integrated into the ray-tracing technique.

VIRTUAL CAMERA

Ray tracing using perspective projection to calculate a raster image proceeds as follows: The image plane is subdivided into an array of rectangular pixels. For each pixel a ray starting at the eye point and passing through the pixel center is cast into a 3D environment to determine the first intersection of the ray with an object if any. If there is an intersection the color of the pixel is calculated by evaluating an illumination model (point light sources, area light sources, ...) and the properties of the intersected surface. Global lighting effects like shadows, reflections and refractions are taken into account by tracing additional rays starting at the point of intersection. For details on ray tracing see Foley[5], Glassner[6], Watt[17]. For each pixel a ray with origin on the image plane and direction given by the pixel position and eye point is processed. The separation of the concept of raster image from the image plane and type of projection allows an easy extension of the camera model (Figure 3).

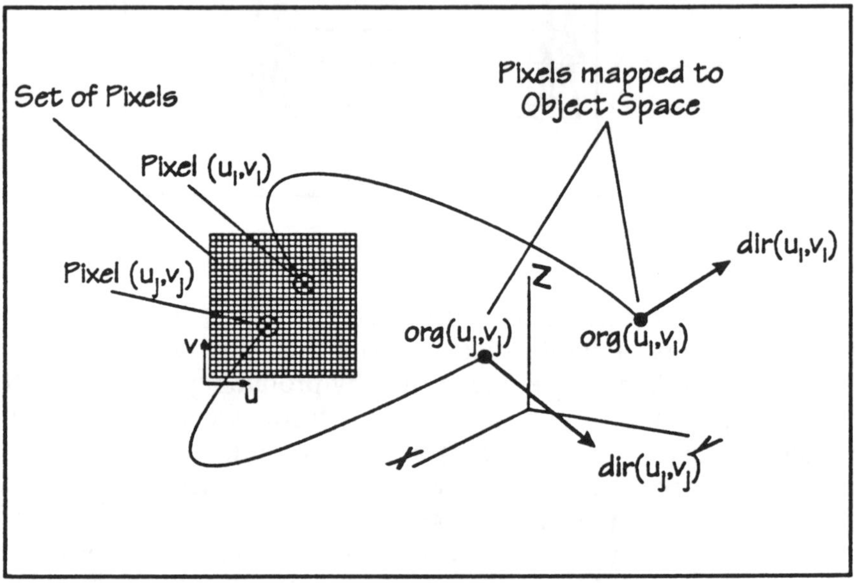

Figure 3: raster image in parameter space (u,v), *org(u,v)*, *dir(u,v)*

We now assume the raster image to be defined in a 2D parameter space, where each pixel is indexed by u and v. In addition two sets, a point set *org* and a vector set *dir*, are given which define a so called virtual camera. *org(u,v)* is the origin and *dir(u,v)* the direction of the ray that is taken into account to calculate a color value for pixel (u,v). The sets *org* and *dir* could be specified as values of a function mapping 2D parameter space into 3D points or 3D vectors (*org*: $R^2 \rightarrow R^3$, *dir*: $R^2 \rightarrow R^3$). With an appropriate definition of *org* and *dir* (points of *org* are lying on a plane, lines defined by the sets *org* and *dir* intersect in one point, i.e., the eye point) the usual perspective projection results (Figure 4).

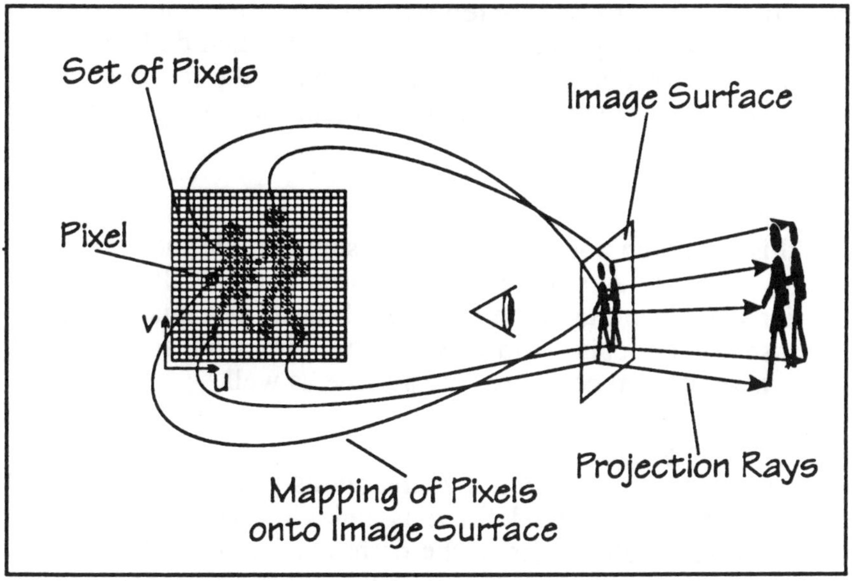

Figure 4: perspective projection defined through sets *org* and *dir*

In the general case the above concept does not pose any restrictions on the sets *org* and *dir*. For example, the set *org* may be an arbitrary point set or may define a non planar image surface (distortion of image plane). Arbitrary sets of *org* and *dir*, however, will usually produce only an arbitrary set of color values. Contrary to our intention no comprehensible image of the 3D environment would be generated in such cases. To achieve interesting results the sets *org* and *dir* should at least vary only smoothly with respect to u and v. Small local changes without discontinuities are preferable.

SOME SPECIFIC FUNCTIONS FOR *ORG* AND *DIR*

In the following some specific types of sets *org* and *dir* are discussed in more detail. Furthermore introducing dependencies between *org* and *dir* allow a simple specification of virtual cameras, e.g., direction *dir(u,v)* could be taken as the normal vector at position *org(u,v)* of the smooth surface given by *org*. Some of the combinations of functions discussed below, e.g., *org* defined by a sphere function and *dir* defined by a constant vector, may produce interesting results, while other combinations, e.g., *org* given as a constant point and *dir* as a constant vector, are not particularly interesting. It has to be pointed out that parametrizations that are used in the following functions are not unique and do have a great impact on the produced image. This is one of the reasons why for certain combinations of the sets *org* and *dir* the outcome is rather difficult to predict.

Plane function

With this approach points of *org* are assumed to define a rectilinear grid on a plane. A simple parallel projection is defined if the directions of *dir* are taken to be parallel. Distorted images result in this case by choosing more elaborate direction sets *dir*. The direction set *dir* can be defined by a plane as well.

Single point, constant vector

org is given as a single point and analogously *dir* can be defined to be a constant vector. *org* defined as a single point and *dir* defined by a plane function results in the common implementation of the perspective projection. With *dir* given as a constant vector all directions are parallel and depending on the specification of *org* an extension to the usual parallel projection is specified. If, for example, *org* defines a curved surface and *dir* is a constant vector then a parallel projection onto the curved image surface is given. Clearly taking *org* to be a single point and *dir* to be a constant vector at the same time does not produce anything interesting (all pixels of the resulting raster image will have the same color).

Dependent vectors

With this approach the directions of *dir* are defined with respect to *org*. If the point set *org* does specify a surface then *dir(u,v)* could be taken as the normal vector (or tangent vector) of that surface at position *org(u,v)*. *dir(u,v)* could be chosen to be a vector with constant deviation from the normal vector at *org(u,v)* as well. With *org* defining a plane this again will result in a parallel projection.

Sphere function

The surface of a sphere is easily parameterized by u and v (longitude and latitude). Therefore *org* may be taken to be part of the surface of a sphere (projection onto a spherical image surface). Using a sphere for the set *dir* is done by taking the vector from the coordinate center to the point on the sphere with parameters (u,v) as *dir(u,v)*. If the sphere does not enclose the coordinate center only a restricted class of directions is possible regardless of the ranges of the u and v parameters. Two different u, v values may produce the same direction vector (if the corresponding points on the sphere are collinear with the coordinate center). This results in interesting mirroring effects along a circle (or part of it) which consists of those points whose connections with the coordinate center are tangent to the sphere. Moving the sphere farther away from the center of the coordinate system produces a zoom effect as the class of possible directions is thereby reduced.

Cylinder function

The points of *org* are chosen to lie on a cylinder surface (or part of it). If the cylinder encloses the objects of the 3D environment and if the vectors of *dir* are taken to be normal to the cylinder surface the usual cylinder projections well known in cartography result. As with the sphere function, also the surface of a cylinder may be used to specify the direction vectors of set *dir*.

Free-form surfaces

From the discussion above follows, that generally any parametric free-form surface (Watt[17]) might be used to specify the sets *org* and *dir*. There are different ways to define such a surface. The free-form surface may be given explicitly by an analytic formula. In another approach a set of control points for a regular array of parameter values (u,v) is given. Interpolation schemes or approximation schemes (e.g., Bezier-, B-Spline, NURBS-Surfaces) then define a smooth parametric surface that either interpolates or approximates the given set of control points. If the control points are not given for a regular array of parameter values (u,v), scattered data interpolation techniques (Hagen[10], Nielson[13], Pottmann[15]) are applicable. Taking a differentiable free-form surface to define the set *org*, dependable vectors (e.g., normal vectors) are an interesting option to specify the set *dir*.

Other functions

Only a small subset of possible functions for the definition of *org* and *dir* has been discussed above (namely those that have been implemented in a test system). One can easily think of a wide variety of other functions and surfaces that are suitable for virtual cameras, e.g., cones, tori, affine transformations of quadrics like ellipsoids, superquadrics and so on.

CENTERS OF INTEREST (COINS)

In this section the concept of centers of interest (coins) is presented, which produces distorted projections of a 3D environment by allowing the user to concentrate on certain regions of object space. Some portions of the 3D environment might be more interesting to the user than other parts and he would like to get a more detailed view of that portion with the remaining 3D environment still present in the final image (something like a local zoom with a continuously varying zoom factor). This can be accomplished through centers of interest (coins). A coin is a position in 3D object space that marks an interesting portion. It acts like a magnet and locally deforms (attracts) portions of the image surface *org* and/or locally biases direction vectors of *dir* (Figure 5).

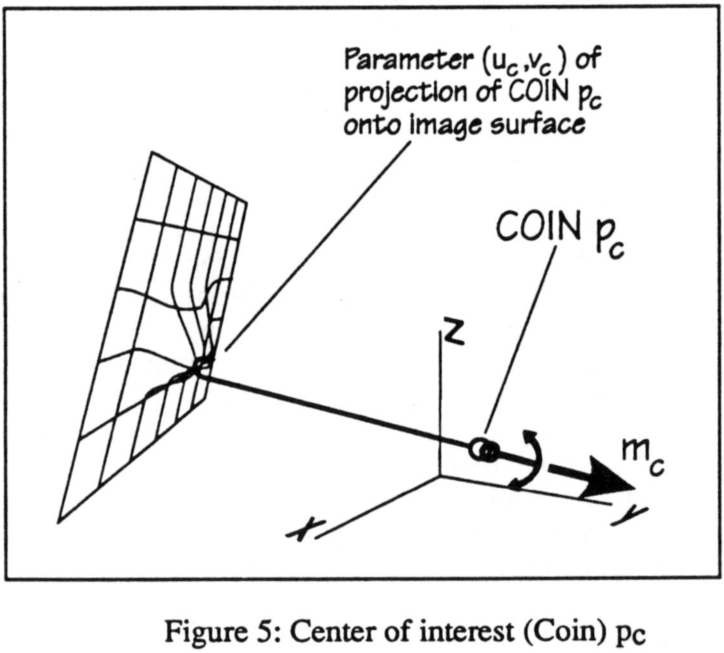

Figure 5: Center of interest (Coin) p_c

The deformation induced by a coin is determined as follows: The coin is projected onto the image surface *org* to calculate parameter values (u_c,v_c) where the influence of the coin is highest. *org(u,v)* and *dir(u,v)* of parameters (u,v) close to (u_c,v_c) are more modified than *org(u,v)* and *dir(u,v)* of parameters (u,v) farther away from (u_c,v_c). Generally the influence of a coin should decline smoothly with increasing distance in parameter space. Given a point p=*org(u,v)* the deformation caused by a coin with position p_c is calculated as follows (p_d is the new position of point p):

$$p_d = p + w(m_c, l, d, s) * (p_c - p)$$

The weight function $w(m_c, l, d, s)$ takes account of the strength m_c of the coin, the distance l of the coin p_c to point p, the parametric distance d between (u,v) and (u_c,v_c) and the stiffness s of the image surface *org*. A low stiffness factor, i.e., a rather flexible image surface is given, allows a significant deformation only very close to (u_c,v_c). The calculation of the deformation of a vector *dir(u,v)* is done analogously. The effect of more than one coin on the image surface is defined as the sum of the individual deformation effects:

$$p_d = p + \sum_i w_i(m_{c,i}, l_i, d_i, s_i) * (p_{c,i} - p)$$

Selecting a weight function w that is non zero only within a certain radius of influence around (u_c,v_c) will increase calculation efficiency. The concept of coins is entirely independent from the functional definition of a virtual camera through the sets *org* and *dir*. Therefore a combination of coins with the previous discussed methods can be done without problems.

IMPLEMENTATION AND RESULTS

The concepts discussed in the previous sections have been integrated into an existing ray-tracing system (Acquisto[1], Stürzlinger[16]). The incorporation of virtual cameras into a ray-tracing system can be done easily without requiring major modifications. Only the camera module of the system that generates origins and directions of primary rays has to be replaced to account for the nonstandard projections of virtual cameras. In general evaluation of the sets *org* and *dir* account only for a small portion of the total image generation time. The sample images (Image 1 - Image 14) were calculated on a simple PC (386, 25 MHz). Calculation time was in the range of 6 to 15 hours. Table 1 lists some statistics for those images.

image #	resolution	type of projection
1	640x480	perspective projection
2	640x480	*org*: plane, *dir*: constant vector, parallel projection
3	1280x320	*org*: cylinder, *dir*: normal vector to *org*, cylinder axis coincides with x-axis
4	1280x320	*org*: cylinder, *dir*: normal vector to *org*, cylinder axis coincides with line through (-1,1,1) and (1,-1,-1)
5	640x480	*org*: point, *dir*: sphere, not concentric with coordinate center
6	640x480	*org*: B-Spline surface, enveloping object half way, *dir*: normal vector to *org*
7	640x480	*org*: plane, *dir*: normal vector to *org*, one coin modifying *org* and *dir*
8	640x480	*org*: plane, *dir*: normal vector to *org*, one coin modifying *org* and *dir*
9	640x480	*org*: plane, *dir*: normal vector to *org*, two coins modifying *org* and *dir*
10	640x480	*org*: plane, *dir*: normal vector to *org*, two coins modifying *org* and *dir*
11	640x480	*org*: B-Spline surface, all control points on a plane elevated above object, *dir*: normal vector to *org*
12	640x480	*org*: as 11, one corner control point elevated, and diagonal opposite control point lowered, *dir*: normal vector to *org*
13	640x480	*org*: as 12, corner points offset by larger value, *dir*: normal vector to *org*
14	640x480	*org*: as 13, corner points offset by a still larger value, *dir*: normal vector to *org*

Table 1:

Aliasing due to the point sampling nature of ray tracing is more severe with virtual cameras as pixels close together may have primary rays with considerably differing points of origin and directions. The nonlinear projections of virtual cameras cause aliasing effects that vary in magnitude locally. Therefore anti-aliasing is done by adaptive sampling (critical regions are sampled more often). It turned out that images produced by using virtual cameras are rather sensitive to parameter selection. Choosing the appropriate parameters (e.g., parameters for the weight function of a coin) might be cumbersome in certain cases.

CONCLUSION

An extension to the usual camera model of ray tracing has been given. For each pixel (u,v) two functions *org*(u,v) and *dir*(u,v) are evaluated. *org(u,v)* is the origin and *dir(u,v)* the direction vector of the primary ray at pixel (u,v). Several functions *org* and *dir* have been discussed. The concept of centers of interest (coins) was introduced. Coins allow to focus on certain portions of the 3D object space (local nonlinear zoom). Coins are point locations in the 3D environment that distort (attract) the image surface and bias directions of primary rays.

In the following some ideas for future work are mentioned. Virtual cameras might be considered for the form factor calculations of non-polygonal objects with the radiosity method. We have not investigated this application in depth so far.

One can think of an interactive system for the definition and modification of virtual cameras. Given a fast hardware, specific parameters of virtual cameras could be animated. How about an animation where the strength of a coin m_C is increased from zero to some finite value? Examples where virtual cameras might be used are arts and virtual reality. If you have the freedom to define your own virtual reality why not take a look at it with your own virtual camera?

REFERENCES

1. Acquisto, P., "Virtual Irreality: Eine Verzerrungskamera", diploma thesis, Technical University Vienna, 1992.
2. Ernst, B., "Der Zauberspiegel des M.C.Escher", Heinz Moser publisher, Munich, 1978.
3. Escher, M.C., "Grafik und Zeichnungen", Heinz Moser publisher, Munich, 1979.
4. Foley, Th.A., Lane, D.A., Nielson, G.M., Franke, R., Hagen, H., "Interpolation of Scattered Data on Closed Surfaces", Computer Aided Geometric Design, Vol. 7, 1990, pp. 303-312.
5. Foley, J.D., van Dam, A., Feiner, St.K., Hughes, J.F., "Computer Graphics: principles and practice", Addison Wesley, 1990.

6. Glassner, A., "An introduction to ray tracing", Academic Press, 1989.

7. Glassner, A., "Graphics Gems", Academic Press, 1990.

8. Glassner, A., "The Theory and Practice of Ray Tracing", Tutorial Note 1, Eurographics'91, Vienna 1991.

9. Greene, N., Heckbert, P.S., "Creating Raster Omnimax Images from Multiple Perspective Views Using the Elliptical Weighted Average Filter", IEEE Computer Graphics & Applications, June 1986, pp. 21-26.

10. Hagen, H., "Scattered Data Methoden", Proceedings of Visualisierungstechniken und Algorithmen, Vienna, September 1988.

11. Haines, K., Haines, D., "Computer Graphics for Holography", IEEE Computer Graphics & Applications, January 1992, pp. 37-46.

12. Max, N.L., "Computer Graphics Distortion for IMAX and OMNIMAX Projection", Proceedings of Nicograph 83, December 1983, pp. 137-159.

13. Nielson, G.M., Foley, Th.A., Hamann, B., Lane, D., "Visualizing and Modeling Scattered Multivariate Data", IEEE Computer Graphics & Applications, May 1991, pp. 47-55.

14. Paeth, A.W., "Digital Cartography for Computer Graphics" in Glassner[7], pp. 307-320.

15. Pottmann, H., Eck, M., "Modified multiquadric methods for Scattered Data Interpolation over a sphere", Computer Aided Geometric Design, Vol. 7, 1990, pp. 313-321.

16. Stürzlinger, W., Tobler, R.F., "FLiRT-Program Source und Dokumentation", Institute for Computergraphics, Technical University Vienna, 1991.

17. Watt, A., "Fundamentals of Three-Dimensional Computer Graphics", Addison Wesley, 1990.

Image 1:

Image 2:

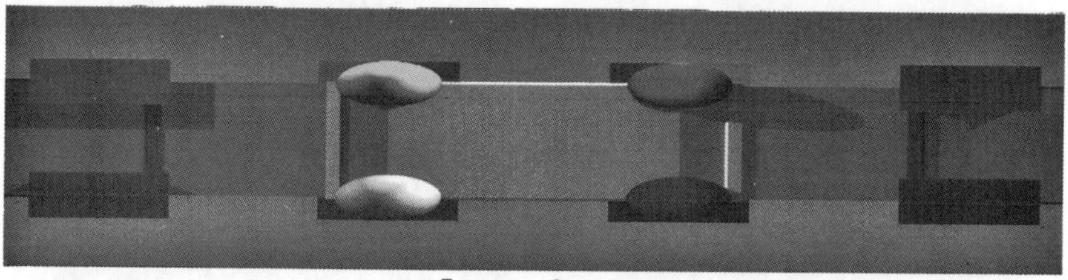

Image 3:

Image 4:

Image 5:

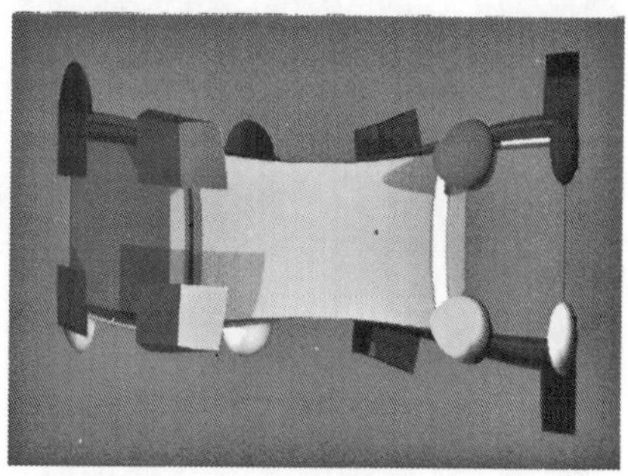

Image 6:

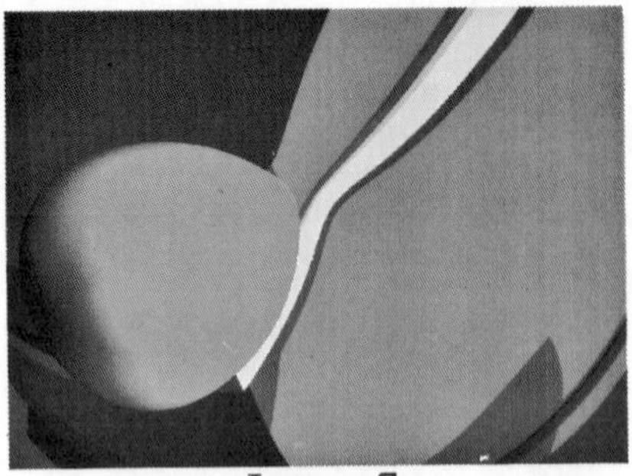

Image 7:

Image 8:

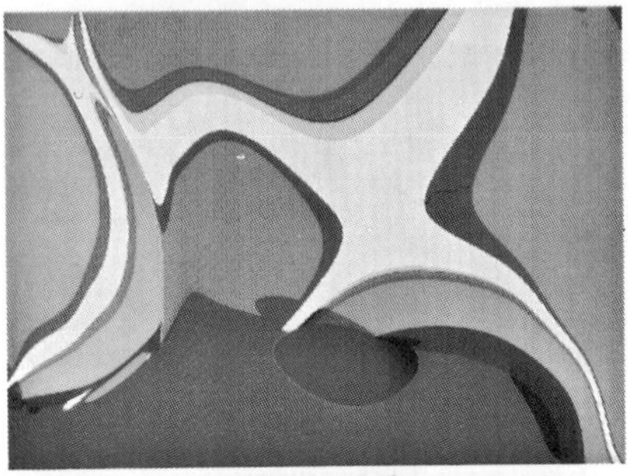

Image 9:

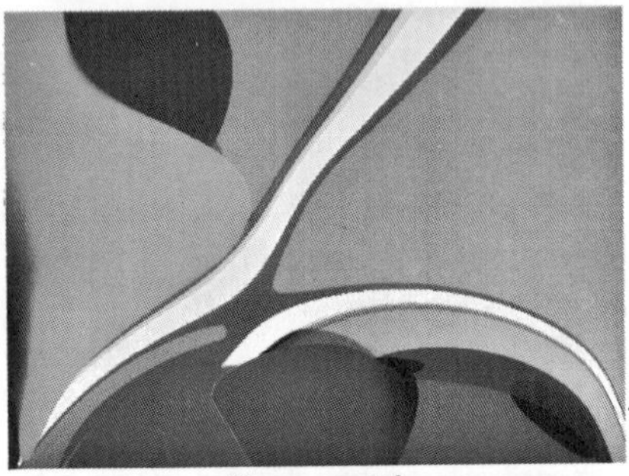

Image 10:

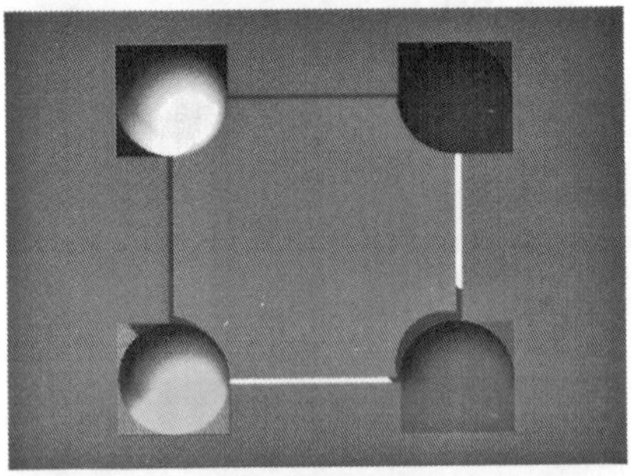

Image 11:

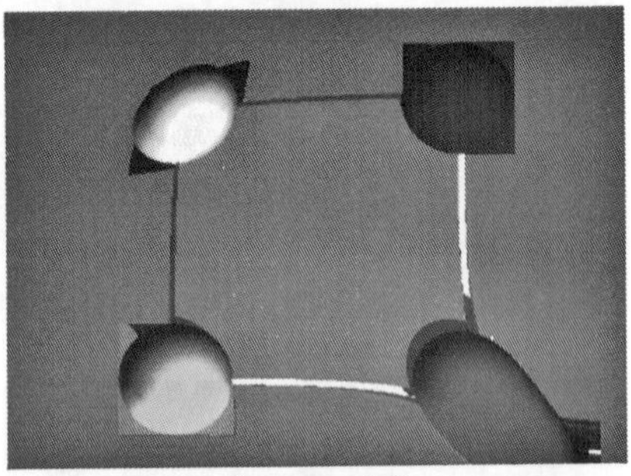

Image 12:

Image 13:

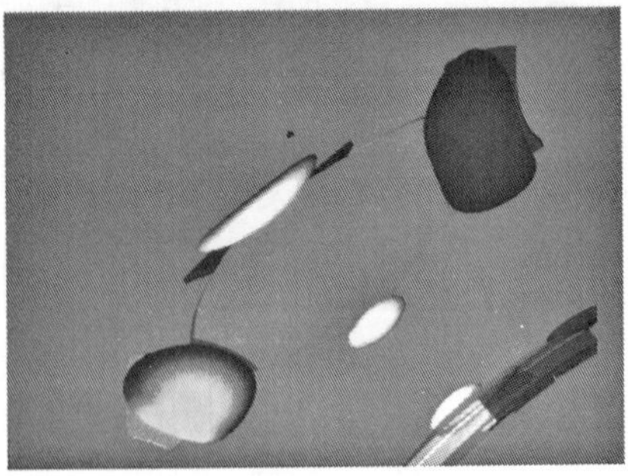

Image 14:

SECTION 2: ANIMATION

Invited Paper

An advanced environment for scientific visualization using cooperative processing for animation

M. Bernaschi, M. Briscolini, P. Santangelo, F. Valentini
IBM ECSEC, European Center for Scientific and Engineering Computing, Rome, Italy

ABSTRACT

This paper describes the IBM AIX **4-D Scientific Data Visualizer**/6000, an advanced tool for the visualization of three-dimensional numerical simulations, where a distributed environment has been used to separate conceptually the numerical simulation from the visualization process. An efficient communication interface let the two processes communicate and exchange data and allows each process to be run on the proper machine. High performance computing devices are used for the numerical simulations and advanced visualizing workstations display the results as they flow through the network. The visualization environment includes i) an interactive tool for the visualization of three-dimensional data, ii) a general purpose communication tool to interface a simulation program to the visualizer and iii) an easy-to-use video server for the production of video animations from numerical simulations. Some application areas where the visualizer has been utilized are briefly indicated.

INTRODUCTION

The visualization environment IBM AIX **4-D Scientific Data Visualizer**/6000 (hereafter referred to simply as **4DSDV**) exploits a simple type of cooperation between two processes where one of them (**the server**) is used to visualize the data which are computed by a numerical simulation program (**the client**). In this environment, the computational client can run on a workstation, on a cluster of workstations, on a mainframe or on any computing device capable of communicating through a TCP/IP network.

The particular visualizer described in this paper is a highly interactive tool for the display of the three-dimensional data used in many scientific and engineering simulations. It is based on the IBM graPHIGS [1] and takes advantage of the three-dimensional capabilities of the graphical accelerator cards. The different types of data managed by the visualizer presently include any combination of the following types: three-dimensional **scalar fields**, three-dimensional **spin-fields** and some primitives used for the representation of **molecular data**.

The video server is the last tool of the environment and is aimed at the production of video animations. It is built on the basis of a high-quality random-access video-disk which allows an exact frame-by-frame video recording. The video-disk is used as the nucleus of a multi-user video recording equipment which allows the simultaneous production of different video animations.

THE VISUALIZER

Our visualization program is based on the IBM graPHIGS [1] and is able to run on any IBM RISC System/6000 workstation equipped with Z-buffered three-dimensional graphics cards. The aim of our visualization program is to offer a simple and interactive environment for the exploration of the three-dimensional data sets, produced by modern scientific and technical simulation programs.

The visualizer is able to handle simultaneously the following types of data:

- three-dimensional **scalar fields**
- three-dimensional discrete **spin fields**
- chemical data including:
  - **stick-and-ball** models and
  - iso-surface representation of molecular field properties.

Work is in progress to add a consistent display of vector fields.

In order to make the interaction with our visualizer as easy and friendly as possible we developed an user interface based on a standard environment: the OSF/Motif and the X Window System [2, 3]. This interface provides on-screen control over the functions of the visualizer through a wide variety of mouse-operated widgets including sliders and buttons for the input of the parameters.

It is also possible to record the complete interaction process into a file which can be edited and used later, as input, for the production of complex animations.

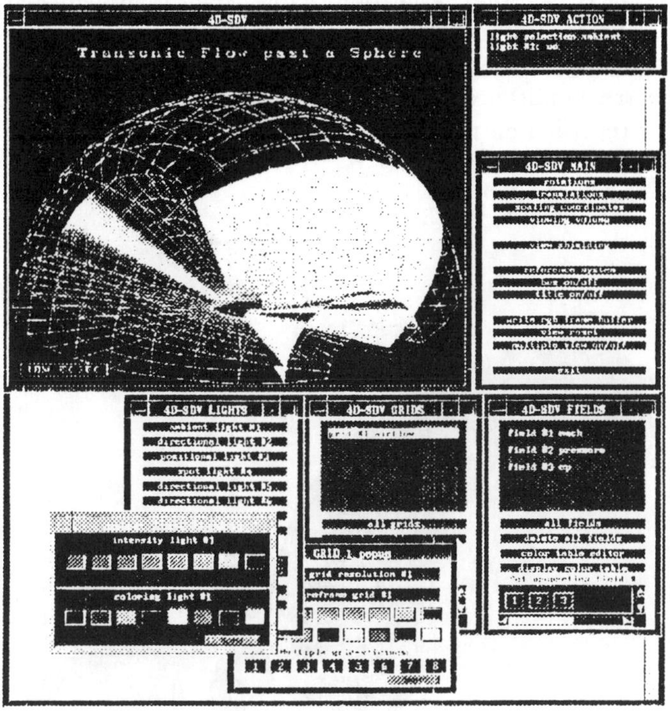

Figure 1: The screen of an IBM RISC System/6000 running the **4DSDV** data visualizer. Note how the availability of the OSF/Motif widgets (push-buttons, sliders, scrolled lists, labels, ...) helps in presenting the results of three-dimensional data obtained from a numerical simulation.

Three-dimensional Scalar Fields

The visualizer can represent a three-dimensional scalar field through one or more iso-value surfaces (the three-dimensional analog the iso-value curves obtained from two-dimensional scalar fields). These surfaces are rendered through a set of triangles that smoothly approximate the desired iso-surface.

The iso-surfaces of a scalar field can be colored interactively according to the value of other fields. As an example, the iso-vorticity surface of a fluid flow can be colored according to the value of the modulus of velocity so that, slowly and rapidly moving components of the surface, can be easily identified.

An interactive color map editor allows to customize easily the RGB (Red, Green, Blue) or HSV (Hue, Saturation, Value) color maps which are used to represent the fields. In this way each field can be colored with the color map that better highlights the field properties. This feature is also useful when many fields are simultaneously shown on the same grid. The customized maps can be recorded on files and used later.

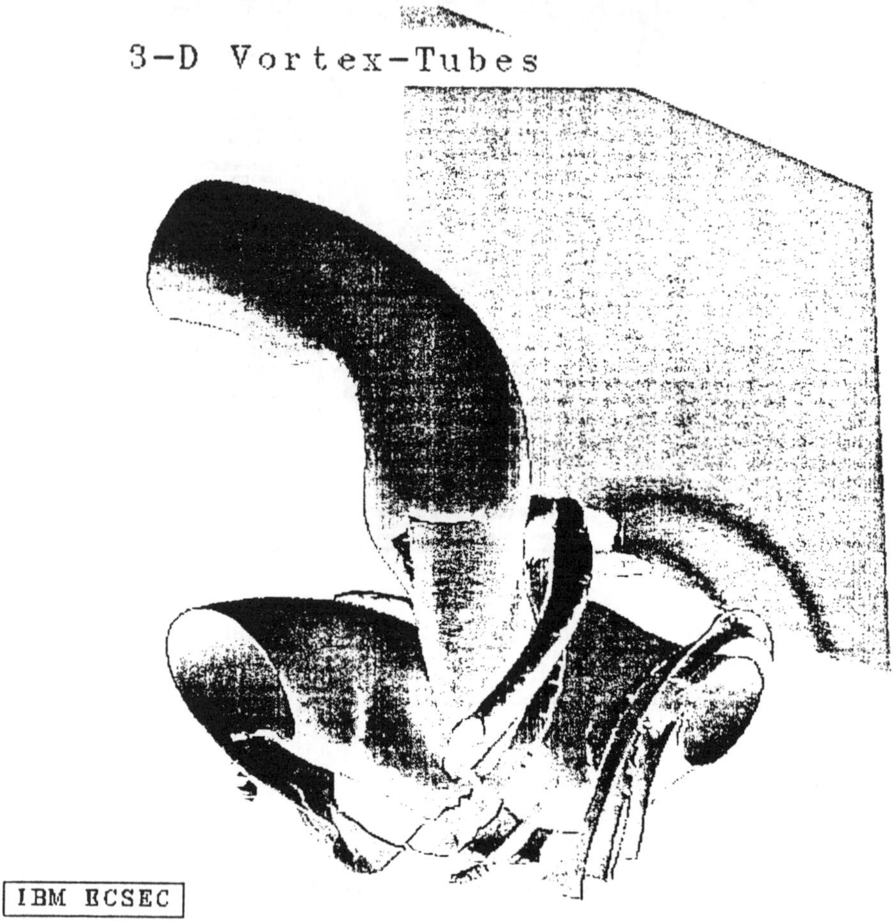

Figure 2: The dynamical evolution of an incompressible fluid. The time evolution of two, initially orthogonally-offset, vorticity tubes produces many small-scale secondary vortex tubes that bridge the gap between the two main distorted tubes [4, 5]. The vorticity iso-surface is *colored* according to the intensity of the local value of the velocity field.

Three-dimensional Grids

The computational grids handled by the visualizer belong to the following types:

- carthesian,
- spherical,
- cylindrical,
- z-deformed,
- fully deformed.

Non-uniformly spaced grids are also allowed.

Some useful functions are provided to display and to manage the computational grids:

- interactive change of grid resolution for the acceleration of intermediate visualization stages and
- concurrent display of different grid-cutting planes (in either shaded or wireframe formats).

Multiblock Grids

The visualizer is able to handle **multi-block** grids. The **multi-block** approach is used in numerical simulations when the geometry of the problem is complex and the generation of the grid is very difficult or when a decomposition of the domain allows a faster execution in a parallel environment. Some industrial codes for Fluid Dynamics use this technique. The **multi-block** approach consist of a preliminary topological subdivision of the domain in a limited number of large hexaedral elements called blocks. Inside each block a smooth grid is produced and finally the flow equations are solved using some well known algorithms developed for single block grids.

PARAGRID [6] is an example of a parallel **multi-block** environment where a large class of *single-block structured* solvers can be easily integrated. With **PARAGRID**, a *single-block* solver can be easily restructured into a *subdomain* solver so that multiple copies of the code can work concurrently on different grid blocks and exploit the computing nodes of a parallel computer.

In our visualizer, each block is considered as a single grid. Using isosurfaces or slicing planes, the flow solution can be shown on the complete computational domain or the representation can be limited to the blocks of interest. Since the number of blocks, needed in problems of industrial interest, is of the order of a hundred or more, the graphical rendering stage requires powerful three-dimensional capabilities.

Figure 3: Output of a **multi-block** aerodynamical simulation obtained using **FLO67P** within the **PARAGRID** parallel environment [6]. The isosurface of the aerodynamic **dtp** field is presented together with the body of the plane (geometry, courtesy of Aermacchi). Note how the **dtp** isosurface clearly indicate the *core vorticity structures* emanating from the wings. The full computational domain comprises 56 blocks and the body of the aeroplane is shaped by the edges of 16 blocks.

Slicing

The representation of the intensity of a scalar field on a plane (**slicing**) often complements the representation through iso-surfaces. For non-orthogonal grids, an extended type of slicing is available. For example, in the case of *spherical coordinates*, slicing can cut the radial coordinate and the corresponding slicing surface is spherical. Analogously, in the case of *fully deformed* grids, slicing can be easily used to show field values on grid-coordinate surfaces.

Field properties on slicing planes, as well as on iso-surfaces, are easily represented either with shaded triangles (or quadrilaterals) or with a set of colored contour lines.

Chemical Data

This type of data represents schematically the molecules through a set of spheres and cylinders. These primitives may be used directly to represent physical properties of the molecules; for example, sphere and cylinder radii may be used for atom sizes and bound strengths respectively. Input data can be either specified directly or using the connectivity matrix. This simple representation of the molecules can be enriched by showing the available field properties of the molecules (e.g., the electron probability distribution function).

The representation of elementary geometrical structures like cylinders or spheres takes advantage of the graPHIGS NURBS structures [1] that give a good rendering with a minimal amount of workstation effort. This is obtained by automatically varying the degree of tesselation as a function of the enlargement so that the surface smoothness appears to be constant.

ANIMATION OF NUMERICAL SIMULATIONS

The Communication Interface

The design of a communication library for handling data communication among processes requires the development of simple high-level subroutines that hide the complexities of the network communication and the differences in data representation due to the different hardware and software environment that participate to the cooperative processing environment.

The library that we have developed is based on the TCP/IP communication protocol. Since this protocol is available on most scientific and technical environments our library is able to handle communications transparently and independently on the operating system chosen to run the simulation program. In our case, the only restriction is the selection of the IBM RISC System/6000 workstation for the visualization environment while client data can be received from any system running a numerically intensive computing simulation.

An important feature of the communication library is the possibility to easily include primitives for data feedback (**steering** of the computing client). An important example for the application of this feature most probably arises from the numerical simulations in Statistical Physics, where the visual analysis of the configurations may help in speeding up the convergence process by changing simple fundamental parameters, (e.g., the temperature of the system).

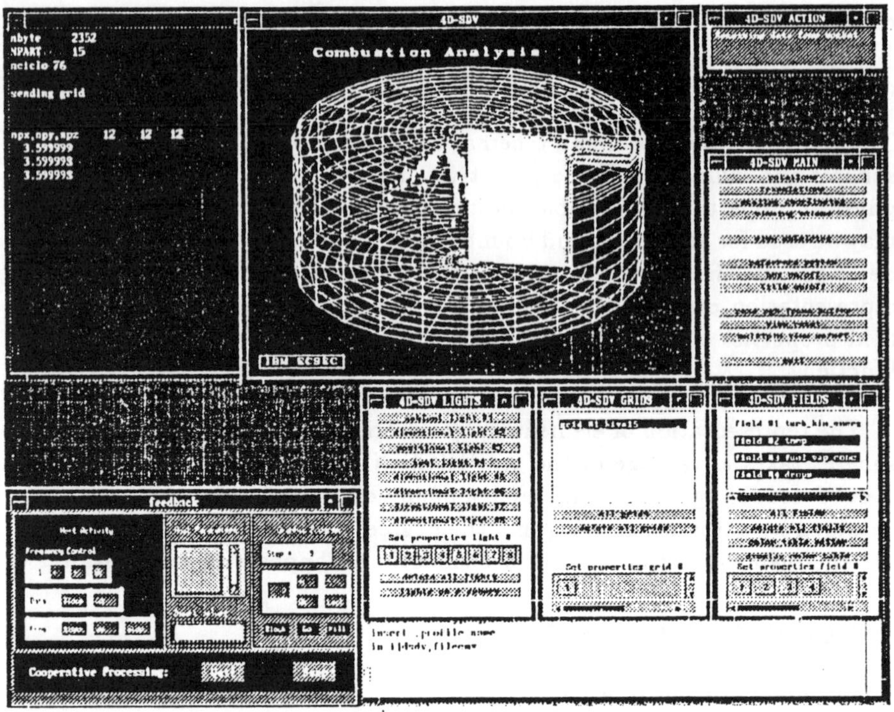

Figure 4: A cooperative session on an IBM RISC System/6000. **a**) In the upper-left window, the KIVA-II combustion code [7] simulates the chemical reaction of fuel droplets transported by a compressible fluid and sends data to the visualizer. **b**) The results are received by the visualization process shown in the upper-right part of the screen; the graphical window shows the combustion chamber and is updated as soon as new data are received from the KIVA-II client [8]. **c**) The **feedback** window is in the lower-left corner

In our system, the **steering** of the computational parameters is done using the **feedback** window which is automatically provided by the communication environment. The same window can also be used to change interactively the many parameters that control the data communication. Here we list some of them:

- temporary halt/resume of the client simulation process,
- temporary halt/resume of the of server visualization process,
- stop of the execution of the simulation client,
- run-time change of some predefined variables of the client process,
- run-time change of the frequency used to send data to the visualizer,

- simple back-stepping through a limited amount of previous time steps,
- temporary disconnection of the visualizer from the simulation client
- monitoring of different numerical simulations.

Figures 4 and 5 and their captions explain in more detail the interactive environment outlined in this Section.

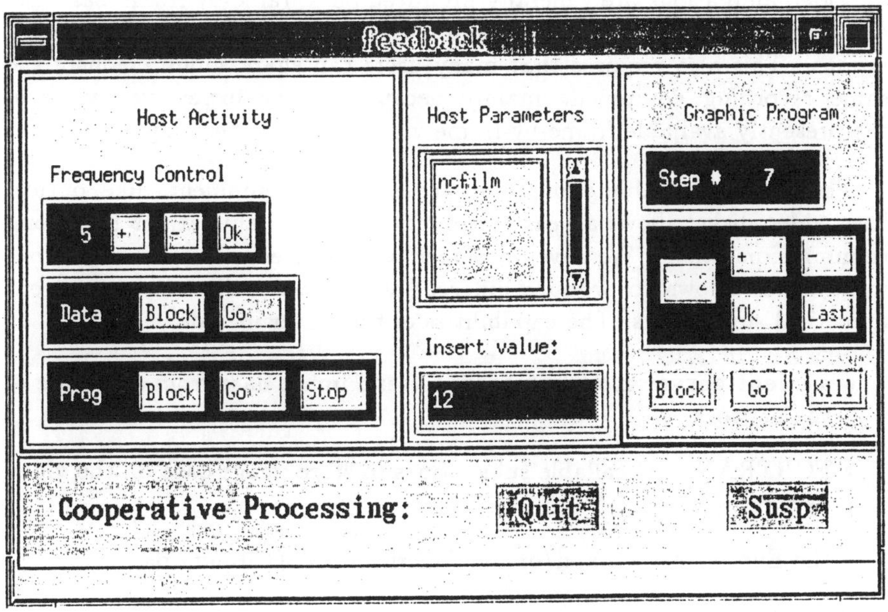

Figure 5: An enlargement of the **feedback** window also shown in the lower left corner of figure 4. Many buttons or input fields can be used to change the properties of the data exchange between the computing client and the visualization server. All parameters can be changed in any order and at any time. Grouped by function, the communication parameters regulate **host activity, computational steering** and the **graphic program**. The last function allows to search for the three-dimensional data corresponding to the previous time steps which are stored in a circular buffer on a first-in first-out basis.

THE VIDEO SERVER

The wide diffusion of high-quality video equipments has clearly indicated that a new way of communicating the results of scientific or technical simulations is the production and the distribution of video tapes. This phenomenon is recognized by an increasing number of scientific magazines [5] which allow the authors to complement the documentation of their work with video tapes, thus easing the task of documenting the numerical results. In fact, a video tape displays the time-dependent component of the simulated phenomena in a way that is almost impossible to represent using a sequence of static figures.

Video animations are generally produced on a frame-by-frame basis because of the long running time required by the modern numerical simulations. This also requires that many users can share simultaneously the same recording equipment mainly because of the expensive nature of the presently available frame-by-frame video recording techniques.

The recent availability of direct-access video equipment, like optical laser video-disks, allows an easy production of video animations from long running simulations. We have developed a multi-user video server environment which allows to produce different animations from many concurrent numerical simulations. The capabilities of the server include:

- a **client-server** model, in order to send image data from any node of the network,
- FORTRAN or C callable subroutines to store image data on the video device,
- the European PAL and/or the American NTSC standards,

Several video disks can be managed contemporarily by the same video server and different instances of the server can run concurrently.

Our video server runs on AIX and is based on the TCP/IP communication protocol. The production of video frames is simply obtained by including a single FORTRAN or C call inside the application that produces the color frames (e.g., the 4DSDV visualizer). The routine sends digital image data (in RGB format and with information about the required positioning on the video media), performs automatically the network connection with the video server or waits until a connection is available.

An easy way to retrieve and playback previously stored video sequences or images is obtained by indexing the video-disks through standard AIX files, hierarchically grouped per user, each one containing the few parameters that uniquely identify the starting and ending points of the corresponding animation.

The hardware equipment that we used to build the video server comprises the following pieces:

- an IBM PS/2 or an IBM RISC System/6000 running AIX as the video server,
- the IBM VCA/A (Video Capture Adapter) card (PAL and/or NTSC) for the digital-to-analog RGB video conversion,
- a SONY LVR/LVS-6000 direct-access laser (PAL and/or NTSC) video-disk complex

and the serial port to drive the video-disk.

The video production equipment that we outlined proved to be a reliable and efficient multi-user tool for the production of broadcast-quality video animations. Other reliable and cheaper frame-by-frame video recording devices in order to devise a cheaper solution.

CONCLUSIONS

An advanced, easy-to-use and integrated environment has been developed in order to obtain visualizations and video animations for large-scale scientific and technical simulations.

The environment comprises three different tools that can be used jointly or separately, in any combination to achieve the desired result.

The first tool is a general purpose visualizer for scientific data which is based on the IBM graPHIGS. The tool can interactively visualize many types of scientific data sets. These data can be read from mass storage or, when the visualizer is used in conjunction with the communication library, can be received continuously from a network computing client.

The communication library is the second tool and is based on TCP/IP. The complexities of socket communications are completely hidden to the user which is only asked to include some FORTRAN or C calls in the computational client code in order to send the relevant data to the visualizer. The design of the communication library is such that other user-written visualization tools, different from the visualizer described in this paper, can be used as well.

The third tool is a simple and powerful multi-user video-server which can be used by any user having access to the TCP/IP network. Being based on

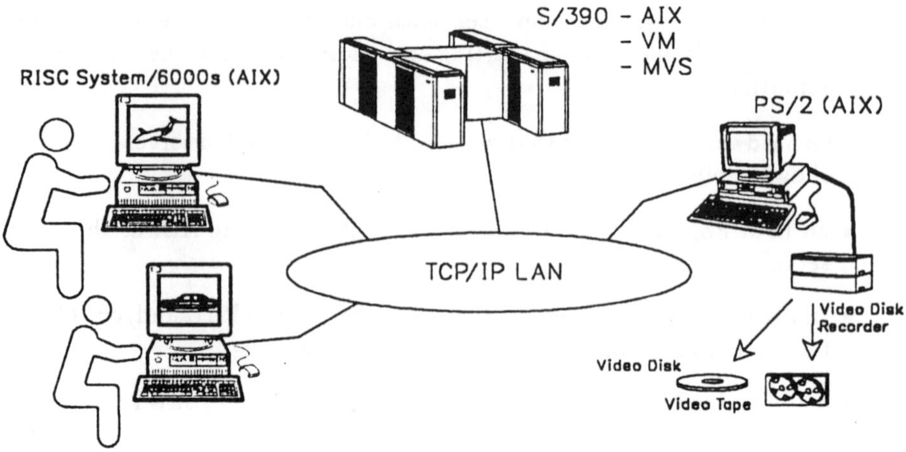

Figure 6: A schematic representation of the visualization environment described in this paper. The user, shown in the upper left corner, uses an IBM RISC System/6000 to visualize the results of a numerical simulation program that runs on a different node of the network.

a random-access laser video-disk device, the tool allows the contemporary production of many different videos; these can be successively reversed on standard video tapes for ordinary playback.

References

[1] IBM, The graPHIGS Programming Interface, Subroutine Reference, SC33-8194, Understanding Concepts, SC33-8191, 1991; available through IBM branch offices.

[2] D.A. Young, The X Window System Programming and Application with Xt OSF/Motif Edition, Prentice-Hall,Englewood Cliffs,New Jersey 07632, 1990.

[3] IBM, AIX Version 3 for RISC System/6000, Call and Subroutines Reference: User Interface, Volume 3, and 4, Order No. SC23-2198, 1990; available through IBM branch offices.

[4] N.J. Zabusky and M.V. Melander, Three-dimensional Vortex Tube Reconnection: Morphology for Orthogonally-Offset Vortex Tubes, p.555, Physica D, 37, (1989)

[5] M. Briscolini and P. Santangelo, Animation of Computer Simulations of Two-Dimensional Turbulence and Three-Dimensional Flows, IBM Journal or Research and Development, p.119, Vol. 35, No. 1/2 (1991).

[6] F. Della Giacoma, S. Paoletti, F. Poggi and M. Vitaletti, Multidomain Computation of Compressible Flows in a Parallel Scheduling Environment, communication presented at the Parallel CFD 92, New Brunswick, NJ, May 1992.

[7] A.A. Amsden, P.J. O'Rourke, and T.D. Butler, KIVA-II, A Computer Program for Chemically Reactive Flows with Sprays, Los Alamos National Laboratory Report LA-11560-MS, May 1989.

[8] F. Papetti, S. Golini, M. Maggiore, S. Succi, P. Gaillard and J.-M. Perez, Internal Combustion Engine Design on IBM Platforms, IBM Systems Journal, p.774, Vol. 31, No. 4 (1992).

Animation of railway simulator results for a mass transit railway

G. Digby, V.J. Strachan
Electrical Engineering Dept., University College, Swansea, SA2 8PP, UK

ABSTRACT

A video animation of the results from a railway simulation is described. The animation techniques used to create the video are discussed. The simulation is for the assessment of the a.c. and d.c. traction power supplies of a large urban mass transit railway.

INTRODUCTION

Power engineering problems involve a large amount of data. The use of computers has allowed the problems that are investigated to become very large and complex. However interpretation of the results from computer analysis of these problems is very difficult unless visualisation techniques are used.

The presentation of results in the form of tables, or even two-dimensional graphs, is no longer sufficient to convey enough meaning about the results. To allow the user to interpret the results quickly and accurately, new techniques are required for presenting them. Computer animation is a possible solution to the problem. The use of animation allows results to be displayed in up to five dimensions, X, Y, Z being three dimensions, time adding a fourth and colour enabling a fifth dimension to be displayed.

A simulator has been developed at University College of Swansea to assess power network design and operation on mass transit railways. At present, the results from this package are presented in the form of tables and two-dimensional graphs.

The paper describes the use of animation techniques to create a video animation of a railway simulation. The textual results were taken from the simulator and post-processed to be displayed as an animation which shows the trains moving with time. Voltage levels and current flows are displayed along with power levels. This allows the interaction between the trains and substations to be observed. The paper outlines the difficulties of interpreting the present textual and 2-dimensional result presentations. The benefits offered by the animation are explained.

The methods used to create the animated video and the problems encountered during the research are summarised.

THE SIMULATOR

A simulator has been developed at University College Swansea to assess voltage controlled rectifier (VCR) operation on d.c. traction schemes, by Kadhim [1], [2]. The simulator models not only the d.c. traction network and the train service but also the a.c. distribution system. The nature, magnitude and geographical location of the loading on a traction power supply changes continuously with time (Mellitt [3], [4], Talukdar [5]). An accurate train service model is therefore essential to any investigation into the power supply arrangements, so that the loading can be clearly predicted. The a.c. and d.c. networks interact in a complex manner, especially when VCR's are used. It is therefore not possible to consider the networks independently if an accurate assessment is to be made. Electrical quantities, such as voltage, current and harmonic levels can be studied in both the a.c. and d.c. networks.

The simulator comprises of three separate but synchronised models:

i) The train service model
ii) The d.c. network model
iii) The a.c. network model

The simulator evaluates the state of these models at discrete time intervals. The inter-sample period is 2 seconds and the complete record length is 200 samples. This equates to 400 seconds of activity on the power system in real-time. The longest headway between trains is 180 seconds. A record length of 400 seconds ensures that the train service model is completed at least twice during the simulation.

Each of the models is now described in greater detail. As an example reference is made as to how the London Underground Ltd Northern Line is broken down for inclusion in each model.

Train Service Model

The Northern Line is broken into five separate track sections or branches, Figure 1 (Allan [6]). These are named:

i) High Barnet
ii) Edgware
iii) Charing Cross

iv) City
v) Morden

The position, speed and power consumption of every train is derived from the train service model. The position and power consumption of the trains are then passed to the d.c. network model. Figure 2 shows the graphical results for a train in the standard output format for the simulator.

The DC Network Model
The d.c. network model comprises of 6 sections, as shown in Figure 3 (Allan [6]). These sections correspond to the d.c. track feeding sections used by LUL at present.

The d.c. network model automatically generates the network to be analysed. The traction substation positions are known, along with the new train positions from the train service model for the current discrete time sample.

From the d.c. network the loading at each traction substation is determined. This information is then passed on to the a.c. network model. Figure 4 shows the graphical results for a substation in the standard output format for the simulator.

The AC Network Model
The a.c. network model represents the 11kV a.c. distribution system (Bletcher [7], Buchanan [8]). The a.c. system is modelled back to the 22kV busbar on the 22kV/11kV transformers in the switch houses.

The 22kV transmission system is not included in the a.c. network model. This means that the a.c. network model breaks down naturally into four sections, each section being fed from a 22kV switch-house, Figure 5. The 22kV busbars are referred to as the Points of Common Connection (PCC) for each section. The four PCC's are the switch-houses:
i) Neasden
ii) Coburg Street
iii) Stockwell
iv) Mansell Street

The a.c. system model allows the electrical quantities at the PCC to be determined. Figure 6 shows the graphical results for a substation in the standard output format for the simulator.

SIMULATOR RESULT PRESENTATION DIFFICULTIES

The simulator has been used in the design studies that have taken place as part of the Northern Line Modernisation Project. The simulator results have been used to provide an insight into the effect of VCRs on the a.c. supply network. During the course of the design studies the authors realised that by presenting the simulator results as simple graphs, it proved difficult to assimilate the information from the results. The difficulties arose in two areas. Firstly the volume of information being generated from the simulation. Secondly the need to present time-based and position-based information at the same instant

The difficulties due to the volume of information were caused because all the individual results from each train, substation and PCC had to be available and visible as conclusions were drawn from each simulation run. To overcome these difficulties necessitated laying out the graphical results over a large amount of desk area. The difficulties increased further when conclusions had to be drawn between the simulation runs.

The conflict between time-based and position-based information arose when it was necessary to consider the interaction between the trains, and the substations and PCC's. This difficulty was caused because the train graphs present the results against train position, Figure 2. Whereas the substation and PCC results are presented against time, Figure 4 and Figure 6.

In light of the difficulties in assimilating the results it was decided to see if the use of animation and video could alleviate some of the problems.

THE ANIMATION

The animation was only performed using results from the Edgware Branch for a single simulation run. The basic picture of the section is shown in Figure 7. The picture is a simple 2-dimensional representation of the railway in order to convey the position-based information. The use of animation allows a 3rd dimension, in this case, time to be displayed. Colour is then used to convey a 4th dimension of information. In this case this 4th dimension is the value of the quantity under consideration, such as power, voltage or current.

Figure 7 shows two straight lines to represent the railway track. Simple block shapes represent the PCC, substations and passenger stations at their relevant positions. Each train is then represented by a block which moves along the track as the animation proceeds. The magnitude of the quantity to be displayed is represented by a change in length of the relevant block. However in the case of the trains the outline of the block remains at a constant size, while the colour within the block is altered. In this animation the colour change is simply between two colours.

The human eye tends to focus on the brightest colour in a scene. Hence attention may be drawn to an area of the scene simply by using a bright colour such as yellow. Similarly a cool colour such as blue can be used to display information in a scene that is not directly relevant but may be useful to the viewer. This technique is used in the animation. The sites of interest, i.e. PCC, substations or trains, are displayed using yellow. The remainder of the scene is displayed using blue.

Figures 8, 9, 10, and 11 show the story-board for the animation of the power results. Initially only the power flow through the PCC is animated, Figure 8. In Figure 9 the substation power flows are displayed, while the PCC is re-coloured in blue. In Figure 10 the train power consumption is visualised, while visual impact of the PCC and substation results is reduced. Finally the complete picture is animated with all the results enabled using the colour yellow, as in Figure 11 .

A similar method is used to display the current and voltage information.

THE ANIMATION AND VIDEO HARDWARE AND SOFTWARE

Figure 12 shows the arrangement of the animation and video hardware. An Atari 1040ST computer was used to generate the animation sequences. These were recorded on to U-Matic low-band format video via a Genlock. A studio quality video camera was used to generate the reference video signal for the computer and U-Matic video recorder to synchronise to via the Genlock. The U-Matic video material was then edited using a video edit suite. Captions were added in the edit suite using an Amiga 2000 system.

The animation sequences were created on the Atari 1040ST using the Cyber Studio software by Antic Software. In particular the Cyber Paint package.

DIFFICULTIES WITH THE ANIMATION AND VIDEO HARDWARE AND SOFTWARE

The objective of this work was to investigate possible display techniques and not actually develop animation software. The software used was unable to accept the results directly from the simulator. Therefore a lot of the images had to be created by hand. This problem would be easily overcome by the development of specific animation software.

The transfer of the computer animation sequences on to video tape resulted in some loss of image detail. This could be overcome by using more expensive video equipment, or only displaying the sequences on the computer.

More significant that the loss of detail was the quality of the saturated colours once transferred on to the video tape. Since video was never designed to cope with this type of colour, the transition between some saturated colours was poor. The difficulties arise as the colours are encoded into the colour difference signals of the video. It can take up to five pixels for the image to recover from one colour to the next. This effect is particularly noticeable if saturated red is used against a black background. The transition between these two colours is very poor causing the red images to appear faded. The effect is worsened by the fact that, to the human eye, red is similar to black in terms of brightness. Improvement in the quality of the video image can be made if the black background contains approximately 30% grey and the red contains a little green. The two colours still appear the same on the computer monitor but the transition between them will be sharpened. For the railway simulation the background was chosen to be an unsaturated colour. Any saturated colours used did not affect the quality of the video image adversely.

CONCLUSIONS

The aim of the work was to investigate the viability of using animation techniques as an alternative to simple graphical plots to display the results of the railway simulation. Hence the results maybe more conveniently interpreted.

The project results clearly show that a better global view of the simulation results is given by the use of the animation. However the fine detail of the graphical plot is lost. Some of this detail could be recovered by using higher resolution animation. But a more cluttered image may result with the consequence that understanding is lost.

It is felt that the use of animation is a valuable tool, but needs to be used in conjunction with the more detailed graphical plots of the results. The animation allows the interaction between the various railway system components to be seen. For example, the animation sequence created clearly shows the impact on substation loading of two trains starting simultaneously at one station. This information is not directly discernible from the graphical plots. It is envisaged that the designers would use the animation to quickly get a feel for the interactions within the system and then study any problem areas using the graphical plots.

The position-based data is only given in 2-dimensions in this work. However if gradient information was to be displayed, then a 5-dimensional animation could be used to show position and height with 3-dimensions, time being a 4th dimension and colour used to display a 5th dimension.

ACKNOWLEDGEMENTS

The authors wish to thank London Underground Ltd., for permission to use the Northern Line Modernisation Scheme data and results as examples within the paper. Special thanks go to Mr. Kevin Payne, Client Engineer (Power Projects).

The authors also wish to thank the staff of the Media Resources Centre, University College of Swansea for their help with the production of the animation video.

REFERENCES

1. Kadhim,R.J., *A Voltage Controlled Converter For DC Traction* Ph.D. Thesis, University of Wales, 1988

2. Kadhim,R.J. Hooper,A. "Voltage Controlled Converters for DC Traction Systems", *Universities Power Engineering Conference Proceedings*, Sunderland Polytechnic 1987.

3. Mellitt,B. Goodman,C.J. Arthurton,R.I.M. "Simulator for Studying Operational and Power-Supply Conditions in Rapid-Transit Railways", *IEE Proceedings*, Vol. 125, No. 4, April 1978, pp 298--303.

4. Mellitt,B. Mouneimne,Z.S. Goodman,C.J. "Simulation Study of DC transit Systems with Inverting Substations", *IEE Proceedings*, Vol. 131 Pt. B, No. 2, March 1984, pp 38--50.

5. Talukdar,S.N. Koo,R.L. "The Analysis of Electrified Ground Transportation Network", *IEEE Transaction on Power Apparatus and Systems*, Vol. PAS 96, no. 1, 1977, pp 240--247.

6. Allan,J. Digby,G. Chan,W.S. McCormick,H.J. "Cost Conscious Design of Power Supply Equipment using a Simulator", *IEE Colloquium Cost Effective Industrial Simulation*, London, 11 June 1990, Digest No. 1990/099, pp. 5/1-5/5.

7. Bletcher,M.J. "Electricity Supply System on The London Underground", *IEE Power Engineering Journal*, November 87, Vol. 1, No. 6, pp. 317-324

8. Buchanan,I.D. "Computer Control of London Underground's Power Network", *IEE Power Engineering Journal*, July 90, Vol. 4, No. 4, pp. 159-163

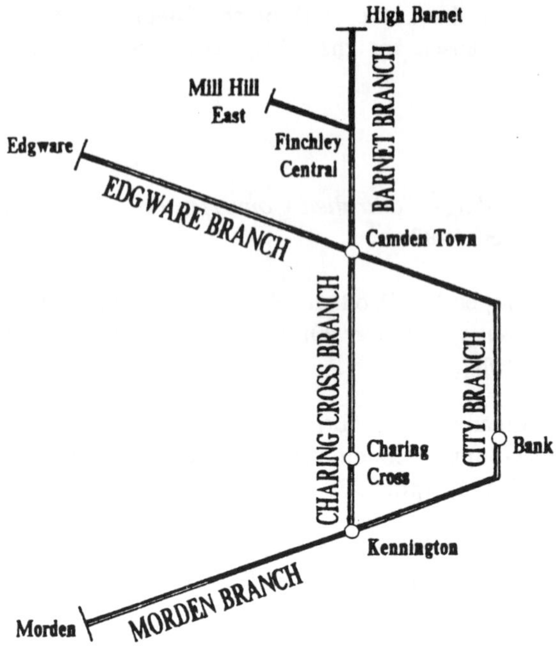

Figure 1 Northern Line track sections

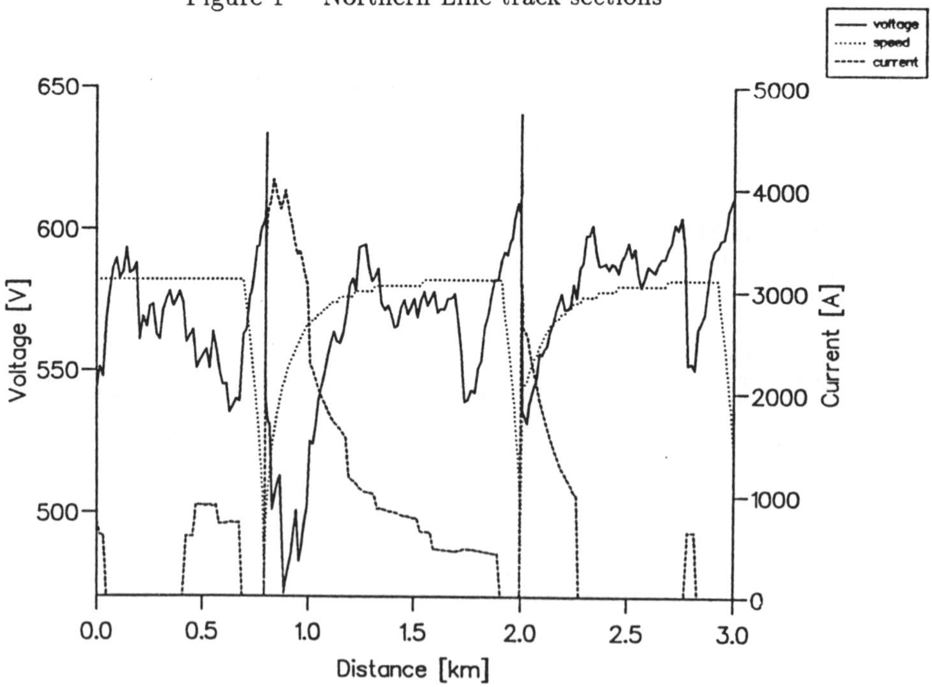

Figure 2 Graphical train results

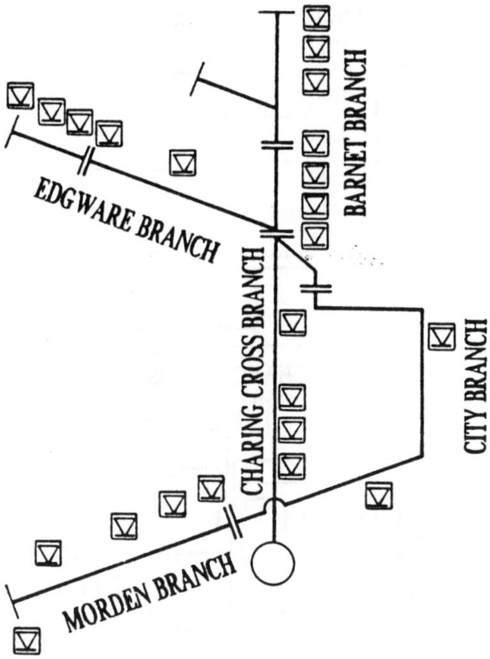

Figure 3 Northern Line DC track sections

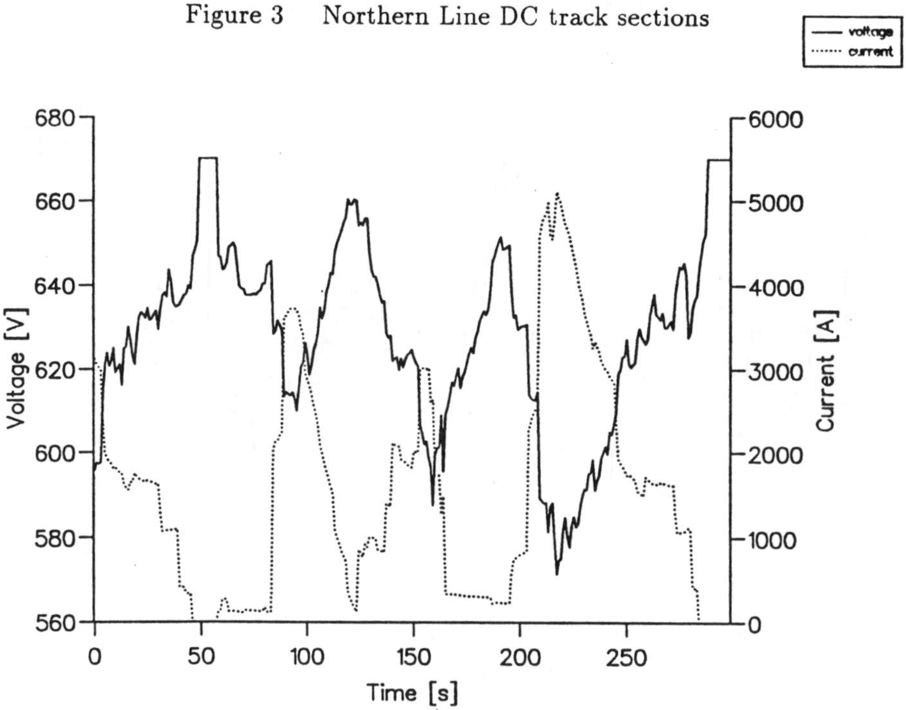

Figure 4 Graphical substation results

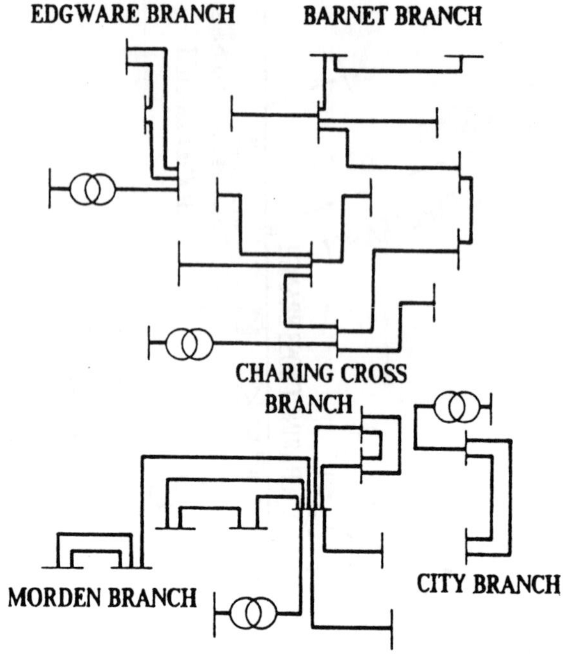

Figure 5 Northern Line AC network

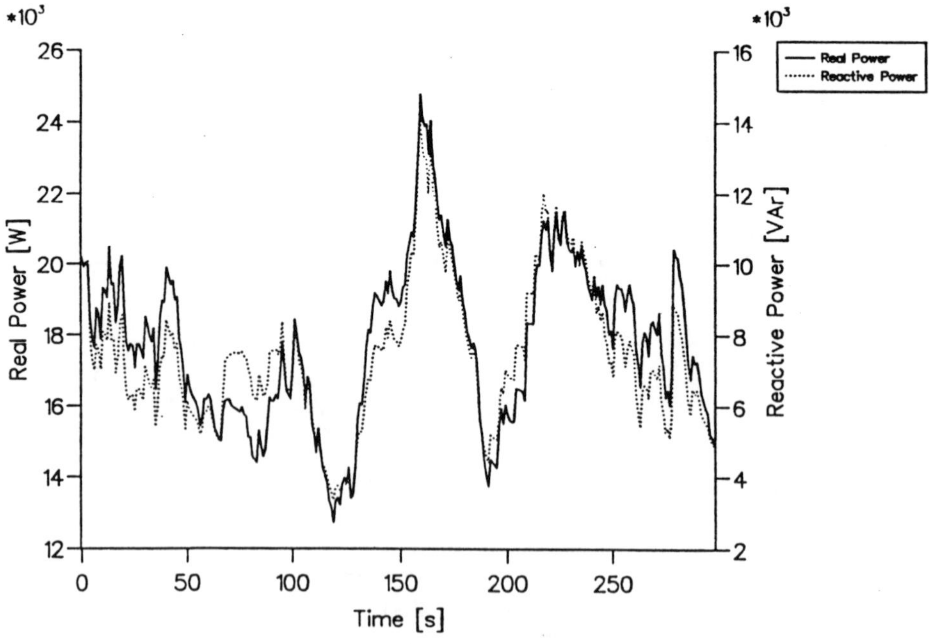

Figure 6 Graphical point of common coupling results

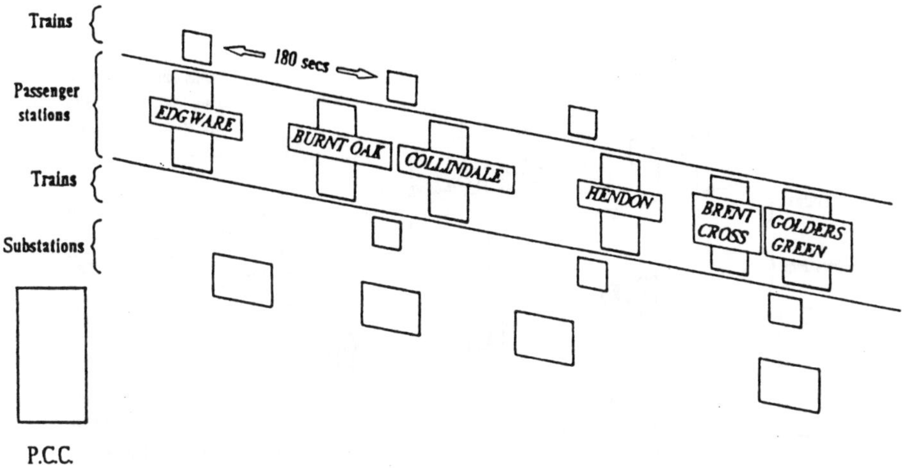

Figure 7 Basic animation picture of Northern Line Edgware Branch

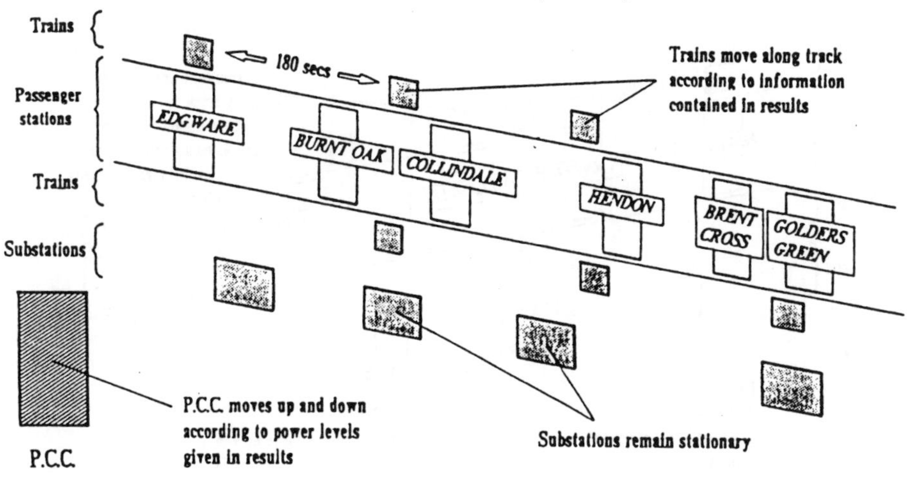

Figure 8 Storyboard for animation results: P.C.C. power

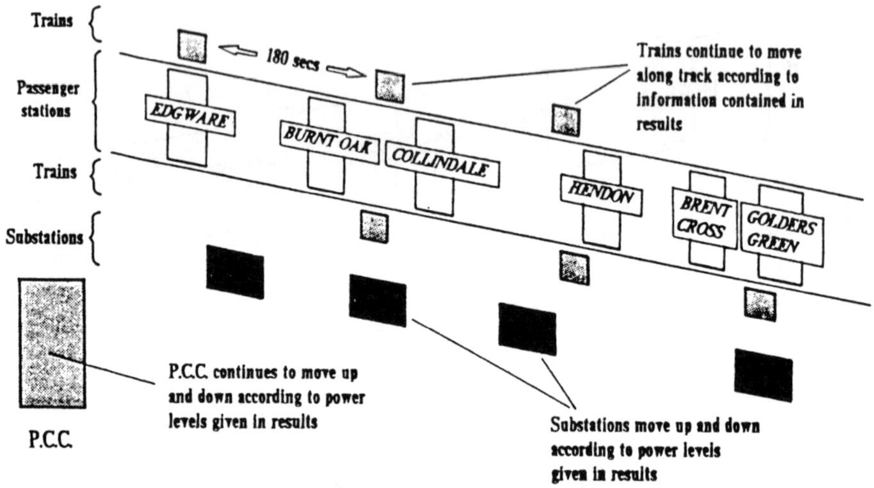

Figure 9 Storyboard for animation results: Substation power

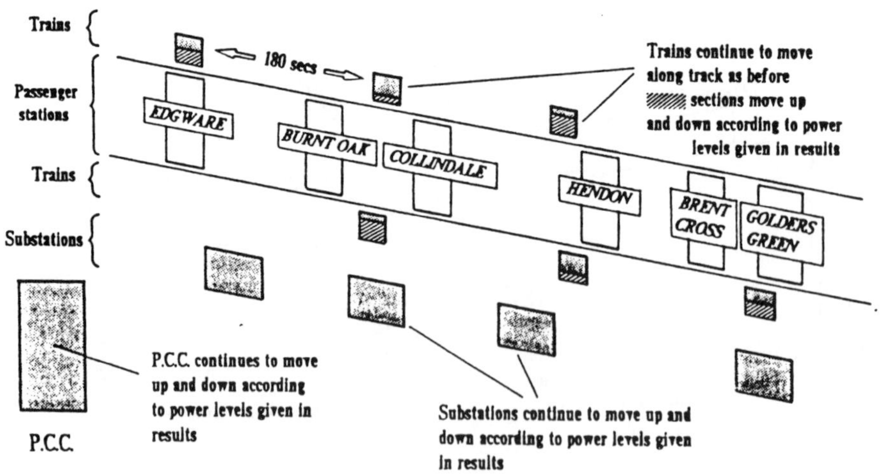

Figure 10 Storyboard for animation results: Train power

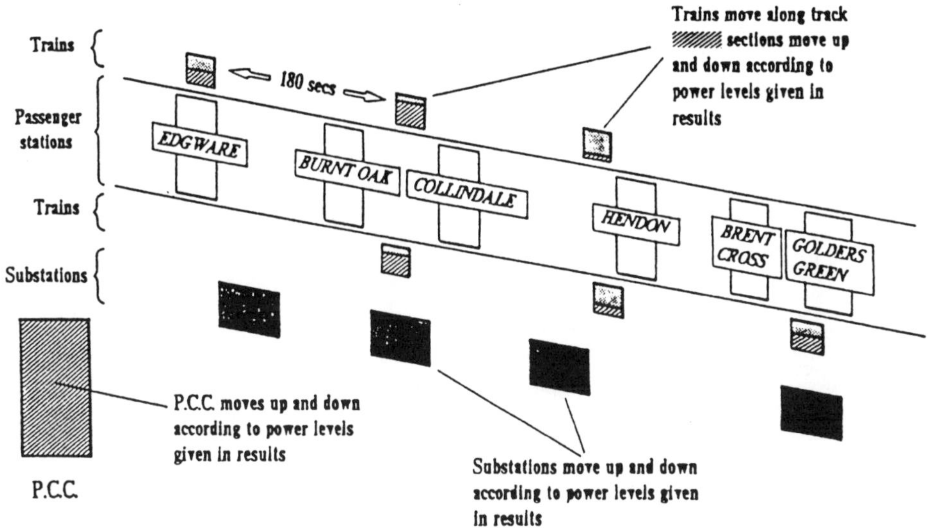

Figure 11 Storyboard for animation results: Total system power

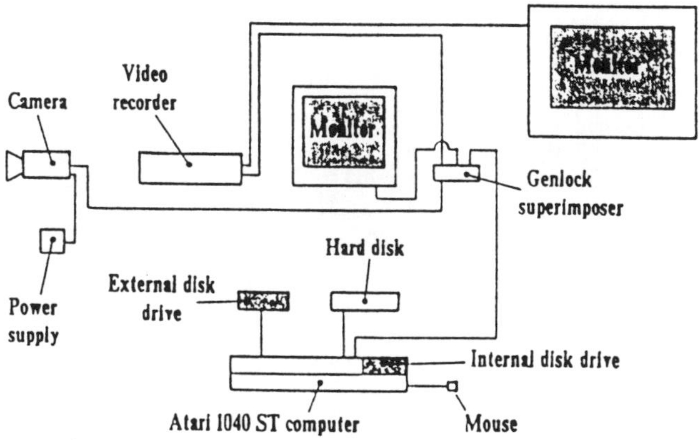

Figure 12 Computer and video hardware arrangement

Animating architectural scenes utilizing parallel processing

F. Van Reeth, E. Flerackers, W. Lamotte

Applied Computer Science Laboratory, Limburg University Center, Universitaire Campus, 3590 Diepenbeek, Belgium

ABSTRACT

The ever increasing demand for realism and complexity in synthesizing architectural animated sequences makes the development performant rendering algorithms and -platforms a necessity. In order to answer this necessity, we take recourse to parallel processing techniques. More precisely, networks of transputers are utilized to implement parallel rendering algorithms. Hence, we can build animation systems which can linearly be expanded in performance, simply by adding new processors to them. Especially for traditionally expensive rendering techniques, a linear performance improvement is measured in systems with up to 50 processors.

INTRODUCTION

Looking at the history of computerized architectural visualizations, one can distinguish several eras. In the beginning, two-dimensional plotter drawings were all one could ask for. Later on, three-dimensional vectorized wire-frame drawings came to be, followed by visualizations in which more sophisticated colored renderings were to be used. Currently, state-of-the-art architectural visualization is requiring the availability of three-dimensional computer animation with an ever increasing quest towards photo-realism. The fact that a lot of the naturalistically looking surface materials which are needed in architectural scenes take up a considerable amount of computing power in order to be generated, makes animation system designers search for more performant rendering solutions. This is especially true when making animated sequences, as the display standards require the production of at least 25 images for every second of animation.

The research group in our laboratory took the decision of going in the parallel processing direction in order to increase the rendering performance. By utilizing parallel processing techniques, it becomes possible to achieve high performance ratings on the one hand, while still staying within software solutions on the other hand (i.e., without requiring special purpose hardware). As many timeconsuming rendering techniques are too complicated (or too expensive) to be implemented in special purpose hardware, this seems a good approach.

In casu, we utilize processors of the transputer family to realize the parallel implementations. On these transputers, a complete three-dimensional computer animation system, incorporating an object modeler, a motion specification subsystem, as well as a renderer, has been implemented. It turns out that many of the calculation intensive rendering techniques can be implemented successfully: procedurally generated 3D texture-, reflection-, bump- and transparency mapping techniques (for rendering types of stone, wood, marble, water, clouds, metals, bricks, etc.), ray tracing algorithms (for generating realistic reflection, transparency and shadowing effects), as well as radiosity methods (for generating realistic images of diffusely reflecting surfaces). An overview of our realizations, together with some relevant references, will be given in this paper.

THE ANIMATION HARDWARE

The existing image synthesis architecture in the laboratory consists of a parallel network of transputers. The transputer of the current generation is a 32-bit parallel processor sustaining 20 MIPS / 1.5 MFlops at 25 MHz. Each of its four serial links allow bidirectional synchronized communication with other members of the transputer family at 20 Mbit / sec. Special purpose logic on chip is provided for rapid context switching in the scheduling and descheduling of the parallel processes running on the processor.

The entire hardware platform of a graphics system in the lab consists of the following components: (i) a PC, utilized as a host for booting the network and for housing the I/O equipment (mouse, graphical tablet & stylus, scanner, video recorder interface, etc.); (ii) host transputer boards (with transputers having up to 32 MB of local memory) used for connection with the PC and for storing the database as well as application specific programs (modelling, motion specification, programming environment, etc.); (iii) a farm or pipeline of transputers having typically 2 to 4 MB of local memory, utilized for executing the calculation intensive rendering algorithms; (iv) two graphics boards, which respectively deliver a non-interlaced high-resolution signal for displaying the user interface and an interlaced true color PAL video signal for displaying the rendered images.

OBJECT MODELLING

In general, the making of an architectural animation can have two very distinct intentions: it can be made for structural analysis purposes, or it can be made for demonstrational purposes. Depending on this purpose, the underlying model representation is of large importance. In the former case, exact mathematical representations through higher order curved surfaces might be necessary, while in the latter case, a polygonal approximation of objects often will suffice. Although we already have been experimenting with bicubic patches to represent curved surfaces exactly (Lamotte [13]), our main attention currently is directed towards representing objects to be animated with a boundary representation consisting of polygonal faces. In our parallel ray tracing algorithm, a procedural representation can also be applied. (Note: As more and more commercial three-dimensional modelling packages become available, it is moreover a necessity to have an "open" file format with respect to object representation, so models created on divers modelling packages can be animated on our animation system; cf. the ray tracing figure later on.)

No matter what representation is chosen, objects should be

represented in such a way that they can be easily created and manipulated by human operators. Appropriate tools have to be offered, enabling animators, designers and engineers to define and shape objects according to their specific wishes. A good user interface, with multiple simultaneous views, is very important in order to reduce the amount of time it takes to create the desired object models: the benefit of having a performant parallel rendering system can indeed disappear completely if a cumbersome user interface on top of it doesn't allow the models to be specified quickly. On the one hand, the available tools should be accessible to a novice animator/user of the system via a graphical interface, while they also should be accessible to an experienced user through a fast "short-cut driven" interface. On the other hand, a continuous visual feedback has to be available alerting the user in case something should have gone wrong. This implies it is necessary to have the ultimate high-quality rendering techniques available in the modelling stage.

Objects in our system are structured as a hierarchical composition of more elementary objects, enabling advanced possibilities during the motion specification. Each of these elementary objects has a reference to its parent object, and consists of a number of groups, being sets of polygons with identical coloring, shading and/or texturing attributes.

At the user interface level, we developed a modeller in which hierarchical objects can be created, deleted, repositioned, shaped, etc. in an interactive fashion. Starting from 2D primitives (polygons, splines, ...) simple 3D objects can be created, e.g., by transformational sweeps. Two examples of elementary transformational sweep methods, are those in which objects are created by rotating and linking primitive surfaces around an axis or by translating and linking primitives in a certain direction. Through parameterization and combination of these primitive transformations (e.g., scaling and/or translating the 2D primitive while rotating around an axis), more sophisticated objects can be created. Although these more sophisticated might not always be used in the creation of buildings, they can conveniently be utilized in the creation of "entourage" objects. Figure 1 gives snapshot of our object modelling screen in which such an object, in casu a car to be used in an animation, is being modelled.

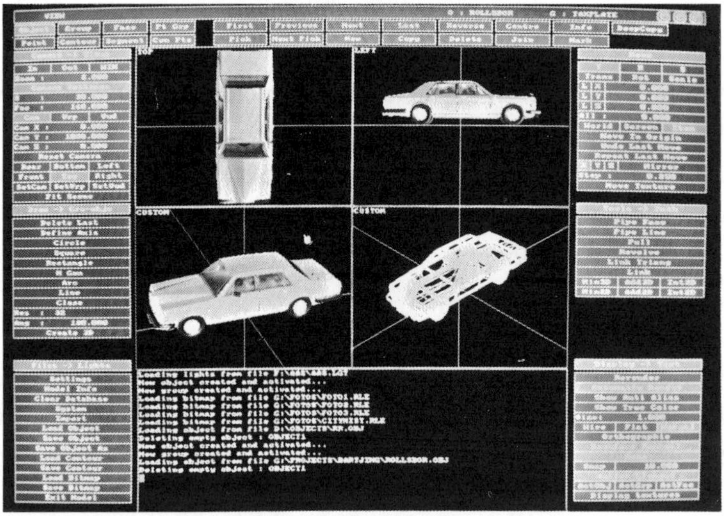

Figure 1. Snapshot of our object modelling screen.

Aside the transformational sweep techniques, contructive solid geometry (CSG) tools can also be exploited very well in architectural modelling. These tools enable creation of models consisting of the union, the difference, or the intersection of other two-dimensional or three-dimensional models. Concretely, we implemented a variant of the polygonal CSG operations found in (Laidlaw [12]).

MOTION MODELLING

Having a two-dimensional screen, controlling the motion of three-dimensional object models in a three-dimensional virtual world is a non-trivial task. Aside the objects, the virual camera itself is often also to be animated: this is especially true in architectural computer animations, as the virtual camera is frequently the main (or even only) moving item.

In order to enable an operator to create sophisticated motion, we developed a user-friendly graphical motion specification system, conform the object modelling subsystem with multiple simultaneous views. Several important key issues support the user-friendliness: (1) the utilization of hierarchies, (2) the manipulation of four-dimensional curves, and (3) real time preview.

By defining objects in a hierarchical structure, it becomes much more easy to specify local motion. We can take the example of a fork-lift: by defining the fork in a co-ordinate axis relative to the vehicle itself, the fork can be moved in the vertical direction of its local co-ordinate system, while the vehicle itself is following a certain path towards its destination in the co-ordinate system of the world.

For specifying smooth motion of objects, lights and cameras over three-dimensional paths, we provided the operator with an implementation of four-dimensional splines. These splines are explicitly represented as three-dimensional objects in the scene, in which editable control points are foreseen for manipulation of position and orientation. Figure 2 shows a screen snapshot of our motion modelling subsystem, in which a fly-through of a radiosity shaded (cf. rendering section) room is prepared. The path itself is represented as a spline curve, while the control points have a cube representation (see the top-left window, as well as the window below it, in which a more distant view is given). The virtual camera is represented as small pyramidal shape. In the bottom left window, two-dimensional editable projections of the splines are represented; besides the x-, y-, and z-components of the motion paths, the superimposed time-dimension can also be edited accurately here. Each desired motion path (and according motion speed) can thus be created by simply inserting and repositioning control points on the curves. An important issue is that the effect of a changement on a curve in any particular window, is automatically displayed in the other windows on the screen.

Last but not least, we implemented the facility to display in real time (25 frames/s) a part of the animation on the screen in any desired rendering technique. This so-called *real time preview* is very important in an animation environment, as it offers the opportunity to get an exact impression of the animation movements without having to make a video recording. The cycle between animation specification and verification can be much faster and more precise in this way. The preview is realized by (i) calculating rapidly a number of frames at lower resolution (one third of the resolution), (ii) storing the frames in the memory of the worker boards, and (iii) finally dumping them in real time to the video RAM via the transputer links.

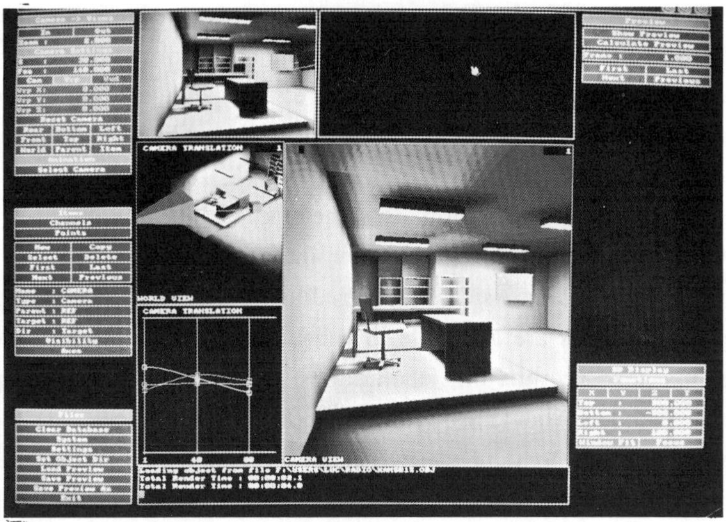

Figure 2. Snapshot of our motion specification subsystem.

RENDERING

General issues

In our rendering system a parallel Z-buffer algorithm as well as a parallel ray tracer are implemented. In the ray tracer as well as in the Z-buffer, parallelism is introduced by dividing the overall screen into screen regions and by distributing the pixel values to be calculated among the available processors. The m x n resolution of the screen regions depends on the rendering algorithm at issue. For ray tracing, this resolution is n x n (with n=4 or 8 or 16), whereas in scan conversion algorithms the horizontal screen region resolution equals the overall resolution. The exploitation of coherence is one of the main factors in the resolution determination.

The amount of work - opposed to the amount of communications - to be done at the pixel level, is crucial for maintaining a linear ratio "performance / number of processors". The load balance in the ray tracer can be kept optimal relatively easily, as (1) rays are fired in the scene independently of one another, and (2) the number and the complexity of the calculations to be performed per pixel are rather large. The more simple the processing per pixel becomes, the more complicated it becomes to keep the procesors usefully busy. Hence, the optimizing of parallelization of a Z-buffer algorithm is already a bit more difficult. Through introduction of some buffering processes a good distribution of the rendering pipeline (transformation - clipping - perspective division - shading - scan conversion - image storage) over the available processors, it is possible to maintain an almost linear expandability.

Within the basic rendering algorithms, some other facilities are integrated also. First of all, parallel algorithms to conquer anti-aliasing (Crow [6]; Van Reeth [18]) are incorporated. Aside the anti-aliasing algorithms, a number of parallelized methods are available to enhance the surface structure of 3D objects via texture mapping techniques (Heckbert [9]). Appropriate illumination models (Hall [8]) are foreseen in the shading process, enabling the simulation of several lighting effects. In the succeeding paragraphs, the most important realizations with respect to architectural visualization are elucidated.

<u>Ray Tracing</u>

In the ray tracing algorithm, parallelism in general can be introduced by dividing the problem in image space, or by dividing the problem in object space. In the latter methodology, object space is subdivided and distributed to different processors. Each of the processors consequently performs the amount of processing necessary for the subdivided space parts which they control. The difficulty with this approach consists of keeping the processors usefully busy. Indeed, processors controlling large parts of non-empty object space receiving relatively little rays may become idle, unless some non-trivial balancing scheme is realized. In the former methodology, the screen is subdivided into rectangular regions, and the work to be done for rendering these regions is distributed among the processors. In order to have a good load balance, relatively small regions should be assigned to the different processors in an interleaved fashion, so the "more time consuming" parts of an image are likely to be encountered by more processors. Given an interleaved distribution, one has a choice between realizing a static distribution in which the processors are always given the same regions, or a dynamic distribution in which the processors are given regions depending on their work load.

We follow the screen space division methodology: the screen is subdivided into rectangular regions of n X n pixels (n = 4, 8 or 16) that have to be ray traced in parallel by the different workers. The master transputer keeps track of the regions that yet have to be processed and distributes the work to be performed. Initially each processor gets a screen region, so each processor starts working. After having rendered the pixels within the region, the processor forwards the result and asks for the next region to be calculated. Hence, processors that coincidently have to process less complicated regions will simply be processing more regions in comparison to the processors that have to calculate more complicated regions (rather than going into an idle waiting state). Appropriate buffering and priority techniques are necessary in order to avoid unnecessary elapsed waiting times (Van Reeth [19]).

Additional optimizations, of which the exploitation of coherence (Kaplan [11]) is the most important, have been incorporated in the ray tracer. (Indeed, the parallelization of an efficient algorithm is always preferable above parallelizing a less efficient one). In casu, we implemented a parallel version of the voxel structure presented in (Amanatides [1]). In this approach, space is subdivided uniformly, and a ray is intersected only with objects lying in the subdivision currently being pierced by the ray.

An important additional issue to take into account when implementing a parallel ray tracing system concerns the distribution of the object data base. As each of our workers has 2MB DRAM at its disposal, it is not uncommon that the entire database organized into the voxel structure can be copied to each of the workers (Note: We expect moreover that in the near future even 4MB or more for every processor will be a very acceptable amount of memory, even when using several dozens of processors). If this is the case, the scene is broadcasted to each of the workers prior to the start of the actual ray tracing and consequently no further scene transmission is necessary. In case the database organized into the voxel structure doesn't fit into the memory of the workers, scene transmission becomes the problem to be tackled. Given the fact that the ray tracing process has a non-uniform work distribution on several levels (on the intersection level, e.g., the intersection of a ray with a bi-cubic patch (Lamotte [13]) takes up far more time than that of ray tracing spheres, whereas on screen level the amount of work varies almost inherently depending on screen content), transmission upon request is the methodology we followed. In order to make this transmission upon request efficient, appropriate data structures have to be defined. Concretely, we utilize the voxel structure, mentioned in the previous paragraph, not only for subdividing space and thus speeding up the intersection process, but also for subdividing the data base and thus speeding up the scene transmission process (as only objects in the neighborhood of the ray are transmitted). Details of this implementation can be found in (Van Reeth [19]). Average anti-aliased rendering timings of scenes (of a few thousand polygons) fitting in the workers' memories on average take less than five minutes to be rendered on a 13 processor network. For scenes not fitting in the workers' memories, a small overhead has to be taken into account. We finalize this paragraph by referring to Figure 3, in which a city model (taken from Autodesk 3D Studio) is ray traced.

Figure 3. A ray traced image of buildings in the mist.

Radiosity

During the last years, a substantial amount of research regarding realistic image synthesis has been focused on the realisation of global illumination models. Two approaches for solving the illumination problem in diffusely reflecting environments can be recognized. The first one is based on extensions to the traditional ray tracing algorithms (Cook [5], Kajiya [10]). The second category incorporates radiosity algorithms (Cohen[4], Goral [7], Nishita [15]). One of the major advantages of the radiosity approach is that it results in a solution independent from the position of the viewer.

Although the early radiosity methods gave rise to the most realistically rendered images of that period, they suffered from a storage and time cost that were $O(n^2)$ - with n the number of elementary surface patches -, so they were not practical to be used for high complexity scenes. With the introduction of progressive radiosity algorithms, however, the performance has drastically improved (Cohen [3], Wallace [20]).

A next step upwards with respect to improving performance is to incorporate parallel computations to solve the radiosities. As the computations of the form factors take up most of the time, it could be worthwhile to parallelize them. Several approaches involving coarse-grained parallelism can be found in the literature (Baum [2], Puech [16], Recker [17]). They are based on parallel computation of the form factors with a hemi-cube like technique to determine visibility and illumination.

We implemented a parallelisation of the form factor computations based on ray tracing methods. The calculation of the radiosities (given the assumption of global Lambertian diffuse illumination) narrows down to solving a set of equations in which the relation of energy leaving each surface element and arriving on each surface element is expressed :

$$B_{dAi}\, dAi = E_{dAi}\, dAi + \rho_{dAi} \int_{j} B_{dAj}\, F_{dAj\text{-}dAi}\, dAj \qquad (1)$$

where B_{dAi} is the radiosity (energy per unit area) of differential area dAi; d_{Ai} is a differential surface area i; E_{dAi} is the energy emission of differential area dAi (i.e. a light source); ρ_{dAi} is the reflectivity of differential area dAi (a value between 0 and 1 indicating which fraction of the arriving light is reflected); and $F_{dAj\text{-}dAi}$ is the form factor from dAj to dAi (a value between 0 and 1 indicating which fraction of light leaving dAj arrives at dAi). As the number of differential areas is infinite, the number of radiosities to be calculated is also infinite. Hence, a discretized version of the above intergral equation is used. In this approach the scene is discretized into finite areas rather than differential ones, resulting in equation (2):

$$B_{Ai}\ Ai = E_{Ai}\ Ai + \rho_{Ai} \sum_{j} B_{Aj}\ F_{Aj\text{-}Ai}\ Aj \qquad (2)$$

Using the reciprocity relationship $F_{Ai\text{-}Aj}\ Ai = F_{Aj\text{-}Ai}\ Aj$ this becomes :

$$B_{Ai} = E_{Ai} + \rho_{Ai} \sum_{j} B_{Aj}\ F_{Ai\text{-}Aj} \qquad (3)$$

Utilizing the ray tracing methodology, one has to compute the form factor from a lighting area A2 to a differential area dA1 at each vertex. For the case in which one subdivides the area A2 uniformly in n parts, the following form factor (including a visibility term δ_i) is to be computed (Wallace [20]):

$$dF_{A2\text{-}dA1} = dA1\ \frac{1}{n} \sum_{i} \delta_i \frac{\cos\theta 1i\ \cos\theta 2i}{\pi\ r_i^2 + A2 / n} \qquad (4)$$

The radiosity B1 at a vertex 1 received from an area A2 is consequently given by (Wallace [20]):

$$B1 = \rho 1\ B2\ A2\ \frac{1}{n} \sum_{i} \delta_i \frac{\cos\theta 1i\ \cos\theta 2i}{\pi\ r_i^2 + A2 / n} \qquad (5)$$

As equation (5) has to be calculated for each vertex in the scene, and this for several dozens of progressive iterations, it is obvious a lot of computation is involved here. Based upon a variant of our parallel ray tracing methodology, these equations can be solved very quickly: for scenes like the one in figure 4, on average a few seconds are needed per iteration on a 13 processor system. We again refer to other material (Lamotte [14]) for the implementational details of this approach, as they go behind the scope of this paper. Figure 4 shows a snaphot of a radiosity shaded office room.

Figure 4. A radiosity shaded office room.

Texture mapping

In the creation of naturalistically looking indoor and outdoor sceneries, the utilization of mapping techniques (Heckbert [9]) is of main importance. Several mapping categories can be defined: (i) texture mapping, for defining structured surface color (e.g. for generating marble); (ii) reflection mapping, for giving a metalistic look to surfaces; and (iii) bump mapping for changing the surface structure of materials (e.g. making it look more roughly). In each of these categories, one can have procedurally defined mappings, as well as a bitmapped approach (in which often scanned-in images are used). We incorporated both mapping methodologies in the parallel shading software. Figures 4 and 5 show respectively an indoor and an outdoor scene in which texture mapping is heavily used. The anti-aliased images took respectively 2.5 and 4 minutes to be rendered on a 13 processor network.

CONCLUSIONS

We have shown that the implementation of parallel rendering techniques can be a viable solution for solving the problem of ever increasing demand for higher realism in the making of architectural animations. It turns out that especially

Figure 5. An indoor scene, rendered with several marble textures

Figure 6. Outdoor scene with texture mapped trees, clouds, gravel and lawn

the complex computational algorithms offer great potential in the area of parallelisation. The next large milestone with regard to demonstrational architectural visualisation will probably be found in the direction of real-time image synthesis; we are convinced that parallel processing technologies can't be omitted in the approach of that milestone.

ACKNOWLEDGEMENTS

The help of all the colleagues in the laboratory who contributed in the realization of the animation- and rendering algorithms is greatly appreciated. This work is partly funded by a grant from the Belgian National Fund for Scientific Research (NFWO).

REFERENCES

1. Amanatides, J. and Woo, A. 'A Fast Voxel Traversal Algorithm for Ray Tracing', pp. 3-10, *Proc. Eurographics'87*, North Holland, 1987.
2. Baum, D.R. and Winget, J.M. 'Real Time Radiosity Through Parallel Processing and Hardware Acceleration', *Computer Graphics*, Vol. 24, 2, pp. 67-75, 1990.
3. Cohen, M.F., Chen, S.E., Wallace, J.R. and Greenberg, D.P. 'A Progressive Refinement Approach to Fast Radiosity Image Generation', pp. 75-84, *Computer Graphics (Proc. SIGGRAPH'88)*, Vol. 22, 3, 1988.
4. Cohen, M.F., 'Radiosity' in *State of the Art in Computer Graphics*, eds. Rogers, D.F. and Earnshaw, R.A., pp. 59-90, Springer-Verlag, 1991.
5. Cook, R.L., 'Stochastic Sampling in Computer Graphics', *ACM Transactions On Graphics*, Vol. 5, 3, pp. 51-72, 1986.
6. Crow, F. 'A Comparison of Antialiasing Techniques', *IEEE CG&A* , Vol. 1 ,1, pp. 40-48, 1980.
7. Goral, C.M., Torrance, K.E., Greenberg, D.P. and Bataille, B. 'Modeling the Interaction of Light Between Diffuse Surfaces', pp. 213-222, *Computer Graphics (Proc. SIGGRAPH'84)* , Vol. 18, 3, 1984.
8. Hall R , *Illumination and Color in Computer Generated Imagery*, Springer Verlag, 1989.
9. Heckbert, P.S. 'Survey of Texture Mapping' *IEEE CG&A,* Vol. 6, 11, pp. 56-57, 1986.

10. Kajiya, J.T. 'The Rendering Equation', pp. 143-150, *Computer Graphics (Proc. SIGGRAPH'86)* , Vol. 20, 4 , 1986.

11. Kaplan, M.R., 'The Use of Spatial Coherence in Ray Tracing', in *Techniques for Computer Graphics,* eds. Rogers, D.F. and Earnshaw, R.A., pp. 173-193, Springer-Verlag, 1987.

12. Laidlow, D.H., Trumbore, W.B. and Hughes J.F. 'Constructive Solid Geometry for Polyhedral Objects', *Computer Graphics (Proc. SIGGRAPH'86)* , pp. 161-170, Vol. 20, 4 ,1986.

13. Lamotte, W., Elens, K., Flerackers, E. 'Surface Tree Caching for Rendering Patches in a Parallel Ray Tracing System', pp. 189 -207, *Proc. Computer Graphics International '91,* 1991.

14. Lamotte, W., Van Reeth, F., Vandeurzen, L. and Flerackers, E. 'Parallel Processing in Radiosity Calculations', paper submitted for Computer Graphics International '93..

15. Nishita, T. and Nakamae, E. 'Continuous Tone Representaion of Three-dimensional Objects Taking Account of Shadows and Interreflection', *Computer Graphics (Proc. SIGGRAPH'85)* , pp. 22-30, Vol. 19, 3, 1985.

16. Puech, C., Sillion, F. and Vedel, C. 'Improving Interaction with Radiosity-Based Lighting Simulation Programs', *Computer Graphics,* Vol. 24, 2, pp. 51-57, 1990.

17. Recker, R.J., George, D.W. and Greenberg, D.P. 'Acceleration Techniques for Progressive Refinement Radiosity', *Computer Graphics,* Vol. 24, 2, pp. 59-66, 1990.

18. Van Reeth, F., Welter, R. and Flerackers, E. 'Virtual Camera Oversampling : A New Parallel Anti-Aliasing Method for Z-Buffer Algorithms', *Proc. Computer Graphics International '90*, pp. 241-254, 1990.

19. Van Reeth, F., Lamotte, W. and Flerackers, E. 'Ray Tracing Large Scenes on a Parallel Architecture', paper to be submitted for the Eurographics'93 conference.

20. Wallace, J.R., Elmquist, K.A. and Haines, E.A. 'A Ray Tracing Algorithm for Progressive Radiosity', *Computer Graphics (Proc. SIGGRAPH'89)*, Vol. 23, 3 , pp. 315-324 , 1989.

Collision detection for objects modelled by CSG

M. Zeiller

Institute for Computer Graphics, Technical University of Vienna, Karlsplatz 13/186, A-1040 Vienna, Austria

ABSTRACT

Physically based computer animation has to deal with the problem of detecting collisions of moving objects and preventing solid objects from interpenetration. This paper discusses a method for detecting collisions among complex objects modelled with the constructive solid geometry paradigm. Bounding volumes in each node of the CSG tree are used to determine as soon as possible whether a collision is likely to occur. Pairs of primitives – one of each object – are determined whose bounds cover common space. Spatial subdivision is used to detect these candidate pairs which have to be tested for collision using a geometric approach. The geometry of the intersection region is used to determine the collision point and the time and type of contact which are required for collision response.

INTRODUCTION

In computer animation objects roll and tumble through space, change their size or shape, alter their visible appearance and many more. When they move around, it may happen that they interpenetrate. To prevent a rigid object moving through another object as if it were not present, such interpenetrations, which are called collisions, have to be detected. To achieve realistic behaviour of the objects, the correct movement of the objects after the impact has to be determined as well. Therefore two tasks have to be performed: collision detection and collision response.

In computer graphics several representations for objects are used. Many modellers are based on the constructive solid geometry paradigm. Therefore CSG objects will be frequently used in computer animation, too. To be able to use this modelling technique in dynamic simulations, collision detection and collision response for CSG based objects have to be developed. However, most of the work on collision detection has focused on polyhedral objects. This paper presents a method for exactly detecting collisions among objects modelled by CSG. Cameron [1] has introduced an approach for detecting intersections among objects defined constructively. Our method is somewhat similar since we also decompose the detection of collisions into three stages. Cameron is only interested whether there is an intersection between two objects or not, since he is using this technique in the robotics field. However, we are dealing with collision detection for computer animation. Therefore we also have to determine the collision point and the time and type of contact which are necessary to be able to perform collision response. We will use Cameron's S-bounds theory to locate the region of a collision if there is any. If this region is not empty then a collision may occur and we have to test those primitives in the collision region whether they collide. To be able to find pairs of primitives – one of each colliding CSG object – we are using spatial subdivision to create an octree-like data structure for the CSG tree. Finally we have to determine the collision point and the time and type of contact of all pairs of primitives if they collide. Collision response will not be dealt with in this paper since analytical penalty methods can easily be applied to CSG objects as well.

The next section will give a survey on work on collision detection and collision response. The following section will present some basics on constructive solid geometry and bounding volumes. Afterwards we give an overview of how collision detection is applied to moving CSG objects. Then we describe the location of the intersection region using S-bounds. In the following section we introduce an algorithm to perform a spatial subdivision of a CSG tree which can be used to determine candidates for colliding primitives when two CSG objects intersect. This is followed by detecting a collision and the collision point between two primitives. Finally we present some ideas for future work.

RELATED WORK

Collision detection is a kinematics problem. Two objects collide if their surfaces intersect. Collisions can be determined by geometric reasoning whether the intersection of the objects is empty or not (e.g. null object detection). An intersection depends on the shape as well as on the position and orientation of the objects. Several algorithms for detecting collisions among rigid objects have been presented. Since the shape of the involved objects greatly influences the applied methods for detecting collisions, these algorithms can be classified depending on the object representation. Most of them concentrate on polyhedral objects. Moore and Wilhelms [2] present a method for detecting collisions among convex polyhedra based upon a clipping algorithm and among flexible surfaces modelled by triangles. Hahn [3] introduces an algorithm suitable for detecting collisions among objects modelled by polygonal surfaces, too. More recent approaches for detecting collisions among moving objects like Bouma and Vanecek [4] as well as Lin and Canny [5] are also restricted to polyhedra. In contrary, Baraff [6] detects collisions among rigid bodies bounded by curved surfaces. Von Herzen et al. [7] present an algorithm to detect collisions between pairs of time-dependent parametric surfaces. Cameron's approach [1] for detecting intersections among objects defined constructively consists of three stages. First he creates a set of bounds called S-bounds around the nodes of the CSG tree. In the next stage he splits the problem into simpler subproblems using spatial subdivision. Finally he generates a set of sufficient points that are classified regarding to the objects being tested. All approaches have in common that the amount of work to detect a collision is quite remarkable. Since these tests have to be performed at each time step in the dynamic simulation, these tests should only be applied when necessary. Therefore optimizations like bounding boxes (Hahn [3]) and octrees

(Moore and Wilhelms [2]), exploiting coherence between two consecutive time steps (Baraff [6]) and even parallel architectures (Bickerstaff and Hellestrand [8]) have been proposed to speed up the computation.

Collision response is a dynamics problem. After a collision has been detected, the time, location and type of the first contact have to be determined. Based on these quantities and the preceding movement, the behaviour of the colliding objects after the impact has to be computed. To achieve a realistic behaviour according to physical laws, the linear and angular momentum has to be preserved. Friction and elasticity of the involved surfaces have to be considered as well. The main problem of collision response is to determine the contact forces between bodies in contact which can be done by analytical methods or penalty methods. Analytical methods have been presented by Moore and Wilhelms [2], Hahn [3] and Baraff [6,9]. Moore and Wilhelms present an analytical linear system solution suitable for rigid and articulated objects that calculates the impulse between two colliding objects. Hahn uses analytical and numerical solution techniques that consider only the states of the involved objects before and after the collision. In [9] Baraff determines forces that prevent a number of non-colliding rigid polyhedral objects from interpenetration by solving a system of equality and inequality constraints, whereas in [6] he determines analytical forces between curved surfaces. On the other hand, penalty methods introduce restoring forces that separate penetrating objects. Moore and Wilhelms [2] present a method based on temporary springs at collision points which can be applied to several kinds of objects.

OBJECTS MODELLED BY CONSTRUCTIVE SOLID GEOMETRY

Constructive Solid Geometry

Constructive solid geometry operates on parameterized instances of solid primitives that are combined by Boolean set operations (Mäntylä [10]). Several kinds of primitive shapes can be used in CSG modelling: spheres, cylinders, cones, cubes, polyhedra and even more complex shapes like torii and superquadrics. Since these primitives are represented by mathematical models, their shapes are exactly modelled.

CSG objects are usually represented by a so-called CSG tree. A CSG tree is a binary tree with Boolean set operations in its intermediate nodes and primitives in its leaf nodes. The set operations used are union ($A \cup B$, OR-nodes), intersection ($A \cap B$, AND-nodes) and difference (A-B, MINUS-nodes). We will assume regularized set operations, i.e. no 'dangling edges' will occur due to these operations. The leaf nodes represent instances of primitive shapes which are combined by the set operations. Each primitive is represented as a type identifier and a sequence of parameters describing its size, position and orientation which might as well be encoded in a transformation matrix. In this paper we will restrict these primitives to be natural quadrics (sphere, cylinder, cone) and regular convex polyhedra (Platonic solids: tetrahedra, cube, octaeder, dodecaeder, icosaeder).

Bounding Volumes

CSG models are used in many fields in computer graphics. Since many of these applications are computationally very intensive, the computations are accelerated by bounding volumes. These volumes enclose the object, but since they have a much simpler shape an approximate solution can easily be obtained. Bounding volumes most frequently used are spheres, axis-aligned boxes, oriented boxes and intersections of infinite slabs. In CSG modelling a hierarchy of bounding volumes is constructed since bounding volumes can be added to each node of a CSG tree. A bounding volume can be easily computed for each primitive and stored in the respective leaf node. Additionally, bounding volumes can be computed in the intermediate nodes by combining the bounding volumes of its child nodes. The bounding volume of the root node encloses the whole object.

In this paper we will use Cameron's S-bounds [1,11,12] to calculate the bounding volumes since S-bounds are usually smaller than ordinary bounds. The upward rule computes the bounding volume of an intermediate node by combining the bounding volumes of the child nodes. The downward rule prunes the bounding volume of a child node due to the bounding volume of its parent node. A sequence of upward and downward rules is applied until a stable configuration is reached. This may even result in nodes with empty bounding volumes. In this case the respective primitive or subtree does not contribute to the entire object. We are using bounding spheres as well as axis-aligned bounding boxes to enclose the CSG objects since these volumes enable fast interference detection.

OVERVIEW OF THE SYSTEM

Dynamic simulation treats objects as masses moving under the influence of forces and torques (Wilhelms et al. [13]). Motion is computed by solving the equations of motion describing the objects. The dynamic state at a given time step is used to compute the dynamic state at a small time increment later. These computations have to be performed at least for each frame, sometimes it is necessary to have time increments less than a frame duration. After each update of the dynamic state of the objects that results in new positions and orientations, collision detection has to be performed. If a collision has been detected, the collision point and the time and type of contact have to be determined. Using collision response techniques, the dynamic state of the affected objects after the impact can be calculated.

Solving the equations of motion results in new positions and orientations of the objects. Based on these parameters collision, detection has to be performed. In an animated scene some objects will exist that do not move. These are static objects that do not change their position and orientation during the animation. Thus we distinguish between static objects and dynamic objects that are moving around. Static objects do not have to be tested for collision among one another. On the other hand dynamic objects may collide with static objects. Therefore each static object has to be tested for interference with each dynamic object which is called static collision detection. Each dynamic object may collide with another dynamic object. Detecting collisions among pairs of dynamic objects is called dynamic collision detection.

At first we perform static collision detection and afterwards we perform dynamic collision detection. Our collision detection method for CSG objects works on two objects. Therefore all possible combinations of two objects (static-dynamic, dynamic-dynamic) have to be created which are then tested for interference. Collision detection can stop if the first pair of interfering objects has been determined. However, if a collision between two other objects has occurred at some instant of time earlier, this will not be detected. Such a collision will probably be detected after the next evaluation of the equations of motion. However, the correct solution is to check the other objects for further collisions as well and to select the earliest one if multiple collisions should occur.

Collision detection for two CSG objects has to be performed in a common coordinate system. Usually objects are modelled in a local coordinate system and then transformed to the world coordinate system. At each frame the dynamic simulation computes new positions and orientations for a moving object which are represented in a new transformation matrix. To be able to perform collision detection all objects have to be transformed to the global coordinate system. However, this results in some problems with bounding volumes. We have decided to use spheres and axis-aligned boxes as bounding volumes because they can be tested for interference most easily. Whether two spheres interfere, can be detected by comparing the distance of the centers with the sum of the radii. Detecting an interference of two axis-aligned boxes reduces to some minimum and maximum operations. Rotating a sphere does not change its shape, but rotating an axis-aligned box will yield a box that is no longer axis-aligned. Thus transformation of a CSG object with axis-aligned bounding boxes will result in an object with non axis-aligned bounding boxes. Since interference detection of boxes that are not axis-aligned is much more complicated than interference detection of axis-aligned boxes, new axis-aligned bounding boxes have to be computed after each transformation. The transformed bounding boxes have to be enclosed by axis-aligned boxes, but unfortunately this will result in larger bounding boxes. We might as well use only spheres as bounding volumes, but boxes usually enclose an object more tightly than spheres. Therefore we have decided to use both of them, although we do not have the tightest bounding volumes that are possible.

After performing the dynamic simulation and computing new bounding volumes for the transformed objects, collision detection is applied. As already mentioned we have decomposed collision detection into three stages which will be treated in detail in the following sections. In the first stage we locate the region of a collision between each combination of two objects. If this region is empty, there is no collision and we can stop. Otherwise a collision may occur and we have to test those primitives in the collision region whether they collide. Therefore in the second stage we are using spatial subdivision to create an octree-like data structure (ECO) for the CSG tree of each object. The ECOs are intersected to find pairs of primitives – one of each colliding object. If no pair can be found the CSG objects do not collide and we can stop. Otherwise we have to determine the collision point and the time and type of contact for any pair of primitives in the third stage. If a collision has been found, collision response has to be performed. Figure 1 gives an overview of the flow of control in this system.

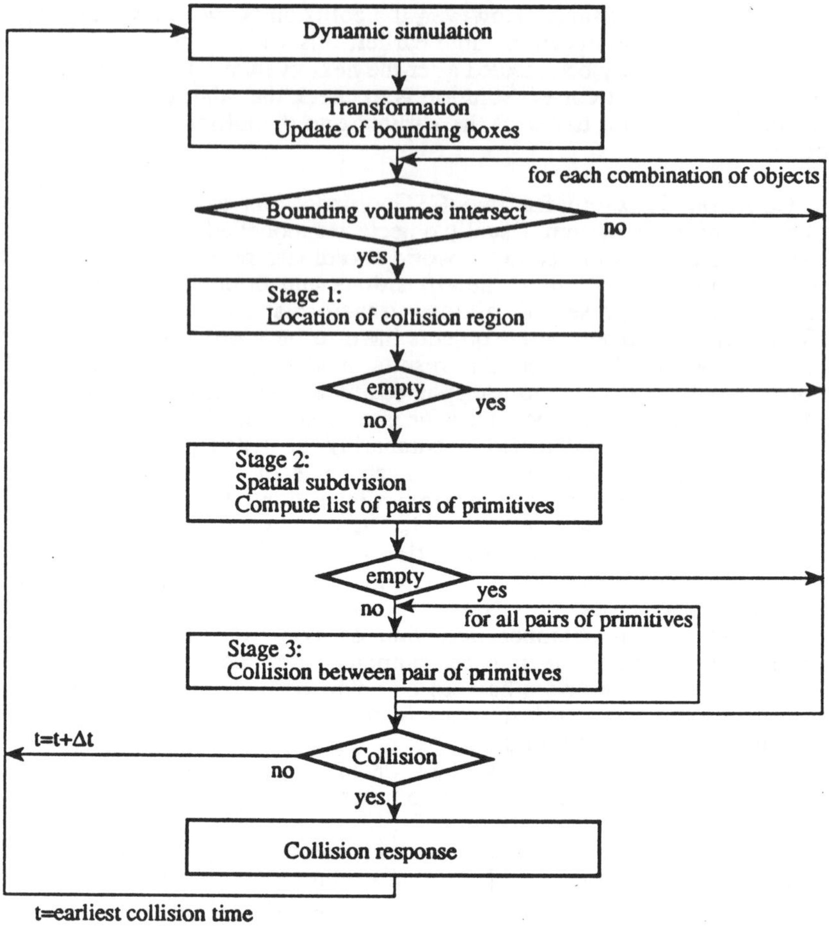

Figure 1. System overview

LOCATING THE COLLISION REGION

If the bounding volumes of two objects do not interfere, the objects which are enclosed by these volumes obviously do not interfere as well. Therefore these two objects do not collide and collision detection stops for this pair of objects. However, if the bounding volumes interfere, further investigations have to be performed since there is at least a good chance that a collision might occur. Detecting collisions among two CSG objects can be accelerated using S-bounds. As Cameron [1,11] already pointed out two CSG trees can be combined by an intersection operation, thus forming a new CSG tree where copies of the original objects are the child nodes of the root node. The bounding volumes of the root node can be computed by combining the bounding volumes of its child nodes. The S-bounds of the new CSG tree are computed by applying downward and upward rules successively until a stable configuration is reached. If the bounding volume of the entire object, i.e. the bounding volume of the root node of the CSG tree, is the empty set, the two original objects do not interfere. Again these objects do not collide and collision detection stops.

If the bounding volume of the root node is not empty, at least the bounding volumes of the CSG objects intersect. A prospective collision will have to occur in the enclosed region and therefore the collision point has to be inside this volume. Applying the S-bounds algorithm to the new CSG tree may have resulted in some nodes (intermediate as well as leaf nodes) with empty bounds although the bounding volume of the root node is not empty. Such nodes and their respective primitives do not contribute to the combined object. Empty bounds will seldom occur in common objects since no animator will model parts of an object that are cut away afterwards. However, since we only have combined two distinct objects symbolically to get the intersection object, the bounding volumes in several nodes will be empty. Hopefully the intersection region is small and many bounds are empty. The redundant nodes will not have to be considered for further collision detection and therefore they can be pruned from the CSG tree.

Since the two objects that shall be examined for a collision have been combined to a single CSG tree to evaluate the bounding volumes for the intersection region, we can as well remove the intersection node again. We end up with two objects which are modifications of the two original objects. Their bounds have been reduced to the region where an intersection of the S-bounds has been determined and where a collision is likely to occur. Those nodes that entirely lie outside the collision region have been pruned from the CSG trees and therefore both objects only consist of nodes that lie at least partially inside the collision region.

PAIRS OF COLLIDING PRIMITIVES

In the previous section we have used the S-bounds theory to reduce object space to the so-called collision region where a collision takes place. If the collision region is not empty a collision is likely to occur. We have to analyze the primitives of the CSG objects to determine whether they really collide. The CSG trees of the objects have been already pruned in the previous stage and thus they only consist of those primitives which at least partially interfere the collision region. Since we use tight bounds we can assume that these pruned CSG trees are not very large and consist only of a few primitives. Therefore we would have to apply collision detection among all pairs of primitives – one from each object. If more than one collision occurs, we have to consider the collision which happens first in the following collision response step. The next section will present some algorithms to determine collisions among such pairs of primitives.

Although the CSG trees have been pruned and their internal structure is simple, the corresponding objects can still have quite complicated shapes. Just think of the 2D object in Figure 2 and its CSG tree. If two primitives have been combined by union they contribute with their whole volume to the object, but if they have been combined by intersection only those parts of their volume contribute to the object which lie in both primitives (Figure 3). The surface of $A \cup B$ and $A \cap B$ is made up of parts of the surface of both primitives. If a primitive B has been subtracted from another primitive A only that part of A contributes to A-B which is not in B. The surface of A-B is made up of some part of A and some part of B, but B's surface is complemented due to the difference operation (Figure 3). Since these combinations are applied recursively in CSG modelling, the resulting volume and surface can be quite complicated. Therefore not every primitive in the CSG tree contributes to the object with its entire volume and surface and applying collision detection to such pairs of primitives would lead to incorrect results.

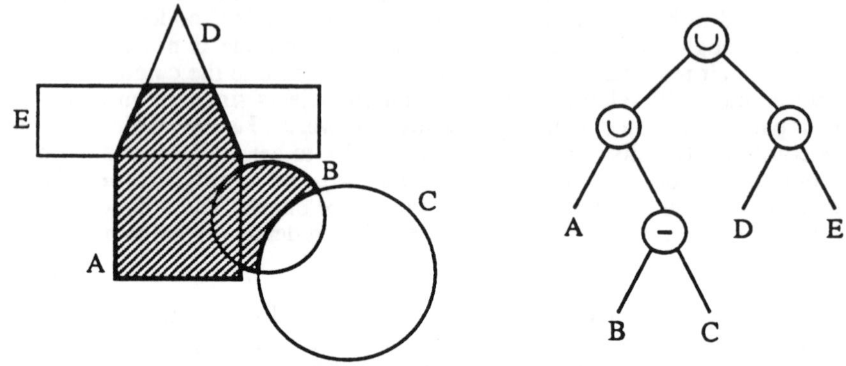

Figure 2. 2D example of a CSG object and its CSG tree

To perform collision detection we need to know the shape of the CSG object. One approach would be to spatially subdivide the objects with each voxel being full or empty (i.e. inside or outside). Collision detection could be easily applied, since if two voxels at the same location in both objects are indicated as full there is an intersection and they collide. In computer animation we have to perform collision response as well and therefore we need to determine the collision point and the type of contact which cannot be calculated from this kind of spatial subdivision. It fails for our needs because the surface of the object has been approximated by voxels, but for collision detection we need to know the exact shape. However, spatial subdivision in itself is quite useful since it helps to reduce the complexity of the object to a collection of simpler parts. In the following subsection we present a spatial subdivision scheme for a CSG object which enables us to determine pairs of primitives which have to be tested for collisions. As another advantage of this subdivision the spatial relationship of the voxels can be used to decide if some of the primitives do not have to be tested for collision.

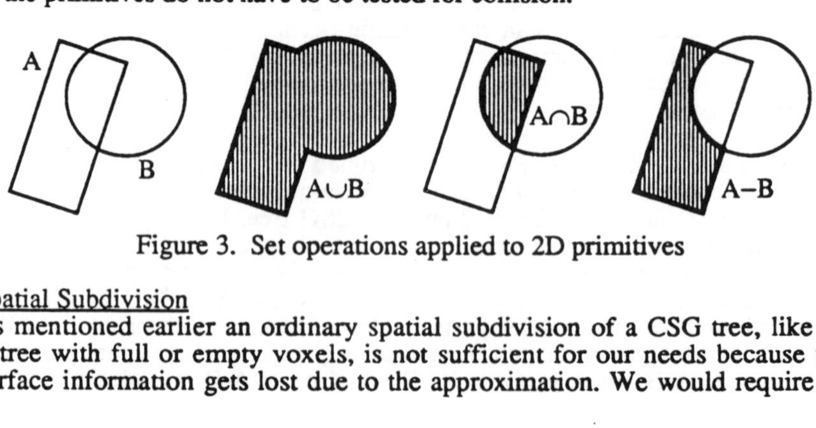

Figure 3. Set operations applied to 2D primitives

Spatial Subdivision

As mentioned earlier an ordinary spatial subdivision of a CSG tree, like an octree with full or empty voxels, is not sufficient for our needs because the surface information gets lost due to the approximation. We would require an octree-like description that exactly represents a CSG tree and therefore also includes information about the surfaces. As pointed out by Samet [14] several approaches exist for exactly representing polyhedral objects in an octree-like data structure that he calls PM octree. A leaf node of a PM octree can be full or empty, but it can also contain at most one face or one edge or one vertex. For the exact evaluation of CSG trees Samet [14] mentions the PM-CSG tree which is based on the work of Wyvill and Kunii [15,16]. Their approach is a variant of the PM octree in the way that their leaf nodes which are not full or empty refer to a single primitive object instead of a face, edge or vertex. They do not use all CSG operations since they restrict only to the subset of union of disjoint operands and difference. The PM-CSG tree distinguishes between five types of leaf nodes: full, empty, boundary between empty space and one primitive object, boundary between full space and one primitive object, and volume of space less than the limit of resolution.

For the aim of collision detection of CSG objects we extend the PM-CSG tree of Wyvill and Kunii [15]. According to the extended octree for representing polyhedral objects (Brunet et al. [17]) we will refer to this extension of an octree to represent an CSG tree exactly as extended CSG octree (ECO) (Zeiller [18]). In contrast to the CSG tree that is used by Wyvill and Kunii the set operations are not limited. Therefore we have to consider union, intersection and difference of two operands when we create the extended CSG octree. In extension to the PM-CSG tree we have to store more information of the underlying CSG object in the octree nodes. Again we have five types of leaf nodes:

1. A *full* node is completely embodied in the object. The corresponding voxel can be contained in more than one primitive (totally or partially). A *full* node refers to a list of primitives in which it is contained.
2. An *empty* node is not contained in the object.
3. A *p-surface* (positive surface) node is partially contained in the object. The full part of it is contained in only one primitive, the other part is empty. Therefore the boundary between full space inside the object and empty space is made up of a fraction of the primitive's surface. The p-surface node refers to this primitive.
4. A *n-surface* (negative surface) node is partially contained in the object. The full part of it is contained in only one primitive, the other part is empty. In contrary to a p-surface node the boundary between full space inside the object and empty space is made up of the surface of another primitive that has been subtracted. The p-surface node refers to the (positive) primitive contributing to the volume and to the (negative) primitive whose surface determines the boundary.
5. A *mixed* node is at the lowest level of resolution. More than one primitive contributes to it and therefore it should be decomposed, but due to the limited resolution it is not further subdivided. A mixed node refers to a list of primitives in which the voxel is partially contained.

Constructing an extended CSG octree is straightforward. The CSG tree is traversed in depth first manner. If a primitive object (leaf node of the CSG tree) is encountered, a single ECO node representing this primitive is created. At an intermediate node the two branches are evaluated and then combined according to the set operation. Tables 1, 2 and 3 represent the decision tables for the union, intersection and difference operation. If a non grey node has to be decomposed into its eight subvoxels the only thing we have to decide is whether the subvoxels are completely inside (full), partially inside (p-surface, n-surface) or outside (empty). Since it is sufficient to perform point tests of the vertices of the subvoxels and to determine the node type from the number of vertices inside the object, this spatial subdivision can be applied to each object made up of any primitive shape where we can decide whether a point is inside or outside.

	grey	*full*	*empty*	*p-surface*	*n-surface*	*mixed*
grey	grey	full	grey	grey	grey	–
full	X	full	full	full	full	full
empty	X	X	empty	p-surface	n-surface	mixed
p-surface	X	X	X	grey	grey	mixed
n-surface	X	X	X	X	grey	mixed
mixed	X	X	X	X	X	mixed

Table 1. Decision table for union of two ECOs (commutative)

	grey	*full*	*empty*	*p-surface*	*n-surface*	*mixed*
grey	grey	grey	empty	grey	grey	–
full	X	full	empty	p-surface	n-surface	mixed
empty	X	X	empty	empty	empty	empty
p-surface	X	X	X	grey	grey	mixed
n-surface	X	X	X	X	grey	mixed
mixed	X	X	X	X	X	mixed

Table 2. Decision table for intersection of two ECOs (commutative)

	grey	*full*	*empty*	*p-surface*	*n-surface*	*mixed*
grey	grey	empty	grey	grey	grey	–
full	grey	empty	full	n-surface	p-surface	mixed
empty	empty	empty	empty	empty	empty	empty
p-surface	grey	empty	p-surface	grey	grey	mixed
n-surface	grey	empty	n-surface	grey	grey	mixed
mixed	–	empty	mixed	mixed	mixed	mixed

Table 3. Decision table for difference of two ECOs (not commutative)

The Extended CSG Octree has to be created for both of the pruned CSG objects that are tested for a collision. Since they can only collide within the collision region the ECO has to be computed for the corresponding region. A bounding cube of the collision region (an axis-aligned box) is created and for each object the ECO is created by decomposing this cube for those parts within the collision region.

Determining Pairs of Primitives as Collision Candidates

The extended CSG octree representation of the two objects is now used to detect whether the objects collide. If they do collide those pairs of primitives are determined which have to be tested for a collision to be able to compute the collision point. Collision detection for two extended CSG octrees is very similar to combining them using an intersection operation. However, we do not generate a new ECO, we only report if they collide and create a list of pairs of (probably) colliding primitives. Table 4 is the decision table of the collision operation applied to two extended CSG octrees.

	grey	*full*	*empty*	*p-surface*	*n-surface*	*mixed*
grey	branches	branches	skip	branches	branches	–
full	X	add pairs	skip	add pairs	add pairs	add pairs
empty	X	X	skip	skip	skip	skip
p-surface	X	X	X	add pair	add pair	add pairs
n-surface	X	X	X	X	add pair	add pairs
mixed	X	X	X	X	X	add pairs

Table 4. Decision table of collision operation (commutative)

In a *branches* case no decision can be made since at least one of the nodes is a grey node. If necessary, the larger node has to be decomposed into its subvoxels and the corresponding subvoxels of both nodes have to be compared. *Skip* means that no action has to be taken. In an *add pair* or *add pairs* case the primitives of both nodes have to be combined to a pair and inserted into the list of collision candidates. Of course, each pair of primitives has to be inserted only once. *N-surface* nodes have to be treated in a special way because they consist of two primitives with one of them complemented. Thus for those nodes an element in the list of collision candidates has to contain both of these primitives of this object and the primitive of the other object. Therefore the list of candidates is made up of three kinds of elements:

- p-pairs with two primitives with positive surface
- n-pairs with one complemented surface and one positive surface
- nn-pairs with two complemented surfaces

The collision operation is applied to the two extended CSG octrees to detect those voxels that are covered by primitives in both objects. If no such voxels can be found, i.e. the combination always ends in a *skip* decision, no primitives will be added to the list of collision candidates. If this list is empty after the collision operation has been finished, no collision has been found and collision detection for these two objects stops. Otherwise we have to perform a third stage where collisions among the primitive shapes are detected.

DETECTING COLLISIONS AMONG PRIMITIVES

In the previous stage we have determined those pairs of primitive shapes that have to be further investigated. In the third stage of our collision detection method we have to consider only those pairs of primitives where a collision is likely to occur. Due to CSG modelling the list of primitive pairs consists of primitives with positive surface and of primitives whose surface has been complemented because a difference operation has been applied. Of course, these primitives have to be treated in different ways. We will first concentrate on collision detection of primitives with positive boundary and explain an equivalent method for complemented primitives afterwards.

Collision Quantities

In computer animation based on dynamic simulation collision detection has to be followed by collision response. Therefore it is not sufficient to detect a collision (yes or no), but the collision point and the time and type of contact have to be determined as well. The collision point can be defined as the point where the two primitives come into contact. However, when we detect a collision the primitives already interfere. A good heuristic in this case is to consider a so-called penetration point to be the point of greatest interpenetration (Moore and Wilhelms [2]). If the penetration point is very close to the surface, the primitives are considered to be in contact. Otherwise the collision point has to be determined by evaluating the geometry of the intersection region (see also Hahn [3] for polyhedral primitives). The time of contact can be calculated as the instant of time when the collision point lies on the surface of both primitives. We also have to determine the type of contact and the plane of collision (Hahn [3]). If quadric surfaces collide, we will gain a planar contact in most cases, since the surface normal can be derived at the collision point. It will rarely happen that two 'edges', an 'edge' and an apex or two apexes will touch. However, we are also using regular convex polyhedra as primitives and when they collide edge-edge, edge-vertex and vertex-vertex contacts may arise. Thus if we can derive the surface normal of one of the involved primitives at the collision point, this normal determines the collision plane. Otherwise the normal of the collision plane is computed by averaging the adjacent surface normals.

Intersection of Primitives

To be able to detect a collision and to compute the quantities mentioned above, the primitives have to be intersected. The primitive shapes we are using are natural quadrics and Platonic solids, i.e. regular convex polyhedra. Therefore we can distinguish between three types of intersections:

- polyhedron-polyhedron
- quadric-quadric
- quadric-polyhedron

For the aim of collision detection, we do not have to determine the intersection curve generated from the two primitives if they really intersect. It is sufficient to calculate the quantities necessary for collision response. Therefore we can apply reduced versions of intersection algorithms.

Polyhedron-Polyhedron Intersection Several methods for detecting collisions among polyhedral objects have been presented in the literature (e.g. Moore and Wilhelms [2], Hahn [3], Bouma and Vanecek Jr. [4], Lin and Canny [5]). We have chosen an approach based on a 3D analogy to Cyrus-Beck Clipping as is has been introduced by Moore and Wilhelms [2], but any convenient method can be used.

Quadric-Quadric Intersection The intersection of two quadric surfaces can be performed algebraically using the algebraic representations of quadrics or by geometric reasoning with the shape of the primitives (Miller [19]). Both methods are using an exact model of the surface, but algebraic representations are based on equations, whereas geometric representations directly store quantities such as size and direction vectors. In the algebraic approach any quadric surface is represented as the set of points satisfying the second degree polynomial equation

$$Ax^2 + By^2 + Cz^2 + 2Dxy + 2Eyz + 2Fxz + 2Gx + 2Hy + 2Jz + K = 0$$

with suitable values of the coefficients A...K. Therefore the intersection can be performed in a surface type independent way. We get an intersection curve which is either a point, a line or one or two curve branches. These branches will either be conic sections or nonplanar space curves of polynomial degree 4 (Miller [19]). However, the drawback of this technique is that is not very easy to compute the collision quantities out of this curve.

On the other hand we can perform geometric reasoning based on the shape of the natural quadrics in 3D space which is determined by their geometric representation. This technique is most often faster, but unfortunately a special intersection procedure has to be implemented for each combination of primitive types. We completely loose the generality of the algebraic approach and have to write several intersection procedures if we want to add a new type of primitive. Despite these drawbacks we are using this technique whenever it is possible. In comparison to the algebraic approach, implementing this method is much more work. Several configurations of the two primitives in 3D space have to be considered and each of them requires special treatment. Figure 4 presents an example of those configurations that arise if a cylinder and a sphere are tested for collision (a. no collision, b. contact, c. collision). Both primitives have been projected onto a plane passing through the axis of the cylinder and the centre of the sphere. After we have derived the configuration we can easily compute the collision quantities. Usually determining the type of configuration can be performed very fast, thus this approach is preferable in terms of computational efficiency.

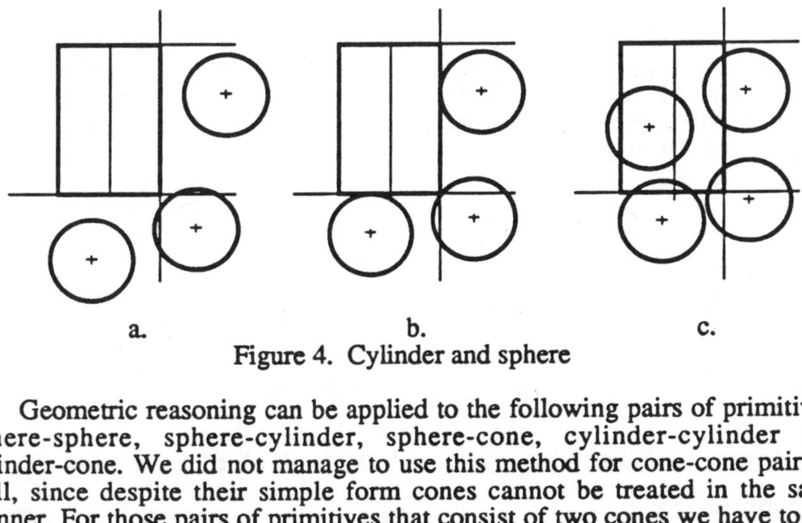

Figure 4. Cylinder and sphere

Geometric reasoning can be applied to the following pairs of primitives: sphere-sphere, sphere-cylinder, sphere-cone, cylinder-cylinder and cylinder-cone. We did not manage to use this method for cone-cone pairs as well, since despite their simple form cones cannot be treated in the same manner. For those pairs of primitives that consist of two cones we have to use a different approach. As in the geometric reasoning method we try to detect the simple configurations of the cones (e.g. an apex is inside or on the surface of the other cone) first. If we do not find such a simple configuration we apply a method that is based on the geometric approaches to intersect quadric surfaces of Miller [20]. The parametric form of one of the cones is substituted into the implicit equation of the second cone and we get the parametric form of the cone-cone intersection curve. Afterwards we have to calculate the so-called critical points on this curve. These points determine the straight lines on the surfaces of the cones with greatest interpenetration which can be used to calculate the collision point.

Quadric-Polyhedron Intersection The intersection of quadric and polyhedral primitives is similar to the intersection of two polyhedral objects. First the vertices of the polyhedron are tested whether they are inside or outside the quadric surface. If at least one vertex has been found which is inside the quadric, the adjacent edges are used to derive the geometry of the intersection region and the collision quantities. Otherwise an edge might pierce the quadric and the edges of the polyhedron are intersected with it. Again the collision quantities have to be derived after an intersection has been discovered. If still no collision has been detected, the bounding planes of the polyhedron have to be intersected with the quadric to determine whether the primitives collide.

Positive and Negative Intersection

So far we have only considered positive primitives that have been added to the CSG object by union and intersection operation. Due to the difference operation we have to deal also with primitives that have been subtracted from the CSG tree thus yielding n-surface nodes in the extended CSG octree. Therefore the list of pairs of primitives consists of n-pairs and nn-pairs as well which contain complemented surfaces. These pairs have to be treated in a special way. Combining the primitives A and B by a difference operation (A-B) means that the resulting object consists of those parts of A that do not lie inside of B. The surface of A-B is made up of some part of A and some (complemented) part of B. If collision detection between a positive primitive C and a complemented primitive has to be performed, a collision has to be reported only if the positive primitive C intersects the surface of the complemented primitive B and that part of C which is outside B lies inside primitive A. Figure 5 demonstrates a 2D example of this relationship (a. collision, b. no collision). Therefore the algorithms for intersecting primitives, that have been presented above, have to be extended to deal with complemented primitives as well.

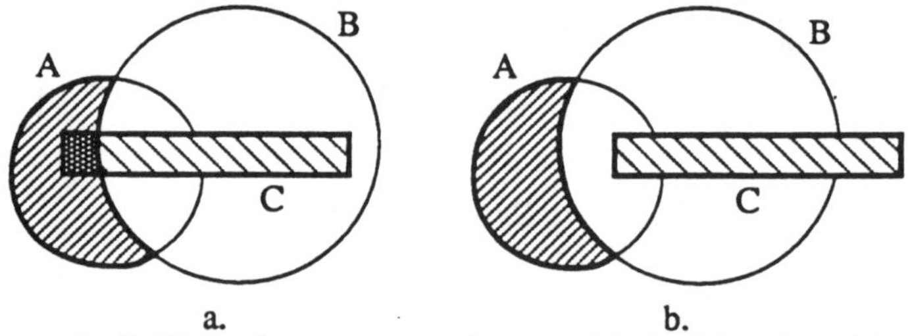

Figure 5. Collision between complemented B (A-B) and positive C

FUTURE WORK

We have restricted the primitive shapes of the CSG objects to be natural quadrics and Platonic solids. We could treat more complex shapes easily in the spatial subdivision stage, but using other shapes would require further intersection procedures for each new combination of primitives in the third stage where collisions among primitives are detected. Perhaps a general and fast algebraic approach can be found which can be used to calculate the collision point and the time and type of contact as well.

Primitive shapes in a CSG tree can be rotated, translated and scaled. Most CSG modelling systems enable different scaling factors in the direction of each coordinate axis. However, this would transform a sphere to an ellipsoid, a cylinder with circular cross-section to a cylinder with elliptical cross-section, etc. Therefore new shapes arise and new geometric intersection procedures have to be developed, too. On the other hand some of them may have to be solved using Miller's geometric approach (Miller [20]) because their shapes are too complicated for geometric reasoning. In the present system scaling factors have to be identical in all directions. Future systems should allow the use of different scaling factors which is another reason for investigating more general intersection methods for various primitives.

In the first stage of our collision detection algorithm we have used bounding volumes to locate the collision region. Out of the variety of possible bounding volumes we have chosen spheres and axis-aligned boxes due to their simplicity in detecting intersections. We have not performed a detailed analysis of using only bounding spheres, only bounding boxes and the combination of both until now. Other bounding volumes, that enclose the object more tightly but are more costly in intersection detection, should be investigated, too.

CONCLUSION

We have presented a method for exactly detecting collisions among objects modelled by the constructive solid geometry paradigm. The detection of collisions between two CSG objects is decomposed into three stages. In the first stage we locate the collision region using bounding volumes. S-bounds with bounding spheres and axis-aligned bounding boxes have proven to be well suited for this problem. If the collision region is empty we can stop collision detection for these objects. Otherwise we have to test those primitives in the collision region whether they collide. We can exactly detect collisions only among two primitive. Therefore we decompose the two CSG objects into pairs of primitive shapes – one of each object. Since CSG objects may have complex shape due to the recursive application of union, intersection and difference operations, we split this problem into smaller subproblems. Both CSG objects are spatially subdivided to create an octree-like data structure for the CSG tree. These extended CSG octrees are intersected to detect those cells that are covered by primitives in both objects. If no such cells can be found, collision detection for these two objects stops. Otherwise those pairs of primitives that may collide are determined. In the third stage we have to detect collisions among all such pairs of primitive shapes. If they collide, we calculate the collision point as well as the time and type of contact.

REFERENCES

1. Cameron, S. 'Efficient Intersection Tests for Objects Defined Constructively' *The International Journal of Robotics Research*, Vol.8, No.1, pp. 3 - 25, February 1989.

2. Moore, M. and Wilhelms, J. 'Collision Detection and Response for Computer Animation' *Computer Graphics*, Vol.22, No.4, pp. 289 - 298, August 1988.

3. Hahn, J. 'Realistic Animation of Rigid Bodies' *Computer Graphics*, Vol.22, No.4, pp. 299 - 308, August 1988.

4. Bouma, W. and Vanecek Jr., G. 'Collision Detection and Analysis in a Physically Based Simulation', pp. 191 - 203, *Proceedings of Second Eurographics Workshop on Animation and Simulation*, Vienna, Austria, 1991.
5. Lin, M. and Canny, J. 'Efficient Collision Detection for Animation', *Proceedings of Third Eurographics Workshop on Animation and Simulation*, Cambridge, UK, 1992.
6. Baraff, D. 'Curved Surfaces and Coherence for Non-penetrating Rigid Body Simulation' *Computer Graphics*, Vol.24, No.4, pp. 19 - 28, August 1990.
7. Von Herzen, B., Barr, A. and Zatz, H. 'Geometric Collisions for Time-Dependent Parametric Surfaces' *Computer Graphics*, Vol.24, No.4, pp. 39 - 48, August 1990.
8. Bickerstaff, M. and Hellestrand, G. 'A Highly Parallel Architecture for Real Time Collision Detection in Flight Simulation' *Computers & Graphics*, Vol.15, No.3, pp. 355 - 363, 1991.
9. Baraff, D. 'Analytical Methods for Dynamic Simulation of Non-penetrating Rigid Bodies' *Computer Graphics*, Vol.23, No.3, pp. 223 - 232, July 1989.
10. Mäntylä, M. *An Introduction to Solid Modeling* Computer Science Press, 1988.
11. Cameron, S. 'Efficient Bounds in Constructive Solid Geometry' *IEEE Computer Graphics & Applications*, Vol.11, No.3, pp. 68 - 74, May 1991.
12. Cameron, S. and Yap, C. 'Refinement Methods for Geometric Bounds in Constructive Solid Geometry' *ACM Transactions on Graphics*, Vol.11, No.1, pp. 12 - 39, January 1992.
13. Wilhelms, J., Moore, M., Skinner, R. 'Dynamic animation: interaction and control' *The Visual Computer*, Vol.4, No.6, pp. 283 - 295, December 1988.
14. Samet, H. 'Volume Data' Chapter 5, *The Design and Analysis of Spatial Data Structures*, pp. 315 - 376, Addison Wesley, Reading Massachusetts, 1990.
15. Wyvill, G. and Kunii, T. 'A functional model for constructive solid geometry' *The Visual Computer*, Vol.1, No.1, pp. 3 - 14, July 1985.
16. Wyvill, G., Kunii, T. and Shirai, Y. 'Space Division for Ray Tracing in CSG' *IEEE Computer Graphics & Applications*, Vol.6, No.4, pp. 28 - 34, April 1986.
17. Brunet, P., Navazo, I. 'Geometric Modelling Using Exact Octree Representation of Polyhedral Objects' in EUROGRAPHICS'85 (Ed. Vandoni, E.), pp. 159 - 169, *Proceedings of EUROGRAPHICS'85*, Nice, France, 1985. Elsevier, Amsterdam, 1985.
18. Zeiller, M. 'Extended CSG Octrees'. In preparation.
19. Miller, J. 'Analysis of Quadric-Surface-Based Solid Models' *IEEE Computer Graphics & Applications*, Vol.8, No.1, pp. 28 - 42, January 1988.
20. Miller, J. 'Geometric Approaches to Nonplanar Quadric Surface Intersection Curves", *ACM Transactions on Graphics*, Vol.6, No.4, pp. 274 - 307, October 1987.

SECTION 3: VIRTUAL REALITY

Invited Paper

Control and autonomy of synthetic actors

D. Thalmann

Computer Graphics Lab, Swiss Federal Institute of Technology, Lausanne, Switzerland

ABSTRACT

This paper discusses the problem of controlling synthetic actors. It presents the development of a general concept of autonomous actors reacting to their environment and taking decisions based on perception systems, memory and reasoning. With such an approach, it is possible to create simulations of situations such as actors moving in a complex environment they may know and recognize, or actors playing ball games based on their visual and touching perception. An animation system with autonomous actors is based on the three key components: the locomotor system, the perceptual system and the organism system.

INTRODUCTION

The long-term objective of our research is the visualization of the simulation of the behavior of human beings in a given environment, interactively decided by the animator [1]. Behavioral techniques make possible the automating of high-level control of actors such as path planning. By changing the parameters ruling this automating, it is possible to give a different personality to each actor. This behavioral approach should be a major step relatively to the conventional motion control techniques [2,3].

There are two ways of considering three-dimensional computer animation [4] and its evolution. The first approach corresponds to an extension of traditional animation methods. The animator uses the computer to assist him in the creation of keyframes and simple motions and deformations like stretching and squashing. The second approach corresponds to simulation methods based on laws of physics, physiology or even psychology. The purpose is not the same: traditional methods allow us to create three-dimensional characters with exaggerated movements while simulation methods are used to try to model a human behavior accurately. High level animation involving human beings and animals may be produced in this way. In fact the behavior of any living creature may be simulated. Reynolds [5] introduced the term and the concept of behavioral animation in order to describe the automating of such higher level animation.

Behavioral animation may be considered as the visual simulation of a behavioral model. A behavioral model means in fact a model of animal or human behavior. Behavior is often defined as the way that animals and human beings act and is also often reduced to the meaning of reacting to the environment. This is not sufficient for computer animation and the definition of the Encyclopedia Universalis is more convenient for us: it is the frame of continuous flow between a living creature and its environment. Behavior is not only reacting to the environment but should also include the flow of information by which the environment acts on the living creature as well as the ways the creature codes and uses this information.

Most of the computer-generated films have been produced using traditional computer animation techniques like keyframe animation, spline interpolation, etc. Automatic motion control techniques [6] have been proposed, but they are strongly related to mechanics-based animation and do not take into account the behavior of characters. As mentioned by Cohen [7], "to provide the creatures with some autonomy, they must have a set of goals or constraints to meet; some of these goals may readily lend themselves to kinematics specification, but others such as a request to move gracefully do not; the subtleties of motion would be influenced by the creature's personality ".

Reynolds studied in detail the problem of group trajectories: bird flocks, herds of land animals and fish schools. This kind of animation using a traditional approach (keyframe or procedural laws) is almost impossible. In the Reynolds approach, each bird of the flock decides its own trajectory without animator intervention. Reynolds introduces a distributed behavioral model to simulate flocks of birds, herds of land animals, and schools of fish. The simulated flock is an elaboration of a particle system with the simulated birds being the particles. A flock is assumed to be the result of the interaction between the behaviors of individual birds. Working independently, the birds try both to stick together and avoid collisions with one another and with other objects in their environment.

Haumann and Parent [8] describe behavioral simulation as a means to obtain global motion by simulating simple rules of behavior between locally related actors. They developed a test-bed used to create a library of physically behaving actors which can realistically reproduce the motion of flexible objects. Lethebridge and Ware [9] propose a simple heuristically-based method for expressive stimulus-response animation. They model stimulus-response relationships using "behavior functions" which are created from simple mathematical primitives in a largely heuristic manner. Wilhelms [10] proposes a system based on a network of sensors and effectors. Ridsdale [11] proposes a method that guides lower-level motor skills from a connectionist model of skill memory, implemented as collections of trained neural networks.

Many authors have proposed methods to implement motion of articulated bodies. Some methods come from robotics like inverse kinematics and dynamics. Inverse kinematics permits direct specification of end point positions (e.g., a hand or foot). There are two major problems in inverse kinematics, finding any solution that will achieve the desired goal and, finding the most desirable solution. Inverse kinematics can be used as constraints with dynamic simulations [12]. Dynamics has been used to automatically generate realistic motion [13,14] that successfully animates the motion of chains, waves, spacecraft, and automobiles. The problem is that force/torque control is not intuitive. The course of a simulation is completely determined by the objects' initial positions and velocities, and by forces applied to the objects along the way. It just solves the initial value problems. How to return some level of control to the animator is one of the most difficult issues in dynamic simulation.

In task-level animation [15,16], the animator specifies what the synthetic actor has to do, for instance, "jump from here to there". Witkin and Kass [17] describe a spacetime constraint system in which constraints and objectives are defined over Spacetime, referring to the set of all forces and positions of all of a creature's degrees of freedom from the beginning to the end of an animation sequence. The laws of physics in spacetime method provide possible constraints on the motion rather than the primary driving force for simulation. The motion of the creatures is derived through a user guided optimization process that minimizes an objective function subject to physical and other user-defined constraints. The process of solving the constraints leads to a series of configurations for the constrained objects, thus creating an animation as a byproduct of the solution process. Cohen [18] takes this concept further and uses spacetime window to control the animation interactively. The subdivision of spacetime into a discrete pieces, or Spacetime Window, over which subproblems can be formulated and solved. The sensitivity of highly non-linear constrained optimization to starting values and solution algorithms can thus be controlled to a great extent by the user. Given visual and numeric feed back about the progress of the convergence towards a desired solution, the user is able to add new constraints, raise or lower the weight of previous constraints, or modify the intermediate solutions. This is followed by further iterations until the animation is deemed satisfactory by the user. Several examples of animation sequences were produced using this approach. For example, a simplified 3-DOF basketball player is simulated; the final sequence includes moving quickly to catch its own throw, and finally to throw the ball into the basket. Girard [19] proposes an algorithm for optimizing limb movement in terms of both kinematics and dynamics-based variables and constraints.

ANIMATION SYSTEM WITH AUTONOMOUS ACTORS

Our main purpose is the development of a general concept of autonomous actors reacting to their environment and taking decisions based on perception systems, memory and reasoning. With such an approach, we should be able to create simulations of situations such as actors moving in a complex environment they may know and recognize, or actors playing ball games based on their visual and touching perception. An animation system with autonomous actors is based on the three key components:

- the locomotor system
- the perceptual system
- the organism system

A locomotor system is concerned with how to animate physical motions of one or more actors in their environment (Fig.1). This is the control part of the system.

A perceptual system is concerned with perceiving the environment. According to Gibson [20], we may consider five perceptual systems: basic orienting system, auditory system, haptic system, taste-smell system, and visual system. The corresponding modes of attention are: orienting, listening, touching, smelling, tasting, and looking.

The organism system is concerned with rules, skills, motives, drives and memory. It may be regarded as the brain of the actor and the autonomous part.

In the next Sections, we explain in more details our three key components.

LOCOMOTOR SYSTEM

Although many techniques have been developed for the control of articulated bodies; they are generally applied to much simpler systems than humans. Moreover, there is no general method applicable to a complex motion like playing football for example. Integration of different motion generators is vital for the design of such complex motions where the characterization of movement can quickly change in terms of functionality, goals and expressivity. This induces a drastic change in the motion control algorithm at multiple levels: behavioral decision making, global criteria optimization and actuation of joint level controllers. By now, there is no global approach which can reconfigure itself with such flexibility.

Fig.1 Several actors moving

Only a combination of various techniques may result in a realistic motion with a relative efficiency. Consequently, our locomotor system based on several integrated methods. As we have already developed several techniques: keyframe, inverse kinematics, direct/inverse dynamics (Fig.2) and biomechanics-based walking (Fig. 3), we integrate them using a blending approach. Production of a natural looking motion based on the integration of all parts of body using different types of motion control methods is certainly a challenge. In our case, a blending module will be associated with the coach-trainee correction method that we have created in the context of walking. This method allows the kinematics correction of joint-space based motion with respect to cartesian constraints [21,22]. In such a way, it is still possible to modify the key frame sequence, a low-level description of motion, for a higher level goal-oriented requirement. We believe that this approach will greatly extends the scope of predefined motions (rotoscopy, specialized model, key-framed etc..). A new methodology emerges for motion conception and editing which is centered on the coach-trainee correction method. The functional diagram in Fig.4 organizes the composition of the input motions around the blending module prior to the coach-trainee correction module.

Fig.2. Dynamics-based motion

Fig.3. Biomechanics walking

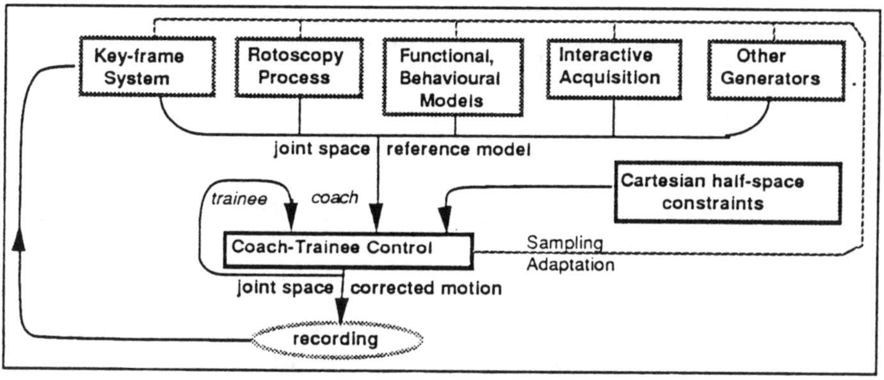

Fig.4 Functional diagram of motion blending

PERCEPTUAL SYSTEM

A perceptual system is concerned with perceiving the environment. Our objective is to integrate only three modes of attention: orienting, touching, and looking. For the orientation, we model the ability to stabilize the posture of the body with respect to the gravitational force. A simulation of the touching system consists in detecting contacts between the actor and the environment. The most important perceptual subsystem is the vision system. The originality of our approach in this wide research area is the use of a synthetic vision as a main information channel between the environment and the actor [23].

Although the use of vision to give behavior to synthetic actors seems similar to the use of vision for intelligent mobile robots, it is quite different. This is because the vision of the synthetic actor is itself a synthetic vision. Using a synthetic vision allow us to skip all the problems of pattern recognition and distance detection, problems which still are the most difficult parts in robotics vision. However some interesting work has been done in the topic of intelligent mobile robots, especially for action-perception coordination problems. For example, Crowley [24], working with surveillance robots states that "most low level perception and navigation tasks are algorithmic in nature; at the highest levels, decisions regarding which actions to perform are based on knowledge relevant to each situation". This remark gives us the hypothesis on which our vision-based model of behavioral animation is built.

Vision simulation is the heart of our system. However, we do not attempt to recreate human vision. Our purpose is to obtain a 2-

3/4 D vision, which means 2D + 1/2D + 1/4D. The 1/2D is due to the fact that for each point of the 2D vision, we know the distance between the eye and the point from which it is the projection. The 1/4D corresponds to the fact that for each point of the vision, we know to which object it belongs (e.g. object 34, which is a table). This 2-3/4D has the advantage of avoiding all problems of pattern recognition involved in robotics vision.

Our system is implemented on IRIS 4D Workstations using an object-oriented extension of the C language. For input, we have a hierarchy containing the description of 3D objects, the environment, and the camera characterized by its eye and interest point. For output, the vision is built as a 2D array of pixels on which the actor's view of the world is projected. Each pixel contains the distance from the eye to the point and the object identifier. The size of the array should be selected in order to have the maximum accuracy without consuming too much CPU time.

ORGANISM SYSTEM

This may be considered as the actor brain. We model the actor brain with a memory (essentially visual) and a limited reasoning system allowing the actor to decide his motion based on information. The movement decision procedure is the central coordinator to determine further movement.

Our approach [25] is based on the use of Displacement Local Automata (DLA), which is similar to the concept of script introduced by Schank [26] for natural language processing. For vision, a DLA is a black box which has the knowledge allowing the synthetic actor to move in a specific part of his environment. Examples are the DLA *displacement-in-a-corridor*, the DLA *obstacle-avoidance*, the DLA *crossing-with-another-synthetic-actor* , the DLA *passage-of-a-door* etc. This concept of DLA has the advantage of being able to increase or decrease the description grain. For example, the displacement in a corridor may be one of the uses of a more general DLA *displacement-between-two-obstructions-not-far-away-one-from-the-other*. An obstruction may be a river, a wall, a road etc. This means that the DLA may be used in a corridor, on a sidewalk or a bridge. Similarly, *to circumvent the swimming-pool* corresponds to going around an obstruction. The DLA system allows for the creation of a kind of assemblage game where, using simple and modular elements, it is possible to react to a lot of everyday simple situations. It should be noted that our DLAs are strictly algorithmic; they correspond to reflexes which are automatically performed by the adult and they only use vision as the information source.

Once we have access to DLAs, how do we use them ? How and when do we activate them and stop them? Our approach consists of using a controller which, in an unknown environment, analyzes this environment and activates the right DLA. In the simpler case of a known environment (i.e. already used), the controller directly activates the DLA associated with the current location during the learning phase. This controller also has to handle the DLA end, the DLA errors and other messages coming from the DLAs. In particular, a DLA may be in a situation where it does not know what to do and the controller should solve the problem. The controller is the thinking part of our system. It makes decision and perform the high-level actions.

In any known environment, the controller should be able to activate the DLA associated with the current location. This means that an environment description has to be stored somewhere. This is the role of the navigator. From information provided by the controller (information that can come itself from DLAs), the navigator builds step by step a "map" of the environment. The navigator then gives to the controller the location of the synthetic actor in the map, and of course, it should plan the path from a known location to another known location (at a high level: e.g. go to the kitchen). Another function of the navigator is to allow the controller to change the initialization of some DLAs (e.g. changing the attention level in the corridor) and anticipate the DLA changes. This description of the navigator functionalities emphasize the role of the map which should be more a logical map than a geographical map, more a discrete map than a continuous map. In fact, human being only remember discrete properties like *the tree is on the right of the car*. This suggests that to give to the behavior of the synthetic actor a maximum of believability and a maximum of developing possibilities, it is essential to give the actor a world representation similar to the representation that human beings have.

We have also improved the behavioral aspect by introducing a local map allowing the actor to detect the movable objects and analyzing their velocity and orientation. This analysis is used for adapting the walking movement in order to avoid (or reach) movable objects.

The purpose of our research is to create behavioral animation of actors moving in a complex environment. The actors should be able to avoid obstacles based on their vision and their experience. This means that the actor are able to remember the environment when they walk in an area already known. Our prototype of vision-based walking is simple and fuzzy: to create an animation involving a synthetic actor automatically moving in a corridor (Fig.5) avoiding other synthetic actors. To simulate this behavior, each synthetic actor was using a synthetic vision as its perception of the world and so as the unique input to its behavioral model. The system was then extended to several actors walking in a rectangular furnished room. The approach has been proved working for simple environments. But, it cannot be extended to a very complex environment like in a problem of walking in a forest. In fact, the method could work when the actor is finding his way for the first time. However, the problem comes when the actor is supposed to know the environment, which means the introduction of an actor memory. Using his vision, the actor sees objects and stores them into his memory. Thereafter, the actor may use this memory for a reasoning process.

Fig.5. Vision-based navigation

The actor visual memory is defined using an octree. An octree is convenient for storing complex surrounding objects detected by the synthetic vision. The octree will be implemented using paging with a hashing function based on the location of the point to be inserted. In summary, the actor will have a synthetic vision, a memory and capacities of learning and reasoning. In order to simulate real situations, we also plan to introduce possibilities of forgetting, corresponding to the deleting of subtrees in the octree, according to criteria to be decided.

The octree has to represent the visual memory of an actor in a 3D environment with static and dynamic objects. Objects in this environment can grow, shrink, move or disappear. In a static environment (growing objects are still allowed) an *insert* operation for the octree is sufficient to get an approximate representation of the world. If there are moving or disappearing objects like cars, other actors, or opening and closing doors, we also need a *delete* operation for the octree.

One procedure to develop will be a recursive algorithm to find a path from the actor to any position avoiding the obstacles based on his memory. The actor should also be able to remember if there is no path at all or if there are loops as in a labyrinth. This requires to implement a backtracking mechanism. Once an actor has found a good path, he may use his memory/reasoning to take the same path. However, as new obstacles could have been added on the way, the actor will use the current synthetic vision to decide the exact path, reacting to the new obstacles.

The algorithm of path finding is based on path searching in a graph. Each free voxel not occupied by an obstacle is interpreted as a node of the graph. All the neighbor voxels are considered to be connected by an edge. So, the octree represents a graph with nodes and edges. The algorithm of path searching uses the principle of backtracking and memorizes all tested nodes in a sorted list. With this list of already tested nodes, circuits can be avoided, and situations without a path from a given source to a given destination can be detected. In a first approach, a path is represented by a sequence of free nodes. To avoid a combinatorial explosion of possibilities in graph searching, we use a heuristic depth first search.

The path finding procedure described above is a mental process of the actor, which is based on the contents of his visual memory (octree). This means that during his reasoning on a possible path, he does not move. Very often, though, he is placed in an unknown environment, and still has not seen and memorized. In this case, he cannot find a path using, for example, some conditional heuristic. So, he is forced to explore his environment guided by his vision and a heuristic. This exploring is an active process and the actor has to walk and memorize what he sees. In this case, the heuristic depth first search step can be used to guide the actor to guarantee that he finds a way if one exists. In this exploring mode he executes the reasoning process of the first step in his real environment.

If the actor, for example, is enclosed in a house, there will be no path to a destination outside the house. In this case, he will explore the parts of the interior accessible to him according to the heuristic. He will finish his search by having memorized the interior of the house and the conclusion that he is enclosed. By using our algorithm, he can avoid turning infinitely in a loop and can recognize that he has checked all possibilities to find an exit.

CONCLUSION

Our goal now is to use the perceptual systems (orienting, touching, and visual) to create a high-level guiding system to drive more complex situations like for example tennis players. We will have the typical situation of a locomotor system driven by perception systems: essentially vision and touching. The locomotor system will consist of kinematics-based and dynamics-based motion of the actor skeleton.

REFERENCES

1. Magnenat Thalmann N, Thalmann D (1990) Computer Animation: Theory and Practice, Springer Verlag, Tokyo, 2nd edition.
2. Thalmann D (1991) Dynamic Simulation as a Tool for Three-dimensional Animation, in: New Trends in Animation and Visualization, John Wiley and Sons.
3. Magnenat Thalmann N, Thalmann D (1990) Synthetic Actors in Computer-Generated Films, Springer-Verlag, Heidelberg
4. Magnenat Thalmann N, Thalmann D (1990) Computer Animation: Theory and Practice, Springer-Verlag, Tokyo.
5. Reynolds C (1987) Flocks, Herds, and Schools: A Distributed Behavioral Model, Proc.SIGGRAPH '87, Computer Graphics, Vol.21, No4, pp.25-34
6. Wilhelms J. Toward Automatic Motion Control, IEEE Computer Graphics and Applications, Apr., 1977
7. Cohen MF (1989) Gracefulness and Style in Motion Control, Proc. Mechanics, Control and Animation of Articulated Figures, MIT, Boston.
8. Haumann DR, Parent RE (1988) The Behavioral Test-bed: Obtaining Complex Behavior from Simple Rules, The Visual Computer, Vol.4, No 6, pp.332-347.
9. Lethebridge TC and Ware C (1989) A Simple Heuristically-based Method for Expressive Stimulus-response Animation, Computers and Graphics, Vol.13, No3, pp.297-303
10. Wilhelms J (1990) A "Notion" for Interactive Behavioral Animation Control, IEEE Computer Graphics and Applications , Vol. 10, No 3 , pp.14-22
11. Ridsdale G. Connectionist Modelling of Skill Dynamics, Journal of Visualization and Computer Animation, Vol.1, No2, 1990, pp.66-72.
12. Isaacs,P. M., and Cohen,M. F. Controlling dynamic simulation with kinematic constraints, behavior functions,and inverse dynamics. Siggraph'87.

13. Armstrong, B., and Green,M. The dynamics of articulated rigid bodies for purposes of animation. In Proceedings of Graphics Interface (May,1986)
14. Wilhelms,J. and Barsky,B. Using Dynamic Analysis to Animate Articulated Bodies Such As Humans and Robots, Proc. Graphics Interface, 1985.
15. Badler NI, Webber BL, Kalita J, Esakov J (1989) Animation from Instructions, in: Make Them Move, Morgan Kaufman, 1989, pp.51-93.
16. Zeltzer D (1987) Task-level Graphical Simulation: Abstraction, representation, and Control, in: Make Them Move, Morgan Kaufman, 1989, pp.3-33.
17. Witkin A, Kass M (1988) Spacetime Constraints, Proc. SIGGRAPH '88, Computer Graphics, Vol.22, No4 , pp.159-168.
18. Cohen MF Interactive Spacetime Control for Animation Siggraph'92, pp.293-302.
19. Girard M (1990) Constrained Optimization of Articulated Animal Movement in Computer Animation, in: Badler NI, Barsky B, Zeltzer D (eds) Make Them Move, Morgan Kaufmann
20. Gibson JJ, The Senses Considered as Perceptual Systems, Houghton Mifflin, 1966, Boston
21. Boulic R, Thalmann D (1992) Combined Direct and Inverse Kinematic Control for Articulated Figures Motion Editing, *Computer Graphics Forum,* Vol. 2, No.4, October 1992, pp.189-202.
22. Boulic, R.,Thalmann, D.,(1992) TRACK a Kinematic Goal-Oriented Animation System for Coordinated Editing of Joint-Space Based Motions, *Eurographics'92, Third Workshop on Animation and Simulation.*Cambridge, UK, September 1992
23. Renault, O., Thalmann,N. M., Thalmann,D. A Vision-Based Approach to Behavioral Animation, Visualization and Computer Animation, Vol.1, No. 1, 1990, pp.18-21.
24. Crowley JL (1987) Coordination of Action and Perception in a Surveillance Robot, IEEE Expert, winter, pp.32-43
25. Renault O, Magnenat-Thalmann N, Thalmann D (1990) A vision-based Approach to Behavioral Animation, The Journal of Visualization and Computer Animation, Vol. 1, No1
26. Schank RC (1980) Language and Memory, Cognitive Science, Vol.4, No3, pp.243-284

Invited Paper

Virtual humans acting in virtual reality

N.M. Thalmann

MIRALab, University of Geneva, Switzerland

ABSTRACT

Several very complex problems must be solved in order to render three-dimensional animation involving virtual humans in their environment. In this paper, we explain several of these problems and present solutions. In the context of interactive animation systems, the relationship between the animator and the virtual humans should be also emphasized. We present an example of facial communication between the animator and virtual humans.

INTRODUCTION

The ultimate reason for developing virtual humans (also called synthetic actors) who seem real is to be able to use them in any virtual scene representing the real world. Anyway, a virtual scene, beautiful though it may be, is not complete without people...virtual people that is. Scenes involving virtual humans imply many complex problems that we try to manage since several years [1]. We slowly come to the point of simulating real-looking virtual humans, taking into account body, face and cloth deformations. In a short future, we will not be able to see any difference between a real person and a virtual one. Any environment could be simulated and consequently, we will be able to experiment in real-time any virtual environment, and to communicate with virtual humans rather naturally. Any simulation will be possible: from the infinite world to the megaworld passing through any daily live situation.

CREATION OF VIRTUAL HUMANS

The synthesis of realistic virtual humans leads to obtain and include the specific features of the character of interest. For the universally known personalities (actors) such as Marilyn, Humphrey, and Elvis, there is a less scope to make mistakes as the deviations will be very easily detected by the spectator. In spite of this ambition to make realism, or better, imitation, this type of realism should not be confused with the photographic or the cinematographic realism.

Sculpting the shape

Creating a body for a virtual human is only the first step, his particular character depends on his body movements and his personality is defined by the subtle changes of his facial expressions and other gestures.

To construct these shapes, we propose the use of an interactive sculpting approach. The surfaces of human face and body are irregular structures implemented as polygonal meshes. We have introduced a methodology [2] for interactive sculpting using a six-degree-of-freedom interactive input device called the Spaceball. When used in conjunction with a common 2D mouse, full three dimensional user interaction is achieved, with the Spaceball in one hand and the mouse in the other. The Spaceball device is used to move around the object being sculpted in order to examine it from various points of view, while the mouse carries out the picking and deformation work onto a magnifying image in order to see every small detail in real time (e.g. vertex creation, primitive selection and local surface deformations). In this way, the user not only sees the object from every angle but he can also apply and correct deformations from every angle interactively.

Typically, the sculpting process may be initiated in two ways: by loading and altering an existing shape or by simply starting one from scratch. For example, we will use a sphere as a starting point for the head of a person and use cylinders for limbs. We will then add or remove polygons according to the details needed and apply local deformations to alter the shape. When starting from scratch points are placed in 3D space and polygonized. However, it may be more tedious and time consuming.

With this type of 3D interaction, the operations performed while sculpting an object closely resemble traditional sculpting. The major operations performed using this software include creation of primitives, selection, local surface deformations and global deformations.

The selection process

To select parts of the objects, the mouse is used in conjunction with the Spaceball to quickly mark out the desired primitives in and around the object. This amounts to pressing the mouse button and sweeping the mouse cursor on the screen while moving the object with the Spaceball. All primitives (vertices, edges and polygons) can be selected. Mass picking may be done by moving the object away from the eye (assuming a perspective projection) and careful, minute picking may be done by bringing the object closer.

Local surface deformations.

These tools make it possible to produce local elevations or depressions on the surface and to even out unwanted bumps once the work is nearing completion. Local deformations are applied while the Spaceball device is used to move the object and examine the progression of the deformation from different angles, mouse movements on the screen are used to produce vertex movements in 3D space from the current viewpoint. The technique is intended to be a metaphor analogous to pinching, lifting and moving of a stretchable fabric material. Pushing the apex vertex inwards renders a believable effect of pressing a mould into clay.

Global deformations

These tools make it possible to produce global deformations on the whole object or some of the selected regions. For example, if the object has to grow in a certain direction, it can be obtained by scaling or shifting the object on the region of interest.

In the construction of a certain category of figures like realistic human bodies, it is often preferable to keep certain irregularities on the surface. A very smooth skin, for example, does not necessarily guarantees a more realistic appearance. In very delicate parts like shoulders or cheeks the imperfections are visible and generally we have to get rid of them right away, but we can keep some others small irregularities in order to create a figure which seems less plastic and robotized and in order to attenuate the feeling we often feel in front of computer generated human bodies. This kind of imperfections in a realistic figure give the impression of a human figure which has not been conceived only in the "designer's" head but has really been observed from the reality.

Simplification of polygonal meshes

The main goal of shape creation is the realism of the final virtual humans. It should look as "human" possible. This constraint implies that the number of faces used to model the body does not come into consideration. What might appear as a small drawback when rendering fixed pictures becomes a problem in animation. Algorithms applied to a virtual human for animation have some others constraints, and among these realism increases computing time the most. If we take the example of an actor dressed with synthetic clothes, one time consuming process in the animation of clothes is collision detection of the clothes with the body. The complexity of these algorithms is dependent on the number of faces of the colliding objects. Moreover, most of the body parts involved in collision detection are covered by the clothes during the animation. There is no need for a very precise body. The solution is to reduce the number of faces used to render the synthetic actor. This kind of reduction can't be applied uniformly to the whole object. If the actor is dressed in a skirt, and short-sleeved shirt, arms and legs need to be as precise as possible since they will appear in the animation.

We need a method to reduce the number of polygons of objects used in animation according to algorithm requirements, thereby reducing computation time and space. The reduction can be locally applied and controlled by the shape information the user wants to keep. We may summarize the whole process of simplification, more details may be found elsewhere.

- Selecting a connexe region of the object by picking the triangles.
- Orthogonal projection of the region onto a plane (defined by an arbitrary point and a vector normal as an average of the normal of all triangles in the selected region). If the projection of a triangle is inside the border of the triangles already projected, it is eliminated from the region.
- Extraction from the 3D region of the elements defining shape of the region. These elements are either points or edges.
- Selection of the elements to keep inside the region.
- Deletion of all other elements.
- Triangulation of the 2D projection using the selected elements to keep and the elements defining the border of the 2D projection.
- Reconstruction of the simplified 3D region from the 2D triangulation. As the border of the region has been kept, there is no problem reconstructing the region and inserting it in the whole object.

Fig.1 shows an example.

MOTION OF VIRTUAL HUMANS

The human animation is very complex and should be split into body motion control and facial animation. Basically a virtual human is structured as an articulated body defined by a skeleton. Skeleton animation consists in animating joint angles. There are two main ways to do that: parametric keyframe animation and physics-based animation. Another complex objective is to model human facial anatomy exactly including its movements to satisfy both structural and functional aspects of simulation.

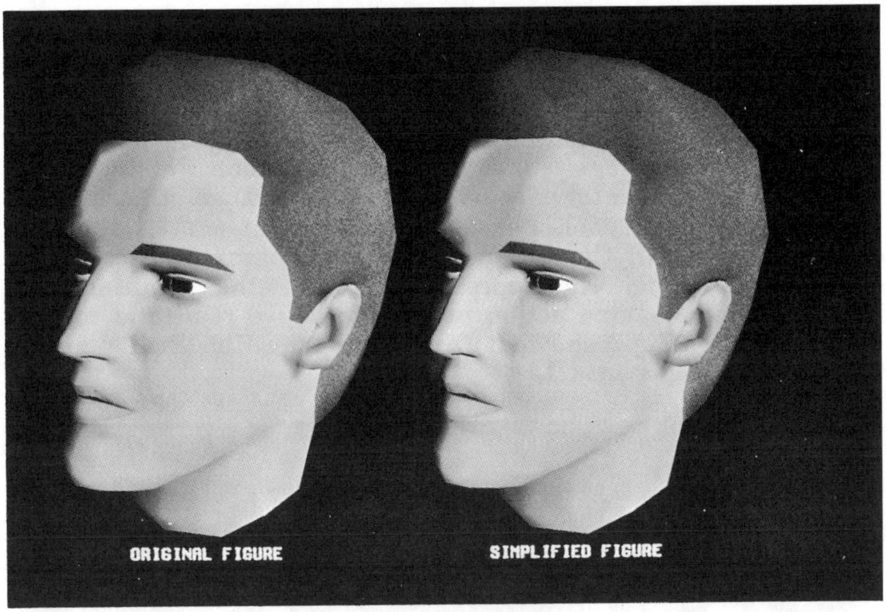

Fig.1 Simplification

Body motion
Basically a virtual human is structured as an articulated body defined by a skeleton. When the animator specifies the animation sequence, he/she defines the motion using a skeleton. A skeleton is a connected set of segments, corresponding to limbs, and joints. A joint is the intersection of two segments, which means it is a skeleton point where the limb which is linked to the point may move. The angle between the two segments is called the joint angle. A joint may have at most three kinds of position angles: flexion, pivot and twisting. Skeleton animation consists in animating joint angles. There are two main ways to do that: parametric keyframe animation and procedural animation based on mechanical laws. For example, to bend an arm with parametric keyframe animation, it is necessary to enter into the computer the elbow angle at different selected times. Then the software is able to find any angle at any time using for example interpolating splines [3]. In the second approach, angles are calculated by inverse kinematics, dynamics or biomechanics.

A high-level approach consists in specifying the animation in terms of tasks. With task level control, the animator can only specify the broad outlines of a particular movement and the animation system fills in the details. In task-level animation, the animator specifies what the virtual human has to do, for instance, "bring me a glass". Task-level animation requires high-level models of human actions.

With task level control, the animator can only specify the broad outlines of a particular movement and the animation system fills in the details. In task-level animation, the animator specifies what the synthetic actor has to do, for instance, "jump from here to there". Task-level animation requires high-level models of human actions. For example, we have developed a human walking model built from experimental data based on a wide range of normalized velocities [4]. The model is based on a simple kinematics approach designed to retain the intrinsic dynamic characteristics of the experimental model. This approach also allows the definition of an individualized walking action in an interactive real-time context in most cases. Fig.2 shows a frame from the film Still Walking using this walking model.

Fig.2. Still walking

Facial animation

Because the human face plays the most important role for identification and communication, realistic construction and animation of the face is of immense interest in the research of human animation. The ultimate goal of this research would be to model exactly the human facial anatomy and movements to satisfy both structural and functional aspects. However, this involves the concurrent solution of many problems. The human face is a very irregular structure, which varies from person to person. The problem is further compounded with its interior details such as muscles, bones and tissues, and the motion which involves complex interactions and deformations of different facial features.

Although all movements may be rendered by muscles, the direct use of a muscle-based model is very difficult. The complexity of the model and our poor knowledge of anatomy makes the results somewhat unpredictable. This suggests that more abstract entities should be defined in order to create a system that can be easily manipulated. A multi-layered approach [5] is convenient for this. In order to manipulate abstract entities like our representation of the human face (phonemes, words, expressions, emotions), we propose to decompose the problem into several layers. The high level layers are the most abstract and specify "what to do", the low level layers describe "how to do". Each level is seen as an independent layer with its own input and output.

For our facial deformations, we have extended the concept of Free Form Deformations (FFD) introduced by Sederberg and Parry, a technique for deforming solid geometric models in a free-form manner. It can deform surface primitives of any type or degree, for example, planes, quadrics, parametric surface patches or implicitly defined surfaces. FFD involves a mapping from R^3 to R^3 through a trivariate tensor product Bernstein polynomial. Physically, FFD corresponds to deformations applied to an imaginary parallelepiped of clear, flexible plastic in which the objects to be deformed are embedded. The objects are also considered to be flexible so that they are deformed along with the plastic that surrounds them. A grid of control points is imposed on the parallelepiped. For any point interior to the parallelepiped, the deformation is specified by moving the control point(s) from their undisplaced latticial position.

As an extension to the basic FFDs, we provide the option of including rational basis functions in the formulation of deformation [6]. These rational basis functions allow incorporation of weights defined for each of the control points in the parallelepiped grid. The advantage of using rational FFDs is that they provide one more degree of freedom of manipulating the deformations by changing the weights at the control points.

Region build-up is accomplished by interactive selection of polygons. These regions on the human face generally correspond to their anatomical descriptions, such as nose, lips, eyes, etc. The selection process can be further hastened by attaching the color attributes of the polygons. The magnitude and direction of the muscle pulled is interactively adjusted by changing the position and weight of the control points on the control-unit of a region.

Certain points in a defined region can be set "anchored" meaning they will not deform when the deformations are applied on the region. The physical characteristics of facial mesh points can also be set interactively.

The free-form deformations correspond to the lower level of our multi-layer system SMILE where, at each level, the degree of abstraction increases. The defined entities correspond to intuitive concepts such as phonemes, expressions, words, emotions, sentences and eye motion, which make them natural to manipulate. The top level requires direct input from the animator in the form of global abstract actions and their duration, corresponding to the intuitive and natural specifications for facial animation [7]. Fig.3 shows an example.

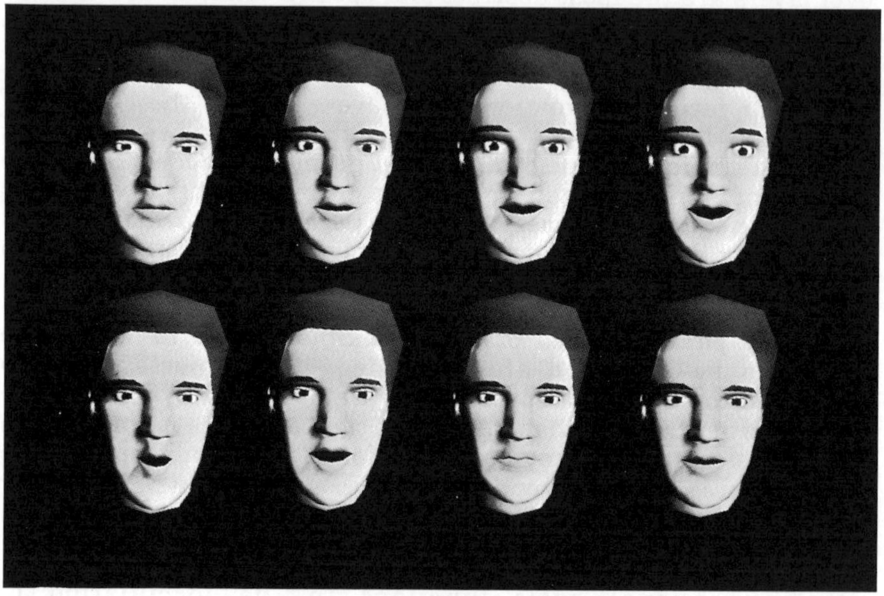

Fig.3. Facial animation

RENDERING VIRTUAL HUMANS

Skin texture

To improve the "Barbie-like" aspect of virtual humans, we propose a technique based on texture mapping of photos of real faces. A separate tool for matching the 3D facial topology on a given picture/photo of a face is developed. Only a few feature points is selected from the 3D model to exactly match the corresponding points on the picture.

Delaunay triangulation is used to connect these points. These points can be moved and displaced on the picture interactively. An interpolation scheme in a triangular domain is used to get the desired texture coordinates. As a result the picture is deformed and mapped on the 3D model. In order to map the entire head, multiple views are needed. These pictures will be projected on a cylinder. Then the corresponding matching will be performed between the cylindrical projected 3D model points and cylindrical projected pictures. By using texture mapping the quality of rendering improves considerably. In addition, it will allow us to put a picture of a specific person on a given 3D model.

It is useful and interesting to be able to generate synthetic textures of different type of skins varying in color and luminance. Samples of skin for different regions of a face are used to generate a synthetic skin texture. Techniques of image treatment are used to obtain characteristic features for the samples in a group, and each group is then assigned to a portion of a face for skin texture.

Hair modeling

In the field of human simulation, hair presents one of the most challenging problems and therefore has been one of the least satisfactory aspects of human images rendered to date. The difficulties of processing hair result from the large number and detailed geometry of the individual hairs, the complex interaction of light and shadow among the hairs, and the small scale of the hair width in comparison with the rendered image.

Rendering an image of hair with our system involves several steps:

- creating a database of hair segments
- creating shadow buffers from all lights
- rendering the hairless objects using all shadow buffers
- compositing the hair on the hairless image

In the hair style rendering module, the process is made step by step. First, the shadow of the scene is calculated for each light source i, as well as for the light sources for the hair shadow. The hair shadows are calculated for the object surface and individually for each individual hair. Finally the hair style is blended into the scene, using all shadow buffers. The result is an image with a three-dimensional realistic hair style rendering where complex shadow interaction and highlight effects can be seen and appreciated. Fig.4 shows an example.

Clothes modeling

In most computer-generated films involving virtual humans, clothes are simulated as a part of the body with no autonomous motion. For modeling more realistic clothes, two separated problems have to be solved: the motion of the cloth without collision detection and the collision detection of the cloth with the body and with itself. A flexible or deformable object is different from a rigid object because it cannot be considered as a whole and its movement cannot be computed from a small set of its points. The flexible object must be divided into small parts and each point is submitted to a set of local and global constraints. These constraints create forces which prevent the violation of these constraints. Solving the dynamic system requires finding an equilibrium between all these forces. Recent research deals with dynamic models for flexible or deformable objects.

Fig.4. Hair rendering

Our work in cloth animation [8] is based on the fundamental equation of motion as described by Terzopoulos et al. [9] with the damping term replaced by a more accurate one proposed by Platt and Barr [10]. When a collision is detected, we pass through the second step where we act on the vertices to actually avoid the collision. For this collision response, we have proposed the use of the law of conservation of momentum for perfectly inelastic bodies. This means that kinetic energy is dissipated, avoiding the bouncing effect. We use a dynamic inverse procedure to simulate a perfectly inelastic collision. Such collisions between two particles are characterized by the fact that their speed after they collide equals the speed of their centers of mass before they collide.

The constraints that join different panels together and attach them to other objects are very important in our case. Two kinds of dynamic constraints [12, 11] are used during two different stages. When the deformable panels are separated, forces are applied to the elements in the panels to join them according to the seaming information. The same method is used to attach the elements of deformable objects to other rigid objects. When panels are seamed or attached, a second kind of constraint is applied which keeps the panel's sides together or fixed on objects. Fig.5 shows examples.

Fig.5. Cloth modeling

INTERACTION WITH VIRTUAL HUMANS

We should also consider the relationship between the virtual human and the rest of the world. We distinguish four basic cases:

1. the virtual human is alone in the scene, there is no interaction with other objects.
2. the virtual human is moving in an environment and he is conscious of this environment.
3. actions performed by a virtual human are known from another virtual human and may change his behavior
4. not only may the animator communicate information to the virtual human but this virtual human is also able to respond it and communicate information to the animator.

Virtual humans are moving in an environment comprising models of physical objects. Their animation is dependent on this environment and the environment may be modified by these actors. Moreover several virtual humans may interact with each other. Several very complex problems must be solved in order to render three-dimensional animation involving virtual humans in their environment. For example, we introduced a finite element method to model the deformations of human flesh due to flexion of members and/or contact with objects [12]. The method is able to deal with penetrating impacts and true contacts. Simulation of impact with penetration can be used to model the grasping of ductile objects, and requires decomposition of objects into small geometrically simple objects. All the advantages of physical modeling of objects can also be transferred to human flesh. For example, the hand grasp of an object is expected to lead to realistic flesh deformation as well as an exchange of information between the object and the hand which will not only be geometrical.

In the context of interactive animation systems, the relationship between the animator and the virtual humans should be emphasized. With the existence of graphics workstations able to display complex scenes containing several thousands polygons at interactive speed, and with the advent of such new interactive devices as the Spaceball, EyePhone, and DataGlove, it is possible to create computer-generated characters based on a full 3D interaction metaphor in which the specifications of deformations or motion are given in real-time. True interaction between the animator and the virtual human requires a two-way communication: not only may the animator interact to give commands to the virtual human but the virtual human is also able to answer him. Finally, we may aspire to a virtual reality where virtual humans participate fully: real dialog between the animator and the virtual human. The animator may now enter in the synthetic world that he/she has created, admire it, modify it and truly perceive it. Finally, computer-generated human beings should be present and active in the virtual world. They should be the synthetic actors playing their unique role in the theater representing the scene to be simulated.

COMMUNICATION BETWEEN THE VIRTUAL HUMAN AND THE ANIMATOR

For the communication between the animator and the virtual humans, we are developing a prototype system. As shown in Fig. 6, this system is mainly an inference system with facial and gesture data as input channels and face and hand animation sequences as output channels.

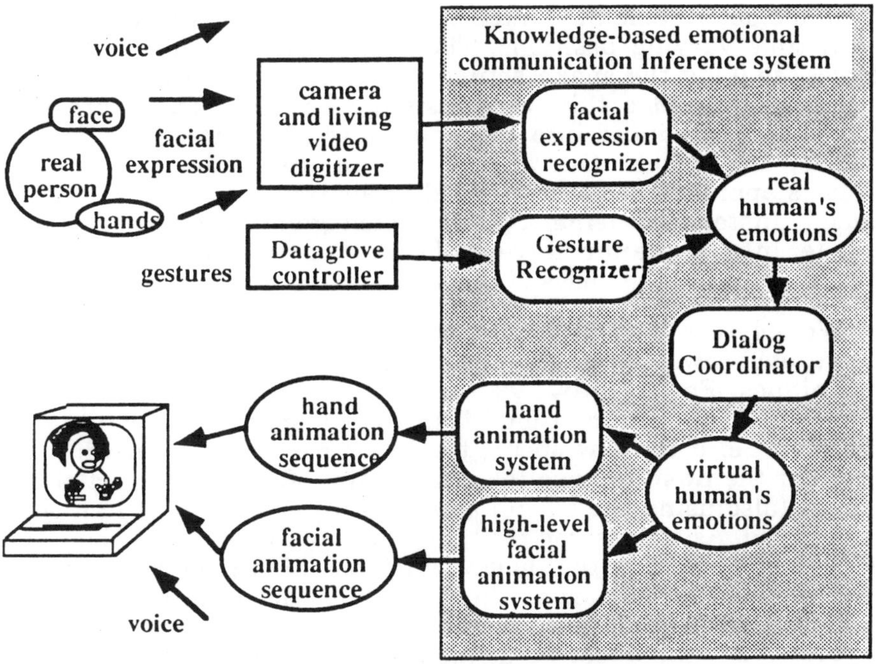

Fig.6 Organization of the proposed system

The development of the inference system is divided into three subsystems:

i) a subsystem for the recognition of emotions from facial expressions, head-and-shoulder gestures, hand gestures and possibly speech
ii) a subsystem for the synthesis of facial expressions and hand motions for a given emotion and speech
iii) a subsystem for the dialog coordination between input and output emotions

This last subsystem is a rule-based system: it should decide how the virtual human will behave based on the behavior of the real human. The dialog coordinator analyzes the humor and behavior of the user based on the facial expressions and gestures. It then decides which emotions (sequences of expressions) and gestures (sequence of postures) should be generated by the animation system. For the design of correspondence rules, our approach is based on existing work in applied psychology, in particular in the area of non-verbal communication.

For the recognition of emotions, our method is based on snakes as introduced by Terzopoulos and Waters [13]. A snake is a dynamic deformable 2D contour. A discrete snake is a set of nodes with time varying positions. The nodes are coupled by internal forces making the snake acting like a series of springs resisting compression and a thin wire resisting bending. The expression forces are introduced into the equations of motion for dynamic node/spring system. Our approach is different from Terzopoulos-Waters approach because we need to analyze the emotion in real-time. Instead of using a filter which globally transforms the image into a planar force field, we apply the filter in the neighborhood of the nodes of the snake.

We only use a snake for the mouth; the rest of the information (jaw, eyebrows, eyes) is obtained by fast image-processing techniques. For the **jaw**, we consider that the lower part of the lower lip (using information given by the snake) is moving with the jaw i.e. if the jaw opens, the lower part goes down with the jaw. For the **nose**, we use the center point of the upper part of the mouth (also using the snake) and we scan upwards until an edge is detected. As we assume that the illumination is very strong, the edge should belong to the shadow of the nose. For the **eyebrows**, we use the same principles as the nose. We start from the forehead and scan downwards until we detect an edge. This should be the eyebrow. For the **eyes**, we define a rectangular region around the eyes (using the position of the nose and eyebrows) and we count the number of white points in the region. If the number of white points is under a threshold value, we consider the eye as closed. Fig.7 shows an example.

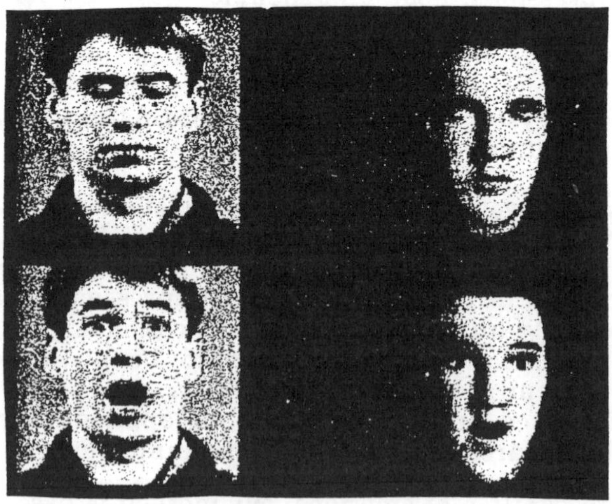

Fig. 7. Animator and virtual human

CONCLUSION

Modelling humans using computers is a very complex task. Several important problems have to be solved to incorporate realistic virtual humans in computer-generated films. However, in a few years, we will be able to recreate any person, living or dead, mix up the past with the present and the future and make this person talk and have emotions.

REFERENCES

1. Magnenat-Thalmann N, Thalmann D (1991) Complex Models for Animating Synthetic Actors, IEEE Computer Graphics and Applications, Vol.11, No5, pp.32-44.
2. Paouri A, Magnenat Thalmann N, Thalmann D (1991) Creating Realistic Three-Dimensional Human Shape Characters for Computer-Generated Films, Proc. Computer Animation '91, Springer-Verlag, Tokyo, pp.89-100
3. Kochanek DH, Bartels RH (1984) Interpolating Splines with Local Tension, Continuity, and Bias Control, Proc. SIGGRAPH '84, Computer Graphics, Vol. 18, pp.33-41
4. Boulic R, Magnenat-Thalmann N, Thalmann D (1990) A Global Human Walking Model with Real-Time Kinematic Personification, The Visual Computer, Vol. 6, No6, pp.344-358
5. Kalra P. Mangili A, Magnenat-Thalmann N, Thalmann D, "SMILE: a Multilayered Facial Animation System", Proc. IFIP Conf. on Graphics Modeling, Tokyo, Japan, pp.189-198.
6. Kalra P, A. Mangili, N. Magnenat Thalmann, D. Thalmann, "Simulation of Facial Muscle Actions Based on Rational Free Form Deformations", Proc. Eurographics '92, Cambridge, pp. 59-69.
7. Magnenat Thalmann N, Kalra P, "A Model for Creating and Visualizing Speech and Emotion", in: Dale R. et al. (eds) Aspects of Automated Natural Language Generation, Lecture Notes in Artificial Intelligence, Springer-Verlag, Heidelberg, 1992, pp.1-12.
8. Carignan M, Yang Y, Magnenat Thalmann N, Thalmann D (1992) Dressing Animated Synthetic actors with Complex Clothes, Proc. SIGGRAPH'92, Computer Graphics, Vol. 26, No 2, Chicago, pp. 99-104.
9. Terzopoulos Demetri, Platt John, Barr Alan, Fleischer Kurt. Elastically Deformation Models. In Proc. SIGGRAPH'87, Computer Graphics, 1987, Vol. 21, No.4, pp.205-214.

10. Platt JC, Barr AH (1989) Constraints Methods for Flexible Models. In Proc. SIGGRAPH'89 , Computer Graphics, Vol.23, No.3, pp.21-30.
11. Barzel R, Barr Alan H. A Modeling System Based on Dynamic Constraints. In Proc. SIGGRAPH '88, Computer Graphics, Vol. 22, No4, 1988, pp.179-188.
12. Gourret JP, Magnenat-Thalmann N, Thalmann D (1989) Simulation of Object and Human Skin Deformations in a Grasping Task, Proc. SIGGRAPH '89, Computer Graphics, Vol. 23, No 3, pp.21-30
13. Terzopoulos D, Waters K (1991) Techniques for Realistic Facial Modeling and Animation, in: Magnenat Thalmann N, Thalmann D, Computer Animation '91, Springer-Verlag, Tokyo, pp.59-74

Applications of virtual reality to computer-aided building design

C. Bridgewater,[a] M. Griffin[b]

[a]*Dept. of Civil Engineering, Imperial College, London, UK*

[b]*Dept. of Cybernetics, University of Reading, UK*

1. Introduction

In this paper, the problems associated with building design and generation within a virtual world environment will be discussed. In this paper the term 'Virtual Reality' has been reduced to 'VR' for the sake of brevity. The field of VR is relatively new in nature and proposes some unique opportunities in building design. The object of this paper is to discuss the problems and the potential solutions to them. Specific reference to the authors' own VR and building design systems will be made.

1.1 General Discussion

The field of Virtual Reality has recently developed out of experience gained in the fields of computer graphics and human/computer interaction. The basic concept is the "immersion" of an operator in a synthetic world, such that sensory information conveyed to the operator enforce the impression that this synthetic world is "real". The model is generated by specialised computing platforms, and suitable displays are used such that the required impressions are conveyed to the user. Wherever possible extraneous impressions from the real world are excluded.

A typical VR system is comprised of a headset system, using liquid crystal displays and suitable optics, so that the wearer can see the images from these displays. A set of headphones supplies auditory information to the wearer (whilst also muffling out unwanted extraneous environmental noise). Finally, a tracking system is also included in the helmet, such that the wearer's head position and attitude may be found. For a typical system, please refer to figure 1, where a picture of the authors' system is presented.

On many of the advanced systems, a "dataglove" system is also often utilised. This is a special pair of gloves that the operator wears. They contain similar tracking systems to the helmet, as well as some form of flexion sensor. From these elements the hand and finger attitude and position may be found. Tactile feedback is very often included in these gloves, either by using piezo electrical "buzzers" in the finger tips, or miniature inflatable air pockets. Thus a user may be given the capacity to pick up and touch objects within a virtual reality.

2. The Building Design Problem

The problem of designing a building is a complex and involved process. In recent years, the development of prefabricated sections, and CAD design suites have enabled more up to date construction methodologies to be utilised. In essence, a building is designed in a similar manner to that of any modern manufactured item. Firstly, the building requirements are decided, i.e. the number of people it has to contain, the volume of space required to fulfil its task, the necessary floor loadings expected. After these have been specified, then the number of work and ancillary areas and their spacial arrangement may be defined.

From these data, the number and positioning of the structural members may be determined. The positioning of the foundations can then be defined and cladding and cosmetic items may be specified and their precise placement determined.

The above approach works best if the complete structure can be specified as a set of prefabricated, manufactured parts which may be slotted together. The construction order can be determined in the design process, such that a complete construction schedule may be created. This, along with a parts list, and placement specification, is sufficient to completely describe the building process.

All the operations above may be generated by a sufficiently complex design tool. Such a tool has been created at Reading, and examples of its output is given in figures 2 to 9 inclusive.

3. The Visualisation Problem

With implementing such a tool, it was noticed that a number of problems enschewed. Firstly, the building requirements have to be tested such that the volumes specified are indeed adequate to the tasks set. Secondly, that the ergonomic layout of these volumes and the necessary ancillary space (toilets and staircases etc), were satisfactory. Thirdly that the safety requirements with respect to fire escape access be satisfactory. Fourthly, that the construction schedule was possible to implement, taking into account the actual site environment. Finally, that the aesthetic qualities of the building be sufficient to satisfy the client's requirements. It can be seen that these problems can be tackled using a sufficiently comprehensive visualisation tool. It was decided to examine the use of VR systems to what potential benefits these could bring.

4. Implementing a Virtual Reality System

In testing the applicability of VR systems on these problems, the Reading University "LOKI" VR system was used. This is a platform based around a 68030 computing system, a headset system and a magnetic based tracking system. A common file format was devised between the two systems, and building geometrical data downloaded into its "Virtual World Modeller". The VR system was then able to generate a realtime graphical representation of the building (in various stages in design) and permit "walkarounds" to be performed.

These walkarounds are effectively realtime investigation of the building structure, giving the observer three dimensional spacial information at any viewpoint within this. Unfortunately, due to limitations on the VR systems software, animation of the construction process was not possible. From these walkarounds, some of the design aspects discussed above could be investigated.

5. Investigating Potential Designs

With respect to determining whether the work volumes defined were adequate to the building's perceived function, there was insufficient information present in the simulation to make accurate assessments. This was judged to be the case because the placement of furniture and people within the scene was not performed. If this had been generated along with the building design, then a reasonable estimate could probably be made.

The ergonomic layout of the volumes could be tested, because there was sufficient information present to come to a judgement. This is again qualified by the fact that no furniture was added to check certain clearance aspects, however, access to staircases and ancilliary areas could be checked.

With respect to safety aspects and access routes, sufficient information could be gained when considering fire escape and stairway access. The visibility of exits and the theoretical placement of exit signs as well as other warning information could be tested and the placement of firefighting equipment. Again, the accuracy of these results is qualified by the lack of furniture detail within the environment.

Due to limitations in the animation system, the construction phase of the operation could not be tested satisfactorily. It was possible, however, to view the building at different stages of construction and make judgements as to where material payloads could be placed without unduly restricting work around them.

Finally, with respect to viewing the final design and presenting this as a proposed solution to the client, this aspect was judged to be excellent. It gave an unparalleled opportunity to examine most of the visual aspects of both the interior and exterior, interactively, such that a clear and unambiguous representation of the building could be brought over to the client.

6. Improvements to the System

From the above experimentation, it was decided that the addition of virtual reality "walkaround" capabilities to the automatic building design system was of considerable benefit. Unfortunately, due to inadequacies in the software and hardware elements of this system, the level of detail needed to test out certain aspects adequately was not present. The next generation of virtual reality hardware and software being developed is targeted at removing these inadequacies.

It is now viewed, however, that part of the building design process should be the specification (to a limited extent) of furniture and person deployment within the proposed structure. This will require a much more detailed brief from the client and possibly an increased, active role in the design stages of the building whilst in this CAD stage. With this information, a much more coherent picture of the buildings' deficiencies may be gained and these corrected.

It was originally envisaged, that the construction phase of the building would be animated, such that the scheduling and materials deployment over the site could be studied in detail. Due to inadequacies in the animator, this could not be achieved. In the next generation VR system, comprehensive animation routines are being generated to address this fault.

Finally, the use of simple file transfer between the CAD system and the VR system is judged to be crude, in that it does not allow for quick design and validation sessions to take place very easily. It is hoped to integrate both systems, such that the building design system is an application which runs on the VR workstation.

7. Conclusions

The use of Virtual Reality in building design to assist in the design and validation of aspects of that design, has been noted. With the relatively crude level of modelling achieved, only a limited number of aspects could be evaluated. This could be increased in number by further refinement of the Virtual Reality system. The use of Virtual Reality suggests a more interactive approach in building design, especially with the capability to visualise the end product more completely. This implies that contact with the client during the design period may have to be increased to take account of this new facility. It is suggested that defining the furniture, or the equipment to be present in the building as well as their placement, should be part of the basic building design.

8. References

[1] Bridgewater C E, "A New Approach to the Design of Buildings for Automated Construction", PhD Thesis, Reading University, Reading, Berks, July 1992.
[2] Griffin, M P, "A Cybernetic Perspective of Virtual Reality", PhD Thesis, Reading, University, Reading, Berks, September 1991.

Figure 1 - The Reading Virtual Reality Workstation

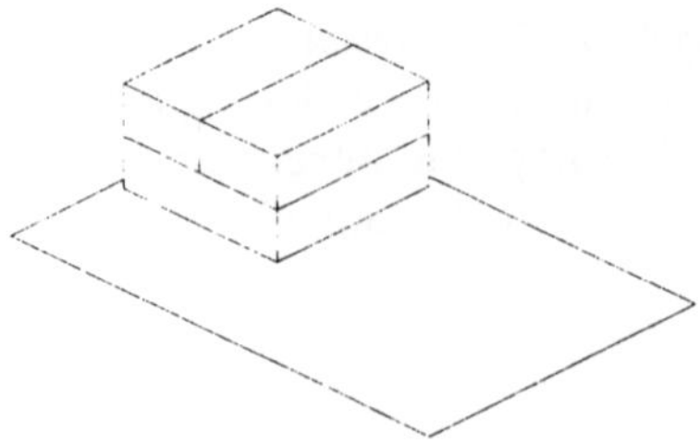

Figure 2 - Defining Initial Volumes

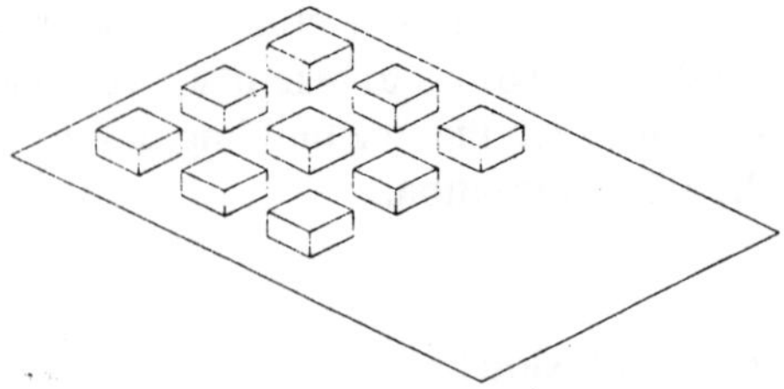

Figure 3 - Laying Pads For Foundations

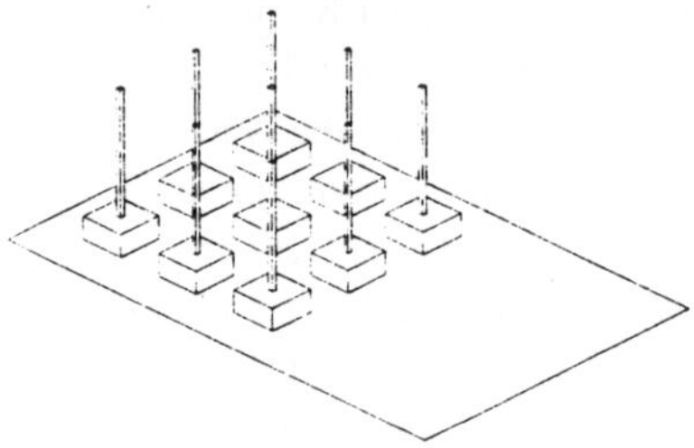

Figure 4 - Placing Columns

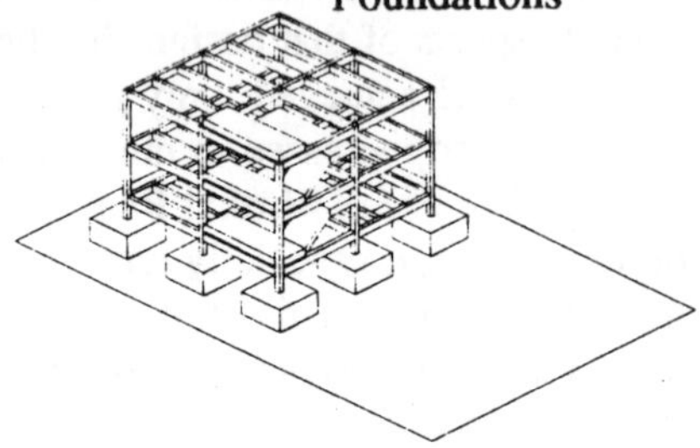

Figure 5 - Placing Stairs and Supports

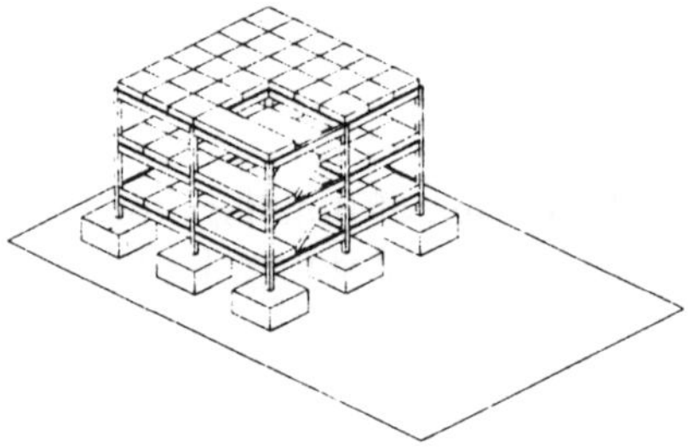

Figure 6 - Placing Floor Panels

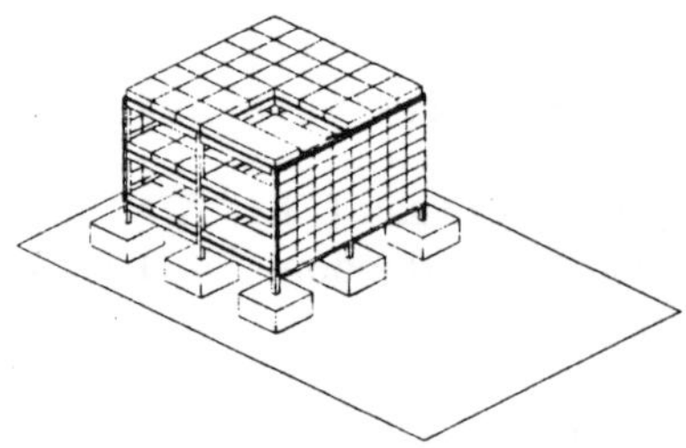

Figure 7 - Placing Wall Cladding

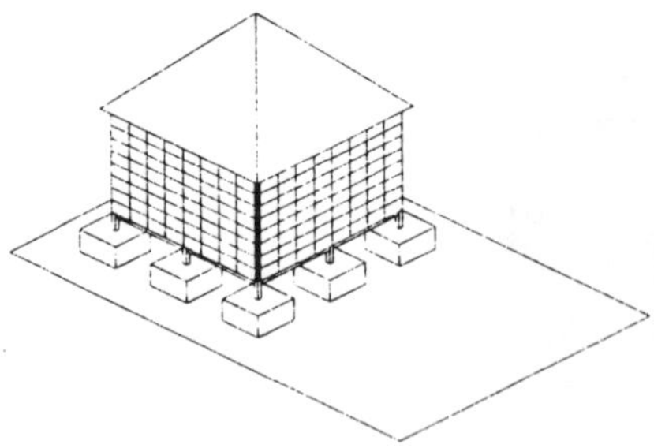

Figure 8 - Completed Building

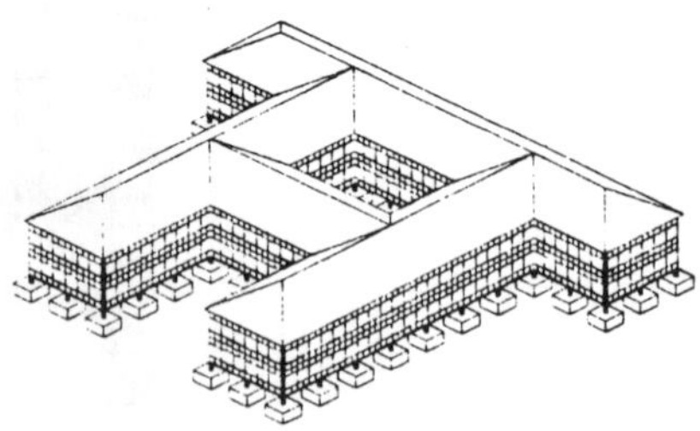

Figure 9 - A More Complex Building

SECTION 4: COMPUTATIONAL GEOMETRY, IMAGING AND VISUALIZATION

Invited Paper

Intelligent decision support of design using advanced multi-media

J.A. Powell

Design Information Research Team (DIRT), Department of Manufacturing and Engineering Systems, Brunel University, Uxbridge, Middlesex, UK

ABSTRACT

Emerging multi-media technologies enable design decision makers to inform themselves and their colleagues more effectively and appropriately than ever before about advances in science, social science, management and technology. This paper is in two parts. Part I discusses the exciting potentials of the multi-media hardware which enables technology transfer to take place in a higly interactive mode. In particular there is discussion of the comparative merits and performance of interactive videodisc, compact disc (CD-DA,CD ROM and CD ROM - XA and CDI) and Digital Video Interactive. Part II is more concerned with softeware issues of media design and, in particular, details a particular attempt to show how such systems could be used to improve busy design professionals' understanding of technical/social issues.

CONTEXT

According to the Management Group Board of the Institution of Civil Engineers (1) "control is **the** key word and management and control of time, cost and quality to meet objectives are **the** key issues" facing future engineering designers. For them the well informed designer is an innovative decision maker in control. This highlights the need for efficient management of all engineering design information. Such information management is becoming increasingly difficult as design methodologies and capital investment programmes themselves become more sophisticated and complex. Now just keeping up-to-date and properly informed, in a manner which does not overwhelm, has become a critical issue. Furthermore design decision makers now have to come up with humanly acceptable solutions, faster than ever before, which are, not only aesthetic sensitivity, but provide value for money. The need is clear - for ready and easy access to relevant and guaranteed information held in a comprehensive data base in a form which **informs** design decision makers - in the sense that such designers can give better **form** to their designs through enhanced knowledge.

Footnote: The author wishes to acknowledge members of DIRT who have made major contributions to the work described here and especially Drs Paul Newland and Chris Creed of Portsmouth University.

This is no longer a trivial task for, as Boardman (2) points out, most existing information technology tools have been "singularly unsuccessful in reducing or coping with complexity".

The national design information scenario is further complicated by the scale and fragmentation of much British industry which is responsible for carrying design solutions through to final products - for instance recent surveys indicate there are still over 150,000 general manufacturing small and medium enterprises (SMEs) and 170,000 private building or civil engineering contractors employing less than 7 people. Furthermore, to add to this complexity, design information itself can be portrayed in many different ways: as printed documents; photographs; on-line bibliographic surveys; Ordnance Surveys; videos; British Standards; spatial information in many forms; data sheets; product information and the like; it can also be held on microfilm or fiche; on vidoes; on 35mm slides; on hard disc; on optical disc; as software; in bubble memory or even be computer hardwired.

The present author believes interactive multi-media could work miracles in providing a cheap and robust single access information medium for designers, that is if the relevant set of comprehensive databases were developed. Interactive multi-media consultants exist to make the necessary databases and optically stored programmes which inform decision making and they now the technology available to develop appropriate systems. All that is needed now is the confidence and the resources from British industry to produce fully developed systems giving the right access to necessary information.

The paper now splits into two parts. Part I discusses hardware issues relating to multimedia. In particular it discusses the vexed problems of systems interfacing and data handling. It then details the potential of optical disc as a suitable medium for storing design information. Part II looks at the software design issues relating to multimedia and gives some concrete examples of successful systems developed by the author. It is important to get the design of the system matched to the needs of the information user if it is to stand any chance of proper communication.

PART I

APPROPRIATELY ACCESSING INFORMING MULTI-MEDIA

According to Aris (3), rigidity of human interfacing, or lack of user friendliness, is a key reason why much of British industry presently fails to capitalise on its general use of IT. *Stability* and *familiarity* of both the human-systems interfaces and the data-bank handling are key components of those information systems which satisfy human acceptability. These factors have therefore become the main design generators for the design of those effective interactive multi-media information systems so far developed. When designed well such systems appear "open", transparent or even semi-intelligent to users, almost recognising their needs and acting accordingly in a sort of dynamically 'bespoke' way.

In this respect, since they suit designers modes of designing, WIMP interfaces(window-icon-mouse-pointer) and hypertext/hypercard based systems have become standard in most of the advanced multi-media information systems used by these decision-making professionals. This is because such tools, techniques and instruments help the inexperienced user, while not hindering the expert. In particular, research by Apple Computers has shown that the mouse, a device for positioning a cursor on the feedback computer screen, is preferred as the information control mechanism by an increasing number of design professionals. Used in conjunction with HyperCard, which portrays data on the screen as a familiar index 'card', the mouse can be used to activate 'buttons' within each card linking it to other cards in a 'stack'. In effect a HyperCard system tries to emulate the free association characteristics of human thought and allows users to create, access and customise information held in digital form on the information system's hard disc. Its success in effective information handling is reflected in its wide and increasing use.

Such semi-intelligent knowledge based interfaces and data-handling systems are beginning to provide the sort of 'openness' all users now demand. In the past the HyperCard based system's major limitation related to the restricted nature of the computer databases open for recall. However, video adaptor cards and associated software - such as the Videologics' DVA 4000 and IBM's M-motion - open HyperCard access to mass storage of text, graphics, animation, stills, moving video, sound, music and numeric elements all held on optical disc. The video adaptor boards digitise the incoming signals, both analogue video and audio, so they can be used as information on the index 'card'. The boards also allow certain special effects so the user can superimpose graphics over video or stills, they can change the size or position of stills or video on the screen, they make a 'button' of a still and use it to call other cards, and they can simultaneously view several different video images. Both video adaptor boards will accept signals from the variety of sources, such as videodiscs, VCR's and compact disc, that could hold the massive databanks of information needed by our industry.

Organisation of and access to these massive databases has become the most fundamental concern of the information systems designers. Unlike a physical library where every book has a spatio-temporal location and competent individuals can easily build and retain a mental map of where required items are, the black box syndrome of an intangible mega optical storage unit can become an offensive Pandora's box. However, many of the new interfaces already mentioned access massive databases in a manner similar to the way professionals think about design and the making of products. For instance, point and click on a component of a product drawing and this 3-D object, held in a CAD system, becomes the search string which retrieves structural characteristics, manufactures' specifications and optional products, etc. In advanced systems such objects are automatically recorded and any updates (e.g. by dynamic links to manufactures' source databases) are communicated to the designer by colour changes in the relevant components.

The above indicates we now have the right technology to **access or interface with appropriately informing massive databases**. The only remaining problem is the precise mechanism required to ensure cheap, reliable and high fidelity storage of a range of infomation sources. The next section concentrates its attention on one suitable medium, namely optical disc technology - a cost effective form of storage capable of mounting the full variety of database formats currently used by British design professionals.

THE POTENTIAL OF OPTICAL STORAGE

All optical storage media allow fast, easy and reliable computer access to a range of high fidelity, but often unstructured, moving video, still images, graphics, audio sound or pages of text. Different systems allow cost effective storage of different types of information. The following sections describe the relative merits of the key optical media currently available to UK industry.

Interactive Videodisc (IV)

Imagine combining 15 hours of audio with over 55,000 still images or 37 minutes of moving video, all in vivid colour. For this is the extraordinary capacity of a single 30cm video disc. All available at random access within seconds or split seconds of demand, comparing favourably with the rate of retrieval from a purely computer-based archive, or material stored on conventional videotape. The videodisc can offer such enhanced retrieval because of the unique method by which video information is accessed. Information is stored in the form of tiny pits 'burnt' into concentric rings on a thin aluminium membrane held within the disc fabric and is retrieved by a well-defined low density laser beam trained onto the disc's reflective surface. A protective tough plastic coating on the disc surface, combined with the friction-free playback mechanism, provides a low maintenance robust system. Eventually this information is distributed to output devices such as a monitor. See Fig. 1 below for diagram of the system.

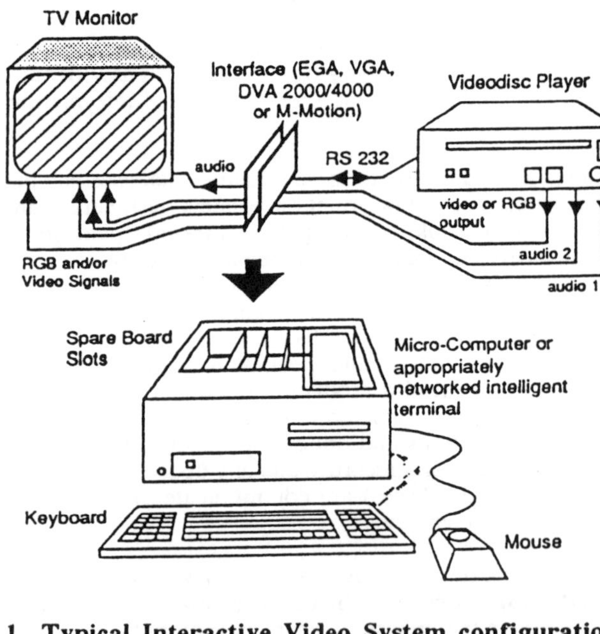

Fig. 1 Typical Interactive Video System configuration

Team the rapid reflexes of the videodisc player with the organising, calculating, text and graphic powers of the computer and you have interactive videodisc - a system that can respons instantaneously to the different needs of the user. All professional videodisc players are equipped with an RS232 serial communication port. This communication facility is the same as that provided on many personal computers to enable them to be used with printers, modem and so on. Basically it allows the information to be sent and received serially as a 'string' of data statements. Theoretically it is possible to use different makes of videodisc player on different computers and make them work satisfactorily. The most common way of achieving this is with a video interface/overlay card/board which plugs into a spare slot in any computer. EGA and VGA cards were the early ways of achieving, on the same screen, a combination of (digital) computer text and graphics with (analogue) video signal. However, as already mentioned, some interface/overlay boards now enable a great deal more than this and both the M-motion or DVA 4000 video adaptor cards have replaced the simpler ones as standard cards. They have done this because they offer many functions which originally had to be undertaken in expensive video editing suites.

Under computer control, the videodisc can take on many roles - simple database, trainer, educator, salesman or entertainer. Quite simply interactive videodisc has the ability to provide access to an almost exhaustive library of audio-visual information shown in dynamic and compelling ways, at the highest quality, without degradation in use. It has so far been used in British industry mainly to provide training programmes in the areas such as the design of safety critical systems, car maintenence, marketing, customer care and the like. It has been shown to reduce training costs from £250 per man day to less than £13 per man day, and this includes the cost of purchasing the original IV hardware. There are also a portfolio of videodiscs available for the general management training on topics ranging from 'assertiveness training', 'telephone answering skills', 'financial management' to 'customer care'.

Videodisc players can now cost less than £1,000, video adaptor card approximately the same and, for multiple copies, disc as little as £20. A weakness of this medium is its analogue form and the fact that it is tied to incompatible broadcast standards adopted by different countries. Thus videodiscs made to the PAL standard used in most of Europe are not compatible with ones made in North America in NTSC, or in France in SECAM. The next optical medium discussed does not have this drawback.

Compact Discs General (CD-DA, CD ROM, CDI and CD ROM - XA)

Compact disc technologies offer another compelling optical disc storage mechanism. These include CD-DA (Compact Disc Digital Audio) for high quality audio, CD ROM (Compact Disc Read Only Memory) for mass data storage, CDI (Compact Disc Interactive) for complete interactivity with pictures, text and data, and also CD ROM - XA (enhanced CD ROM) to enable storage and retrieval of standard images, graphics and sound as well as text. Like videodiscs, compact discs are based on the reflective optical disc. However, they are only 12cm in diameter and the data is stored on the disc in digital rather than analogue form. Fortunately, there is a common standard between all forms of CD, so such optical discs can be used anywhere throughout the world. Unfortunately, the problem with existing forms of CD is that it is impossible to store more than a very small amount of moving video on each disc. To produce high quality video images, data must be displayed on a screen at a rate of nearly 15 Mb per second. Such a video transfer rate is well beyond the reach of most low cost systems. As we shall see later some CD systems try to overcome this using compression and decompression devices.

CD-DA (Compact Disc Digital Audio) is the most familiar form of CD technology. It delivers up to 72 minutes of broadcast quality stereo audio on a relatively cheap and durable player. It is remotely produced in a factory environment to a physical format conforming to standards set out in the "Red Book". Interactive CD-DA players can cost as little as £200 and individual discs only £2 when cut in multiple copies.

CD ROM (Compact Disc Read Only Memory) have evolved out of CD-DA and have the same physical characteristics and specification, as laid down in the "Yellow Book". Unlike CD-DA, CD ROM store only large quantities of digital data - up to 600 megabytes - of various forms. Each disc can store the equivalent of up to 200,000 pages of text, enough storage to contain the information on all 18 volumes of the telephone directories of Switzerland. Like videodisc it is 'read only memory', being made in a factory environment. It can be easily interfaced with existing databases, on most computers, and is now being increasingly built in as a standard facility in the higher level PCs. Now, at a cost of less than £500, it is becoming the preferred storage technique for distributing mass software.

CDI (Compact Disc Interactive) simply extends CD-DA and CD ROM into a multimedia platform. It offers the same data or audio storage as its forbears, but can also handle up to 6000 high quality still images, 20 minutes of quarter screen broadcast video or 20 minutes of 42% screen VHS quality video. The technology is not yet sufficiently advanced to provide full screen full motion video. The higher levels of audio or video produce the best results but take up more disc space. The Phillips "Green Book" sets down a complete systems specification including how data is laid down on the disc, the disc content and the CDI player itself. Once again players will retail for around £500 in the UK. CDI is mainly intended as a consumer product.

CD ROM - XA (enhanced CD ROM) extends the High Sierra specification of CD ROM to add interleaved audio and video. It will not add much to the cost of a conventional CD ROM drive, but will slightly extend certain features of CDI. It will include quarter screen motion video and graphics with simultaneous audio tracks. According to Bayard-White (4) 'CD ROM - XA achieves the bridge with CDI via two audio processors on the CD ROM drive interface card which compresses audio data. The XA standard also overcomes the problem of playing sound by interleaving blocks of audio data at regular intervals between blocks of text, image or programme data. This means that it is possible to synchronise a text display with any audio output - to see and hear them simultaneously - critical for example in developing foreign language learning programmes. Like CDI, XA offers different audio quality from mono speech to stereo sound and two audio channels. The physical form of CD ROM - XA is a plug-in adaptor card to a PC mainly aimed at the professional world. Aside from the conventional PC with VGA display and an ordinary CD ROM drive, the only hardware needed is the XA interface. The XA card has been specified to allow existing CD ROM software to run on XA hardware and CD ROM-XA software to run on CDI. As an intermediate technology XA lies just short of the more complex full motion video systems, although even this constraint may not remain as XA has been designed as a subset of other systems including digital video interactive'.

Digital Video Interactive (DVI)

Digital Video Interactive is similar to CDI but offers 72 minutes of full-screen full motion video, 20 minutes of motion video with 7 hours of voice over and 5,000 high resolution still, or 2 hours of quarter screen video with thousands of textures and 3D objects. It also offers 360 degree panorama views. It has been developed for the PC market and can make use of CD ROM, Hard disc or WORM storage. Bayard-White again, 'DVI is an evolving software based process. At its heart is Intel's 1750 chipset, the most powerful custom processor chipset yet developed for the PC. Perhaps most importantly, the chipset is programmable from software. In short, users will therefore not have to purchase new hardware with every new version of DVI. It provides compression and decompression of digital data (see above) on any IBM compatible computer, whatever the storage medium'.

Many new applications are being developed for this extremely new optical retrieval mechanism. In comparison with Interactive Videodisc it still has a problem in coping with fast acting sequences especially with large colour changes. However, its flexibility, versatility and usability - it will be able to handle multiple video signals - give it an advantage over other optical media, including IV. Eventually, I expect the developers to overcome its existing fast data handling problems. So far there are few practical applications of this device in the UK and none in the construction industry.

The Benefits of Optical Storage

The benefits of optical storage for British industry are myriad. First and foremost they can store all forms of data and information, to a consistently high standard at any level of detail. The medium is extremely versatile and highly durable. The quality of audio-visual presentation from massive databases are always clear and cogent, to broadcast standard or any lower standard required. Presentations can appear personalised, or factual and can be presented at a convenient time and place, to meet a variety of needs. Images from the system are so good they grab attention and are very persuasive. Hardly surprising, therefore, that Poulter Communications have convinced over 2,500 manufacturers to sell their products by means of optical storage. Relatively speaking optical storage is also an extremely cost effective way of storing high quality images of many different sorts. In conjunction with the right computing technology, both control and interface hardware: they can act extremely patient, but remote and private, teachers; they can provide consistent messages for remote training in the use of standards and codes; they can provide remote help in an intensely human way; they can provide a realistic context for role playing simulations; they can provide a quality sales environment and many more benefits. Finally, they seem to motivate people to want to use them, what better reason for incorporating them into all new information systems.

Informing systems based around highly interactive, user-friendly multi-media databases can be an instrument for beneficial change, if design professionals are motivated to understand their true potential.

PART II

INTELLIGENT DESIGN DECISION SUPPORT - THE EDUCATIONAL SOFTWARE OF MULTI-MEDIA DESIGN

Context

As hinted at earlier the real design problem is usually to know what the real problem is, Rittel and Webber (6). In this respect the new informing technology can be conceived of as a tool which helps designers understand the nature of problems and provides a context to facilitate solutions which enhance the experiences of the individual. To this end the present author, and his colleagues in the Design Information Research Team, have been investigating the role of interactive media in the creation of appropriate experiential learning environments for use in the area of continuing professional design education. Early evidence suggests such self motivating interactive learning technologies promotes confidence in the acquisition of useful skills and decision making knowledge.

Repeatability and the Loss of Creativity - A Theory of Improved Communication

Replication may be a high priority in maintaining the consistency of a manufactured product, but the present research questions such an approach when to task is to find the most effective way of purveying understanding to unique individuals. Unfortunately, learning by 'rote' all too often finds its way into professional education where it fails to influence designers. A good demonstration is a much better communicator for them since it makes explicit the decisions made in the course of the activity. Thus a good demonstration shows a learner what not to do as well as what to do. A skilled performance makes these same decisions invisible. This insight from Olson and Bruner (7) hints at two crucial factors my studies have found to be missing when inappropriate understanding results from traditional pedagogic training and 'rote' learning, as opposed to emotive experiential learning.

Much skilled professional performance, especially complex decision-making, has components of actor, stage and script and is essentially a set piece. As with acting, we believe introducing any form of training development for naive professional learners which simply instigates mere imitation of set pieces, has two fatal flaws with respect to the knowledge learned:

i. no account can be taken of either change on the stage (context) or in the script (content)
ii. no account can be taken of the individual's dignity and how their previous knowledge invokes a truly personal perspective (timing).

Let us maintain the theatrical analogy of learning contexts just a little longer hopefully to aid understanding. Imagine a naive actor (learner), who has simply 'rote' learned a script, being faced with one of those magical revolving stages whose sets she had not seen before. A ninety degree rotation of the stage would produce a totally new set, a novel situation with which our naive (learner) actor would at least feel uneasy. If the script no longer matched the setting what would they do, since their 'rote' learning would have left them without, either the confidence to create a new script for the novel set, or the experience to ad lib around this anomaly until the previous set returns. Imitation of the skilled performance leaves our (learner) actor bereft of truly useful tacit knowledge of how to act in the sort of 'wicked' contexts Rittel and Webber (4) that so often frequent, not only the stage, but especially the real world.

Furthermore, such imitation ignores the personal history of knowledge already held by our naive actor. For sure she may be naive with regard to the nature of the present information, but asking her to mimic without acknowledgement of her personal viewpoint dismisses a wealth of creative thinking. For, learning through imitation negates the development of adaptive ability necessary for real world decision-making of the professional and spurns the possibility for creative self-discovered knowledge. In the end knowledge has to be the acceptance of co-creative distinctions, not the rigid enforcement of another's views, if all individuals are to be given opportunity for creative action in response to novel circumstance.

Seeking to use multi-media technology to elliviate these flaws has been the goal of the remaining investigations reported here.

<u>First Steps - Individuals and Their Learning</u>
In particuler the research presently reported first tackled the problem of individual preference for the assimilation of knowledge. Various studies including my own, Pepper (8), Thompson (9), Newland et al (10) had already looked at the approaches different individuals take in order to acquire knowledge - that is in order to act viably in their professional arena. The field of learning has been extensively investigated by Kolb (11) who summarised the various approaches open to an individual by a learning cycle, shown in Fig.2. Kolb's model portrays an ideal situation where the learner has immediate experience of the world, steps back to reflect on this feeling in context, thinks up an abstract model to explain the phenomenon, which is then tested by further active exploration. These doing actions cause the learner to re–sense the world and so begins a new learning cycle.

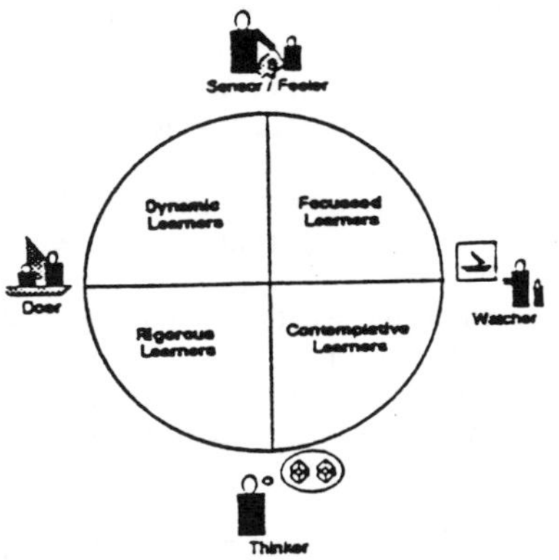

Figure 2 Kolb Learning Cycle

It might be thought that well balanced individuals would be those who follow this idealised learning circle. For this should give them a well balanced view of the world. However, Thompson's (9) work indicates that such a fully rounded view may be impossible and even counter productive to real world living, learning and especially complex problem decision-making. His view is based on the observation that in the real world people are not simply individuals learning by themselves, rather they are social animals in continual negotiation with each other as they learn. He, and now many others, have found that learning is clearly as much a socialisation process as a means of acquiring factual knowiedge. Socialisation imposes severe constraints on learning, leading different people to learn in different or preferred ways. This itself eventually drives particularly professionals to adopt one of the four different and often incommensurable world views which we portray in summary below and in detail elsewhere Newland et al (10).

In order to cope with pressures of time and money, decision-makers will tend to stick to their preferred learning approach, their chosing of appropriate information and their corresponding action style while practising their profession. This gives them a secure basis from which to make decisions and they rarely change their approach to learning new ideas – it is not only too much effort, it also drains their confidence. In summary, it appears that all individuals have a predisposition to learn, then think about the world as they see it and finally act in it, using one of four strategic action patterns, each is different, independent and recognisable. Powell (12) has called the types of learner-actors who portray these underlying traits Dynamics, Focussed, Rigorous and Contemplatives. Let us briefly discuss each of these in turn.

i. Dynamics have an approach to learning which is centred on doing and sensing. They are entrepreneurial and innovative, acting for the moment, sensing its potential and doing something about it immediately. It is the dramatic active events in the world which give them motivation and allow them to register understanding. To learn they must be personally involved and if this necessity is satisfied then they are eager for challenges and novelty. By being assertive and individual in their actions they can swiftly switch tack, being entrepreneurial and initiating centres of profit at one moment and, at the next, rapidly creating innovative designs. They seek acclaim for their work. Their risk–taking attitude makes them good copy for glossies and the professional, financial and management pages of national newspapers.
ii. The Focussed concentrate their attention on decisive action. They operate in the world with a down–to–earth approach which is above all practical. Their need to do can often lead them into being campaigners and expounders of alternative technologies and conservation policies. Their small is beautiful philosophy may make them reticent to try untested approaches. Instead these people stay with the safe, known approach.
iii. The Rigorous understand the world by creating abstract models of it. These models give the rigorous predictive power over the environment. Through assumed knowledge of the underlying patterns from nature these professionals have secured rules that guide them in all their decision making. They stand as the guardians of professional standards and aim to produce regularly competent solutions.
iv. Contemplatives with their emphasis on reflection, may feel no great urgency in real world decision making. Their reluctance emanates from a desire to create the all encompassing solution. To this end they will thoroughly analyse all the data which is at hand and then try to retrieve more in an effort to ensure every angle has been covered.

Appropriate Interaction and Tone

The preceeding fourfold of strategic action patterns have guided DIRTs design and production of a number of information or learning systems which appear to have enabled more ready acquisition of useful professional knowledge. The effectiveness of these systems has been tested by practicing designers using them in their own offices. Therefore the underlying theory upon which the systems have been developed has been validated and other useful conclusions can drawn which are detailed below.

DIRT's systems seem to work well because of the special ways in which they handle the interaction with the learner and through the tone of the information portrayed to the learner.

In the first respect particular emphasis in the developed systems has focused on designing an acceptable acquisition mechanism for the stored information. I have found that when an information system exactly matches a preferred style of learner engagement, or has the right "interactive sensitivity", it communicates particularly well. Thus, information presented to the appropriate "parish" of learning and in the right configuration stands a better chance of communicating than one that does not. For example the visual portrayal of a topic could be either: by a set of still photographs with text annotation; or via a film with an audio commentary; or through self-paced questions and answers; or as audio/visual sequences etc. The mix and style of media presentations provided by any system like film, audio and text will facilitate different self-informing strategies (Jonassen (13), Salomon (14)). The diagram below indicates, in summary form, the sort of interactive sensitivity required by different design information users.

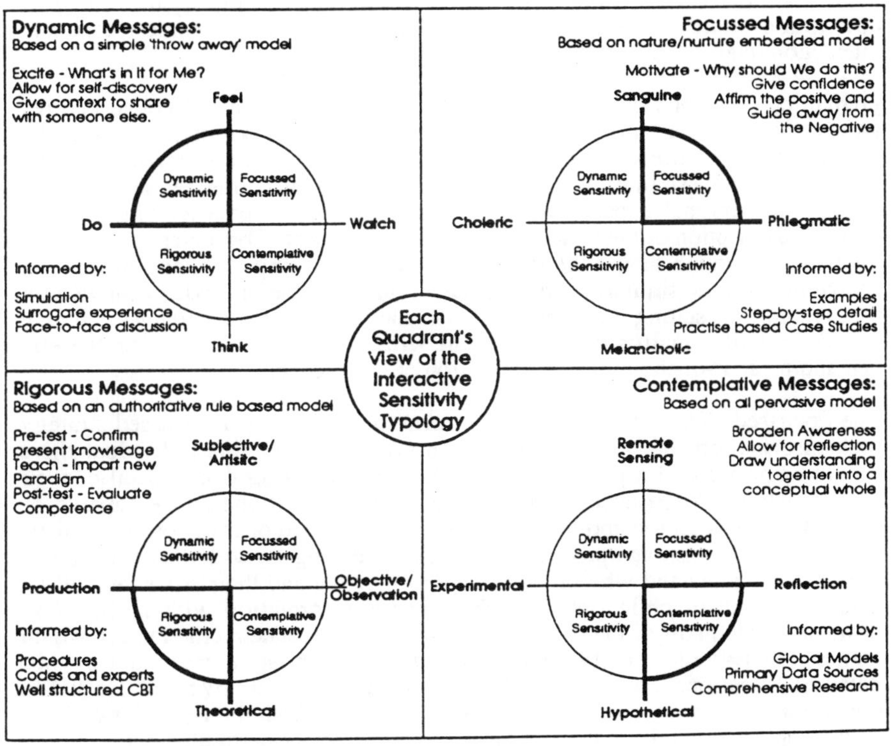

Summary of Messages Preferred by different Designer Learners

Concentration on better presentation of information alone, contrary to previous research, has been shown to be insufficient to remove the barriers to communication if the underlying interactive sensitivity is inappropriate.

While interactive sensitivity forms the basis for good communication, of equal importance is the need to get the tone of knowledge in the system right as a means of achieving resonance between any individual and the imagery, learning, interpersonal interaction preference, philosophical basis and social assumptions they naturally seek. I have coined the term "semiotic credibility" to define this important matching of content characteristics with a learners belief structures and culture. Identical information protrayed by a system can be semiotically credible to some learners while being semiotically incredible to others. The latter group find information transfer much more difficult than the former even if the information is presented with the appropriate interactive sensitivity.

"Interactive sensitivity" and "semiotic credibility" work together in a complementary way so as to improve technology and information transfer, and help to promote deep learning in infromation systems users. Finally the following descriptions indicate the general characteristics of media presentation, tone and interaction relevant for each strategic action patterns.

Dynamic. The tone of information needed for a Dynamic individual centres around the notion of 'information transfusion'. In terms of portraying a particular topic, a sense of transfusion is only invoked when a topic is made tangible. Therefore, for these individuals, any topic of information transfer should draw heavily on personal experiences or shared experience events leaving them to weave a description of the topic into their own viewpoint. Dynamics portray their knowledge through their actions, and to be of a matching tone, an information source needs to be similarly active. I have found one useful mode of interaction for Dynamics is setting up a framework of challenges. Such challenges can give a near real life quality to involvement with an information source thereby demanding spur-of-the-moment acquisition of knowledge.

The Focussed. Those who choose to inform themselves by a Focussed strategy seek information that has a 'tone of transposition'. For this individual there should be provision for practical knowledge both at a general and specific level. For knowledge to be acceptable to these individuals it should be capable of grabbing their attention and being naturally incorporated or transposed into their own normal professional processes. The Focussed engage with the environment in order to understand why processes are as they are and their's is essentially a small is beautiful philosophy which demands information with a clear cut, down-to-earth quality. These individuals would appreciate an interaction which makes it possible to readily interrupt the documentary style case studies when detailed questions are raised. Such interrupts would give the Focussed opportunity to: review the material presented; enquire about the general usefulness of the topics covered for their present concerns; select particular points in the material for which greater detail could be requested.

The Rigorous. These attain viability through the principle of 'transcribing' codes into practise. Individuals who find this strategy appropriate are those who seek information which reinforces the validity of the rules they already use to guide their action. Such rules are recognised as emanating from authorities. In a similar fashion information which is seen to have been generated by these authorities will also be given credence as well. Therefore, they would be assured by information whose tone implies it is a code or standard practice. Procedural information would also be welcomed for its strength in providing straightforward, economical solutions.

Contemplatives. My research shows that those who choose to adopt a Contemplative strategy are interested in a broad spectrum of information. For them a comprehensive montage is necessary to adequately support their desire to accumulate a mass of knowledge which is sufficient to allow them to transcend a unifying principle. These individuals are given confidence through having access to a wide range of primary information sources. In essence these individuals are seeking the means to undertake an integrative analysis of a topic and a welcomed information source would need to divulge uncensored data and incorporate guidelines to unify the known facts.

DEMONSTRABLE ENVIRONMENTS

In developing the above systems I have investigated the range of multi-media information technology mentioned earlier. Each has the potential to facilitate educational environments which allow response on a bespoke basis to at least four different types of user. I this section I briefly describe two of the systems DIRT has already developed on multi-media information systems using the aforementioned theory.

The first of such system presents information on energy conscious design from four perspectives and aims at engaging individuals of any strategic action pattern preference (Newland et al, 5). This is accomplished by use of interactive videodisc a medium which allows a pictorial base of film and stills to be instantly sequenced and re-sequenced by the learner. The context from which such visual imagery is presented can be easily manipulated, extended by multi-track audio, integrated with text and combined with an overlay of computer generated graphics.

More recently DIRTs learning/information system has been enhanced using Apple Macintosh series II computers themselves extended by the addition of transputers and real-time video digitising boards. These hardware improvements have enabled us to make more responsive educational environments to support specific strategic action patterns. The following more detailed description concerns a second educational environment developed using this refined system.

A Case Study of QUALITY: My Definition - Architectural Video Package

The Problem. In order to gain an understanding of different peoples strategies for attaining quality in design, it is important for those managing a design project to have a clear, reciprocal awareness of what each active participant involved in the design process means by quality. At present quality is a noun looking for appropriate adjectives. In the round table design discussions normally associated with most early phases of designing each participant of a design team, and even the client, can bring their own particular and idiosyncratic view of quality to bear. In many cases they will not be sure themselves what they mean by quality. The project manager should have a clear understanding of that quality before the notion of quality simply becomes a catch all noun to represent many facets and contradictions that can be left undefined.

The Solution. Through the use of a MacIIx equipped with a video digitising board and CD-ROM/ videodisc storage sources a one screen mouse controlled environment was created giving the learner access to a montage of information on designed artefacts portraying very different aspects of quality. This provided a changing array of menus gives access to the views, thoughts and expressions on quality from architects, engineers, and product and fashion designers using nine key artefacts as a suit of common parlance. Filmic vignettes of impressions of quality and a database of Design Council award products were also open to interaction. The design professional using the system was asked to agree, or otherwise, with a series of views portrayed on quality. These judgements were continuously stored by the systems as participants negotiate their own perception of quality. This is summarised for them when they have explored suffient of the quality information package. Designers and clients can therefore explore what each means by the term quality giving the opportunity to explore perspectives and hopefully gain reciprocity. The information environment can also act as database of raw information appropriate for a more Contemplative individual.

The Effectiveness. Students within Portsmouth University's Environmental Studies Faculty together with industrialists, managers and designers have found their use of the package illuminating and rewarding. All find they are familiar with at least one of the key artefacts and the range of comments accessible for each artefact is often surprising and an inducement to constructive discussion with colleagues when they have finished using the system. They leave with a clearer understanding of their own perception of quality and many comment that the experience encourages them to seek out the opinions of others on this concept. We have also seen that participants become more aware of the nature of language and the creation of definitions as a social construct. The design process itself then benefits from this heightened understanding that negotiation of as many viewpoints as possible to achieve an agreed design objective(s) is a fundamental basis for quality design.

CONCLUSION

I believe an edge in competitiveness is achievable if individuals can forge their understanding in experience rather than having it formatted as training. Use of DIRT's multi-media informing systems seems to aid understanding which in turn improves decision making skills in design professionals: their capacity for immediately understanding the systemic nature of a problem and in knowing the right course of actions to take when faced with novel situations is undoubtedly enhanced.

REFERENCES

1. Institution of Civil Engineers. Management Research and Development in the Construction Industry, Report of a Working Party of the Management, ICE, 1988

2. Boardman. J.T., A theory for decision support in technology management, ACME Final Report, SERC, 1992

3. Aris, J. et al Survey of IT User Needs, Pub. DTI (private communication), 1990

4. Bayard-White, C., Multimedia Notes, Chrysalis Interactive Services, London, p.53, 1990

5. International Standards Organisation, Draft STEP proposal, ISO TC184 SC4/ WG1, 1988

6. Rittel, H.W.J. and Webber, M.M, Policy Sciences, 4, 155-169, 1973

7. Olson, D.R. and Bruner, J.S,The 73rd year book of the National Society for

the Study of Education (ed) Olson D.R, The University of Chicago Press, Chicago, 1974

8. Pepper, S.C, World hypotheses, University of California Press, Berkeley, California, 1942

9. Thompson M, Beyond self interest, a cultural analysis of risk, WP-81-17 IIASA, 1981

10. Newland, P.M, Powell, J.A. and Creed C, Design Studies, 8, 1, 1-17,1987

11. Kolb, D.A,The learning style inventory technical manual. MacBer, Boston, USA. 1976

12. Powell, J.A, Is design a trivial pursuit, Design Studies, 8, 4, 1987

13. Jonassen, D.H, Learning strategies:- a new education technology. Programmed Learning and Educational Technology, 22, 1, 26-34, 1985

14. Salomen, G, The use of visual media in the service of enriching mental thought processes. Instructional Science, 9 327-339, 1986

Invited Paper

Visualization and generation of iterated function systems

H. Jones,[a] A. Campa[b]

[a] *School of Mathematics, Statistics and Computing, Middlesex University, Bounds Green Road, London N11 2NQ, UK*

[b] *Centre for Advanced Studies in Computer Aided Art and Design, Middlesex University, Cat Hill, Barnet, EN4 8HT, UK*

ABSTRACT

Iterated Function Systems (IFSs) have been developed by Michael Barnsley for the generation of a class of fractal objects. The authors show how IFSs can also be used to generate three dimensional solid objects. A method for visualizing such systems is presented, using the standard Computer Graphics depth cueing and z-buffer algorithms. The method has potential for adding naturalistic features, such as trees, shrubs and clouds, to architectural drawings and some of the forms generated could be inspirational to architects and engineers in the generation of innovative structures.

ITERATED FUNCTION SYSTEMS

Iterated Function Systems (IFSs) were developed by Michael Barnsley [BARNSLEY88] and have been used by him to illustrate organic structures such as the ferns and leaves; figure 1 shows a leaf (after Barnsley) generated using an IFS. A feature of such systems is that complex structures can be generated using a straightforward algorithm applied to relatively simple data sets, giving major potential for compression of the data required for description of visual images. This feature is currently being investigated by Barnsley and his colleagues as a potential method for compression of High Density Television (HDTV) images, enabling more efficient transmission of such images in a narrower bandwidth than that normally used.

Most applications of IFSs generate images within a two dimensional space, generating objects that have fractal dimension between 1 and 2 [MANDELBROT82]. The authors have extended and illustrated a set of fractal objects lying in three dimensional space, based on the Platonic solids [JONES91b]. The Sierpinski tetrahedron and the Menger sponge were already well known, but the method of illustration using IFSs is novel and gives more correct illustrations of such fractal objects than the methods normally used.

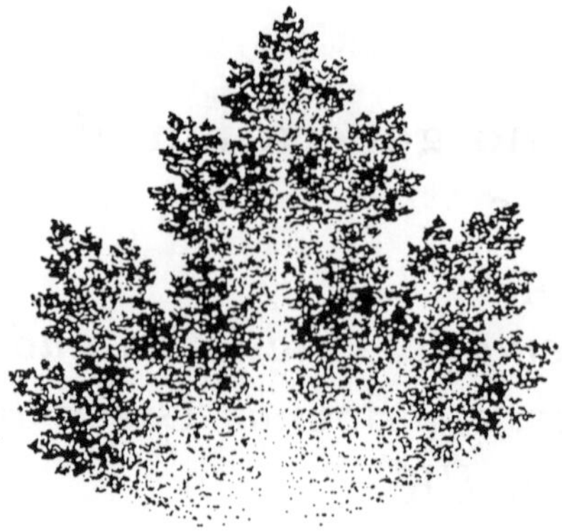

Figure 1. A leaf drawn using a four transform Iterated Function System

An Iterated Function System is defined by a set of n contractive affine transformations $\{T_1, T_2, \ldots T_n\}$ applied to some space and n associated probabilities $\{\pi_1, \pi_2, \ldots \pi_n\}$, such that the sum of the probabilities is unity,

$$\sum_{i=1}^{n} \pi_i = 1.$$

By definition all the π_i are non-negative as they represent probabilities.

An affine transformation is one in which lines are transformed into lines, parallel lines and only parallel lines are transformed into parallel lines [SELKIRK91]. Affine transformations can be created as combinations of the three standard Computer Graphics transformations of translation, scaling and rotation, which are implemented in most Computer Graphics systems using matrix operations on homogeneous coordinate representations of space [FOLEY90][FERRANTE91]. A contractive transformation diminishes the area of a two dimensional object or the volume of a three dimensional object. This can be tested by showing that the determinant of the square matrix used to represent the transformation has value less than one.

There are two major methods for depiction of an IFS [BARNSLEY88]. One involves recursive application of the set of transformations to a specified region of the plane. The points remaining after many iterations of the transformations fall into the attractor of the IFS, for example the leaf shape of figure 1. Different forms of attractor result from different sets of transformations. Interactive methods have been developed to enable designers empirically to change transformations to produce different effects in the image of the attractor [HORN91]. Barnsley describes how 'tiling' of fractal structures can be used to generate the required transformations. For most developers, the generation of transformations to produce particular shapes is an 'art'; Barnsley's research is to make it into a science, so that the process can be automated. This will enable the rapid algorithmic reduction of complex image data into a compact set of transformations.

The second form of depiction is the one that will be used in this study. The simple algorithm for such depiction is given below (algorithm 1). This generates an apparently random sequence of points, P_0, P_1, P_2, ..., which is attracted towards the shape defined by the IFS. If the first point chosen, P_0, is not in the attractor defined by the IFS, the first few points plotted will lie outside the attractor. However, convergence is usually rapid and this problem can be overcome by not plotting the first few points generated, say the first 20. The overall number of points needed for a reasonable result will generally depend on the precision of the display surface. With a medium precision VDU, no added detail should be noticed after about 20 000 points. If the probability π_i associated with the transformation T_i is proportional to the contraction ratio of the transformation (the area into which a unit area is transformed under T_i), the random scatter of points produced will have uniform density with regard to the different transformations throughout the image generation. However, if generation continues for a very large number of iterations with probabilities π_i being otherwise assigned, the whole of the attractor will eventually be visited, although the density of depiction will be uneven for the early stages of the image generation.

```
Given a set of transformations {T1, T2, ... Tn} and a set of
   probabilities {π1, π2, ... πn} ;
Define an initial point P0;
For k = 1 to a large number
      Select a transformation Ti with probability πi;
      Apply the transformation Ti to point Pk-1 to create point Pk;
    Plot Pk;
End k loop.
```

Algorithm 1: A general IFS plotting algorithm

A particularly straightforward family of IFSs uses transformations consisting only of translations and uniform scalings with factor $\frac{1}{2}$ (this means the same scale factor is applied to all Cartesian coordinates). Such scalings can be represented as projections with scaling $\frac{1}{2}$ towards a given centre of projection. Figure 3 illustrates the affine transformation that shrinks the square ABCV to half size so that the vertex V is unchanged. This can be performed by applying the formula

$$P' = \frac{1}{2}(P + V)$$

to any point P in the original shape, generating the 'image' of P under the transformation as P'. Here, P', P and V are taken to be position vector representations of the points concerned, so that the formula should be applied separately to all Cartesian coordinates of these points.

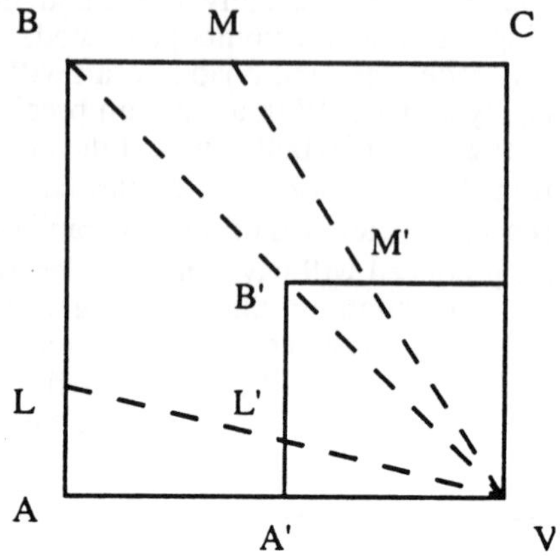

Figure 2. Effect of half scale shrinking of a square towards one of its vertices

The algorithm for generating such IFSs can now be reduced to the following form, called the 'chaos game' by Barnsley.

Given a set of vertices $\{V_1, V_2, \ldots V_n\}$;

Define an initial point P_0;

For k = 1 to a large number

Select a vertex V_i with probability $\frac{1}{n}$;

Set $P_k = \frac{1}{2}(P_{k-1} + V_i)$;

Plot P_k;

End k loop.

Algorithm 2: The 'chaos game' algorithm

If this algorithm is applied to the vertices V_1, V_2 and V_3 of an isosceles triangle, the result is as in figure 3, generating the 'classical' fractal shape of the Sierpinski triangle or gasket [SIERPINSKI16].

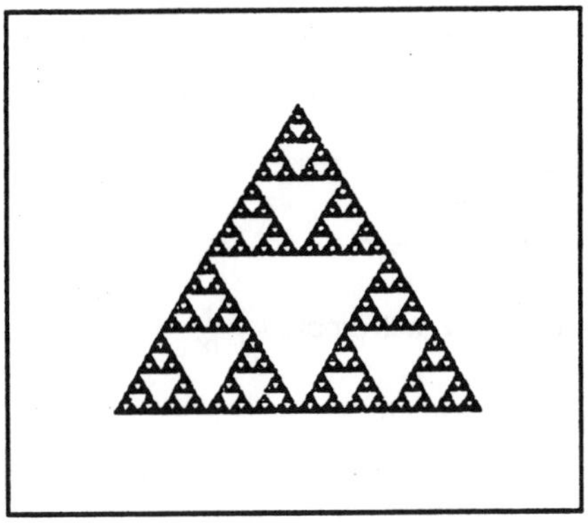

Figure 3. Chaos game depiction of a Sierpinski triangle

Sierpinski's original paper describes two ways of producing this shape. One involves an infinite recursion, drawing smaller and smaller triangles, the other fills the shape with a curve whose length approaches infinity and has nowhere a well defined tangent, as it is generated by an infinite set of zero length line segments. Both these methods must be drawn to a finite precision and they impose, at their limited precision, the shapes of triangles and straight lines where there should be none. The IFS method is intrinsically better for the representation of such fractal forms, as it imposes no superfluous information other than the shape of a pixel, the basic drawing element of a display surface. Thus the depiction is automatically to the greatest precision possible on the display surface used. Lines are seen in figure 3 only as the result of generating a scatter of points that happen to lie in the line. Completeness of the depiction depends on the number of repetitions used in the drawing algorithm; after a certain number of repetitions, all possible pixels that lie within the shape will have been visited and no further change to the image will be observed.

Figure 4. The 'chaos game' algorithm applied to the vertices of a square

Jones [JONES91a] has illustrated that the same algorithm, when applied to the four vertices of a square, evenly fills the square with a uniformly scattered set of random points (figure 4), as proved by Jarrett [JARRETT90]. This shows that an IFS can be used to 'fill' regions of two dimensional space.

```
Given a set of vertices {V_1, V_2, ... V_n} and a shrink factor f;

Define an initial point P_0;

For k = 1 to a large number

    Select a vertex V_i with probability 1/n;

    Set P_k = fP_(k-1) + (1 - f)V_i;

    Plot P_k;

End k loop.
```

Algorithm 3. The adapted chaos game algorithm

Figure 5 shows the effect of applying algorithm 3 to the vertices of a regular pentagon and to a regular hexagon. In the case of the pentagon, the shrink factor is set to $\frac{1}{2+\rho}$, where ρ is the 'golden ratio' [CUNDY61], $\rho = \frac{1+\sqrt{5}}{2}$. For the hexagon, the shrink factor is $\frac{1}{3}$. These factors are calculated using simple trigonometry to enable the subportions of the figures to touch, but not to overlap.

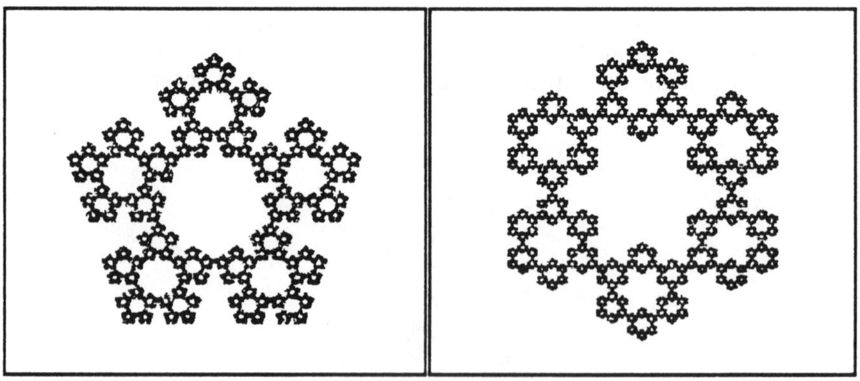

Figure 5. Algorithm 3 applied to a regular pentagon and hexagon.

ITERATED FUNCTION SYSTEMS APPLIED TO SETS OF THREE DIMENSIONAL VERTICES

Without changing the algorithm, the vertices and points generated can be interpreted as three dimensional Cartesian coordinate triads, (x, y, z), with formulae being applied to each of these coordinates in turn. Visualization of objects generated in this way can be enabled by using two algorithms commonly used in Computer Graphics, depth cueing and z-buffering [FOLEY90].

In many Computer Graphic systems, an intermediate stage of generating images of objects placed in a three dimensional Cartesian universe redefines the Cartesian space so that the origin is at the viewing position (equivalent to the eye of an observer or the lens of a notional camera). The z-coordinate of this 'view coordinate system' points away from the observer towards the scene to be viewed and thus represents the 'depth' of a point from the observer. Depth cueing shades points or lines according to this depth. To generate an image of an IFS using one of algorithms 1, 2 or 3 applied to a three dimensional set of vertices, points closer to the observer are given high colour intensity, the intensity diminishing for points further away. This device enables a real observer to distinguish the depths of points in the illustration generated. The authors have set the highest possible intensity of colour allowed on the display surface for points in the objects generated closest to the observer, diminishing linearly to the lowest possible intensity for points in the object furthest from the observer. This gives the largest possible intensity difference within the object's depth.

Images are synthesised on the 'raster' device of a graphics display screen. This contains a regular grid of 'pixels', or picture elements, each of which can be allocated a colour value and intensity. Projection from the three dimensional description of a point to the equivalent pixel location that will represent that point forms part of the process of image generation. A point known to occupy a particular pixel position should only be displayed if the pixel does not currently illustrate a point closer to the observer. Use of a z-buffer enables this check. A z-buffer is a dedicated section of computer memory containing one real valued variable for each pixel of the display device. A memory location in a z-buffer holds the depth (or z-coordinate value in view coordinates) of the object currently illustrated in its pixel. The z-buffer is initiated to hold a large 'background' z distance, beyond which nothing will be depicted. As elements are added to the image on the raster display, values in the z-buffer are upgraded to keep pace with the image. A new point generated is not drawn if its pixel address is already occupied by a point closer to the view position (one with lower z value). Otherwise, the new point is displayed, overwriting the previous occupant of the pixel, with the correct depth cued intensity, and the z-buffer is adjusted accordingly to contain the z-depth of the new point.

This technique has been applied to the vertices of regular polyhedra (or Platonic solids) [CUNDY61] to generate another family of fractals [JONES91b][JONES91c]. Shrink factors are calculated so that the shrunken versions of the object concerned will just touch each other when fitted inside the original object. Considerations of solid geometry give shrink factors of $\frac{1}{2}$ for the tetrahedron (figure 6) and octahedron (figure 7), $\frac{1}{(2+\rho)}$ for the dodecahedron (figure 8) and $\frac{1}{(1 + \rho)}$ for the icosahedron (figure 9), where ρ is the 'golden ratio' defined above.

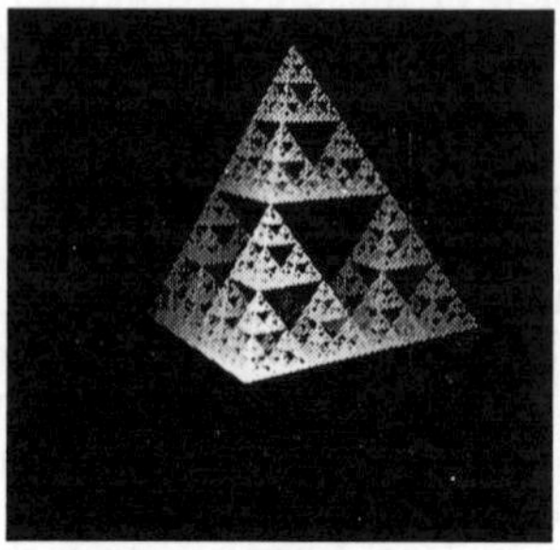

Figure 6. A Sierpinski tetrahedron illustrated using an IFS

Figure 7. An octahedral based fractal , generated using an IFS

Note that the tetrahedron generated is the classical Sierpinski tetrahedron. Once more, the method of generation and depiction used here is purer than the normally used methods of recursively redrawing small tetrahedra. Recursive redrawing of tetrahedra imposes surface normals on the fractal object where none should exist, unless the subdivision is continued so that the sub-polyhedra depicted occupy less than the space of a single pixel in image space. The adapted chaos game method shows the object to the full precision of the display device without imposing any unnecessary and invalid surface information. The octahedral based form has a complex double-pyramid structure. The other forms illustrated have almost organic, coral like forms.

Figure 8. A dodecahedral based fractal, generated by an IFS

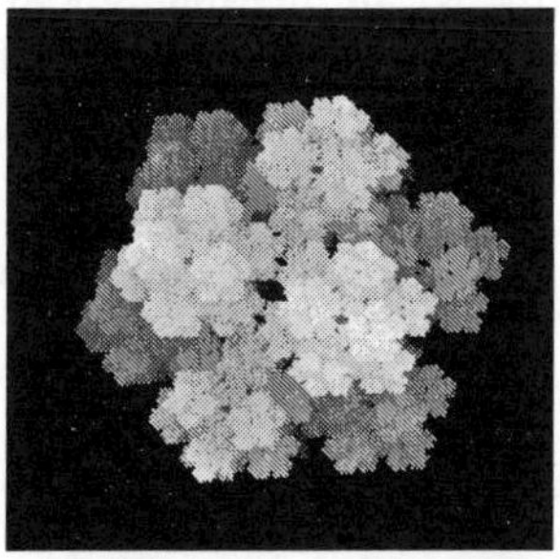

Figure 9. An icosahedral based fractal, generated by an IFS

Further interesting figures have been generated by taking narrow sections through this collection of objects, producing fractal patterns sometimes displaying several forms of symmetries, otherwise with some non-symmetric but self similar patterns. Figure 10 shows a cross section through an icosahedral based fractal that displays pentagonal symmetry. Other slices through the same object display different forms of symmetries. The lay outs of such patterns could, perhaps, have desirable properties for town planners; the objects based in three dimensions could give architects and designers a new source of potential shapes to explore.

Figure10. A slice through an icosahedral fractal

The algorithm does not generate a fractal object from the vertices of a cube or hexahedron when the natural shrink factor of $\frac{1}{2}$ is applied, as the points generated evenly fill that original cube, in a manner analogous to the filling of a square described above. Different cube-based distributions of vertices with different shrink factors can, however, produce interesting effects. An IFS with 32 vertices based at the eight corners of a cube and at $\frac{1}{3}$ and $\frac{2}{3}$ distances along the cube's twelve edges with shrink factor of $\frac{1}{3}$ generates a fractal object similar to the classical Menger sponge, [MANDELBROT82] (figure 11).

Figure 11. A new cube based fractal, the 'Jones/Campa cake'

This new form has been named the 'Jones/Campa cake' by John Lansdown [LANSDOWN91]. The Menger sponge itself can be generated by using the same shrink factor of $\frac{1}{3}$, but with 20 vertices at the cube's eight corners and at the mid points of each of its twelve edges [figure 12].

Figure 12. The Menger sponge illustrated using an IFS

In theory, the block shapes so formed have zero overall volume. It would be interesting to analyse if a limited precision approximation of these structures could produce light but strong building blocks. There is already an analogy of a light, but strong, structure in the similarity of the Sierpinski tetrahedron (figure 6) to the 'pylons' that carry high tension electrical transmission lines. A similar 'pylon' shape can be generated by applying a shrink factor of $\frac{1}{2}$ to the vertices of a square based pyramid. Could such a shape be of interest to structural engineers? Do the fractal cube structures illustrated have potential as the basis for building designs? The 'Défense' development in Paris shows that such structures are not totally outrageous for modern architectural development.

It has been demonstrated above that IFSs can be used to generate filled regions such as squares in two dimensions and cubes in three dimensions. Another organisation of vertices can be used to generate an apparently solid tetrahedron. Using a shrink factor of $\frac{1}{2}$ on eight vertices consisting of the corners and edge mid-points of a regular tetrahedron gives the image of figure 13. In this

Figure 13. An apparently solid tetrahedron generated using an IFS

figure, points generated by attraction towards a corner vertex were coloured in yellow with points generated by attraction towards an edge mid-point vertex were coloured red, showing up as different tones in the monochrome reproduction shown here. The exterior surface of the tetrahedron is filled, although if the object generated is sliced through, interior holes are found. Combinations of cubes, tetrahedra and other IFSs of 'solid' appearance could, conceivably, be used to illustrate models of more complex solid objects. The combination of depth-cueing and z-buffer algorithms gives a relatively simple rendering method for such shapes.

It is worth considering the implications of the rendering method used. No surface information is presented to the observer. The objects generated are illustrated as clusters of points scattered on a two dimensional surface. The only form of shading is to vary the intensities of the shading of the points to match a concept of distance from an observer. There is no attempt to reproduce such lighting features as inverse square laws. However, the images are interpreted as having surfaces as in figure 13. This feature is also evident in fractal images based on the tetrahedron (figure 6), octahedron (figure 7) and cube (figures 11 and 12), where co-planar features can be distinguished. It is a matter of fascination to the authors that the eye and brain interpret the images as objects existing within a three dimensional space on such relatively minor clues.

The half-toned figures illustrated above are monochrome reproductions of images from a Silicon Graphics Personal Iris Workstation, with the capability of displaying up to 16.5 million colours. The process does not depend on equipment of this degree of sophistication to deliver useful results. Figure 14 gives an example of an IFS generated on a PC by David Scott, with a limited range of colours [SCOTT92]. This fern differs from those normally shown in

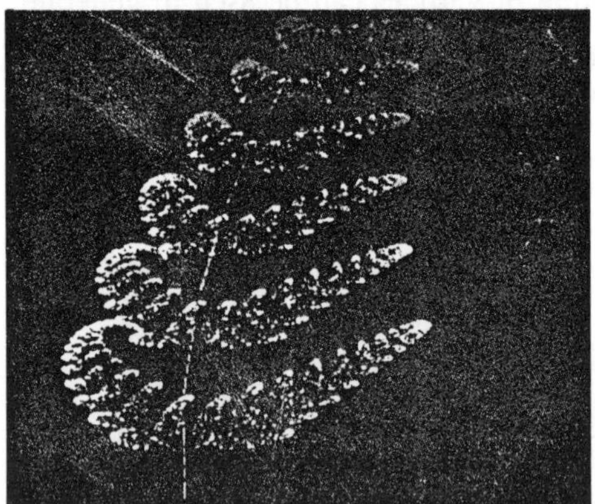

Figure 14. A three dimensional IFS fern

that it is a representation of a three dimensional structure based on a three dimensional IFS. Choosing different viewing directions will generate different views of the fern. On more limited equipment, a little subtlety in development of the display method can do away with the need for a z-buffer, which may be too costly in storage space for equipment with limited memory resources. Each pixel holds implicit information on the depth of the point currently illustrated in that pixel, as the depth cueing has tied the intensity of pixel illumination to the z-depth. Using a 'read pixel' command to compare the current pixel intensity to that which would be allocated to a new point can enable the decision whether the new point should be displayed, or whether it is behind (less intense than) the point currently illustrated within the pixel.

Figure 15. An IFS generated three dimensional cloud

Other natural forms such as clouds [LUCKINS92] and trees [SCOTT92] have been successfully illustrated by this method on devices with limited memory and limited display capacity. An example of an IFS generated cloud is shown in figure 15, although this picture used the full capabilities of the advanced Workstation cited above. Here the vertices used for algorithm 3 are scattered within an ellipsoid and the value of f is empirically determined to give a satisfactory image. The cloud generated here is a representation of a set of particles suspended in three dimensional space and is therefore different from the majority of computer generated cloud simulations [SAUPE88], which are shown as painted objects on a two dimensional backdrop sky. This form of illustration could be used in flight simulator systems, giving a flight officer an experience similar to that of flight through a cloud.

SUMMARY

The concept of generation of Iterated Function Systems has been developed, mainly within the context of a limited sub-class of such systems. A relatively straightforward method of visualizing IFSs that are based within three dimensional Cartesian space on a two dimensional display surface has been discussed. Although no real surface information is given in this method, images generated are interpreted by observers as having spatial properties such as co-planar sets of points and even planar surfaces. These even allow the depiction of apparently solid objects and could be used for generation of images of a wide class of objects, including realistic simulations of naturally occurring phenomena. The new fractal forms generated through IFSs could be inspirational to designers. IFSs have the property of encapsulating the forms of complex objects in a relatively limited data set and a simple reconstruction algorithm.

The monochrome images displayed in this publication do not do justice to the precision of the full colour images generated on the authors' workstation and displayed as colour slides.

REFERENCES

[BARNSLEY88] Barnsley, M.F., *Fractals Everywhere,* Academic Press, San Diego, 1988

[CUNDY61] Cundy, H.M. and Rollett, A.P., *Mathematical Models* (2 ed), Oxford University Press, Oxford, 1961

[FERRANTE91] Ferrante, A.J., Moreira, L.F.R., Boggio Videla, J.M. and Montagu, A., *Computer Graphics for Engineers and Architects*, Elsevier/Computational Mechanics Publications, Amsterdam, 1991

[FOLEY90] Foley, J.D., vanDam, A., Feiner, S.K. and Hughes J.F., *Computer Graphics, Principles and Practice* (2 ed), Addison Wesley, Reading, Mass, 1990

[HORN91] Horn A.N., *IFSs and the Interactive Design of Tiling Structures*, in Fractals and Chaos, Crilly A., Earnshaw R. and Jones H. (eds), 119 – 144, Springer-Verlag, New York, 1991

[JARRETT90] Jarrett, D., Personal correspondence, Middlesex University, 1990

[JONES91a] Jones, H., *Dürer, Gaskets and the Chaos Game,* Computer Graphics Forum, 9 (3), 327 – 332, 1991

[JONES91b] Jones, H. and Campa, A., *Fractals Based on Regular Polygons and Polyhedra*, in Scientific Visualization of Physical Phenomena, Patrikalakis, N.M. (ed), 299 – 314, Springer-Verlag, Tokyo, 1991

[JONES91c] Jones, H. and Campa, A., *Fractal Polyhedra – New Shapes from Old Figures*, in Compugraphics'91, Santo, H.P. (ed), 47 – 56, Conference Proceedings, Sesimbra, Portugal, September 1991

[LANSDOWN91] Lansdown, R.J., *Serendipity, Not only computing – also art*, The Computer Bulletin, Vol 3, Part 7, 12 – 13, September 1991

[LUCKINS92] Luckins, S., *Three Dimensional Cloud Modelling Using the Chaos Game*, MSc Applied Computing Technology Dissertation, Middlesex University, 1992

[MANDELBROT82] Mandelbrot, B.B., *The Fractal Geometry of Nature,* W.H.Freeman, 1982

[SAUPE88] Saupe, D., *Algorithms for Random Fractals*, in Peitgen, H.-O. and Saupe, D. (eds), The Science of Fractal Images, Springer-Verlag, New York, 1988

[SCOTT92] Scott, D. *Three Dimensional IFS Generation and Rendering of Botanical Images*,MSc Applied Computing Technology Dissertation, Middlesex University, 1992

[SELKIRK91] Selkirk, K., *Longman Mathematics Handbook*, 236, Longman, Harlow, 1991

[SIERPINSKI16] Sierpinski, W., *Sur Une Courbe dont tout point est un point de ramification*, in W.Sierpinski: Oeuvres Choisis, Tome II, PWN - Polish Scientific Publishers, Warsaw, Poland, 1975, first published 1916

Formex configuration processing

D. Hadker, D. Tzourmakliotou

Space Structures Research Centre, Department of Civil Engineering, Univeristy of Surrey, Guildford, Surrey GU2 5XH, UK

ABSTRACT

Remarkable advances in computer aided design, have made it possible to have effective techniques for organisation of graphics as well as data generation. In particular, the fields of architecture and structural engineering may greatly benefit from the concepts of formex algebra which is a mathematical system providing a convenient basis for generating and modifying configurations. Formian is the programming language of formex algebra which handles problems of data generation and computer graphics with ease and elegance.

The concepts of formex algebra have found applications in many disciplines. However, this paper describes some of the basic concepts of formex algebra in relation to a variety of space structures. Some composite transformations have been illustrated in the paper, that allow the creation and manipulation of a number of families of surfaces used for representing membranes, pneumatic structures and lattice shells.

The paper concludes by describing a convenient technique for projecting patterns on various surfaces such as spheres, ellipsoids and hyperbolic paraboloids. Also, a function is described for evolving configurations based on polyhedra. This function constitutes the kernel of the problem handling strategy for visualisation and generation of data for geodesic domes.

INTRODUCTION

In the fields of architecture and structural engineering, advancements in computer aided techniques have made it possible to design and realise more and more complicated structural forms. Especially, in the field of space structures, the role of computers for visualisation, design and analysis is becoming indispensable. The term "space structures" in this context, refers to three dimensional systems such as spaceframes as used for single and double layer grids, vaults and domes, tension structures such as prestressed cable nets and membranes, pneumatic structures and hybrid space structures.

The concepts of formex algebra and its associated programming language Formian provide a convenient tool not only for computer graphics, but also for generating data for analyzing space structures. The rudimentary ideas from which formex algebra has emerged were evolved in the early seventies. These concepts were published in 1975, Nooshin [1]. The first textbook giving a comprehensive account of the current ideas on the subject was published in 1984, Nooshin [2].

FORMEX CONFIGURATION PROCESSING

In a very broad sense, the term "configuration" is used to mean an "arrangement of objects". Although the term configuration has been used in this context to refer to assemblies of structural elements, in general, even parts that form an electrical network or a finite element idealisation of a solid medium may be regarded as a configuration. The most common usage of the term configuration is in reference to a geometric composition consisting of points and/or lines and/or surfaces. Such geometric composition may itself be the subject of study or it may represent another arrangement of objects. When digital computers are involved, the internal representation of configurations is in terms of numerical models. The term "configuration processing" is used to mean "creation and manipulation of numerical models representing configurations".

Some examples of space structure configurations commonly encountered by architects and engineers, are shown in Fig 1. Graphical visualisation of these structures and preparation of data for their analysis are very important requirements in the design process. The analysis requires information regarding various aspects of the structural system including the elements of the structure, the positions of supports, the nature of loads and the manner in which the elements are connected together.

THE BASIC APPROACH

To begin with, consider the dome shaped configuration in Fig 2. Let it be required to generate data for the

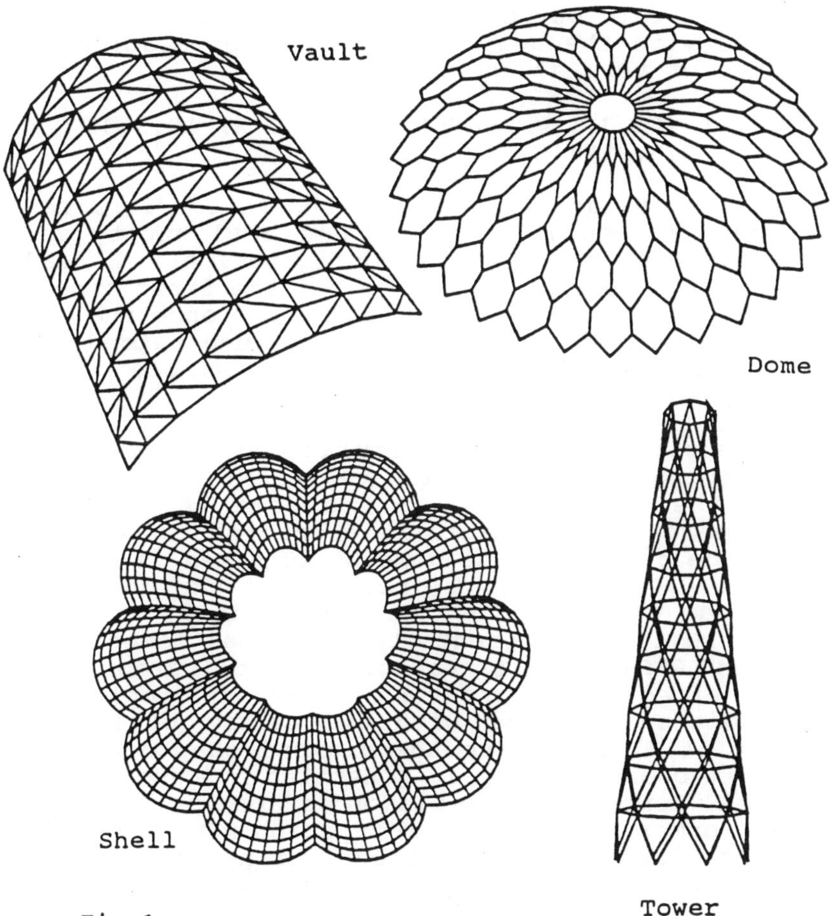

Fig. 1

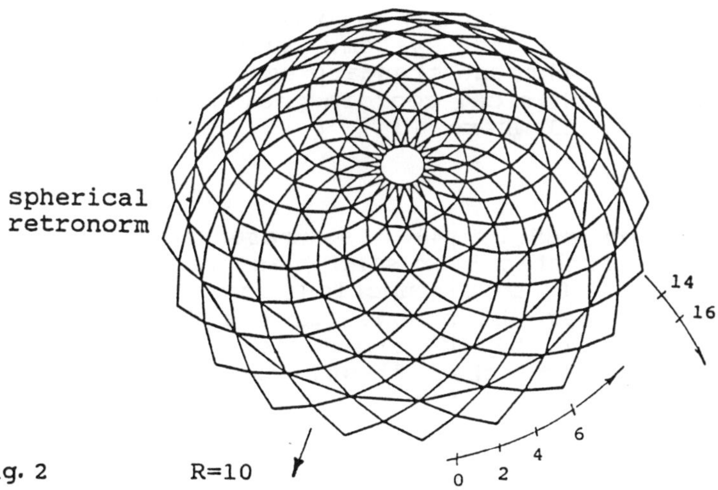

Fig. 2

analysis of this structure. The configuration may be described with the help of a formex formulation in a few lines of text as shown below

```
E1=[10,1,1; 10,3,1]#ROSAT(1,2)|[10,0,2; 10,1,3]
E2=RINIT(20,6,2,2)|E1
```

The entities E1 and E2 are referred to as "formices". A "formex" (singular) is a mathematical entity which may be used to represent a configuration. A reader encountering the above formices for the first time, is bound to find it a little confusing but once the basic principles of formex algebra are understood, the formex approach is found to be a simple way of describing a configuration.

To explain the above formulations, one may begin by describing formex E1 which represents only a small part of the configuration. The dome is formulated with respect to a suitable reference system shown in Fig 2. The radius of the dome R, is taken as 10. The construct [10,1,1; 10,3,1] represents the horizontal member near the crown of the dome and the construct [10,0,2; 10,1,3] represents the diagonal member. The "function" ROSAT implies rotational replication. This function effectively, creates a rotational replication of the diagonal member with the centre of rotation given by (1,2) to create the diamond shaped pattern. The entire configuration is generated by repeating the horizontal member and the diamond shaped pattern using another formex function. The function RINIT(20,6,2,2) is used to imply 20 replicational translations in the second direction and 6 in the third, with a step of 2 units in each direction as represented by formex E2.

The configuration may also be described relative to the x-y-z coordinate system. Such a transformation may be achieved through standard equations that relate Cartesian and spherical coordinates. This is obtained through formex E3 where

```
E3=BS(1,9,4)|E2
```

The function BS(1,9,4) is referred to as a "retronormic function" or simply as "retronorm". The general form of this function may be written as BS(b1,b2,b3) where BS stands for the "basispherical" retronorm and b1, b2, b3 are scale factors relating, respectively, to the first, second and third directions of the reference system.

In most design procedures it is often necessary to generate a number of shapes before arriving at a satisfactory configuration. By changing the scale factors, one can achieve this with ease and convenience. To show the effects of changes in the scale factors, the plots of formices

E4=BS(1,9,7)|E2

and

E5=BS(1,9,2)|E2

are shown in Figs 3 and 4 respectively.

When using Formian, the formex formulation representing a configuration is typed in through a keyboard and the result is obtained as a graphical output through the VDU and/or a plotter. Formian has a number of constructs through which images may be created on the graphical output media. Depending on the viewing requirements, plans, elevations and perspective views may be obtained using different viewing specifications available in Formian, Nooshin and Disney [3]. The advantage of using the formex approach is that the same information can be used as input data for a structural analysis package.

The process of data generation is aided by a number of formex functions. Although the scope of the present paper does not allow the detailed description of all formex functions, an elaborate explanation is available in Nooshin and Disney [3]. However, some functions such as the cardinal and tendial functions which are in constant use have been described in the sequel. Cardinal functions have been described in Table 1. A brief description of each cardinal function is accompanied by a sketch showing the effects obtained by using the function.

Tendial functions are commonly used combinations of cardinal functions. For example, in formex E2, a tendial rindle function was used where

E2=RINIT(20,6,2,2)|E1

was used instead of writing

E2=RIN(3,6,2)|RIN(2,20,2)|E1

The suffix IT implies double action in the second and third directions. The suffix ID is used to imply double action in the first and second directions and the suffix IS is used to imply double action in the first and third directions.

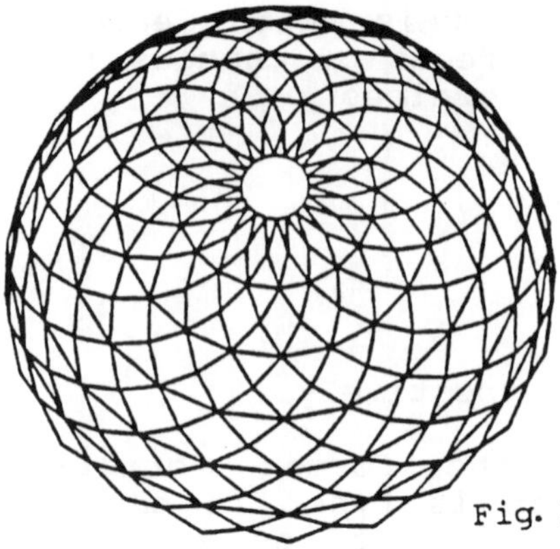

Fig. 3

Fig. 4

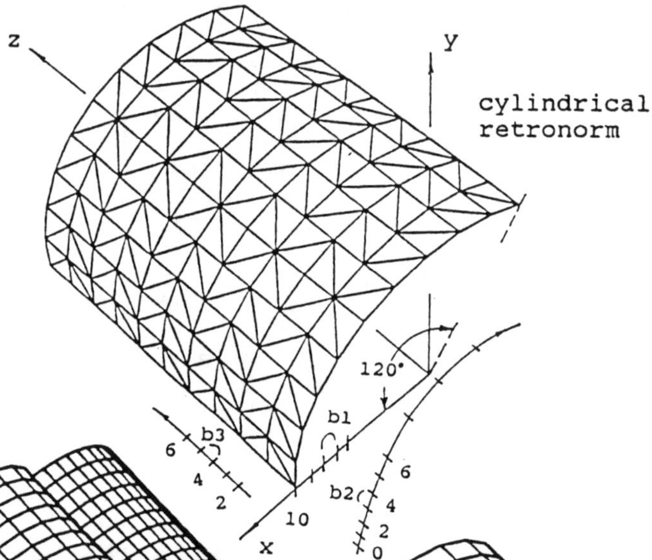

Fig. 5

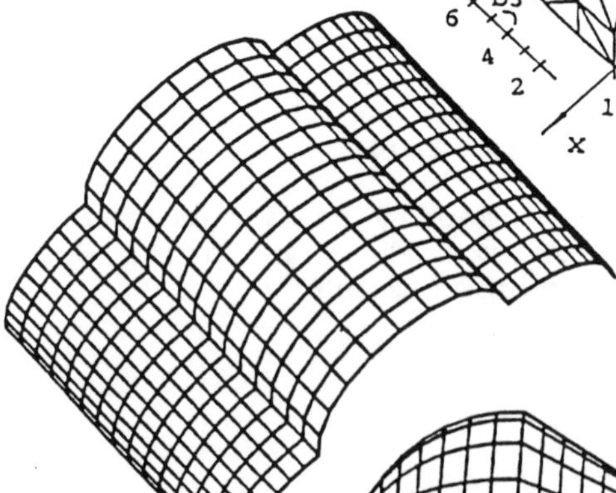

Fig. 6

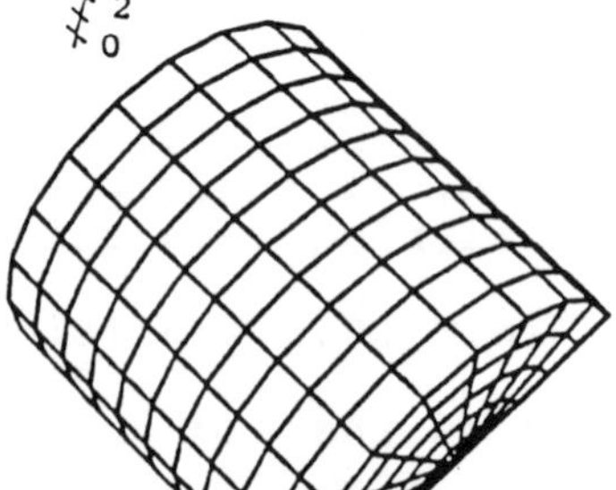

Fig. 7

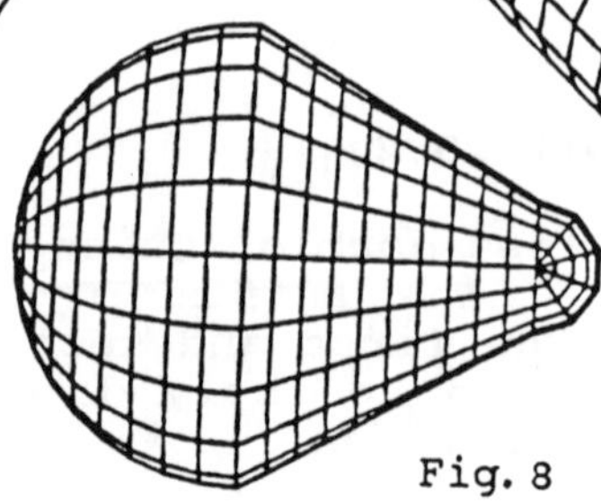

Fig. 8

TABLE 1 : CARDINAL FUNCTIONS

TRANSLATION G=TRAN(h,q)|E h gives direction of translation q gives amount of translation 	RINDLE G=RIN(h,s,p)|E h gives direction of replication s gives number of replications p gives amount of replications 
REFLECTION G=REF(h,q)|E h gives direction of reflection q gives position of the plane of reflection 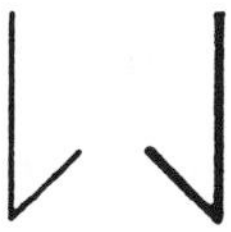	LAMBDA G=LAM(h,q)|E h gives direction of reflection q gives position of the plane of reflection 
VERTITION G=VER(h1,h2,q1,q2)|e h1 and h2 define the plane of rotation q1 and q2 give coordinates of the centre of rotation 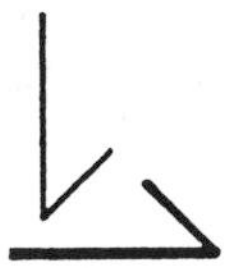	ROSETTE G=ROS(h1,h2,q1,q2)|E h1,h2 directions define plane of rotation q1,q2 give coordinates of the centre of rotation 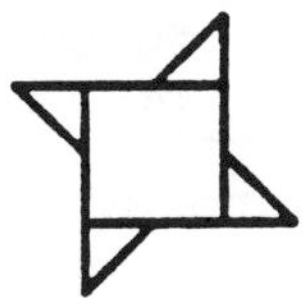
PROJECTION G=PROJ(h,q)|E h gives direction of projection q gives position of the plane of projection 	DILATATION G=DIL(h,q)|E h gives dilatation q gives factor of dilatation 

In the formulation for the dome of Fig 2, constructs were combined using an "operator". The symbol # is an operator, referred to as a "duplus symbol" which effectively combines two or more constructs or formices.

STANDARD RETRONORMS

Standard retronorms are the ones which are already available in Formian. The basispherical retronorm was discussed in the earlier sections of the paper and the "basicylindrical" retronorm is shown in Fig 5, along with a cylindrical reference system. A formex formulation for this configuration may be written as

```
D1=[10,0,0;10,2,2]#ROSAT(1,1)|[10,2,2;10,2,0]
D2=PEX|RIN(3,5,4)|LAMIT(2,10)|RIN(2,5,2)|D1
```

In the process of representing the configuration, a number of extra elements are created. Although these may not be visible in the graphical output, it is very important to eliminate them to avoid problems in the analysis of the configuration. In order to overcome this problem, the formex function PEX is used, which stands for "pexum". The function has the effect of eliminating the doubly represented members.

Formex D2 represents the whole configuration relative to the cylindrical reference system. The same configuration may be written in terms of the x-y-z coordinate system using the relation,

```
D=BC(1,120/20,1)|D2
```

The function BC stands for the "basicylindrical retronorm". The general form of the retronorm may be written as BC(b1,b2,b3), where b1, b2 and b3 are scale factors.

The retronormic functions whose names begin with "basi" imply simple scaling. There are certain retronormic functions whose names begin with "metri" and imply accelerated and decelerated scaling in accordance with a rule based on geometric progression, Nooshin [2].

Using standard retronorms, segments of basic surfaces such as spheres, cylinders and cones may be combined to generate more complex forms. Some examples of configurations obtained using combinations of parts of surfaces obtained from cylindrical and spherical retronorms are shown in Figs 6 to 8.

A number of possibilities arise when more than

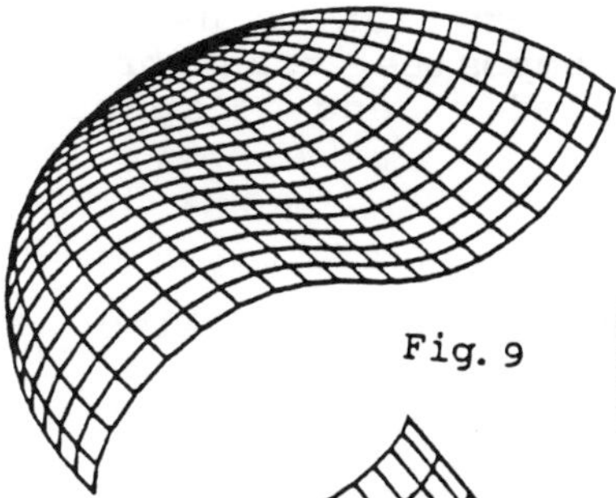

Fig. 9

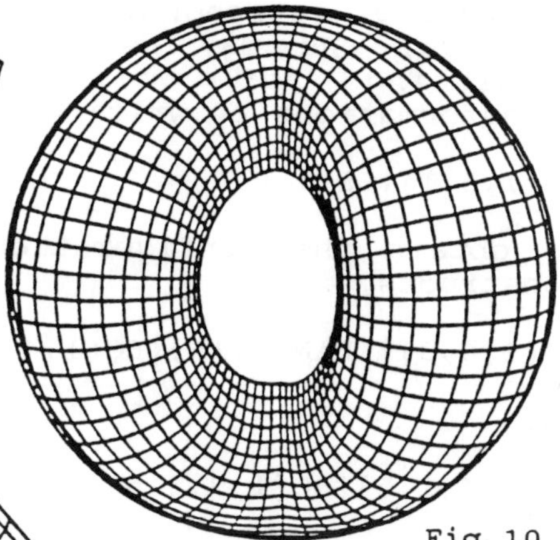

Fig. 10

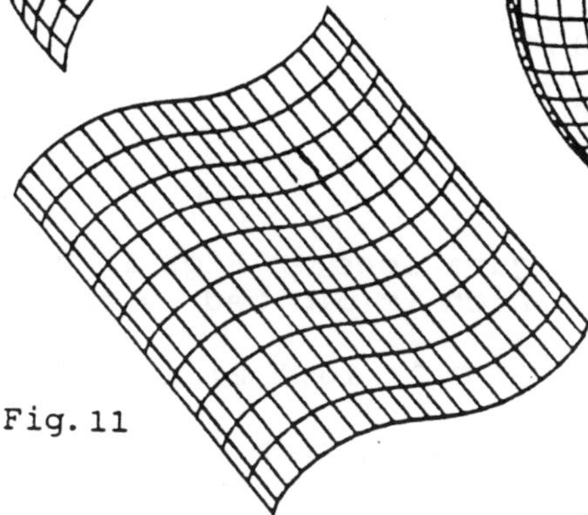

Fig. 11

Fig. 12

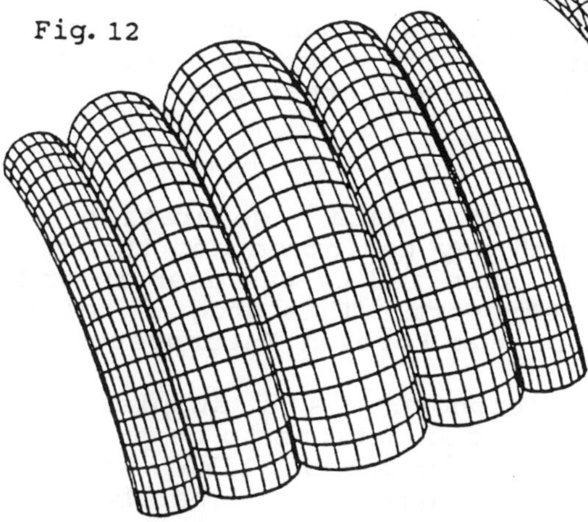

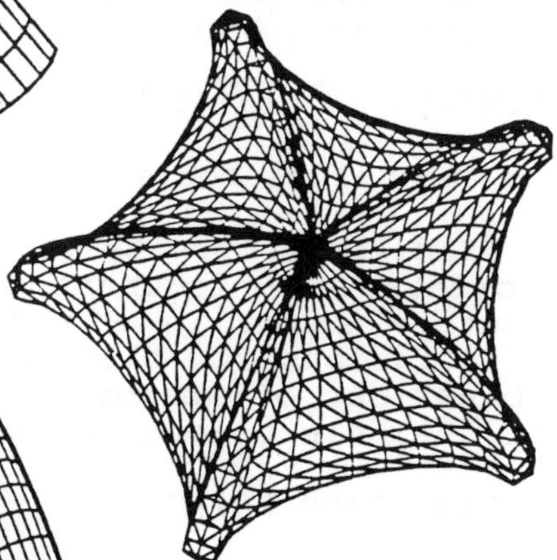

Fig. 13

Fig. 14

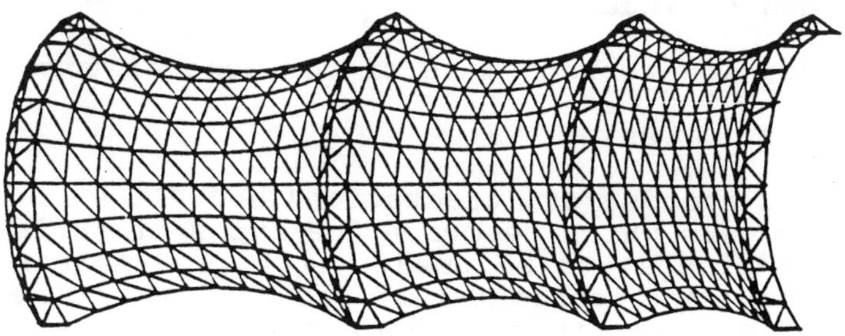

one type of retronormic transformation is used. For instance, one may use a composite transformation consisting of cylindrical and spherical transformations by writing

```
BC(1,5,1)|BS(1,6,3)
```

to obtain the configuration in Fig 9. Also, the constructs,

```
BS(1,6,3)|TRAN(1,15)|BC(1,4,1)
```

and

```
BC(1,4,1.5)|BC(1,9,1.5)
```

will give rise to the configurations in Figs 10 and 11 respectively. The formex which is plotted is given by

```
F=RINIT(m,n,1,1)|[10,0,1; 10,0,2; 10,1,2; 10,1,1]
```

with the values for m and n varying from case to case.

Similar composite transformations have been used to create a configuration suitable for a pneumatic structure shown in Fig 12. Figs 13 and 14 show configurations for hybrid space structures which are combinations of spaceframes and membranes, generated with the help of composite transformations.

SUPPLEMENTARY RETRONORM

During the process of data generation, a "supplementary retronorm" may be introduced through a program segment which is supplied by the user in order to use a non-standard retronorm. This program segment is linked to the body of the Formian Interpreter.

The "tractation retronorm" is a supplementary retronorm which enables configurations to be projected on different types of surfaces like spheres, ellipsoids, paraboloids, hyperbolic paraboloids, planes and cylinders.

The tractation retronorm is of the general form

```
TRAC(P1,P2,...,Pn)
```

where TRAC stands for "tractation" and P1, P2,..., Pn are parameters which describe the types of surface(s) and projection(s) to be used. If E is a formex that represents a configuration then a formex F representing the projection of the configuration on the specified surface(s) is given by an equation of

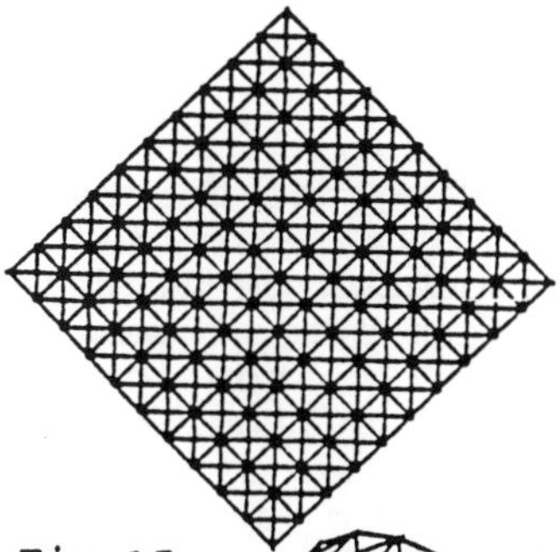

Fig. 15

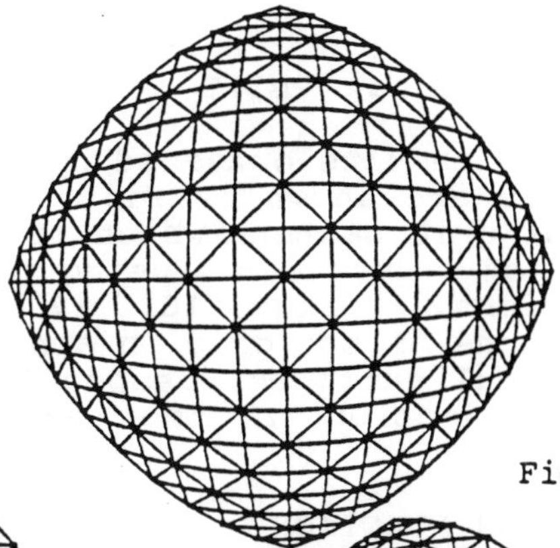

Fig. 16

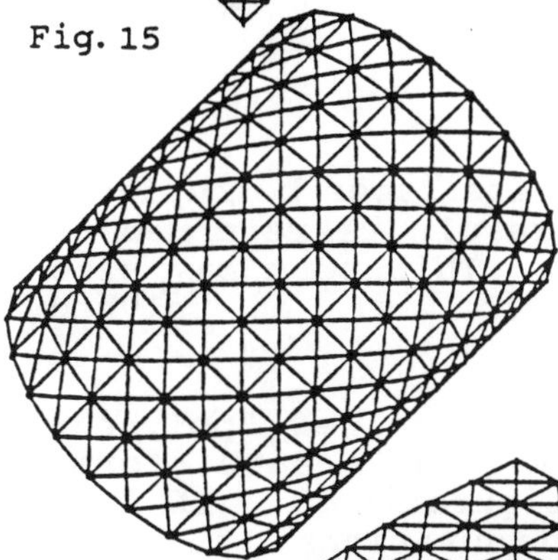

Fig. 17

Fig. 18

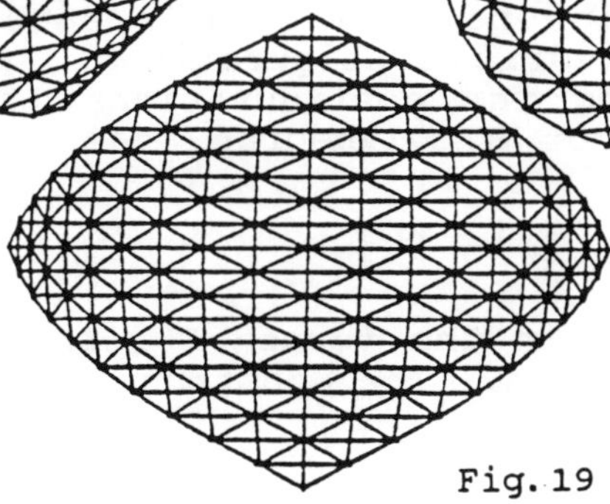

Fig. 19

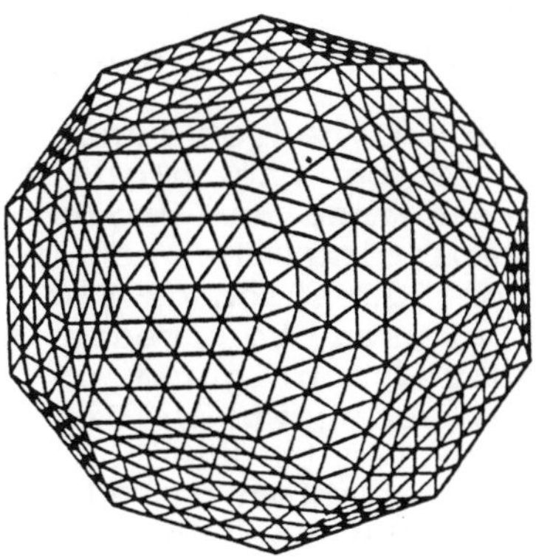

Fig. 20

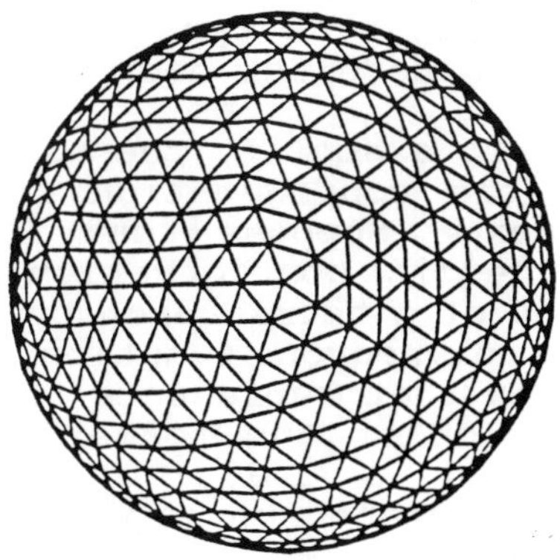

Fig. 21

the form

F=TRAC(P1,P2,...,Pn)|E

Some results obtained by applying the tractation retronorm are shown in Figs 16 to 19. Fig 15 shows the flat grid pattern which is projected on different types of surfaces using different types of projections. Figs 16, 17 and 18 show central projections of the grid on spherical, cylindrical and ellipsoidal surfaces, respectively. Fig 19 shows an axial projection of the same grid on a spherical surface.

POLYHEDRON FUNCTION

Consider a polyhedron and let its faces be replaced by a given configuration. The result is referred to as a "polyhedric" configuration. This term may also be used to refer to a portion of a polyhedric configuration.

There is a formex function called "polyhedron" function that can be used to create formices representing polyhedric configurations. If E is a formex representing the configurations that are to replace the faces of a polyhedron, the construct of the form

F=POL(P1,P2,...,Pn)|E

can be used to obtain the formex F representing the required configuration. The construct

POL(P1,P2,...,Pn)

is the "polyhedron" function, where POL stands for polyhedron and where P1, P2,..., Pr are parameters specifying the type of polyhedron, the radius of the circumsphere for each layer involved, ... , etc.

To exemplify the use of the polyhedron function, let the configuration in Fig 22 be the plot of a formex. Suppose that it is required to produce a polyhedric configuration by replacing the top four faces of a tetrahedron by the configuration of Fig 22. The polyhedron function may be used to create the formex that represents the required configuration and the plot of the formex will be as shown in Fig 23. Moreover, a similar operation applied to the top part of an octahedron produces the plot shown in Fig 24. Using the same configuration on the top half of an icosahedron produces the plot shown in Fig 25.

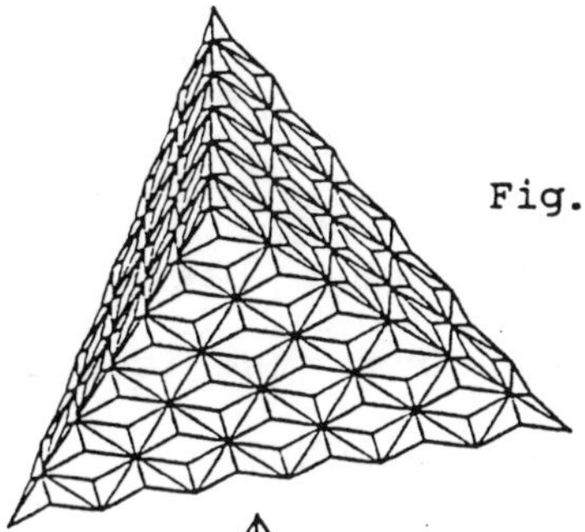

Fig.23

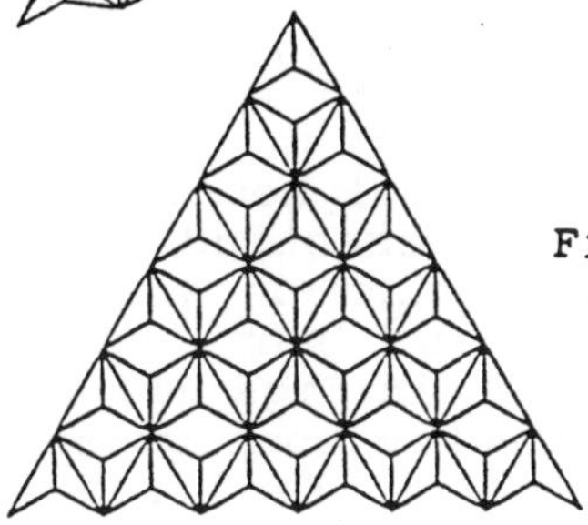

Fig.22

Fig.24

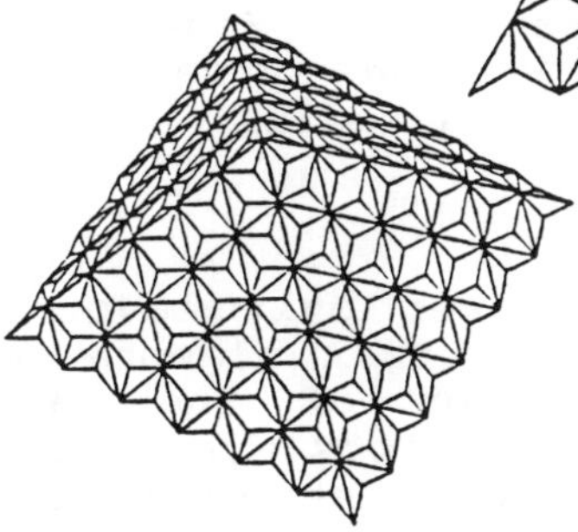

Fig. 25

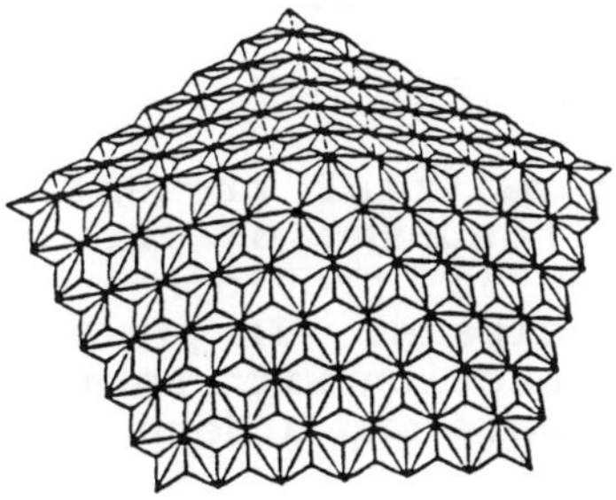

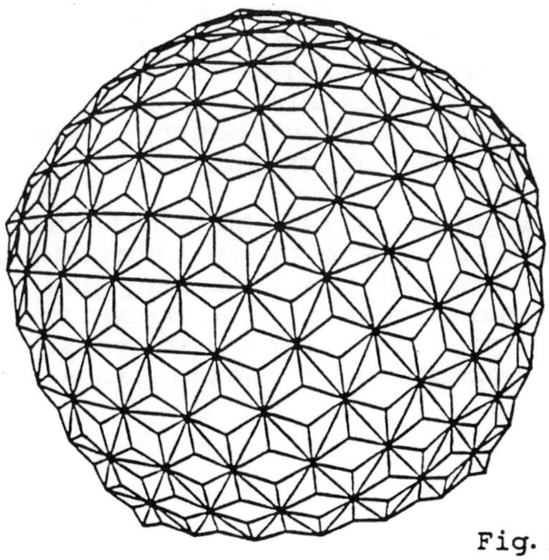

Fig. 26

If the configuration to replace the top faces of the three polyhedra is as shown in Fig 27, then the repetition of the above operations will result in the configurations shown in Figs 28 to 30.

So far, polyhedric configurations based on Platonic polyhedra have been shown. However, the configuration represented in Fig 20 is based on an Archimedean polyhedron, the icosidodecahedron.

CREATION OF GEODESIC DOMES

The combination of polyhedron function and the tractation retronorm can be used to produce a wide variety of geodesic forms. The procedure is simple. To begin with, the polyhedron function is used to create a formex representing the projection of the polyhedric configuration on a suitable surface. The above procedure has been used to produce the configurations shown in Figs 21, 26 and 31. These are based on the polyhedric configurations of Figs 20, 25 and 30 respectively. The projection surfaces are spherical and the projection type is radial.

CONCLUDING REMARKS

The material presented in the paper only covers a small subset of the concepts of formex algebra but aims at providing an appreciation of the basic principals. In practice, these concepts are used in conjunction with Formian which is a suitable computing software and a description of this is given in, Disney [5].

The concepts of formex algebra compliment the human imagination and allow mentally visualised configurations to be expressed in terms of concise numerical models. This aides in creating a versatile medium for communication between people in data generation and computer graphics.

Formex algebra, like other mathematical systems grows like a living entity in response to practical needs. It is anticipated that many new pioneers will join the present research workers to explore the exciting new possibilities that present themselves through the concepts of formex algebra.

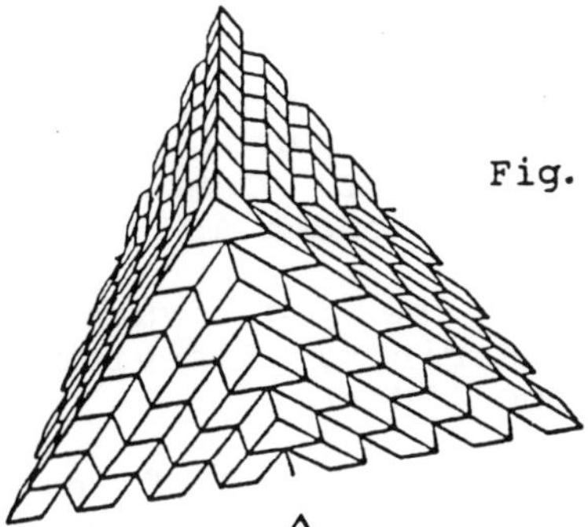

Fig. 28

Fig.27

Fig. 29

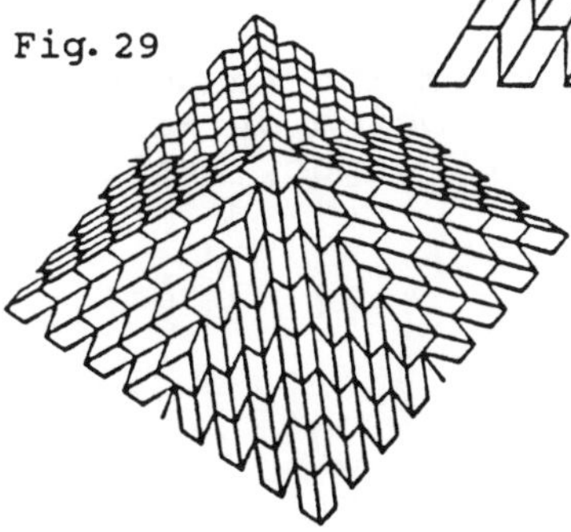

Fig. 30

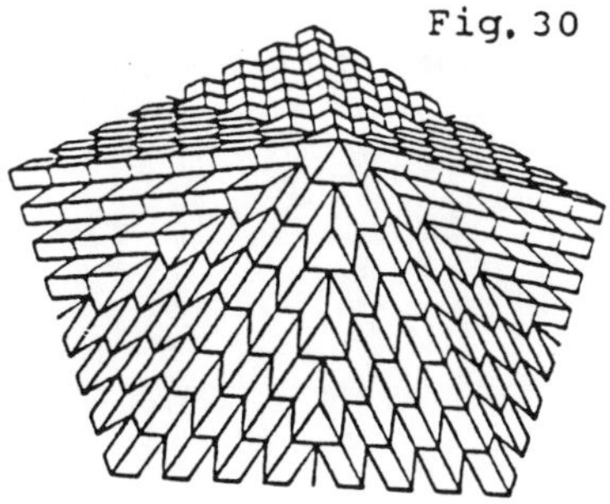

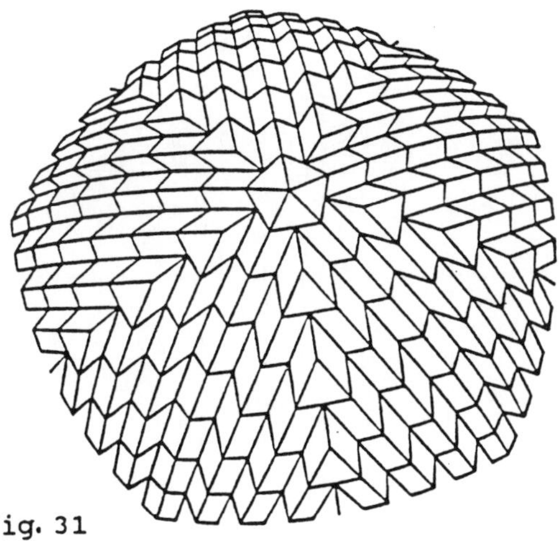

Fig. 31

REFERENCES

1. Nooshin, H. "Algebraic Representation and Processing of Structural Configurations", *International Journal of Computers and Structures*, Vol. 1, pp. 119-130, London 1975.

2. Nooshin, H. "*Formex Configuration Processing in Structural Engineering*", Elsevier Applied Science Publishers, London 1984.

3. Nooshin, H. and Disney, P. "Elements of Formian", *International Journal of Computers and Structures*", Vol. 41, No 6, pp. 1183-1215, London 1991.

4. Nooshin, H., Disney, P., Hadker, D. and Tzourmakliotou, D. "Some Aspects of Formex Configuration Processing", pp. 310-333, *Proceedings of the First International Seminar on Structural Morphology*, (Ed. Motro, R. and Wester, T.), Montpellier, France, September 1992.

5. Disney, P. "Formian : The Programming Language of Formex Algebra", *International Symposium on Membrane Structures and Space Frames*, Osaka, Japan, September 1986.

Highly interactive 3D geometric modeling

J. Liang, M. Green

Department of Computing Science, University of Alberta, Edmonton, Alberta, Canada, T6G 2H1

ABSTRACT

This paper presents an effort to build a 3D modeling system that can be used for 3D quick prototyping. We focus on providing a highly interactive user interface for 3D direct manipulation, and allowing for the efficient creation and modification of 3D shapes. The main features of the system are (1) the use of kinetic depth effect to provide effective 3D visualization on a traditional CRT screen, (2) the use of a hand-held six degrees of freedom input device (a bat) for direct manipulation of 3D objects, and (3) a set of modeling operations based on the new input device. The system allows a designer to quickly create, modify, place, orient, and organize 3D objects in a natural and intuitive manner. Preliminary results show that such a highly interactive interface has the potential of increasing the efficiency of 3D geometric modeling by a large factor.

1 INTRODUCTION

Three dimensional geometric modeling has been a challenging topic for computer graphics and computer-aided design [11]. Although many modeling systems have been developed to facilitate 3D geometric modeling, few have penetrated into the early stage of the design process, that is, the conceptualization or sketching phase [4]. The major reason for this is the generally low efficiency of interaction. Using a commercial system, it usually takes several hours to model a mechanical part of moderate complexity [6]. This amount of turn-around time makes it virtually impossible to "sketch" with a computer [5].

We believe that this difficulty in quick prototyping is largely due to poor user interface design. In particular, the traditional mouse-based interaction style is a bottleneck to 3D geometric design, as it forces the designer to decompose a 3D task into a series of 1D or 2D tasks.

Based on this observation, we have designed a new highly interactive user interface for 3D geometric design. In our system, the designer is equipped with high degree-of-freedom input devices. They allow the designer to directly manipulate objects in 3D space, without having to map the task into 1D or 2D space. The naturalness and intuitiveness of such interactions can greatly enhance the efficiency of geometric design, thus making it possible to sketch with a computer. Using our system, we show that the prototyping time can be reduced to about one tenth of that required by commercial systems.

In the following section, we discuss the problems with mouse-based interaction techniques and examine some experiments on using high degree-of-freedom input devices for modeling tasks. In section 3, we discuss the overall setting of our system, and several basic design issues. In section 4, we discuss the use of the bat in modeling operations. Some preliminary results are presented in section 5.

2 PREVIOUS WORK

Mouse-based 3D manipulations have two major limitations. First, simultaneous translations along all three directions are not possible, due to the 2D nature of the mouse. This is the fundamental obstacle barring mice from being useful in 3D quick prototyping. When a designer wants to create a box-shaped object, he/she wishes to do it in a single stroke, as in the case of creating a rectangle with a 2D drawing tool. If a system forces the designer to break such a task into sub-tasks (such as creating a rectangle, and then adding the third dimension), the designer will have to think in terms of subspaces, rather than focusing on the 3D nature of the object. In other words, the level of abstraction will be lowered, and the result is a low interaction efficiency.

Second, 3D rotations are not effectively supported. Most techniques limit the rotations to around one of the major axes at a given time. To effect a general 3D rotation, three consecutive rotations have to be performed. The fact that rotations are not commutative makes the situation worse. As a result, it is difficult to orient a 3D object in space.

To overcome these problems, several experiments have been carried out in using high degree-of-freedom interaction techniques to support 3D modeling. Of them, three are of great importance.

In mid 1970's, James H. Clark built an experimental system for designing surfaces in 3D [2]. It used a head-mounted display to provide a stereoscopic view to the designer, and let the designer use a mechanical 3D "wand" to drag control points in 3D. The system offered the designer a strong kinaesthetic correspondence which gave it a great advantage over the more common 2D techniques of viewing and designing curves and surfaces.

Emanuel Sachs et al at MIT built the 3-Draw system for interactive 3D shape design [9]. The system was based on a pair of Polhemus Isotrak six degree-of-freedom tracking devices. The designer held one sensor (in the form of a palette) in his left hand to specify a moving reference frame, and used another sensor (in the form of a stylus) to draw and edit 3D curves in space. A graphics display was used to visualize the scene from a virtual camera position.

Butterworth et al built the 3DM system at University of North Carolina [1]. The system uses a head-mounted display to place the designer in the modeling space. The input device is a Polhemus 3Space Isotrak mounted in a hollowed-out billiard ball with two buttons on it. The designer used the hand-held tracker to select functions from a toolbox, and to create/manipulate objects in 3D space.

Both 3-Draw and 3DM offered easy and intuitive ways of modeling 3D shapes, and demonstrated that 3D direct manipulation interfaces are superior to mouse-based ones. However, the lack of ergonomic considerations and support for precision in modeling made them unusable for engineering purposes.

3 SYSTEM OVERVIEW

Our system is based on an IRIS 4D/310 VGX graphics workstation and a pair of Polhemus Isotrak six degree-of-freedom tracking devices. We use one tracking device to dynamically monitor the designer's head position, and provide the kinetic 3D display effect [10] [8]. We use another six degrees of freedom tracker as a hand-held bat [12]. It is the main input device for our system.

3.1 3D Display and Input

To enable 3D direct manipulation, a system must first provide enough spatial information to the designer. Since we don't have true 3D display techniques yet, we have to rely on 2D displays to visualize the 3D scene.

In addition to common depth cues such as perspective projection, hidden surface removal, and shading, we also supply kinetic depth effect by tracking the designer's head position. When sitting in front of the graphics workstation, the designer wears the sensor of the tracking device on his head. Through a series of transformations, the approximate eye position with respect to the display window is obtained and used as the

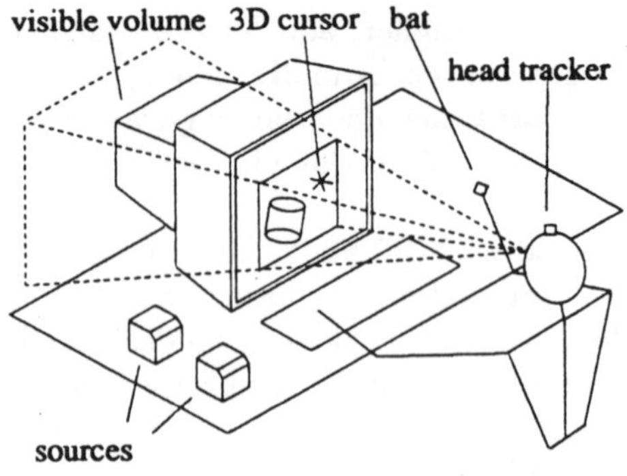

Figure 1: A typical setting of our system.

center of the perspective projection. Since the display is updated at 15 to 60 frames per second, the designer has a strong spatial perception as if he/she is looking into a real 3D space. Similar experiments with kinetic depth effect were also made by other researchers [10] [3].

This setting provides a platform for 3D direct manipulation.

The main input device for our system is a bat — a hand held Isotrak sensor. The position and orientation of the bat is monitored in real time, and is echoed on the display as a 3D cursor (in the form of a jack). The designer uses the bat to interact with objects and to navigate through the design space.

Many design issues have to be carefully treated before one can make good use of such an environment.

3.2 Coordinate Transformation

In order to achieve correct kinaesthetic correspondence between hand movement and the 3D cursor motion, the tracking data from the bat has to go through a series of transformations.

First, the measurement from the Isotrak is usually noisy and bears a certain lag [7]. So we need to filter it first.

Second, since the tracking data are given in the device coordinate system D, they have to be transformed into a useful reference frame T, in our case, that of the graphics terminal.

Third, since the Isotrak is an absolute locator, a clutch mechanism is necessary for re-mapping its physical location into a logical location. In addition we found two kinds of relocation operations to be very useful: (a) a "homing" operation which brings the 3D cursor to the center of the visible volume, (b) a "masking" operation that disables the passing of either positional or orientational change from the bat to the user interface. With these mechanisms, the bat can be operated with the same ease as a relative locator (such as a mouse).

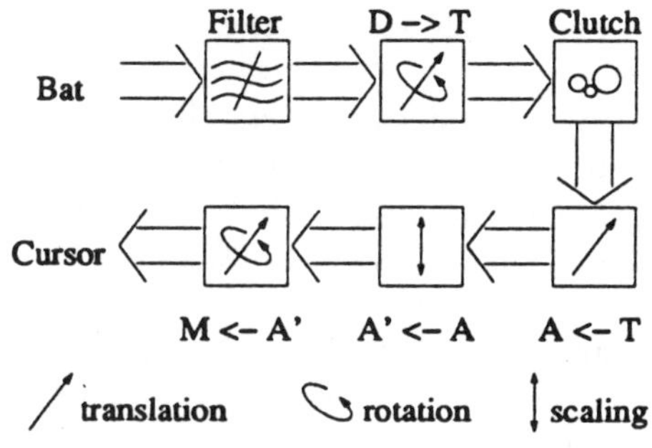

Figure 2: Transformations from bat to 3D cursor.

Fourth, certain 3D manipulations require a reference frame other than the terminal or the modeling coordinate system. For example, to inspect an object and its relationship with the neighboring objects, one may grab it with the bat and simultaneously rotate/translate it. Such operations require a reference frame whose origin is near the center of the visible volume. Zooming operations also have a similar requirement. For this purposes, we define an *apparent coordinate system* A, which is centered at the visible volume. The translation from T to A is dynamically determined by the viewpoint (and also by the window size).

Fifth, any zoom factor should be applied to the positional part of the bat data before it is transformed into the modeling coordinate system M. This ensures a constant ratio between hand movement and the apparent cursor motion. So, at a zoom factor of 10, the 3D cursor still seems to move 1 cm to the designer when the hand moves 1 cm, although the distance it travels in the modeling coordinate system M is actually 0.1 cm.

3.3 Navigation in Space

Since the designer has a space of virtually unlimited volume to work with, but at any given moment can only see a part of it (i.e. the visible frustum), some forms of navigation are necessary.

We found the following four types of navigation useful.

First, a "grabbing" operation lets the designer move the entire model with the bat. This mechanism, when used in conjunction with zooming, provides a natural way of bringing any part of a model into the visible volume.

Second, a "focus" operation brings a selected object to the center of the apparent coordinate system, i.e. the visual sweet spot, so subsequent manipulations on the object become convenient. In order to maintain visual continuity, we use an animation sequence to provide a smooth transition. The system first determines the minimum translation and/or rotation needed for the focus operation, then divides them into 10 parts and carries them out in about half second (the update rates of the system is about 20 frames per second).

Third, a "fly-with-bat" operation lets the designer move around in space, and inspect different parts of the model. This is useful for tasks such as architectural walk through. Since the apparent flying speed is zoom-invariant, one can effectively change the pace by adjusting the zoom factor.

Fourth, the designer can mark a particular view setting by putting down a "spacemark". Later, he/she can come back to that setting by traversing the chain of spacemarks. Also, the designer can always bring the model to its canonical positional by a "homing" operation.

These navigation mechanisms are very useful when the model is large and complex.

4 MODELING OPERATIONS

The most important advantage of a bat over a mouse is the ability to create and make changes to objects by 3D direct manipulations. The fact that the designer is in the same space as the geometric model relieves the designer from having to break down 3D tasks into 2D or 1D operations, and thus makes the interface "transparent". We will discuss the design of basic modeling operations in the following sections.

4.1 Creation by Rubber-Banding

A primitive solid can be created in a single stroke using 3D rubber-banding. The designer first chooses a shape with the bat using a "daisy" menu [8], then presses a key to set the current position of the 3D cursor as one corner of the bounding box, the other corner of the bounding box follows the 3D cursor as it moves with the bat. When the solid is stretched to the right shape, the designer releases the key to finalize the solid.

For shapes with three axial measurements (such as a box), the mapping from the bounding box to object geometry is trivial. However, certain shapes have non-axial measurements (such as a circular shape) or more than three measurements (such as an extruded trapezoid). In those situations, proper mappings have to be devised.

Initially, we tried to employ the data of all six degrees of freedom simultaneously. It proved to be the wrong choice. We observed three facts. First, controlling all six degrees of freedom simultaneously tends to overload the hand. Second, orientation appears intrinsically more difficult to control than position. Third, mappings from an orientation change to a position change are usually difficult to understand and use.

Based on these observations, we adopted the following two rules.

First, if the shape in question has three measurements, but some of them are not axial in nature, we associate each one with an axial change.

For example, a truncated cone has three measurements: the height, the diameter of the top, and the diameter of the base. We map the width of the bounding box to the top diameter, and the length of the bounding box with the base diameter. (The height of the bounding box obviously maps to the height of the truncated cone.) In this way, we made good use of the ability of the designer to control 3D positions.

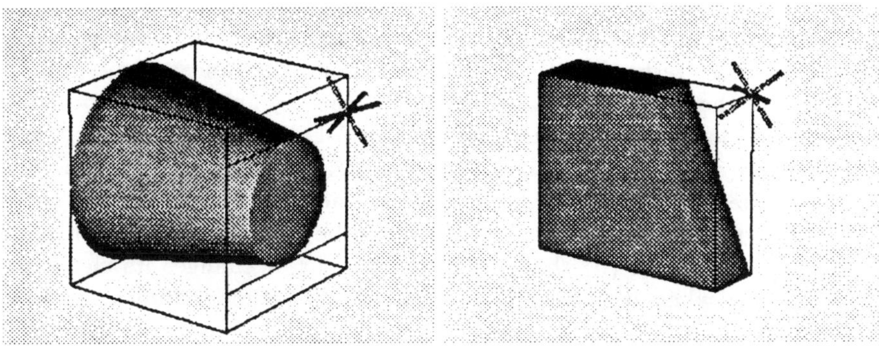

(a) A truncated cone. (b) An extruded trapezoid.

Figure 3: Rubber-banding creation of primitive objects.

Second, if the shape in question has more than three measurements, we only directly control three of them, and compute the rest with a predefined proportion. In this way, the designer can quickly get the object in roughly the desired shape and then come back to fine tune it. For example, an extruded trapezoid has four measurements: the width, the height, the bottom length, and the top length. We map the length of the bounding box to the bottom length, and keep the top length at a fixed proportion to it. We use the golden proportion as a default for these cases. The linear mapping makes it equally easy to set either the top length or the bottom length to a desired amount.

4.2 Region-Based Reshaping

Modification to a model is at least as important as its creation, and usually occurs much more frequently than creation. We designed a set of reshaping operations that take advantage of the bat.

In an ordinary 2D drawing system, the reshaping operation is often based on a set of "handles" on the object. The user drags a handle with the mouse to enact a certain type of reshaping. Although this method can be directly extended into 3D, we found it less effective than in the

2D case. The measurement noise and lack of sufficient depth cues often makes it difficult to place the 3D cursor at an exact handle point.

We designed a region-based method to facilitate the reshaping operation. The idea is to extend the handles into 3D regions, so the precision requirement can be lowered. The space around the object is divided into different regions, each of which corresponds to a different type of reshaping operation. After the designer selects an object, he/she moves the 3D cursor into one of the regions before commencing the reshaping action, to specify the type of the reshaping action. As a visual feedback, the 3D cursor takes different shapes in different regions. When the designer presses a key to start reshaping, the system attaches certain measurements of the object to the the 3D cursor. Releasing of the key finalizes the operation.

The following figures illustrate the region-based reshaping of a cylinder. Figure 4 (a) shows that the 3D cursor takes the shape of a two-ended arrow when it is near the center of the cylinder end, meaning that should the designer start to reshape, he/she will be changing the length of the cylinder. Figure 4 (b) shows that the 3D cursor changes its shape to a bow-arrow when it is near the middle of the cylinder, meaning that a reshape operation will change the radius of the cylinder. Figure 4 (c) shows that the 3D cursor has both shapes on it when it is near the "rim" of the cylinder, meaning that both the height and radius of the cylinder may be changed simultaneously. Figure 4 (d) shows a reshaping operation changing both the height and radius of the cylinder. At the beginning of a reshaping operation, the 3D cursor is first snapped to the corresponding point on the object, so the object will not undergo a sudden change in shape.

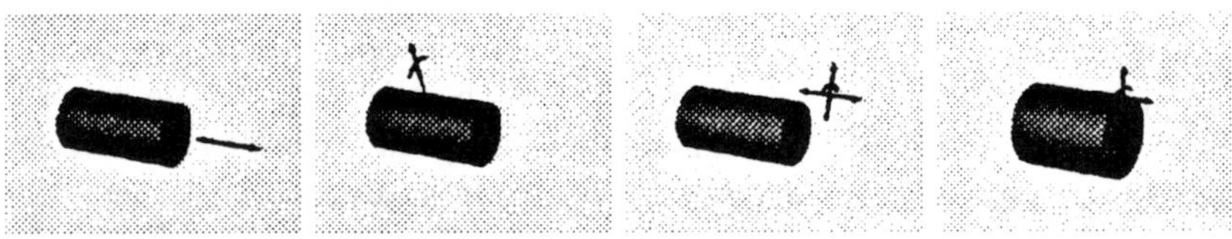

(a) Ready for height change

(b) Ready for radius change

(c) Ready for height and radius change

(d) During reshaping

Figure 4: Region-based reshaping.

Composite objects can be reshaped via their bounding boxes, in one, two, or three dimensions. If the starting position of the cursor is near the center of a face, then the object will be stretched only along the dimension perpendicular to that face. If the cursor started near the center of an edge, then only the two dimensions perpendicular to that edge will change with the cursor. Similarly, the designer can pick a corner and change all three dimensions at the same time.

4.3 Placement, Alignment, and Others Operations

Translations and rotations of an object with the bat are straightforward. The designer first selects the object, then presses the key for translation, rotation, or both, and the object moves with the 3D cursor as the bat moves. Releasing the key finalizes the position and orientation of the object.

To defeat the measurement noise of the bat and the unsteadiness of the hand, an optional snap grid can be turned on for position and/or orientation.

The bat is also used to specify an axis for certain operations. Since the shape of the bat makes its X-axis very distinct from the other two, we use it as the direction of the bat. The designer points the bat at roughly the direction of the desired axis, and presses a key to instruct the system to find the nearest axis to that direction. The system echoes the axis it chose as a thin rod, and the designer can start a constrained translation, rotation, reshaping, or alignment between selected objects, with respect to that axis. This natural way of specifying an axis eliminates the need for the designer to know the names of the axes.

Once an axis is specified, the designer may also perform 'conformation operations' to change the size of an object according to some reference objects. Such operations relieve the designer of repeatedly entering the common dimensions in a design.

For visual continuity, all the changes in position, orientation, and size are visualized with animation. Each command results in a 10-frame animation sequence, and is carried out in about half a second. This eliminates the "jumping" of objects, which often require the user to mentally re-assimilate the scene.

The designer may also select several objects and group them together or perform CSG operations. A hierachical model can be built in this way.

5 CONCLUSIONS

In this paper, we proposed a new user interface for interactive 3D geometric modeling. The aim is to remove the bottleneck between the designer and the modeling system by using high degree-of-freedom input devices and novel interaction techniques. We combined a kinetic 3D display scheme with hand held bat input to provide the designer with a 3D direct manipulation interface. We discussed coordinate transformations for achieving the correct kinaesthetic correspondence, which is vital in such an environment.

Although only at a preliminary stage, the system has shown to be very efficient in 3D quick prototyping. Compared with commercially available systems, it usually takes only one-fifth to one-tenth of the time to build a prototype. For example, the simple mechanical assembly in figure 5 can be modeled in about half an hour.

Figure 5: A simple assembly modeled in about half an hour.

Our next steps are to incorporate geometric constraints in this modeling environment, to further increase the efficiency in modifying 3D models, and to experiment with free-form surface design using the bat.

ACKNOWLEDGEMENTS

The first author wishes to thank Chris Shaw and Dani Beaubien for many discussions and comments.

REFERENCES

1. Butterworth, J., A. Davidson, S. Hench, and T. M. Olano, '3DM: A Three Dimensional Modeler Using a Head-Mounted Display,' *Proc. 1992 Symposium on Interactive 3D Graphics,* Cambridge, Massachusetts, March 29–April 1 1992, pp. 135–138.

2. Clark, James H., 'Designing Surfaces in 3-D,' *Communications of the ACM,* Vol.19, pp. 454–460, 1976.

3. Deering, Michael, 'High Resolution Virtual Reality,' *Proc. ACM SIGGRAPH '92,* Chicago, Illinois, July 26–31, 1992, pp. 195–202.

4. Encarnação, J., 'CAD Technology: A Survey on CAD Systems and Applications,' *Computers in Industry,* Vol.8, 145–150, 1987.

5. Hatvany, Józef, 'Can Computers Compete with Used Envelopes?' *Proc. IFAC 9th Triennial World Congress,* Budapest, Hungary, 1984, pp. 3373–3375.

6. Hodskinson, M., and R. Patel, '3-D Modeling in a Teaching Company Scheme,' *Computer-Aided Engineering Journal,* Vol.9, pp. 177–180, 1989.

7. Liang, J., C. Shaw, and M. Green, 'On Temporal-Spatial Realism in the Virtual Reality Environment,' *Proc. ACM Symp. on User Interface Software and Technology,* Hilton Head, South Carolina, November 11–13, 1991, pp. 19–25.

8. Liang, J., and M. Green, 'Quick Prototyping for Engineering Design,' to appear in Proc. ASME Symposium on Computers in Engineering, Houston, Texas, January 31 - February 4, 1993.

9. Sachs, Emanuel, Andrew Robers, and David Stoops, '3-Draw: A Tool for Designing 3D Shapes,' *IEEE Computer Graphics and Applications,* Vol.11, pp. 18–24, 1991.

10. Schmandt, Christopher, 'Spatial Input/Output Correspondence in a Stereoscopic Computer Graphics Work Station,' *Computer Graphics,* Vol.17, pp. 253–261, 1983.

11. Sproull, R. F., 'Parts of the Frontier Are Hard to Move,' *Computer Graphics,* Vol.24, p.9, 1990.

12. Ware Colin, and Danny R. Jessome, 'Using the Bat: A Six Dimensional Mouse for Object Placement,' *Proc. Graphics Interface '88,* Edmonton, Alberta, June 6–10, 1988, pp. 119–124.

Hardware platforms for computer graphics visualization

F. Gareeboo

Datapath Ltd., Alfreton Rd., Derby DE2 4AD, UK

ABSTRACT

The purpose of this paper is to present the hardware platforms currently used at various levels for computer graphics visualization. The main parameters that characterize the different hardware platforms are described and methods explored to help in the selection of the most appropriate platform for particular visualisation tasks. The paper concludes with a methodology for the selection of appropriate hardware platforms for visualization.

INTRODUCTION

Visualization embraces both image understanding and image synthesis. That is visualisation is a tool both for interpreting image data fed into a computer, and for generating images from complex multi-dimensional data sets. It studies those mechanisms in humans and computers which allow them in concert to perceive, use and communicate visual information. Visualization unifies the largely independent but convergent fields of computer graphics, image processing, computer vision, computer-aided design, signal processing and user interface studies.

McCormick et al [3]

History.

Computer visualization probably started in 1950s, with the use of point-display screen on the MIT Whirlwind computer. The seminal work in the field was Ivan Sutherland's 1963 Ph.D dissertation 'Sketchpad'.

In the mid 1960s, CRT displays were expensive ($50,000 - $100,000) but in 1968 Tektronix launched an inexpensive graphics terminal which encouraged expansion of the market.

In the 1980s, the emergence of raster displays (as opposed to the previous generation of displays which was vector based) led to even cheaper solutions and further expansion of the market.

Since the publication of the NSF Panel on Graphics, Image Processing and Workstations report "Visualization in Scientific Computing", there has been a very high level of interest in both the computer industry and the user community in Visualization.

Some current uses of computer visualization are :

Discipline	Use
Planetary Science	by NASA JPL to accurately define optimal flight paths and camera positions of spacecraft.
Pharmacology	Drug Design
Mathematics	Fractals, Iterated Function Systems
Medicine	Volume visualization in diagnostic radiology
Science in general	To help define, compute and analyze simulations

Table 1. Uses of Computer Visualization

VISUALIZATION PROCESSES

An understanding of the processes used to achieve computer visualization is vital for developing an appreciation of hardware requirements.

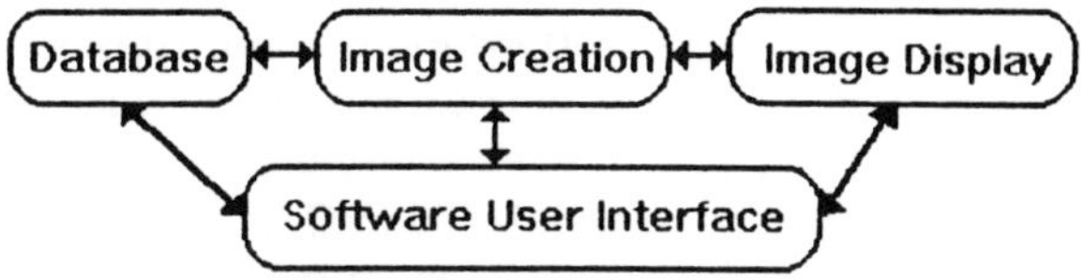

Figure 1. Main components of a visualisation system.

In this paper, we focus on the image creation and image display parts of the process.

Simple Raster Display system

In the most common type of display system (Figure 2), the image is created from a 'Raster', consisting of a matrix of pixels representing the entire screen. The matrix is scanned by the video controller sequentially from top to bottom (displaying the image concurrently on the monitor) continuously. (The number of times per second that the whole raster is scanned by the video controller is called the refresh rate).

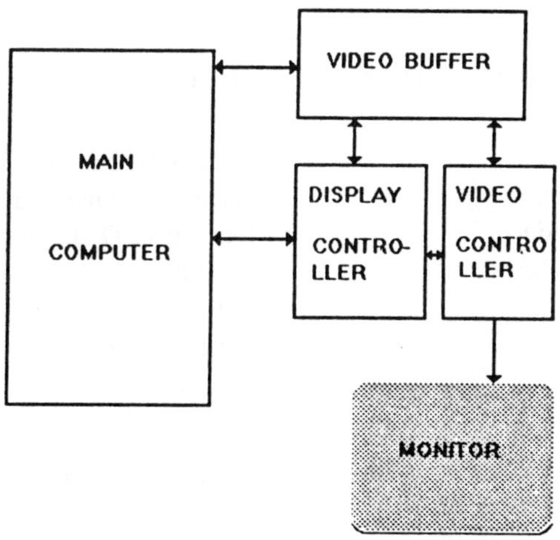

Figure 2. A Raster display system

Table 2 below lists the main characteristics of raster display systems.

Parameter	Description
Resolution	Greater resolution of the display allows you to see more detail. However, it also requires more expensive hardware and monitors. Typical resolutions of PCs for example is 640x480 (VGA), but 1280 x 1024 is more commonly used for visualization.
Refresh rate	For a static picture, the human eye requires a refresh rate of 70 frames per second to avoid eye strain due to screen flicker.
Colours	Number of colours displayed varies between 16 to 16.7 million. This is an important parameter for volume visualisations, because of the use of colour and intensity to represent data variables.

Table 2. Main characteristics of raster display systems.

There are a number of different types of visualization; as listed in Table 3 below.

Visualization type	Description
2D lines	Uses only 2 dimensional lines
3D lines	Uses 3 dimensional lines
3D shaded	Displays 3D shaded objects/scenes
3D rendered	Displays 'realistic' 3D objects/scenes uses textures, transparencies, etc
Volume	Displays contents of 3D objects; not just surface.

Table 3. Types of Visualization

Different processes are used to create the image in the video buffer, depending on the type of visualization being used. For example, for 3D shaded and 3D rendered visualization, Table 4 below illustrates the stages in the image creation process. (The cost of each stage is characterized in terms of operations per vertex. 3D shading and rendering is typically applied to a collection of 3D polygons. See Foley,Van Dam,Feiner & Hughes [1] for more details.)

Image Creation Stage	**Operations/ vertex.**
Database Traversal	
Modeling Transformation	25 x; 18 +
Trivial Culling	18 x; 14 +
Lighting	12 x; 5 +
Viewing Transform	8 x; 6 +
Clipping	
Perspective Transform	2 x; 2 +; 3 /
Rasterization	

Table 4. The Image Creation Process

The end result of the image generation process is a 'picture' in the display buffer.

Using a simple PC, it is technically possible to generate images of unlimited complexity and realism given enough time (and memory). However, computer visualization is no longer a practical option if it takes too long to generate the visuals (the length of time that is 'too long' depends on a number of factors eg importance of visualization, seniority of user, etc).

The three main bottlenecks in the image generation process are

- Stages 2-6 require extensive Floating-point processing. This is measured in MFLOPS (Million Floating point Operations Per Second). Most common hardware platforms have low MFlops ratings compared to MIPS rating.

- Stage 8 requires extensive Integer Pixel processing capability. This is a function of the Mips rating of the hardware .

- After Stage 7, we start to write the pixels that compose the picture into the display buffer. The Video-buffer memory bandwidth limits the speed at which the 'picture' can be composed.

The performance of a visualization system can be improved by using:

- **display processors**, which are specialised for displaying graphics,

- **integrated processors**, which are general purpose processors with integrated support for graphics operations,

- **multiple processors**; and splitting the graphics tasks between them using pipelining or parallelism.

A display processor (eg TI34020) is an inexpensive method of increasing the display performance of a system. Increased performance is achieved by purchasing a 'plug-in expansion card'. An integrated processor (eg Intel i860 which supports 3D graphics) is a more expensive option, but delivers a much higher performance. The most expensive option, (and highest performing) is the use of multiple processors (specialised or general purpose) for processing the visuals.

The last two options usually require the purchase of complete high-performance visualization workstations. (Except for the Merlin plug-in card which uses the i860 as a display processor).

HARDWARE PLATFORMS CHARACTERIZED

The previous sections of this paper outlined the types of visualization currently used, and the processes used to implement them. We also looked at the performance bottlenecks in the visualization process, and hardware solutions to those performance problems. In this section, we look at the hardware platforms currently available and their suitability for computer visualization.

Hardware platforms are usually characterized by their manufacturers in terms of peak processing power expressed in MIPS, MFLOPS, etc. Those figures can only hint at the visualization potential of those platforms, since a number of crucial layers come between the raw hardware performance, and visualization power :

> implementation of graphics libraries that use the hardware power to the maximum, and,
>
> availability of suitable software to perform visualization

To have a feel for the performance requirements, we look at a number of typical drawings (I choose some models shipped with AutoCAD as examples). Table 5 below lists some of those models, and the number of lines and number of triangles that they are made up of.

Model	Lines	Triangles
Airplane	2852	2480
Site-3D	7670	5424
Hook	12155	11060

Table 5. Some typical models and sizes

A medium complexity data set has of the order of 10000 lines/triangles. On average, the lines may be assumed to be 10 pixels long, and the triangles to have an area of 100 pixels. To display different views of this data set in real time (10 updates per second) would require performance of the order indicated in Table 6 below.

WireFrame	Smooth Shading
15 MFLOPS	34 MFLOPS
5 MIPS	33 MIPS
2 MFBAS	51 MFBAS

Table 6. Performance required to visualize average model
(MFBAS stands for Million Frame Buffer Accesses per Second)

Table 7 below lists some of the currently available hardware platforms and their performance characteristics. The table lists only the Mips, Mflops ratings.

Platform	Mips	Mflops
IBM PC 486 50MHz	40	3
Merlin (i860)	40	80
Mac (Quadra 950)	25	2.9
Next	25	2.9
Sun-SparcSation 10 Model 41	96.2	17.2
SG (Indigo R4000)	85	16
HP9000/730	76	22

Other platforms : Tektronix, Evans and Sutherland,Convex,Stellar, Cray

Table 7. Characteristics of Current Platforms

As can be seen from Table 7 above, current 'low-cost' platforms can deliver reasonable solutions for 2D and 3D wireframe visualization. However, for 3D shading, rendering and volume visualisation, the cost of adequate hardware platforms starts to rise dramatically.

Apart from raw graphics performance, there are a number of characteristics of hardware platforms that are relevant for visualization. These are listed below :-

Display quality - Resolution and refresh rates.

These characteristics are defined earlier. For adequate visualization, the minimum requirements are 1024 x 768 resolution, and 72Hz refresh rate.

Openness of the system.

This is an important consideration to safeguard any software development investment that you may have to make in the system. The issue of openness can be considered at various levels;

hardware	:	eg use of industry standard bus architecture like ISA,EISA,VL, VME,etc..
operating system	:	use of 'industry standard' operating system; POSIX compliance.
Graphics API	:	use of standard Application Programming interface to display graphics (eg PHIGS3D,PEX,GL)

Connectivity

The ability to connect to other 'identical' stations, or different ones electronically is an important consideration. This may be particularly important if the visualization facilities are to be made available to a large number of users, and if the data is generated from another station and need to be viewed in 'real-time'.

Software solution - Features

The visualization hardware on its own is completely useless without usable software that gives you access to the functionality (unless you are prepared to develop your own visualization software).

Video compatible computers

The ability to record the visualization sequence(s) onto video tape is of relevance to a number of applications.

SELECTION OF HARDWARE PLATFORM

The first step in the selection of a hardware platform for Computer Visualization (and for any other application for that matter) is a precise definition of your requirements. These must include the data, the form(s) of output, functionality, ease of use, support, upgrade paths, and budget.

Visualization Software

Most of the requirements are satisfied by the software component(s) of the system rather that the hardware. See Appendix A for a list of commonly available visualization software, and reference 4 for an evaluation of some of them.

In most scenarios, the need for visualization arises after the user has already made significant investments in hardware, so that the search for appropriate software that runs on existing machine is the most important task. In fact, the origination of some of the current crop of visualization software from Academia means that some of them (eg Khoros) are available free of charge.

When there is an opportunity to pick both the software and a new hardware platform, some of your requirements may have important implications for the hardware platform (See Table 8 on the following page).

ITEM	ASPECT	IMPLICATION
Data	Origin -	
	eg on Tape	Tape drive
	Format	Is it readable by platform ?
	Real-time	Network, HIPPI channel
	Size	Hard-disk size Memory size
Output	Type	Raw performance
	Update speed	Performance
	Hardcopy	Driver for hardcopy device
	Video	Video interface
Budget	Tightness	Prevents us from getting what we really want !

Table 8. Requirements and hardware implications

FUTURES

The nature of the computing industry is such that the future holds out some very exciting prospects. For example, more powerful chip-sets allowing better price-performance ratios are being announced now, and we can look forward to desktop stations of £8,000 with a rating of 200 MIPS and 100MFLOPS by the middle-end of next year. (A 486 50MHz and Merlin currently delivers 80MIPS and 80MFLOPS for £8,000). This will, for example, allow volume visualization (the most processor intensive type) on an affordable desktop platform.

New and exotic displays and input devices are also appearing over the horizon. For example, Deering[7] describes a 3D holographics station integrated with a 3D pointing device in his paper on "High resolution Virtual Reality" . Stereoscopic specs are also available from a number of manufacturers. The physical display side of Visualization has now seen as much change as for example the hardware processors, but this may change soon with better LCD technology.

REFERENCES

1. Foley James D., van Dam Andries, Feiner Steven K., Hughes John K. Computer Graphics Principles and Practice. Addison-Wesley 1990.

2. Defanti Thomas A, Brown Maxine D. Visualization in Scientific Computing. Advances in Computing Vol 33 1991.

3. McCormick, B.H., DeFanti, T.A. and Brown, M.D., eds Visualization in Scientific computing. Computer Graphics 20 (6) (1987) .

4. Advisory Group on Computer Graphics Technical Report Series, Evaluation of Visualization Systems No 9, May 1992.

5. Rae A. Earnshaw (Ed) Fundamental Algorithms for Computer Graphics, Springer-Verlag (Section 6 - Hardware Architecture and Algorithms) 1985.

6. Ian Hirschsohn, Personal Supercomputing, Dr. Dobb's Journal, June 1992.

7. Michael Deering, High Resolution Virtual Reality, SIGGRAPH '92.

APPENDIX A

List of currently available visualization software and graphics libraries.

Alias's 3D software		
AVS	-	AVS Inc.
MOVIE.BYU	-	Brigham Young University.
DISSPLA	-	Computer Associates.
PC-MGL II	-	Datapath Ltd.
Visualization Data Explorer	-	IBM.
PHIGS+		
RenderMan and CubeTool	-	Pixar.
PV-WAVE	-	Precision Visuals.
Iris Explorer	-	Silicon Graphics.
Open GL	-	Silicon Graphics.
DORE	-	Stardent.
MatLab	-	Stardent.
apE	-	TaraVisual Corporation.
FIGARO	-	Template Graphics System.
Khoros	-	University of New Mexico.
	-	Wavefront Techology .
Mathematica	-	Wolfram Associates.

Visual composition of concurrent applications

A.F. da Silva

Universidade Nova de Lisboa, Grupo de Eng. Sistémica, 2825 Monte de Caparica, Portugal

ABSTRACT

This paper describes an object-oriented visual programming system. The system manages a dataflow framework for the composition of concurrent applications expressed as active visual objects. A multidimensional data structure is used to manage interface related operations. This structure supports the management of the dataflow architecture and the scheduling of applications. The dynamic properties of the graphical interface are enhanced by the specification of replacement behaviours associated with the active objects. Together with the dynamic properties of the underlying data structure, active objects simplify the composition of applications and allow for the easy reconfiguration of the user interface.

INTRODUCTION

Visual language systems have been the subject of intensive research in a large number of different fields. The term visual language has been used to refer to all systems that use graphics, whether the objects to be dealt with have or not an inherent visual representation. The classification of visual language systems in visual programming and program visualisation systems was proposed by Myers [1]. In visual programming systems there are images or pictorial objects that are associated to construct a program or logical interpretation. In program visualisation systems the program is specified in a conventional textual manner being the graphics used as a complementary tool. The lack of a formal specification for visual languages, somehow equivalent to the BNF's formulation of textual languages, makes visual programming difficult since each graphical language requires a different environment and specific visual utilities. Moreover, low-level graphics programming contribute to the non-portability of visual programs. Several visual programming environments based on an all-visual approach have been developed for specific application domains. However, for general-purpose programming textual languages are generally incorporated in the visual system to circumvent the shortcomings of current visual programming languages. User interface toolkits define an application area where visual languages have been successfully applied. The widespread adoption of the object-oriented paradigm appear to have contributed decisively to enhance user interface design.

In the field of scientific visualization several systems have been proposed based on the dataflow model (Felger et al. [19]). These systems tend to restrict the composition of applications to the process of configuring a set of modules (filters) linked by pipelines for visualization purposes. This approach has several limitations. The modules do not support semantic interaction, since they have rigid built-in behaviour and interfaces. The transfer of information is based on the flow of raw data due to the absence of abstract data types. Finally, the composition of applications is static since run-time reconfigurations are not supported by the interface. Some recent proposals have been put forward to address these problems; e.g. de Mey et al. [18], Chen et al. [20]. These approaches tend to conceive the application architecture as a composition graph defined in an object-oriented environment.

In this paper, an object-oriented visual programming system following similar guidelines is introduced. Nevertheless, we focus on the problem of dynamic reconfiguration in the framework of a strongly-typed object-oriented language such as C++. Two mechanisms are introduced to support the dynamic properties of the interface: concurrency and replacement behaviour. The interface manages a dataflow environment for the composition of concurrent applications expressed as active visual objects. A quaternary tree data structure is used to support all the required operations for managing the dataflow diagram layout and the scheduling of applications. Together with the dynamic properties of the underlying data structure, the active objects simplify the composition of applications and allow for the easy reconfiguration of the user interface.

MODEL SPECIFICATION

The basic motivation in developing the present environment has been the need to simplify software development and promote dynamic reconfiguration. For this purpose three main guidelines have been followed. First, the object-oriented approach has been used for the construction of a graphical interface for interactive composition of application programs [2,3]. Graphics are used to represent the high-level structure of the global application. Second, by using a coarse-grain dataflow language supporting modification of program graphs a systematic way of building programs is supported. To this aim a visual framework has been developed in which particular programs, written as sequential procedures, are perceived as modules associated with nodes of a flow graph. Third, a concurrent programming toolkit has been introduced for the development of multitasking applications. The toolkit facilities are intended to integrate well with existent C concurrent libraries by using standard C++ compilers and no language extensions. A set of C++ constructs has been devised to effectively enable the integration of the concurrent model in the object-oriented framework. The above considerations led to the development of a programming environment with the following additional features:

Visual programming

One important graphic tool used in many visual systems is the dataflow diagram [4,5]. In the dataflow approach the program is perceived as a directed graph in which nodes are computation primitives and arcs represent data dependencies between nodes. Typically, each node has a visual signature and encapsulates a function which receives input data and generates output data. An application is supposed to be written as an encapsulated program or module which contains data and the associated procedures to operate on the data. Particular applications are viewed as icons. A graphical editor enables the visual composition of applications in order to build up the global flow diagram. After being individually tested and compiled modules can be integrated at run-time in the environment by interacting with the graphical editor. A C++ object-oriented approach in developing the visual user interface was adopted. Some additional tools support overlapped windows and class browsers [6].

Dynamic composition

Interactive node manipulation such as insertion, deletion, and relocation of nodes is enabled. A quaternary tree data structure in which the leaf nodes reference icons is used as the building block supporting dynamic operations. The dynamic redefinition of icon connected functions is largely supported.

Concurrency

Recently, concurrent object-oriented languages have been developed which support concurrency within objects [7,8]. Frameworks for parallel programming have been proposed as well [9,10]. Our objective is to set up an object-oriented framework to enable concurrent programming of applications. In our present model concurrency within objects is supported under restricted conditions: concurrency is limited to a single thread and only specific member-functions may generate active objects. Individual modules implement specific user defined applications. Each module expresses a sequential set of actions and is written as a procedure intended to be run as a task. The term task is used to denote the activation of a procedure for concurrent execution. The term is synonymous with lightweight process and thread of execution. These terms denote a task which runs in co-operation with other tasks in a single address space. Contrary to heavyweight processes used in multitasking operating systems (such as UNIX) lightweight processes have the advantages that they can be scheduled very rapidly and that they can share access to global variables within their common address space. A shared-memory model of computation has been used. Objects become active objects when some member-function may be run as a task. In order to ease integration with existent concurrent programming facilities objects must be declared active objects in order to run as tasks. Moreover, only those member functions associated with a user-defined task may be scheduled for execution. This corresponds to a coarse-grain model of computation for active objects.

ARCHITECTURE

Managing the visual composition of applications involves three main responsibilities: architecture management, behaviour management and task scheduling management. Next, the design options adopted for each one of these problems are considered.

Architecture Management

Dataflow diagrams have been used in many visual programming systems. Some systems allow the designer to specify the user interface graphically while others require textual representations. A problem common to many graphical representations is the generation of a suitable layout. An optimal path is one with the shortest possible length and having the fewest number of corners. Finding the optimal path for a connection between two icons can be very difficult. Another source of complexity deals with the number of input and output links for connecting icons. Trying to manage these problems in the most general way tend to yield unattractive visual designs and impair the system dynamic capabilities. Our design strategy tries to emphasize the dynamic features ruling the user interface. For this purpose, several restrictions were imposed. The layout locations are restricted to be coincident with the nodes of a two-dimensional grid. The icons after being interactively located in the screen by the user are automatically attracted to the nearest grid position. An occupied icon location may not be overridden unless the user explicitly activates the icon substitution command. On the other hand the links between icons are restricted to be horizontal or vertical lines. When dragging inter-icon links, repeated clicks of the mouse button toggles between upper-oriented routing and down-oriented routing in links requiring chained adjoining lines. Rubber-band lines visually inform the user of the routing that will be set up. The number of adjacent nodes in the graph are restricted to be four, one for each side of the rectangular icon box. Therefore, the number of one-way communicating lines per icon is also restricted to be at most four.

As regards the underlying data structure many visual systems rely on simple data structures to store icon information. Hash tables have also been used (SunPict for instance [11]). Multidimensional data structures have, however, been widely used in computer graphics and image processing. Data structures such as the quadtree/octree have emerged as convenient structures for operating on multidimensional data [21]. Two features of these structures are particularly interesting: their hierarchic structure based on the principle of recursive decomposition and their efficient dynamic capabilities. As exemplified in Figure 1, a modified pseudo point quadtree structure [12,13] was implemented to store the 2D icon handle points. These points are represented as leaves in the quadtree but they reference quadrant subdivision as well. Figure 1a illustrates the assignment of points to quadrants, including those lying on a quadrant border. Points on a horizontal border are assigned to the upper quadrant. Points on a vertical border are assigned to the right quadrant. The implementation of the dynamic operations in this data structure closely follows the techniques described in [12]. The leaves of the quadtree structure do not store the icon objects themselves but hold pointers to arbitrary data types which can be customized by the application. The quadtree structure was defined as a generic class [17]. The member functions implement the required dynamic operations of node insertion, node deletion, node searching, etc.. The generalization of the class to objects of any type is implemented through macros [14]. An external user-defined function may be passed to the quadtree structure. This feature enables the dynamic (run-time) redefinition of behaviour associated with the quadtree nodes. In order to represent the dataflow diagram we have to define the icon objects themselves. The resulting data structure is the framework for all subsequent icon related operations. Most of these operations implement tree walking algorithm. The tree data structure may also be used by the task manager to schedule icon tasks.

<u>Icons as Agents</u> An icon object can be perceived as an agent able to provide some specific services. The agents exist in a network controlled by a manager. The network provides communication channels between agents. Figure 2 represents the overall structure linking the manager´s structure to the network of co-operating agents. Icons are modelled as active objects. In [17] some of the responsibilities associated with the icon class were defined.

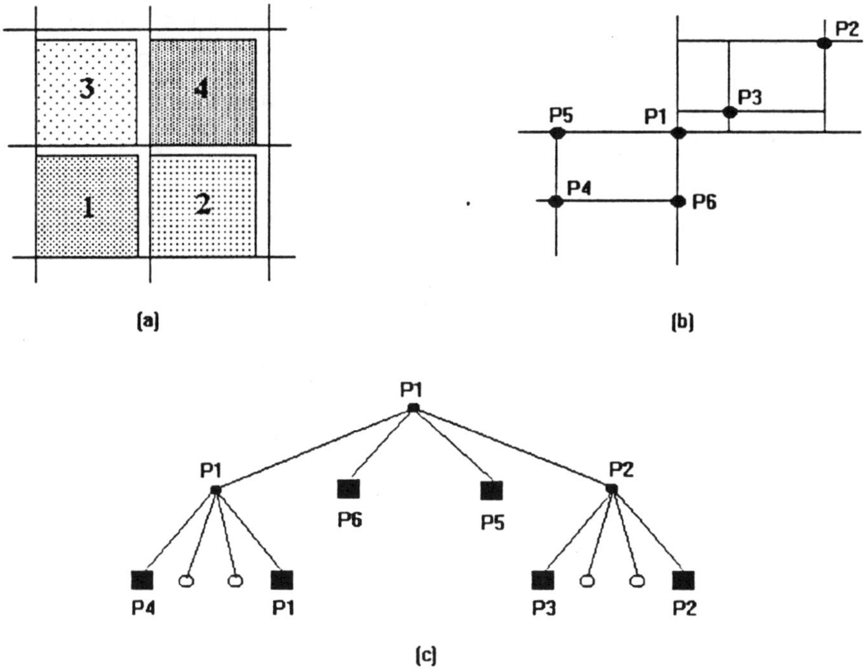

Figure 1. (a) Quadrant assignment. (b) Points to insert. (c) Data structure.

Each icon object has the possibility of managing an attached window object. The window object is an interactive tool at user disposal and a peephole into the flow diagram data and activities. It permits the visualization of icon-specific operations as well as the querying of icon related information. In particular, the user may inquire about active ports, verify the definitions of objects to be communicated, and browse through the icon function. On the other hand, derived window classes may easily be incorporated for proper customisation of user requirements and preferences in the context of the icon activities. The icon-window is valuable for monitoring dynamic phenomena as well. A set of member-functions provide support for attached window objects, namely creation, deletion, and suspension. The core of all icon activities is concentrated in the external function which is passed to the icon object. This function can be run as a task in accordance with the schedule set up by the task manager. Since the icon function can be substituted at run-time the dynamic capabilities of the environment are enhanced. Moreover, this function can be substituted by the global icon manager whenever necessary.

For instance, to perform some action common to all icons the icon manager may assign a common function to all icons and then walk down the tree requesting each icon services. A key design decision was to separate specification of the communication and synchronisation from specification of the units of computation. In this manner, the external function may be oblivious of the global setting and may therefore be tested independently. However, for intertask communication purposes each external function manipulates input and output objects which must be taken into account in a proper flow diagram setting. In order to separate communication from computation the icon as the possibility of enqueueing and dequeueing objects which are not directly interpreted by the icon but passed back to the function manipulating them.

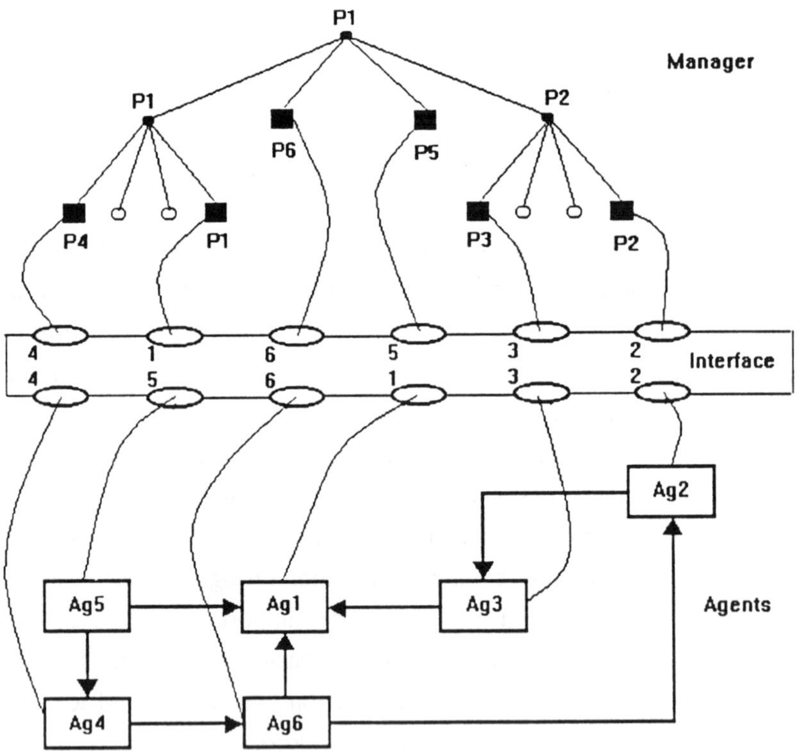

Figure 2. Structure linking the manager to the agents´ network

Behaviour Management

In order to build a dataflow application a default table of icons is presented to the user. A diagram may be constructed by moving instances of those icons to specific screen locations and dragging lines between them. Initially each icon has a pre-defined function assigned to it. As we have seen, dynamic function management is a fundamental issue in our design approach. In keeping with the visual approach, the redefinition of the icon functions (or simply their substitution) must be supported by the interface. Redefining an icon function involves two aspects: the redefinition of the textual code associated with the icon function, and the redefinition of its visual (iconic) content, presentation and textual-to-visual correspondence. For dealing with these matters a class has been defined. The class manages operations on an array of pointers to functions. Operations such as add, remove and replace care for the maintenance of a set of basic functions. An external user defined function, for instance, may be added to the basic function set and thus be made available to the designer. The class deals with the textual to visual mapping as well. Each icon function as a unique visual signature which evokes its code contents. The function's visual signature must then be incorporated in the icon table to be presented to the user.

To deal with the manipulation of information as described above, the tools provided by the main interface for text and icon editing may be used. In most cases, however, the library of functions and icons required for composing the application is specified beforehand. Since there is separation at the interface level between communication and synchronisation on one hand and computation on the other, the functions may be externally coded and compiled. For icon functions having common scope, i.e. those functions which may be meaningfully applied to any other icon object no matter their location in the diagram, at least one icon location is reserved in the visual layout. This icon location (or locations) holds signatures of frequently replaced visiting functions having common scope. This location is used by the icon manager to perform common dynamic operations and updates. Actually, the objects of the class responsible for managing the set of functions are used by the icon manager to dynamize the interface.

Task Management

The concurrent capabilities of the interface relies on a concurrent programming toolkit. The toolkit is a C program library allowing the user to build programs as a set of co-operating lightweight processes, which are threads of execution that share a single address space within a single application program [15]. A variety of facilities, commonly available in equivalent systems referenced in the literature [16], have been implemented. In particular, there are facilities for creating and destroying tasks, for constructing intertask communication, for scheduling tasks and selecting the scheduling policy, and for handling exceptional conditions. A running task may be suspended or killed. Suspension occurs in three situations: when the task blocks awaiting the arrival a resource from another task, when the task is preempted by another task or interrupt handler, or when the task sleeps for the desired time interval. A suspended task may resume execution after the appropriate condition is satisfied. When a task is first created it enters the ready state. When a task is scheduled for execution, according to its priority and scheduling policy, it enters the running state. If a running task is preempted it returns to the ready state awaiting future scheduling. A sleeping task returns to the ready state when the specified time interval expires or when it is awakened by another task. Two task scheduling methods are supported: event-driven scheduling and real-time scheduling. In the former the system time automatically advances when all tasks are either blocked or sleeping, and the task scheduler exits when no task exists or can be awakened by advancing the time. This method is useful in computer simulations. In the latter, one can advance the system's clock using an interrupt source in order to synchronise it with the passage of time. This is useful in real-time applications.

Tasks communicate through shared memory. Several facilities for structuring intertask communications are provided. They allow tasks to cooperate by synchronising their actions and passing data to each other. The main facility used by the interface for intertask communication is the blocking queue (Figure 3). Each task automatically receives its own built-in blocking queue implemented as a doubly linked list. Objects can be passed between tasks by placing them into queues. A task can enqueue objects into another task's blocking queue using the task's identifier. Objects are passed by reference. That is, the queue maintains references to the enqueued items. The blocking queue has the capability of synchronising the activities of multiple tasks. If a task attempts to dequeue an object from an empty blocking queue, the task suspends execution awaiting the arrival of an object. When another task places an object into an empty queue it unblocks the waiting task, which then dequeues the object after resuming execution. On the other hand, a single task may serve a blocking queue to which requests are submitted by other tasks. A time-out interval may be defined for tasks kept waiting in the blocking queue. To deal with active objects a set of C++ classes has been built on the above toolkit [17]. All objects transmitted between tasks are instances of the same Item class. They serve as standard carriers which reference the actual data objects containing the information to be communicated.

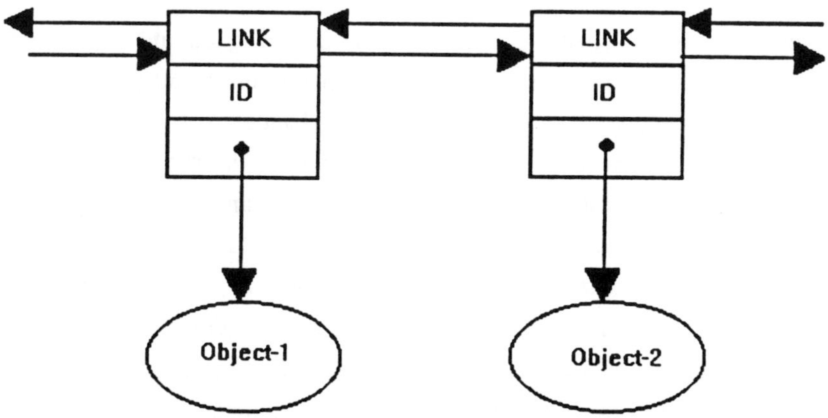

Figure 3. Queue link structure.

BUILDING APPLICATIONS

A dataflow diagram is built by dragging icons from the icon-table into discrete screen locations and linking the icons afterwards. Figure 4 shows two views of the graphical user interface. The Composition menu item references operations for layout management, namely insertion and deletion of icons and links, icon related queries, selective updating, and modification of drawing options. These services are controlled by the manager. Agents request the manager´s services when they need to perform some dynamic operation. For instance, an agent may request the manager to set up a new agent in the network and be connected to him. The manager´s menu item enables the suspension of the network´s operation in order to access the browse tools and windowing system described elsewhere [6]. Once the network is established the menu item `Run_Task_System´ may be selected. In this mode, the agents´ services are scheduled according to their priority, and the flux of information follows the network´s connections. Icon windows may be selectively open to visualize dynamic phenomena ocurring as a result of the agents´ activities. In debug mode the execution of the agents´ processes is traced. In particular, information is generated to trace the scheduling of tasks.

CONCLUSION

An interactive object-oriented framework has been designed and implemented for the integration of programs specified as separate units and able to run as lightweight processes. By normalising the channels of communication between units and their synchronisation mechanisms, the separation between computation and communication has been greatly enhanced. This separation, in conjunction with the definition of abstract data types for function manipulation, contributed to the dynamization of the interface. For composing coarse-grain applications the system provides a visually clean interface supporting behavioural re-specification. Several improvements are currently under consideration. The icon´s visual content may be enriched. The hierarchic nature of the data structure seems adequate to activate multi-layered diagrams. Real-time scheduling may be applied to discrete-event simulation and animation. These issues constitute directions for future research.

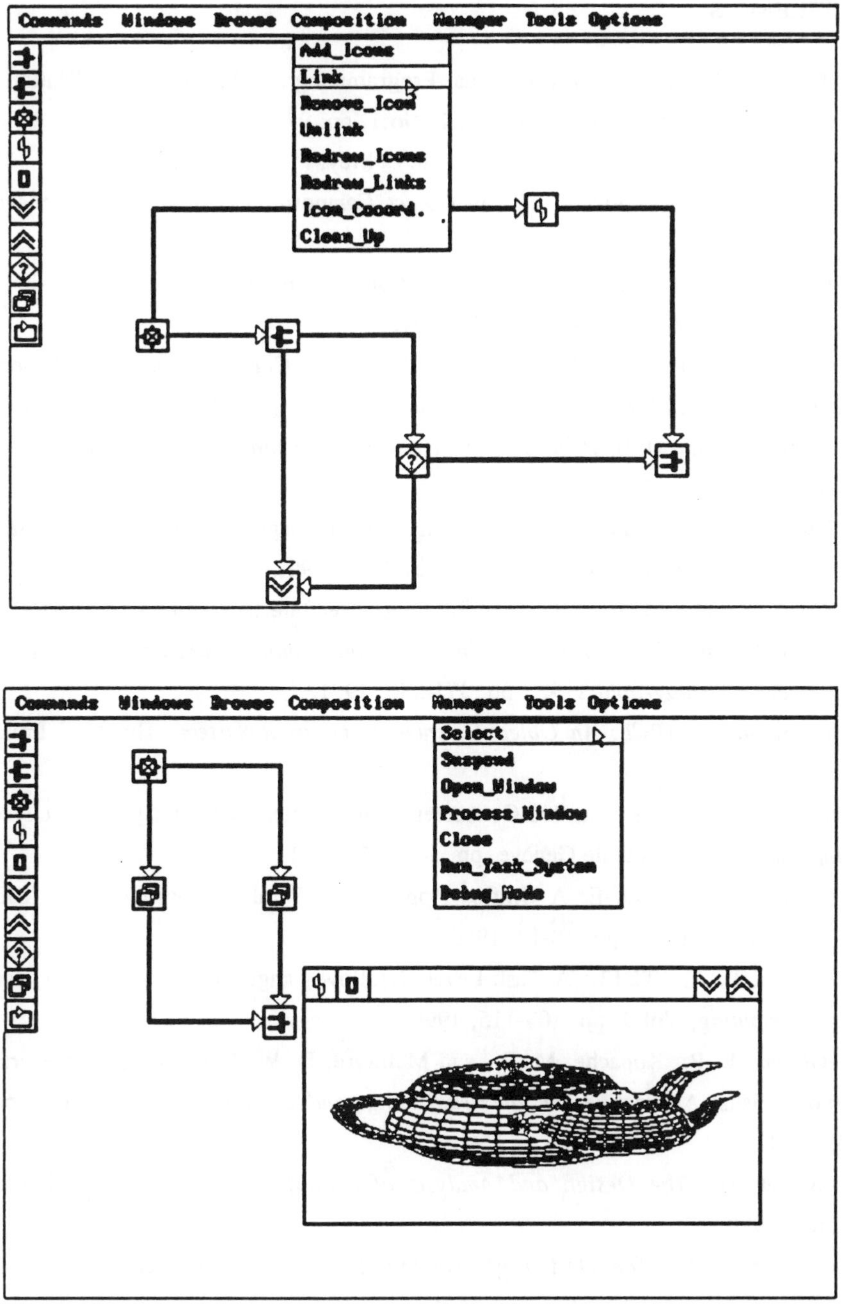

Figure 4. Aspects of the user interface.

REFERENCES

1. Myers, B. A. `Taxonomies of Visual Programming and Program Visualization´, *J. of Visual Lang. and Computing*, Vol.1, No.1, pp. 97-113, 1990.

2. Nierstrasz, O., Tsichritzis, de Mey, Vicki, Stadelmann, M. `Objects+Scripts=Applications´, in *Object Composition*, Université de Genève, pp. 11-29, 1991.

3. Harbert, A. et al. `A Graphical Specification System for User-Interface Design´, *IEEE Software*, Vol.7, No.4, pp. 12-20, 1990.

3. Wirfs-Brock, R. and Johnson, R. E. `Surveying Current Research in Object-Oriented Design´, *Comm. of the ACM*, Vol.33, No. 9, pp. 104-124, Sept 1990.

4. Chang, Shi-Kuo (Ed). *Principles of Visual Programming Systems*, Prentice-Hall, 1990.

5. Tanimoto, S. L. `VIVA: A Visual Language for Image Processing´, *J. of Visual Lang. and Computing*, Vol.1, No.2, pp. 127-139, 1990,.

6. Silva, A. and Ramos, H. D. `Building User-Interface Support for Object-Oriented Design´, *Proc. of the First Int. Conf. on Comp. Graph. and Visualization Techniques, Compugraphics´91*, pp. 371-380, 1991.

7. Yonezawa A. *ABCL: An Object-Oriented Concurrent System*, The MIT Press, 1990.

8. Nierstrasz, O. `The Next 700 Concurrent Object-Oriented Languages´, in *Object Composition*, Université de Genève, pp. 165-187, 1991.

9. Browne, J. C. `CODE: A Unified Approach to Parallel Programming´, *IEEE Software*, Vol.7, No.4, pp. 10-19, 1989.

10. Suhler, P. A. `TDFL: A Task-Level Dataflow Language´, *J. of Parallel and Dist. Computing*, Vol.9, pp. 103-115, 1990.

11. Glinert, E. P., Kopache, M. E., and McIntyre, D. W. `Exploring the General-Purpose Visual Alternative´, *J. of Visual Lang. and Computing*, Vol.1, No.1, pp. 3-39, 1990.

12. Samet, H.. *The Design and Analysis of Spatial Data Structures*, Addison-Wesley, 1990.

13. Overmars, M. *The Design of Dynamic Data Structures*, Lecture Notes in Computer Science No.156, Springer-Verlag, 1982.

14. Ellis, M. and Stroustrup, B. *The Annotated C++ Reference Manual*, Addison-Wesley, 1990.

15. Axford, T. *Concurrent Programming: Fundamental Techniques for Real-Time*

and Parallel Software Design, John Wiley & Sons Ltd., 1989.

16. Anderson, T. E., Lazowska, E. D. and Levy, H. M. `The Performance Implications of Thread Management Alternatives for Shared-Memory Multiprocessors´ *IEEE Trans. on Computers*, Vol.38, No.12, pp. 1631-1644, 1989.

17. Silva, A. `Acting on Inheritance Hierarchies´, *Proc. of the 3th Eurographics Workshop on Object-Oriented Graphics*, pp. 83-96, Oct. 1992.

18. de Mey, Vicki, Breitneder, C., Dami, L., Gibbs, S. and Tsichritzis, D. `Visual Composition and Multimedia´, *Computer Graphics Forum*, Vol.II, No.3, pp. C-9 - C-22, 1992.

19. Felger, W. and Shroder, F. `The Visualization Input Pipeline - Enabling Semantic Interaction in Scientific Visualization´, *Computer Graphics Forum*, Vol.II, No.3, pp. C-139 - C-149, 1992.

20. Chen, M., Townsend, P. and Wang, C. Y. `A Development Environment for Constructing Graph-Based Editing Tools´, *Computer Graphics Forum*, Vol.II, No.3, pp. C-345- C-355, 1992.

21. Silva, A. and Ramos, H. D. `On Integrated Finite Element Modelling´, Proc. of the 6th Mediterranean Electrotechnical Conference, MELECON´91, Ljubljana, Vol.II, pp. 965-968, 1991.

SECTION 5: FLUID FLOW SIMULATION AND VISUALIZATION

Invited Paper

Scientific Visualization in computational fluid dynamics

H.-G. Pagendarm

Deutsche Forschungsanstalt für Luft- und Raumfahrt, Bunsenstr. 10, D-3400 Göttingen, Germany

While visualization of flows in experimental situations has a long tradition, numerical flow solutions may be presented using easily understandable images only since fast computer graphics techniques are available. During the recent past a number of software systems appeared, some developed at universities or research facilities, others being commercially available. When studying those systems different designs become obvious which make a given system more or less suited to a particular application. An overview about some of the most relevant software packages illustrates their differences and advantages.

Most of the available visualization software offers a common set of visualization techniques. The methods used by established software systems largely determine the way how numerical flow simulations are presented today. Nevertheless some activities may be identified which try to develop new methods for flow visualization in computational fluid dynamics (CFD). These new techniques will add more variety to the methods for creating images of flows and thus allow a better communication of results, concepts and understanding of fluid dynamics.

In many cases easy and fast visualization is a necessary condition for the discovery of structures and phenomena in flow fields. The reason for this are the vast amounts of data which are produced during a numerical simulation of flows. By visualizing these data one can make efficient use of the capabilities of the human brain to detect structures even in partly obscured or incomplete images. Once a particular, may be unexpected structure has drawn the attention of the researcher, he may follow this trace using a fast interactive visualization system until finally he eventually discovers a new unknown feature. Automatic data reduction algorithms, however, would find only structures for which they are designed to detect.

FEATURES OF CFD DATA

Numerical simulation of flows becomes easier to understand when data are mapped to graphical presentation. Since these flow calculations are created using computers the use of computer generated graphics is highly desirable. Thus early visualization of numerical results frequently presents the state of the art in computer graphics. The technique chosen to illustrate the data depends on the methods which are easy to implement in computer codes and the output is determined by the quality of existing hardware. While in experimental fluid mechanics a significant amount of research work is put into the development of visualization apparatus for specialized applications, a specialized flow visualization computer does not exist. Software in use offers visualization methods which are generally applicable in the wide area of scientific visualization rather than tackling the special problems of fluid dynamics. Direct comparison between the visualization of real flows and their numerical simulation is rarely possible.

Data Types

Numerical simulations of flows essentially produce a limited set of data types which are very frequent in engineering sciences. This makes it easy to apply mapping techniques which are commonly used. Such visualization will often be sufficient and satisfying., In some cases however, additional methods dedicated to fluid mechanics will be required. Typical data types are:

- location, a vector quantity which defines a location for a data value in a space or in a space-time frame
- scalar quantity which describes a selected physical quantity
- vector quantity, for example the flow velocity
- tensor quantity

Data structure

Besides the data type, the structure of how the data are accessed within the computer's memory plays an important role for the visualization. There are mainly two reasons for that. Often visualization software will be designed to operate on the same data structures without any necessary conversion or generalization of the data in order to benefit from possible advantages in processing speed. On the other hand the user may want to avoid any data conversion or interpolation in order to avoid artefacts introduced by the visualization scheme being visible on the display, while they are not present in the data. Some flow data may be reduced to single numbers or discrete representations of functions. Since computational fluid dynamics today deals with two-dimensional data fields at least, most often however with three-dimensional data or even time-dependant data in 3D-space, these data are organized using computational meshes. These meshes define a certain ordering of the location in space and time where the governing partial differential equations of the problem are solved numerically. All flow data is stored at these discrete locations. Frequently these locations are visualized by connecting them with straight lines which gives the illusion of a mesh. The data may be stored at the node of the mesh, in some cases as well in the centre of the cells of the mesh. Four major types of meshes or grids are used in CFD:

- structured grids
- multi-blocked structured meshes
- unstructured meshes
- hybrid meshes

The unstructured mesh is the most general but at the same time most memory consuming way to store data.

COMPUTER GRAPHICS TECHNIQUES

Some data requires only moderate visualization techniques which are readily available on most computers today. This holds for example for the problem of visualizing discrete functions of one independent variable. Such data are frequently displayed using x-y-charts. Besides a large number of commercial software packages which serve this purpose there are some excellent programs to be found in the public domain. One example is the well known xvgr tool by P. Turner [1].

In the two-dimensional domain the use of various techniques is known for the visualization of scalar quantities:

- iso-contours
- pseudo-colours
- height maps

All these methods are easy to understand for most users since they are known from cartography and serve well for their purpose.

For the display of vector quantities two techniques are most common in CFD:

- arrow or hedgehog plots
- streamlines or particle paths

In order to create an arrow plot the direction of the arrow encodes the direction of the vector quantity, the length of the arrow will encode the magnitude of the vector quantity. To avoid visual clutter the magnitude may be encoded by the colour of the arrow as well while all arrows are drawn using an equal length. If different visual idioms [2] rather than classical arrows are used these plots are commonly called hedgehog plots.

Streamlines are a numerical simulation of the experimental flow visualization techniques which results from the recording of particle paths through a flow field. It is important to notice that particle paths turn out differently if a flow field is considered independent of time or not. A review of streamline tracking algorithms was recently given by Kenwright and Mallinson [3]. While in fluid dynamic streamlines, that are trajectories of the velocity vector field, may be the most common use of this visualization technique, these methods are not restricted to velocity vector fields. The same technique may be applied to the wall shear stress vector or any other vector field. Nevertheless the computation of trajectories in the velocity field of a flow in order to display streamlines or particle paths is probably the most often used visualization technique which tries to mimic the visualization of real flows.

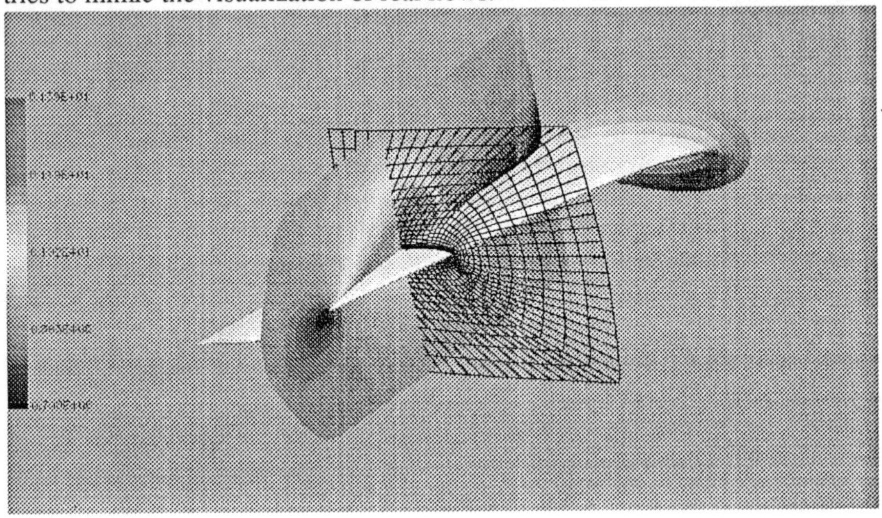

Figure 1. A planar cut through a numerical flow field. The Mach number distribution is displayed in pseudo-colours on the planar section.The computational mesh which holds the data is presented in black. The planar section cuts through the cells of the mesh in an oblique fashion

In three-dimensional space the visualization of flows becomes much more difficult since the techniques known from the 2D case are no more applicable or misleading. The use of height maps, iso-contours and pseudo-colouring is not directly usable in a 3D field of scalars. However these techniques may still be applied to surfaces in a 3D-domain. Therefore they can often be found in visualization of CFD results in 3D space. This requires some data reduction process to have taken place beforehand. Typical data reduction methods to be applied in order to create surfaces in a three-dimensional domain for which then 2D visualization techniques may be applicable are:

- subsampling of surfaces according to identifiers or grid indices
- interpolation of data upon planar cuts
- thresholding of surfaces to limit visibility to a scalar interval

After this data reduction, scalar quantities in space may be visualized on selected significant surfaces using height maps, iso-contours and pseudo-colouring. Of course, it is possible to display arrow plots of vector quantities or streamlines on surfaces, however the resulting images are often misleading and if at all need to be used with special caution. Streamlines, if displayed as simple lines, are not really easy to understand in a three-dimensional domain since the depth information is easily lost.

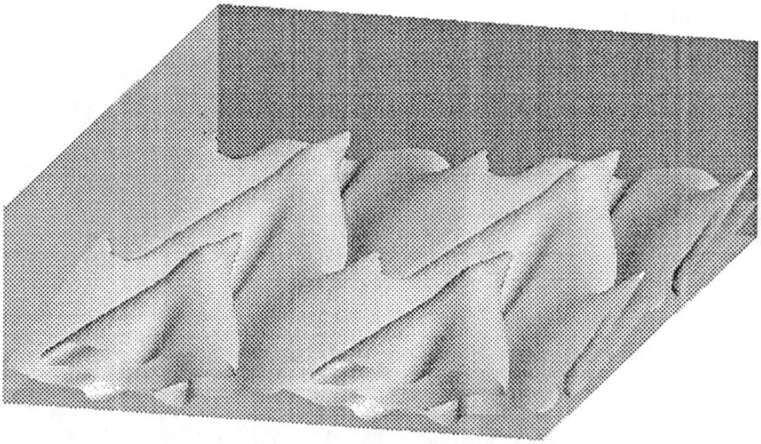

Figure 2. Iso-surface of the vorticity within a wall shear layer during transition from laminar to turbulence. This visualization of a numerical simulation exhibits the lambda shaped structures which are known from experimental visualizations of such flows.

As there is no effective method for rendering scalar quantities within a volume of space, much emphasis needs to be put on the interactive selection or creation of surfaces inside the volume onto which the desired quantities may be mapped for display. If the data is stored using structured grids, surfaces of constant index often provide a significant location such as a wall, a plane of symmetry or a domain boundary. In other cases, it may be necessary to create surfaces using a special data processing and to use these artificial surfaces for display. Any quantity which is mapped onto such a surface may be visualized at that location.

Frequently, surfaces within a volume of data are created using an iso-surface method. This powerful mechanism allows to spot locations where a selected quantity takes a certain value. In arbitrarily shaped data meshes the same algorithm may be used to create planar cuts through a volume. In such meshes a planar cut is nothing but an iso-surface where a linear combination of the coordinates takes a constant value.

Iso-surfaces are a very powerful technique to illustrate internal structures in data fields. In order to be able to create meaningful services which reveal some information from the data, one needs to interactively adjust the parameter which defines the surface carefully. Since this process may require a large number of loops, an effective algorithm for the calculation of such surfaces is needed. Due to the effect that shaded surfaces can exaggerate any kink or unsmoothness iso-surfaces are extremely sensible to numerical wiggles or noise in the data. Therefore they may be of great help when looking for the quality of numerical simulations.

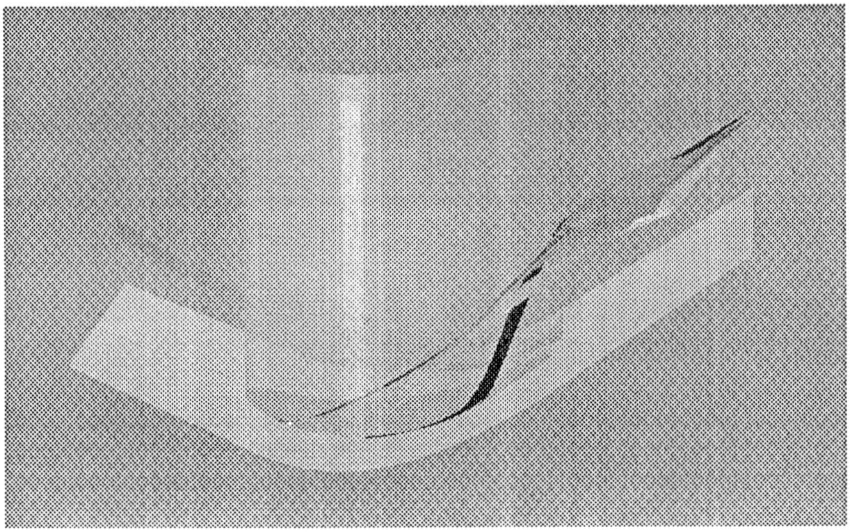

Figure 3. Streamribbons illustrate the vortical flow near a turbine blade. Increasing width of the ribbons visualizes the flow divergence. The transparent blade geometry allows to visualize the flow on both sides of the blade.

Many systems have implemented an algorithm suggested by Lorensen and Cline [4]. This algorithm which was originally designed to process medical imagery data has been thoroughly discussed in the literature and improvements have been suggested (see for example: [5]).

For the visualization of vector quantities within a volume of data several techniques are widely used:

- streamribbons
- stream surfaces
- particle traces
- particle advection
- topological analysis of vector fields

There are various methods to generate the surface elements of stream ribbons from the information of particle traces. Some modern extendable software systems allow to implement a choice of methods [6]. A simple technique is achieved by connecting adjacent streamlines by surfaces obtained from triangulation. This method may be used to create streamsurfaces as well.

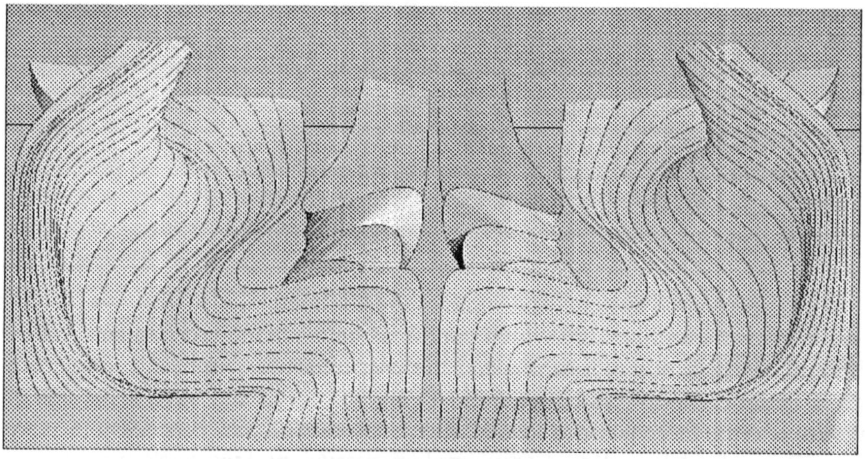

Figure 4. Streamsurfaces in a wall shear layer. In order to illustrate how the surface is created by connecting adjacent streamlines, these lines are drawn upon the surface. While the left and right streamsurface is selected properly to display the vortical behaviour of the flow, the surface in the centre was intentionally composed of a diverging set of lines to illustrate the need for a careful selection of lines in order to avoid unrealistic folds.

Even more information about the velocity field may be seen in images of particle systems which move along with the flow. These images can display not only the local flow direction but also the velocity magnitude as well as the local shear in the flow field. This becomes visible in particular by using animations. Some very interesting research work concerning motion of particles has been described by Stolk and van Wijk [7]. Another technique which is useful especially for vortical flows and three-dimensional flow separation is the extraction of topology information from the flow field. The visualization system FAST [8] which is in use at NASA has incorporated some algorithms reported by Helman and Hesselink [9] for that purpose.

SURVEY ON VISUALIZATION SYSTEMS

A survey on visualization systems conducted recently at DLR examined more than 30 systems which are commercially available or known to be in use at various universities or research establishments. About one third of these systems was installed on our computers for a serious testing. Other systems have been tested at different sites of cooperative partners. A technical report [10] is available which summarizes the results of this survey.

Most of the systems offer a good selection of the present state of visualization techniques as presented in the previous chapter. However the particle system technique as well as the extraction of topology information from the flow field is still not available in these systems.

Architecture of visualization systems

All systems examined during the study can be ordered according to their internal architecture into three groups.

- turn-key systems with traditional single program architecture
- application builders with data flow driven user interfaces for visual programming
- modular systems

Most of the research systems and some of the commercial visualization systems have a software architecture which combines data storage mechanisms, all the different visualization algorithms and the user-interface within one single program. These turn-key systems are most easy to design and have some advantages due to the fact that once loaded into the system all data is accessible at any time. This advantage allows easy implementation of feed-back mechanisms such as probes or picking devices to select data within the image and trace back to the related input data. This architecture as well has advantages in processing speed as long as the amount of data to be processed does not exceed the available memory of the computer by far.

Unfortunately CFD tends frequently to tackle problems which are large even for the biggest available computer. Therefore the advantage of processing speed is often lost due to the need of swapping data to the disk which is in general not done with an optimal strategy.

Another severe disadvantage of this architecture is the well known tendency of such programs to become very complex. This makes them difficult to maintain and extend. Even if the source code is available to onlya limited amount of people, if not only the designer of the program, they will have a chance to customize the system and implement new visualization technologies.

Several commercial systems such as AVS [11], apE [12], Khoros [13] and Explorer offer a visual programming interface. This interface allows to select algorithms from a toolkit and program the data flow between those modules visually by drawing connecting lines. Further customizing a system to deal with a given application becomes easy. To a certain extend new algorithms may be implemented by using a combination of existing modules. Some systems like AVS offer facilities to create user designed modules and therefore allow in principle to include any future visualization technique. In practice it may turn out that these extensions require significant knowledge about the system and unexpected effort [14].

Due to the fact that all modules communicate using pipes or socket connections they are loaded into the computer simultaneously once they are required for a programmed sequence. Since they all allocate memory to hold their data a significant duplication of data is unavoidable, which makes such systems very memory consuming.

Whereas the turn-key systems are a good choice in a production environment where a given class of visualization has to be produced frequently, the visual programming systems have their advantages when developing new visualization technology or processing very different types of data.

Modular systems, the best of two worlds?

The third class of visualization systems are the modular systems. SUN's visualization and image processing package SUNVISION [15] may be considered to be in this group as well as the system designed at DLR named HIGHEND [16]. As our own software is mentioned here the heading of this paragraph may be provocative, indeed. In fact, modular systems can combine some of the benefits of the two other known architectures but they certainly show their own particular drawbacks. Note, that depending on the computing environment and the application any of the three architectures may turn out best suited for a user.

The modular system HIGHEND consists of a large number of programs which each perform a comparatively small task of the visualization process. Data is passed between those modules by writing or reading files. These files are created and accessed automatically. Therefore, the user does not have to bother with this data system. In order to keep things comfortable all modules have a similar, consistent user interface. In the modular system SUNVISION the user interface is even combined in one special program which communicates with all modules using a parameter management system.

For both systems holds that by invoking modules in a desired sequence many different visualization tasks may be performed. Much like in the visual programming systems even new effects which have not been considered during the design of the system are possible due to the flexible arrangement of the modules. Since the data may be easily retrieved from the intermediate files, new algorithms may be added to extend the system. This is even possible if the source code of the visualization system is not available. By creating a simple program which may be used in between two of the existing modules data from the visualisation may be captured and undergo additional user-defined processing. Afterwards it will be feed back into the visualization system.

The HIGHEND system presently offers a large selection of state of the art visualization technology. Besides generally applicable algorithms a number of features is dedicated explicitly to fluid dynamics. Presently the following features are available:

- structured, body-fitted grids
- multi-blocked grids
- unstructured grids in 2D and 3D
- display of gridlines
- display of surfaces, flat shaded and Gouraud shaded
- display of poly-lines and single points
- extraction of subvolumes from unstructured grids according to markers
- extraction of surfaces of constant grid index including subregions
- extraction of surfaces from unstructured grids according to markers
- extraction of lines from unstructured grids according to markers
- calculation of aerodynamics quantities on all data structures
- location of shock positions
- gradient calculations
- 3D interactive rotation with mouse input
- OPEN LOOK compliant user interface
- reusable layout definitions
- pseudo-coloured encoding of scalars on any surface or line
- hedgehog plots
- isocontours, isosurfaces
- planar cuts through structured and unstructured grids
- calculation of streamlines, trajectories, vorticity lines
- streamribbons, streamsurfaces
- triangulation of surfaces between sets of lines
- triangulation of arbitrary polygons
- thresholding of polygons and lines
- conversion of data formats, import and export converters
- interactive control of all processes and parameters
- multi-lingual user-interface selectable at run-time
- open data interfaces for user designed extensions
- definition of keyframes for calculation of animations
- on-line help facility
- modular structure
- remote visualization and cooperative working facilities

This makes HIGHEND a competitive system even compared to commercial visualization systems. In particular, HIGHEND turns out to be very effective in environments with relatively poor computer hardware and large amounts of data. Due to the build-in need for disk I/O the system is less effective, once the problem size fits well into the computer.

FUTURE DIRECTION OF VISUALIZATION RESEARCH

Improved rendering technology

Modern hardware provides the necessary processing speed for enhanced rendering techniques. These allow to include visual aids such as:

- shadows
- atmospheric attenuation
- transparency
- texture mapping

Such visual aids will provide a more rapid and more intuitive understanding of images to persons who were not involved in the process of the visualization or which are not trained in understanding and interpreting simplified computer graphics. Visualized data will be presented in a way much closer to human viewing experiences and thus be less difficult to perceive.

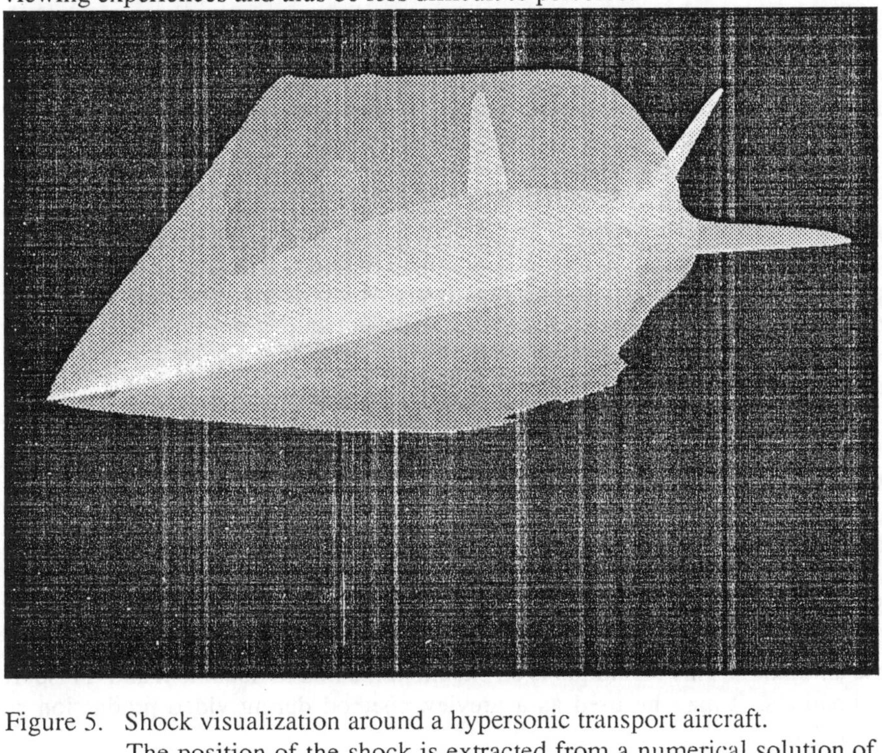

Figure 5. Shock visualization around a hypersonic transport aircraft. The position of the shock is extracted from a numerical solution of the flow field. The transparently rendered shock surface allows to see the details of the complex shock pattern near the wing.

Improved data reduction and mapping techniques

Visualization requires to transform data which is essentially not visible into some kind of visualizable objects. Since these objects frequently do not have much in common with real objects the human mind has to draw conclusions from often very abstract structures. Once visualization techniques are readily available additional effort may be spent by customizing visualization towards specific applications. Therefore,some of the pattern recognition task which are normally performed by the human mind may be transferred into the computer. Structures may be localized and identified within the data and then displayed on the screen using a dedicated method for those structures. This is why today visualization systems in fluid mechanics are equipped with topology analysis systems [9] or shock wave analysis algorithms [17].

The shock wave localization method provides a polygonal representation of the three-dimensional shock systems in supersonic flow fields. These may be displayed in a way which is quite close to some experimental visualization techniques in real flows. Researchers which have experience with optical methods of flow visualization in experimental setups will thus more easily compare their images to numerical results.

Another final goal may be to simulate experimental visualization methods on numerical data in order to obtain a direct comparability of images.

Animation and Digital Video

Today, workstations offer sufficient processing capabilites to display animated graphics. However in general, the amount of information within a scene as well as the choice of rendering methods is limited to those techniques which are supported by the specialized graphics hardware. High quality visualization may on the other hand require more effort. Still, excellent animations may be created on small computers by calculating single images in advance which are afterwards assembled to make up the required animation scenes.

There is a low cost possibility to record such scenes frame by frame to a video tape [18]. However new technology developed for multi-media applications offers opportunities to create and store even long animations on a disk in a compressed fashion. Real-time decompression software is then used to run the animation on any workstation screen. The new SUN XIL [19] software for example was found to be fast enough to perform this task even on a low cost SPARCstation II computer. This software offers decompression without loss of image quality. The processing speed is even fast enough to include audio with the animation. This technique is an excellent method to document and transport animations. It may be used as a preview method during video production as well [18].

Remote visualization and cooperative visualization

Computer networks provide a mechanism to visualize data remotely from a distant site. This makes it easy to bring different data to a display for comparison. Multi-media technology today offers the possibility for researchers at distant locations to visualize the same data simultaneously and discuss about the images using telepointers, audio and video communications [20]. All these communication mechanisms may be integrated within one single software system which uses a standard computer network to transfer necessary information. The feasibility of such cooperative system was already demonstrated on local networks. A wide area connection will be built in the near future in the coarse of a european project.

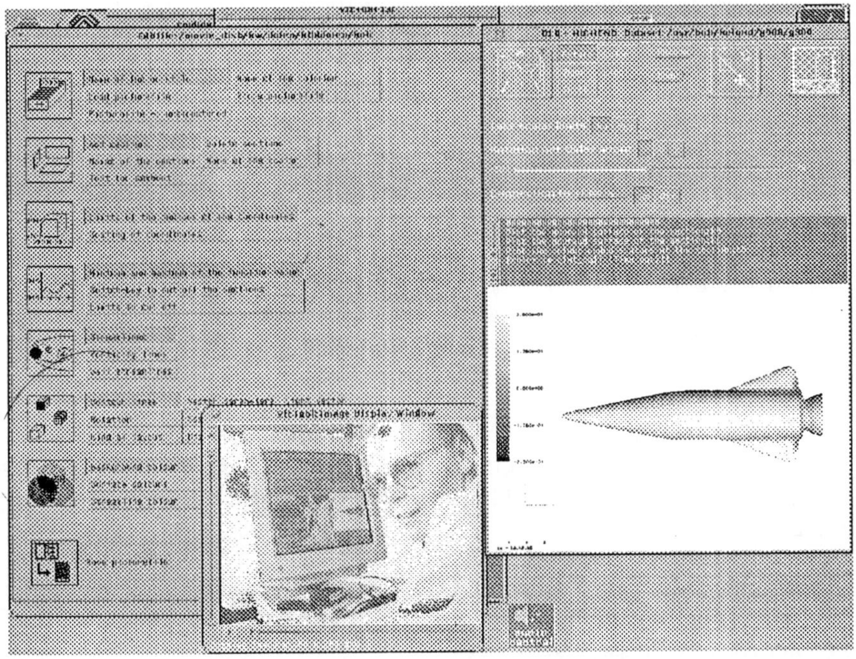

Figure 6. The cooperative visualization facility built upon the HIGHEND visualization system
Besides providing all standard visualization techniques the system offers a bi-directional audio communication, a low speed video connection over standard Ethernet networks and a remote pointer to enhance human discussion about the data in display.
For demonstration purposes video cameras were directed towards the computer screen in order to illustrate the synchronous visualization on both workstations.

By the use of networked visualization systems, scientific visualization faces the chance to become an integrated part of communication in modern science.

CONCLUSION

The lack of visualization software about 6 years ago led to a large number of software systems being developed in European research establishments. Today there is a choice of systems as well as several commercial visualization packages available. These software may be sorted into three groups depending on the software architecture. Each type of architecture turns out to best suited for a certain application environment. Turn-key systems work best for production purpose, visual programming provide high flexibility for varying visualization demands. Modular systems while still very flexible show smaller appetite for computer memory and processing speed.

A common set of standard visualization technique may be found in most systems. Future developments direct towards improved rendering quality, enhanced data abstraction and recognition of structures and phenomena in the-data as well as the use of multi-media techniques for animation and communication purposes.

REFERENCES

1. Turner,P. '*ACE/gr - graphics for explanatory data analysis, Manual-Pages, Sun Release 4.1*', Oregon Graduate Institute, 1992

2. Haber, R.B. and McNabb, D.A. 'Visualization Idioms: A Conceptual Model for Scientific Visualization', *Visualization in Scientific Computing* eds. Nielson, G.M., Shriver, B.S. and Rosenblum, L.J., pp. 74-93, IEEE Computer Society Press, Los Alamitos, 1990

3. Kenwright, D.N. and Mallinson, G.D. 'A 3-D Streamline Tracking Algorithm Using Dual Stream Functions' *Proc. of Visualization '92*, pp. 62-68, IEEE Computer Society Press, 1992

4. Lorenson, E. and Cline, H. E. 'Marching Cubes: A High Resolution 3D Surface Construction Algorithm' *ACM Computer Graphics*, Vol. 21, pp. 163-169, 1987

5. Wu, E.H. and He, R.R. 'A Solution of Disambiguation in Iso-surface Generation', *Proceedings of the 3rd Eurographics Workshop on Visualization in Scientific Computing*, Viareggio, Italy, 27-19 April 1992

6. Currington, I. 'Visualization of Fluid Flow Data' *Proc. of Computer Graphics '91* London, Nov.5-7, 1991

7. Stolk,Jj. and van Wijk, J.J. 'Surface-Particles for 3D Flow Visualization' *Advances in Scientific Visualization* eds. Post, F.H. and Hin, A.J.S. , Springer Verlag, Heidelberg, 1992

8. Bancroft, G.V. et al. 'FAST, A Multi-Processed Environment for Visualization of Computational Fluid Dynamics' *Proceedings of 'Visualization '90* ed. Kaufman, A., Oct. 23-26, 1990, San Francisco, IEEE Computer Society Press, 1990

9. Helman, J. and Hesselink, L. 'Representation and Display of Vector Field Topology in Fluid Flow Data Sets', *Visualization in Scientific Computing* eds. Nielson, G.M., Shriver, B.S. and Rosenblum, L.J., pp. 61-67, IEEE Computer Society Press, Los Alamitos, 1990

10. Seitz, B. and Pagendarm, H.-G. '*Software for Scientific Visualization of Fluid Dynamics Data - A Survey-*', Internal Report DLR-IB 221 92-A-03, Institute for Theoretical Fluid Mechanics of the DLR, Göttingen 1992

11. Upson, C. et al. 'The Application Visualization System: A Computational Environment for Scientific Visualization' *Computer Graphics and Applications*, Vol.9, No.4, July 1989

12. Wettering,M.V. 'Visualization Tools, apE 2.0' *Pixel* Nov./Dec. 1990, (see also: *Introduction to Volume Visualization* ACM-SIGGRAPH 91 Course Notes 7, Las Vegas 28 July- 2 August 1991)

13. '*KHOROS Manuals, User's Manual*', The Vision Lab, Department of Electrical and Computer Engineering, University of New Mexico, Albuquerque, NM 87131, USA, 1990

14. Pagendarm, H.-G. and Choudhry, S.I. '*Visualization of Hypersonic Flows - Exploring the Opportunities to Extend AVS*' paper submitted to the 4th Eurographics Workshop on Visualization in Scientific Computing, Abington, UK, April 21-23, 1993

15. Sun Microsystems Inc.: '*Sunvision - Sun's Visualization Platform , Technical White Paper*', MountainView, 1990

16. Pagendarm, H.-G. 'HIGHEND, A Visualization System for 3d Data with Special Support for Postprocessing of Fluid Dynamics Data' *Visualization in Scientific Computing* eds. Grave, M., LeLous, Y. and Hewitt, W.T. Springer-Verlag, Heidelberg , 1992

17. Pagendarm, H.-G. and Seitz, B. 'An algorithm for Detection and Visualization of Discontinuities in Scientific Data Fields Applied to Flow Data with Shock Waves' *Proceedings of the 3rd Eurographics Workshop on Visualization in Scientific Computing*, Viareggio, Italy, 27-19 April 1992 (will be published by Ellis Horwood, 1993)

18. Choudhry, S.I. and Pagendarm, H.-G. 'A Low Cost Approach to Animated

Flow Visualization Using High Quality Video' *Advances in Scientific Visualization* eds. Post, F.H. and Hin, A.J.S. , Springer Verlag, Heidelberg, 1992

19. Sun Microsystems '*Solaris XIL 1.0 Imaging Library Beta Release Notes*', Mountain View, 1992

20. Pagendarm, H.-G. and Walter, B. '*A Prototype of a Cooperative Visualization Workplace for the Aerodynamicist*' paper submitted to EUROGRAPHICS '93, Barcelona, Sep. 6-10, 1993

Invited Paper

Interactive visualization techniques in CFD#

D. Vucinic,* J. Decuyper, D. Keymeulen, Ch. Hirsch**
Department of Fluid Mechanics, Vrije Universiteit Brussel, Pleinlaan, 2, 1050 Brussel, Belgium

Abstract

In this paper we discuss the application of the Object Oriented Programming (OOP) paradigm to the development and implementation of a software system for the interactive visualization of CFD data. We show that OOP provides both a powerful framework to analyze and design the system as well as an efficient implementation tool.In particular we show how a high degree of interactivity and user- friendliness can be achieved through the use of class libraries encapsulating graphics, networking and interfacing. We illustrate our approach with examples taken from the development of CFView, which is an advanced interactive visualization tool capable of handling both structured and unstructured meshes. We include a discussion of how OOP facilitates parallelization.

1. Introduction

Problems in CFD[1] are commonly solved with numerical techniques implemented on powerful computers. The amount of data produced during such a solution may be very large and not directly suggestive to the engineer who is looking for qualitative features. A possible technique for interpreting such large amounts of data is to visualize them on a graphical workstation and to use advanced computer graphics to enhance interesting features.

Although very useful, visualization techniques may not be sufficient by themselves to uncover interesting features. For example, features may be visible or obscured depending on the choice of a viewpoint or on the choice of a colour map used to convert a range of values into displayed colours. The main difficulty here is that the proper choice of a viewpoint or of a colour map depends itself on the data to be visualized and hence may not be available when the visualization is set up.

Part of this work is sponsored by EspritIII under the contract PASHA-7074 and by RACEII under the contract PAGEIN-R2031.
* Research Assistant , Member AIAA.
** Professor, Head of Department, Member AIAA
[1] CFD stands for Computational Fluid Dynamics

One way to overcome this problem is to make visualization an interactive process. By offering the user the possibility to change dynamically all the important parameters of a visualization, he may explore the data and discover the interesting features without having to know in advance where to look for them.

CFView[2] is a visualization package for CFD, which allows the interactive visualization of large amounts of data supported by structured or unstructured topologies [Vucinic et al., 1992]. It offers the engineer a set of powerful operators, e.g. viewing data on an arbitrary cutting plane or tracing virtual fluid particles. It is fully interactive in the sense that the user can change the parameters of a visualization at run time and observe the result directly on the screen (figure 1).

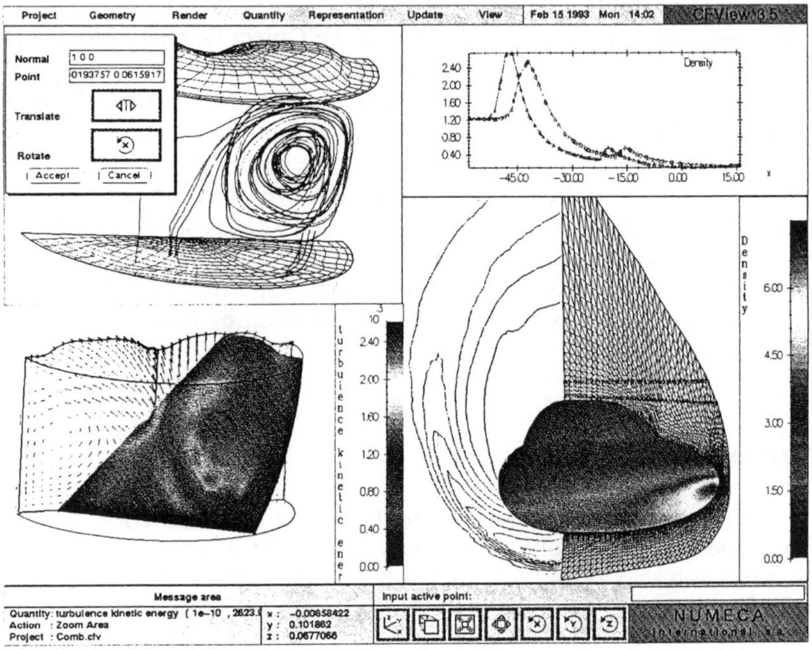

Figure 1. Simultaneous views generated with CFView.

Figure 1 shows a Particle Trace (top left) and Cutting Plane (bottom left) for a combustion chamber, a Cartesian Plot (top right) and an IsoSurface for a double ellipsoid (bottom right), all obtained with CFView. The views on the l.h.s of the figure are generated starting from a structured mesh while the r.h.s views involve unstructured meshes.

In order to provide many different functionalties in an interactive way, CFView integrates several types of routines. For example, the graphical user interface, which provides interactivity, uses a dynamic windowing environment together with several I/O facilities and a set of graphical display primitives. The visualization itself and the operators which are provided to the

---

[2] CFView is a trademark of NUMECA International s.a.

user require routines from computational geometry and from combinatorial topology.[Wilhems, 1990]. Furthermore, routines of different type may interact with each other through function calls or by exchanging data of different types.

The diversity of routines and the tight connection between them are two factors which add considerable complexity to the task of developing and maintaining a software system like CFView over an extended period of time. To deal with this complexity, CFView has been fully developed using the Object Oriented Programming (OOP) approach. All the different data types and routines are encapsulated in objects. These objects are grouped into classes covering the different domains involved in the interactive visualization of CFD data. The classes themselves are organized into class hierarchies which reflect both the internal connections within each domain and the interconnections between different domains. In this way the OOP approach provides a natural decomposition of the software into manageable modules, software components.

In this paper we illustrate the application of the OOP to interactive visualization of CFD data. In section 2 we present the class decomposition underlying CFView. We show how concepts from CFD, from computational geometry and combinatorial topology, together with concepts from computer graphics suggest a hierarchy of objects and classes that implements a full fledged interactive visualization tool. In section 3 we take a closer look at one of the main mechanisms in OOP: polymorphism. We show how polymorphism can be used to implement 2-D and 3-D versions of the same operator and also to unify different visualization operators. Finally, in section 4, we briefly discuss how the OOP approach supports in an elegant way the modification of CFView in view of using parallel hardware for some of the more computationally intensive operations .

2. Class decomposition

As stated in the introduction, the class hierarchy in CFView mirrors the conceptual structure of the domains involved in the interactive visualization of CFD data. There are three main domains: CFD, computational geometry and combinatorial topology and finally computer graphics. The CFD domain provides the application related concepts. These are specific to CFD and are directly meaningful for the engineer who is identifying, extracting and manipulating geometrical subregions. Computational geometry and combinatorial topology provide a set of operations needed to carry out the visualization process [Preparata, 1985],[Henle, 1979]. These operations can be thought of as specialized mathematical operations over the CFD data sets. Finally computer graphics provides concepts related to the user interface and to the graphical primitives that actually display data on a computer screen (figure 2).

Interactive visualization

From the viewpoint of the CFD engineer, visualization is a postprocessing on data representing the numerical solution to a CFD problem. The goal of this postprocessing is to extract relevant features from the numerical data. What features are relevant depends largely on the context posed by the CFD problem as well as on the nature of the CFD field: scalar or

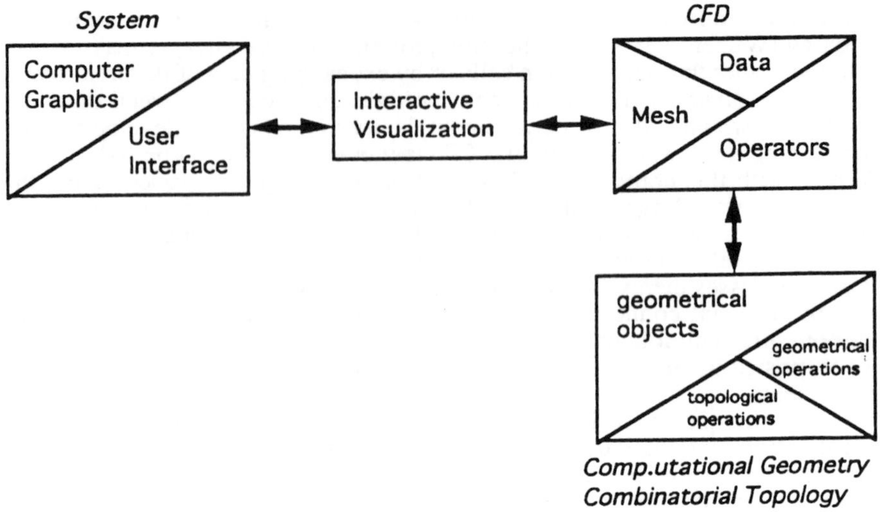

Figure 2:The different conceptual domains involved in interactive visualization

vector. Examples of interesting features are high pressure zones in the airflow around a spacecraft, the flowlines of fluid particles in a turbulent flow or the regions of space where the mach number has a constant value. In order to detect such features, the CFD engineer must be able to focus on different regions of the CFD domain and to manipulate the data in these regions.

CFD data can be divided into two types. The first type concerns the geometry of the CFD problem, i.e. the domain in space to which the data are attached, the boundary of this domain and the way the domain is discretized. The second type of data are the values of the scalar, vector or tensor fields that characterize the fluid under study.

The interactive visualization of these data is based on a set of operators. For example, the Local Value operator allows the user to select a particular point in the domain and to retrieve the value of the quantity fields in that point. Similarly, the Cutting Plane operator provides the possibility to view the field values on an arbitrary plane defined by the user. Note the interactive character of these operators: they adapt to the users choice and hence allow him to explore the flow fields in a customized way. These operators, together with the different types of data provide a first set of objects for the implementation of an interactive visualization tool.

Computational Geometry and Topology

The different operators considered in interactive visualization can be analyzed in terms of the mathematical operations they require. For example, the Local Value operator may be decomposed into two steps. First, the particular point of the domain corresponding to the selection of the user must be determined. This step involves the intersection of a line, through the viewpoint of the user, with the surface on which the CFD data are specified.

Next, the value of a selected field must be computed at this point. This step involves some form of interpolation because the selected point does not generally correspond to one of the points in the discretization of the domain.

The Local Value example illustrates the need for mathematical computations such as intersections and interpolations. These computations belong to the field of computational geometry. Other visualization operators require still a further set of mathematical operations. For example, if a cutting plane operator is applied to a structured mesh, the result is a secondary *unstructured* mesh covering a portion of the plane. Depending on the shape of the original mesh and on its relative position and orientation with respect to the cutting plane, the secondary mesh may be more or less complicated. It may happen for example that the secondary mesh is composed of two disconnected components or that there are some holes in it. These properties need to be known in order to generate a correct visualization on the cutting plane.

Determining the number of connected components or the number of holes in a mesh pertains to the domain of combinatorial topology. Combinatorial topology, together with computational geometry provides the necessary mathematical underpinning for making visualization consistent and performant. Hence they are natural candidate domains to provide useful objects to encapsulate the data and operations needed by CFView.

Computer Graphics and Interfacing

The final product of interactive visualization is a picture displayed on a computer screen. The generation of such a picture requires a special set of operations such as the drawing and colouring of polylines and polygons or the shading and rendering of surfaces in order to show their three dimensional extent. All these operations belong to the domain of computer graphics.

Interactivity calls for fast and efficient access to the controls of the visualization process. This can be achieved by using advanced interfacing techniques based on user-friendly tools such as buttons and dialog boxes.

Computer graphics and Interfacing constitutes a third source of objects for the implementation of an interactive visualization system. The translation of concepts and operations to objects is direct if implemented through class libraries. CFView uses a combination of Fiagaro+ and PEX functionalities encapsulated in class libraries for graphics and Interviews which is a class library for X-window functionalities.

The four domains, CFD, computational geometry, combinatorial topology and computer graphics correspond to important chunks in the class hierarchies underlying CFView. The class libraries of CFView consist of objects encapsulating all the relevant data and operators of the different domains together with their interrelations [Vucinic et al., 1992].

The center part of the diagram shows the *controller* class which is responsible for the activation of the different objects involved in one visualization and for the communication with the user. The *controller* accesses the CFD objects through an object of type *project* which contains a *mesh*. A *mesh* comprises one or multiple *domains* each having a *boundary* which consists of *segments*. The CFD data are attached to the domain and to its boundaries as *field sets*.

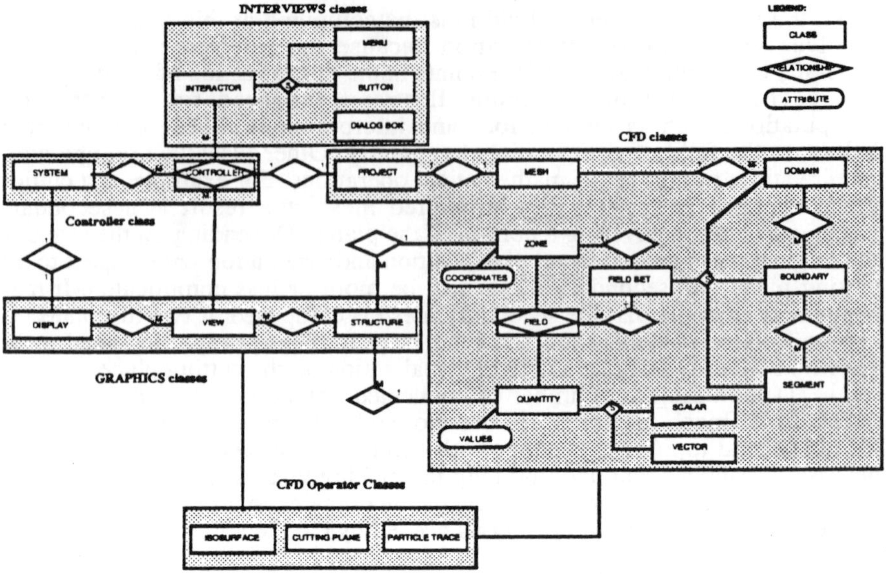

Figure 2 Class decomposition of CFView

On the other side, the controller also activates the objects which implement the graphics part of the visualization. To this end the controller is connected to an object of type *view* which takes one or multiple *structures* and posts them onto the *screen*. The *display* controls multiple views such as illustrated in figure 1. Each *structure* consists of both a geometrical object: a *zone*, and of a *quantity* object which holds the quantity data of the flow being visualized. Once the CFD object is created, it is treated as a graphic primitive which can be manipulated using only computer graphics class libraries.

The operations of computational geometry are activated by the *structure* object each time an intersection on a *zone* object is needed together with the corresponding interpolation operation on the attached *quantities*. The *structure* object is also responsible for activating the operations from combinatorial topology which are not explicitly shown in this graph.

3. Use of inheritance and polymorphism

The class hierarchy in figure 2 only presents a top level view on the whole hierarchy underlying CFView. Each of the classes is further subdivided into more specialized classes. An example of this is the class Zone which is the generalization of geometrical classes. Its hierarchy consists of four subclasses: point, curve, surface and body corresponding to zones in different dimensions (figure 3). Examples of Zones considered in visualization are constant mesh lines or particle paths, which are Zones of dimension 1, or cutting planes and isosurfaces, which are Zones of dimension 2.

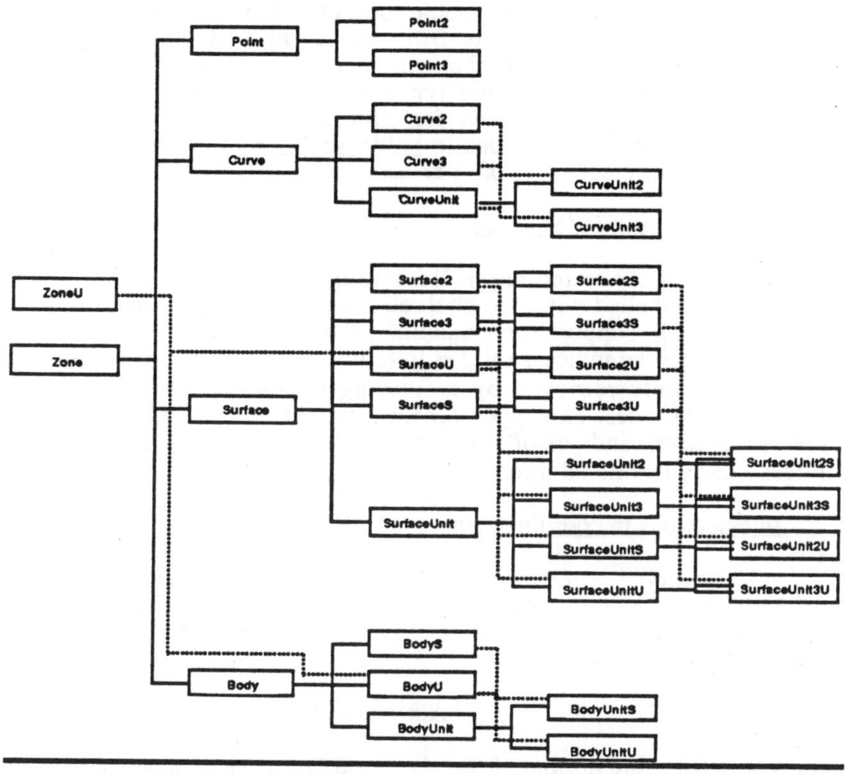

Figure 3. Geometrical class hierarchy

The different possible Zones considered in visualization induce a variety of visualization operators. For example, CFView provides operators such as LocalValue, Cartesian Plot, Isolines, IsoSurfaces, Cutting Plane and Particle Trace[Lorensen, 1987] [Mondiano, 1989]. At first sight all these operators may seem different. A closer inspection however reveals a considerable overlap. For example, the four operators Cartesian Plot, Isolines, IsoSurface and Cutting Plane are in fact all instances of a same operator. This correspondence becomes apparent by analyzing these operators in terms of the operations of computational geometry they involve. The prototype visualization operator consists of two different operations:

Compute the geometrical support for the visualization

Compute the values of the field on this support

The first step typically involves some type of intersection between the given mesh and a geometrical object (line, plane) specified by the user. Outcoming intersection data are obtained by a geometrical filtering process.

The second step typically involves some form of interpolation to determine the values in the intersection points from the computed mesh values. The differences between the visualization operators follow mainly from differences in the type of intersection. Three types of intersections are encountered: intersections in index space (only for structured meshes), intersections in coordinate space and intersections in value space. Following we present an example for each of these types of intersections.

Intersections in index space.

Consider a structured mesh. Depending on the dimension of the mesh the vertices are labelled with two or three indices usually denoted by (I,J) or (I,J,K). A possible support for visualizing data is obtained by specifying a condition over the the indices, e.g. $J = J_0$ (constant). The corresponding values can be immediately retrieved by a filtering operation over the CFD data attached to the vertices. This type of intersection is not used with unstructured meshes because the indices of the vertices nor the labels of the constituting simplices* carry positional information.

Intersections in coordinate space.

A second type of intersection is based on the coordinate representation of a mesh. Suppose for example an unstructured mesh in 3 dimensions (a set of adjacent tetrahedra) and suppose the user has specified a plane by the equation:

$$\mathbf{A}.\mathbf{x} + D = 0$$

where **A** stands for the normal of the plane and D for the distance of the origin to the plane. If this plane is intersected with all the tetrahedra of the mesh the result is a set of adjacent triangles covering a part of the plane. The vertices of these triangles correspond to the intersections of the egdes of the tetrahedra with the user defined plane.

Intersections in value space.

A third type of intersection is obtained by considering the mesh together with a scalar value attached at each vertex as a mesh in a space of one dimension more than the mesh itself. For example, vertices of a 3 dimensional mesh are characterized by three coordinates (X,Y,Z). To each vertex a scalar value may be attached, say V(X,Y,Z). The three space coordinates of a vertex together with the scalar value attached to it define a point in a 4-dimensional space. All the points obtained in this way from the vertices of the 3-dimensional mesh form an analogous mesh in 4-dimensioanl space. This mesh may now be intersected with a plane in 4-dimensions, for example the plane specified by the equation:

$$V = V_0$$

---

* The Simplex concept is used here according to its definition in combinatorial topology and differential geometry.

The result of this intersection projected along the V-axis constitutes an isosurface. It consists of points (X_i, Y_i, Z_i) where the V-field has value V_0.

Based on the type of intersection and on the dimension of the mesh four of the main visualization operators can be grouped together as instances of one more general operator. This abstraction is illustrated in the table 1:

Type of mesh / Type of intersection	**2-D mesh** ---------------- **simplex:** Tetragon (structured) Triangle (unstructured)	**3-D mesh** ---------------- **simplex:** Hexahedron (structured) Tetrahedron (unstructured)
Quantity	IsoLines	IsoSurfaces
Geometry	Cartesian Plot	Cutting Plane
Results	Segments and 2 Values $\{P_1,P_2,V_1,V_2\}$	Faces and 3 Values $\{P_1, P_2, P_3, V_1, V_2, V_3\}$

Table 1: Classification of visualization operators.

The OOP philosophy permits to exploit the underlying similarities of the visualization operators. The operators are encapsulated as messages that can be sent to objects of type Zone. The generic part of the operators can be implemented inside the Zone object itself. This part corresponds to the intersection of an edge with the user defined plane. The specific part of the operator may be implemented inside the different objects Surface and Body. These parts involve the particularities of intersecting the simplexes of different types constituting a Surface or a Body. For example, intersecting a triangle with a plane requires a scan of the three edges making up the triangle while intersecting a tetrahedron requires an additional scan over all the faces making up the tetrahedron.

The mechanism used for encapsulating the intersection operators is called polymorphism. There are two types of polymorphism [Vucinic, 1991]. In *object polymorphism,* the same message sent to different objects results in different computations. In *message polymorphism*, the same message but with different arguments sent to the same object results in different computations. Polymorphic objects can be used to implement 2D and 3D versions of the same visualization operator. For example, if the message *IsoZone* is sent to an object of type *Surface* it generates an *IsoLine* computed on this surface. If the same message is sent to an object of type *Body*, an *IsoSurface* is generated on this body.

The implementation of *IsoZone* is distributed over different objects. The intersection of an edge with a plane is common to both *IsoLine* and *IsoSurface*. It is encapsulated as a method for the *Zone* object. The different scannings over the edges of a simplex are stored within *Surface* and *Body*. In this way, conceptual transparency is achieved together with a reduction of the implementation effort because the common part of two computations can be attached to the *Zone* class from which the two objects *Surface* and *Body* are inheriting.

Polymorphic messages may be used to push the unification of the visualization operators one step further. Both the Cutting Plane operator and the *IsoSurface* operator involve the intersection of segments connecting two vertices with a plane. In the case of the *IsoSurface*, the segment connects two points in 4-dimensional space whereas for the Cutting Plane the segments connect points in 3D space. Table 2 summarizes the differences.

	IsoSurface	Cutting Plane
vertex	X, Y, Z, V(X,Y,Z)	X, Y, Z
plane	$V = V_0$	$\mathbf{A} . \mathbf{X} + D = 0$

Table 2: Differences between IsoSurface and Cutting Plane.

Consider the case of a Cutting Plane and define for each vertex **X**, an auxiliary field by

$$W(\mathbf{X}) = \mathbf{A}.\mathbf{X}$$

then it is obvious that a Cutting Plane corresponds to an IsoSurface where the intersecting plane is now defined by

$$W = -D.$$

This generalization can be implemented with a polymorphic message, i.e. a message taking arguments. In one case, the *IsoZone* operator can be defined as

IsoZone(Quantity, Threshold).

When *IsoZone*(V, V_0) is sent to an object of type *Body* (i.e. a 3D mesh) then it generates the *IsoSurface* corresponding to the value V_0 for the field V. If on the other hand the *IsoZone*(**A**, -D) is sent to the same object of type *Body* then a *Cutting Plane* will be generated. By combining a polymorphic message sent to polymorphic objects, all four of the visualization operators in table 1 may be unified into one operator.

4 Parallelism and the OOP

So far we have shown how the OOP supports the decomposition of a visualization software system into modules which cover conceptually different domains. We have shown how classes for CFD, for computational geometry and combinatorial topology can be integrated with class libraries for computer graphics and interfaces. We now show how this class hierarchy can be modified to support the use of parallel hardware[Aggarwal et al., 1988] [Akl, 1993].

The approach we have followed starts from the assumption that only the most computation intensive routines need to be executed on a parallel machine. We assume that the graphical part of the visualization can still be carried out on the dedicated graphical workstation. From this point of view it appears that the calculation of intersections is the most likely candidate to benefit from parallelization. Because this type of calculation is very well localized within the object hierarchy; it requires only few modifications to the original code in order to parallelize it.

The parallel implementation of the *IsoZone*(quantity, threshold) operator introduced in the previous section involves the following steps.

for all vertices of the mesh
calculate a label depending on the quantity and threshold specified in the IsoZone operator

for all the cells in the mesh
label the cell according to the labels of all its vertices. This cell label specifies where and how the cell is intersected.

for all the possible labels of cells
calculate the intersection with the individual edges of the cell and generate the appropriate simplices of the intersection

The three steps involved in the parallel implementation explicitly mention over which data the computations are parallelized. In the first step all the vertices are handled in parallel. The labelling operation consists in determining whether the value of the quantity attached to the vertex is lower than or equal to the threshold. Which quantity needs to be considered and what threshold it needs to be compared with is specified in the arguments of the IsoZone operator.

The second step in the intersection is executed for all the cells of the mesh in parallel. It involves the classification of a cell according to the labels of all its vertices. The main idea behind is that if the two vertices of an edge have a different label then this edge is intersected by the plane specified by the user. Depending on the type of cell (see table 1) and on the number and position of intersected edges several auxiliary computations are needed to construct the intersection. These computations are listed in a table, called the intersection table. The label of the cell constitutes the index into this table.

The third step is parallelized over the different types of cells. For all the cells of a given type the intersection points on the edges specified in the intersection table must be obtained. Finally, the intersection points must be connected in order to form the simplices making up the intersection of the mesh with the user specified plane.

An interesting side effect of the parallelization is the fact that the distinction between structured and unstructured meshes may be dropped. Looking back to the three parallel steps outlined above, it is clear that the only difference between structured and unstructured meshes will reside in the type of cell (hexahedron or tetrahedron for example) and in the associated table of intersection. Apart from this, all the operations are completely identical for structured and unstructured meshes.

Parallelism and the local view on visualization

Because parallelism enforces the local view, i.e. bound to a cell, it provides an additional argument to the OOP. In particular, the unification of the visualization operators as proposed in the previous section becomes very natural when viewed at the cell level. Recall that we have proposed an IsoZone operator which generalizes Cutting Planes and IsoSurfaces in both 2D and 3D. We can go one step further in the generalization process by combining the intersection and interpolation operations within one cell into one operator. This combination leads to the following specifications:

Cutting Plane:

Inside a cell of a mesh the Cutting Plane requires the intersection of an edge and the value of a field in the intersection point by linear interpolation (figure 5).

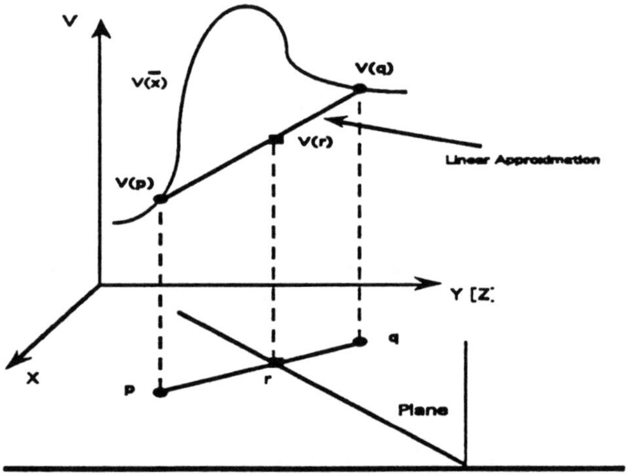

Figure 5. Intersection for a Cutting Plane

Formally, suppose **p** and **q** are two vertices in a mesh. The segment (**p**, **q**) is given by the points **x** such that

$$\mathbf{x} = \mathbf{p} + \lambda\,(\mathbf{q} - \mathbf{p})$$

The plane defined by the user has the equation:

$$\mathbf{A} \cdot \mathbf{x} + \mathrm{D} = 0$$

The intersection point corresponds to the point of the segment (**p**, **q**) with

$$\lambda_{\mathrm{int}} = -\,[(\mathbf{A} \cdot \mathbf{p} + \mathrm{D}\,) \,/\, \mathbf{A} \cdot (\mathbf{q} - \mathbf{p})]$$

By linear interpolation the value of the field V in the intersection point, $\mathbf{x}_{int}$ is given by:

$$V(\mathbf{x}_{int}) = V(\mathbf{p}) + \lambda_{int} (V(\mathbf{q}) - V(\mathbf{p}))$$

IsoSurface

The IsoSurface corresponding to the threshold V_0 is defined by the points **x** which satisfy the equation:

$$V(\mathbf{x}) - V_0 = 0$$

The IsoSurface operator executed on a mesh has to compute the intersection of this IsoSurface with the elements of the mesh(figure 6).

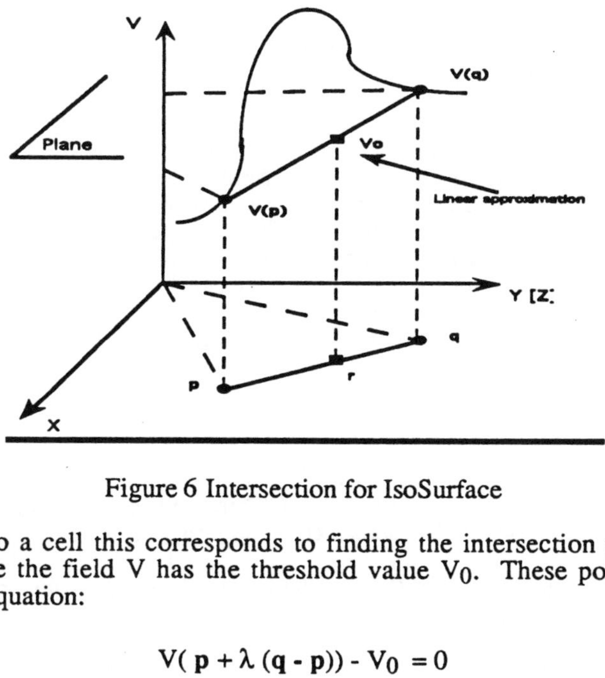

Figure 6 Intersection for IsoSurface

Restricted to a cell this corresponds to finding the intersection point of the edges where the field V has the threshold value V_0. These points have to satisfy the equation:

$$V(\mathbf{p} + \lambda (\mathbf{q} - \mathbf{p})) - V_0 = 0$$

If the field V(**x**) is approximated linearly inside a cell we find for the intersection point on edge (**p**,**q**):

$$\lambda_{int} = - [(V(\mathbf{p}) - V_0) / (\mathbf{grad}\ V. (\mathbf{q} - \mathbf{p}))] + \Theta(|\mathbf{q} - \mathbf{p}|^2)$$

It follows that the IsoSurface operator can be viewed locally, i.e. inside a cell as an example of a Cutting Plane with equation:

$$\mathbf{grad}\ V.\ \mathbf{x} - V_0 = 0$$

This identification is nothing but the expression that the IsoSurface may be locally approximated by a plane. In the parallel implementation of both operators this correspondence can be fully exploited because operations are parallelized over the cells of a mesh. A preprocessing step may be used to extract the gradient of the field V in every cell. This gradient defines the plane which plays the role of the user defined plane in the case of the Cutting Plane operator.

5. Conclusion and future work

In this paper we have illustrated the application of OOP to the design and implementation of an interactive visualization system for CFD. In particular we have shown how different conceptual domains such as computational geometry, combinatorial topology, computer graphics and interfacing may be encapsulated into class libraries. These libraries provide a high level abstraction of the computational processes and as such they represent reusable software components for a variety of applications.

We have also discussed how the OOP supports the parallelization of the software system. In particular we have shown that some distinctions between structured and unstructured meshes can be removed in the parallel implementation if the computations are viewed locally, i.e. bound to a cell of the mesh. It follows that the main source of parallelism is to be found at the cell level.

References

Aggarwal, A., Chazelle, B., Guibas, L., O'Dunlaing, C. and Yap, C. (1988). Parallel Computational Geometry. *Algorithmica*, 3:293-327.

Akl S.G. and Lyons K.A. (1993). *Parallel Computational Geometry*. Prentice-Hall, Englewood Cliffs, New Jersey 07632.

Lorensen, W. and Cline, H. (1987). A High Resolution 3D Surface Construction Algorithm. *Computer Graphics*, 21(4):163-169.

Henle, M. (1979). *A Combinatorial Introduction to Topology*. W.H. Freeman and Company, San Fransisco.

Modiano, D., Giles, M., and Murman, E. (1989). Visualization of three-dimensional CFD solutions. In *AIAA*, number 89-0138.

Preparata, P.F., Shamos, M.I. (1985). *Computational Geometry, An Introduction*. Springer-Verlag, 175 Fifth Avenue, New York 10010, U.S.A.

Vucinic, D., Pottiez, M. Sotiaux, V., and Hirsch, C. (1992). CFView - An Advanced Interactive Visualization System based on Object-Oriented Approach. In *Proceedings of the AIAA 30th Aerospace Sciences Meeting Exhibit*, Reno, N.V. AIAA, 370 l'Enfant Promenade, S.W. Washington, D.C. 20024. number 92-0072.

Vucinic, D. (1991). Object Oriented Programming for Computer Graphics and Flow Visualization. In *VKI Lecture Series on Computer Graphics and Flow Visualization in CFD*, September, Brussels, 1991.

Wilhelms, J. and Van Gelder, A. (1990). Topological considerations in isosurface generation. *Computer Graphics*, 24(5):79-86.

Invited Paper

An automatic digital image processing system for whole field flow measurement

T. Kobayashi, T. Saga

Institute of Industrial Science, The University of Tokyo, 7-22-1, Roppongi, Minato-ku, Tokyo 106, Japan

ABSTRACT

An automatic image processing system for measuring instantaneous, whole field of velocity vectors was developed. This system is connected with the classical flow visualization technique (tracer injection method) and sophisticated image processing technique. The measurement system consists of a laser light source, a TV camera, a laser optical disk recorder and an image processor interfaced with a 16 bit micro-computer. The tracer particles injected into the liquid flow are illuminated using the pulsed laser light sheet and are tracked by the TV camera. The images recorded on the laser disk are later replayed and sent to the image processor, and instantaneous velocity vectors of whole flow field is automatically calculated. To explore the ability and usefulness of this system the complex flow fields have been investigated.

INTRODUCTION

An important subject of modern experimental fluid mechanics is the development of techniques for the measurement of whole, instantaneous velocity. Conventional point measurement techniques such as Hot-Wire Anemometry (HWA) and Laser Doppler Velocimetry (LDV) have been widely used for various flow measurements with sufficient accuracy. However, the difficulties of the calibration of multisensor hot-wire probe and the optical arrangement of laser doppler velocimeter become to decline the practicability of their techniques. It is expected to develop a new technique which overcomes this basic problem.

Particle-Imaging Velocimetry (PIV) enables the dynamic measurement of flow velocity, by the tracking in the real time the movements of fine particles injected into a flow [1],[2],[3]. Several outstanding characteristics have generated the great advantage in PIV : ① two-dimensional and three-dimensional information can be logged in real-time for multi-points over a whole field of the flow, ② from this information the flow structure can be interpreted, ③ the need of the velocity calibration is eliminated, ④ the method is applicable to complex flows, ⑤ measurement is no contact, ⑥ data can be automatically processed with ease.

With these advancements of PIV, we have been developed an automatic image processing system for the flow measurement. The utility of this system for detecting the flow structures is demonstrated.

MEASUREMENT TECHNIQUE AND INPUT IMAGE

The flow velocity is calculated by tracking the particles injected into the flow field. Motions of particles are observed simultaneously by TV camera. By processing the consecutive TV frames, instantaneous particle positions are reconstructed, and then the path of each particle over the several time steps is identified. The flow velocity vector is calculated from the displacement of each particle during a short time interval.

MEASUREMENT SYSTEM

Shown in Fig.1 is the measurement system designed specifically for PIV, consisting of a illumination source (SPECTRA PHYSICS 2016 ; 4W Argon laser), a charge-coupled devise camera (SONY XC 007 ; NTSC video format), a laser optical video recorder system (SONY LVR,LVS-5000 ; 48000 frames capacity), an image processor with 16 MB external image memory (NEXUS 6400) and a 16 bit micro-computer (NEC PC9801 VX). The image processor and the video recorder are interfaced with the host computer which controls whole operations in this system. Image data from the flow field are logged directly by the TV camera and analyzed automatically. For an automatic image processing, consecutive images from the TV camera are recorded on the optical video disk. By the command from the host computer to

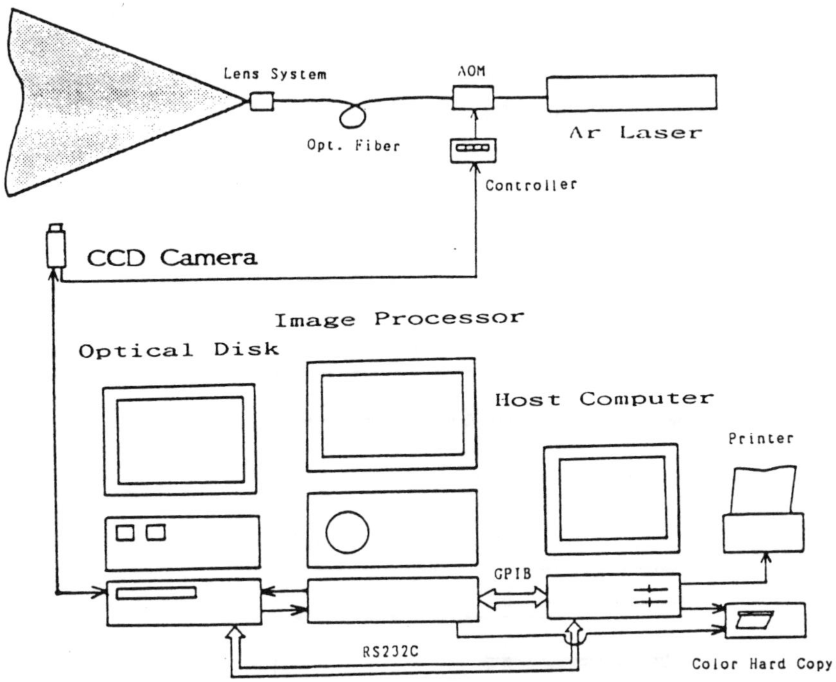

Fig. 1 Measurement system

the video recorder, the images of the disk are selected and are sent to image processor. The monochrome analog video signal of NTSC video format are digitized into 256 gray levels over 512 X 480 pixels by the image processor. The digitized images are then thresholded by an adequate gray level to extract the images of particles from those of the background. Finally, the gravitational center of each particle image is calculated by the image processor and transferred to the microcomputer. Both image acquisition and data reduction are executed automatically by the original computer programs.

CODED ILLUMINATION

For covering wide range of flow speed measurements, a suitable coded pulsing illumination is introduced into the measurement system. The illumination system, shown in Fig. 1, consists of an Argon laser source, an acousto-optic modulator (AOM) which connected with the TV camera through a pulse controller, an optical fiber and a lens system. The laser beam passes through a series of optical components to produce a modulated sheet of laser light. The temporal light illumination is obtained by using the scattered light of the first order which is produced by the AOM. The illumination is synchronized with the vertical and horizontal blanking signals of the TV camera. Figure 2 shows the two typical illumination codes for the low and high flow speed measurements, and the input images obtained by using those

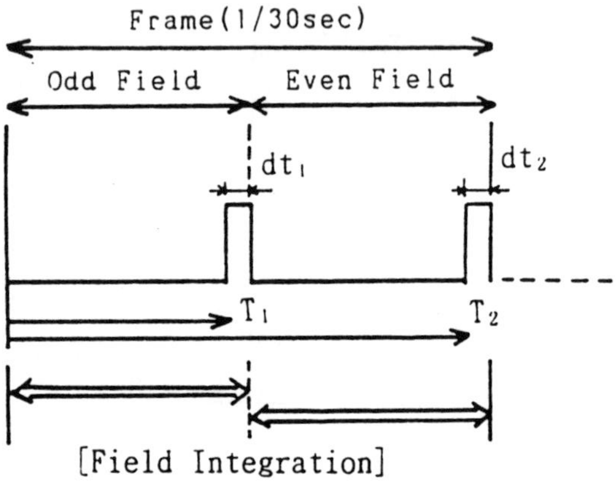

(a) Illumination code for low-speed flow

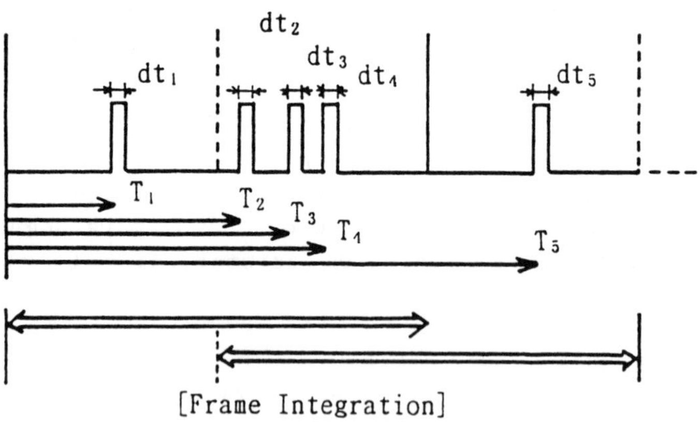

(b) Illumination code for high-speed flow

Fig. 2 Illumination code

codes are shown in Fig.3. In order to adapt to various flow conditions, the timing of illumination Ti at the video frame and the pulsing time dti can be adjusted respectively by using unit time period (1H = 63.56 µs) in this system. The illumination times of the laser light also can be changed. At the low speed illumination code, the flow field is illuminated once at each field (Fig. 2(a)). In the case of

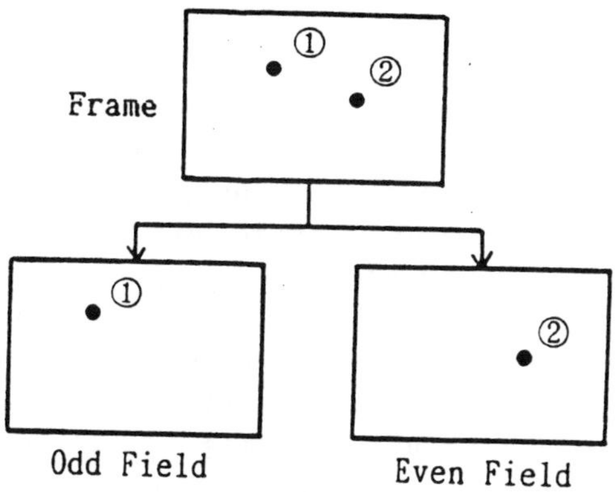

(a) Input images by low-Speed illumination code

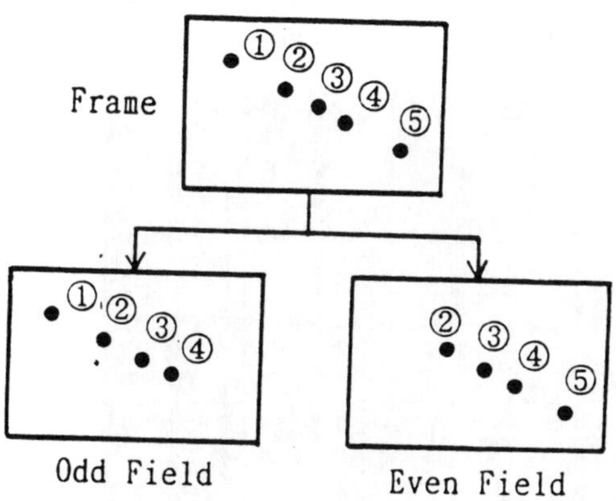

(b) Input images by high-Speed illumination code

Fig. 3 Frame image and its' divided field images

the high speed flow measurements, the illumination code is settled as shown in Fig.2 (b). In this mode, the flow field is illuminated one time at the odd field and three times at the even field. Furthermore, the electric charge mode of the TV camera is changed between two illumination codes. The field integration mode is used for the low speed code and the frame integration mode is used for the high speed one. By selection of the illumination code and the electric charge mode, the particle images during several time steps are photographed on a frame image, as shown in Figs.3 (a) and (b).

INPUT IMAGE

Since the information at each field image is required for the present particle tracking, the frame image is divided into the odd field and the even field before the particle tracking. The input image of NTSC video format, consisting of 525 horizontal scan lines, is dissected into two fields, the odd field and the even field. Each field has 262.5 scan lines, which are combined and interlaced to produce a complete frame. At first, the frame is divided into two field images by using a image dividing mask. Next, to fill up the left pixel densities of both field images by this frame divide, the pixel densities are interpolated by using the surrounding pixel densities. Thus two new field images are produced from a frame image. Figure 4 shows a typical field image produced by using the low speed illumination code. This process is continuously performed on consecutive frame images before the particle tracking.

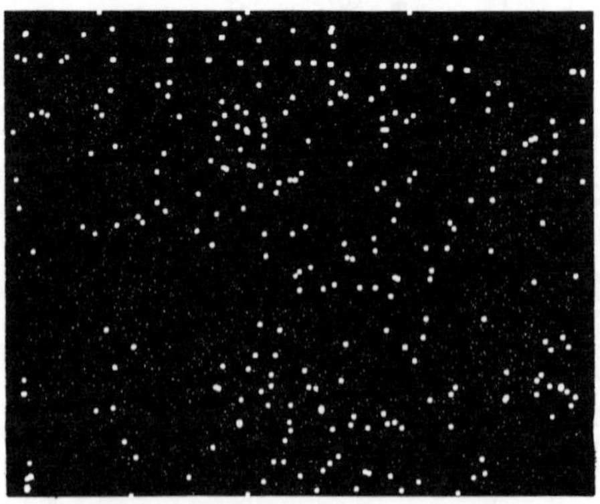

Fig.4 Instantaneous tracer particles of the produced field image by using a low-speed illumination code

IMAGE ANALYSIS

Calculation procedure of velocity vectors from the produced field images consists mainly of three steps, the extraction of particle position, tracking of particle and determination of particle path. There are several proposals of tracking motions of tracer particles in video image, but those have the same fundamental principle : a series of images of particles during four time steps are superimposed to determine the distance which the particle has travelled and thence the velocity vector. The most crucial task performed by the system software is to identify the same particle over series of images. The details of tracking and identification algorithm is discussed in [4],[5],[6].

At the low speed illumination code, instantaneous particle positions at each illumination time Ti are obtained on the consecutive each field image, independently (Fig 3(a)). In this case, a set of four consecutive field images are used for the particle tracking and the determination of the particle path.

In the case of the high-speed flow measurements, particle positions of same particle during the time T_1 to T_5 were superimposed on a frame image by selection of the high-speed illumination code and the frame integration charge mode as shown in Fig. 3(b). However, these particles have no information concerning the time history, it is needed to get some information of the time from the input image. Here, to extract the initial position of particles, the images of the odd and the even fields are subtracted from each other. By this operation, all of the initial particles with positive values of image density and the terminal positions with negative values are identified in the frame image. Once the initial position is determined, the particle tracking is accomplished by the prediction of displacement and the direction of the particle during four time steps.

EXPERIMENTAL FACILITY AND PROCEDURE

To explore the ability and the usefulness of the present technique, the system is applied to two kinds of complex flow fields. The flow measurement in a rectangular channel with a backward-facing step was focused on the evaluation of the measurement accuracy of present technique.

In the measurement of a casting model which is scaled down by 1/7 casting in metallurgical process, we emphasis on the view to confirm the applicability of the system for the complex flow measurement.

The working fluid is water, and the fluid motion is visualized by fine particles injected into the fluid. The particles are made of Nylon 12, specific density of which is 1.02 and is closely equal to that of water. The mean diameter of particles is about 100 μm.

BACKWARD FACING-STEP

The experiment is carried out for the two water models as shown in Figs. 5 (a) and (b). The transparent rectangular channel has a 50 mm height of backward-facing step. The expansion ratio is 1.5, and the channel width is 2h. The free stream velocity U_0 at the 5h upstream of step is 0.8 m/s, and the Reynolds number based on U_0 and h is 4×10^4. The 2 mm thick of coded illumination is provided into the mid-span plane of the channel, and the particle motions in the plane are observed by the TV camera. Measuring region is taken over 3h upstream of the step and 12h downstream. It is determined that the measuring domain of each experiment is 3h × 3h × 0.04h to consider the measurement uncertainty. In this case, the image resolution is 0.31 mm / pixel.

CASTING MODEL

The main components of the casting model are a curved and transparent rectangular parallel channel and a transparent injection pipe (Fig. 5(b)). The channel has dimensions of 214 mm width, 34 mm depth, 3000 mm length, and the pipe with two injection nozzles is placed at the channel center (; the inner diameter of the pipe is 10 mm and the nozzle diameter is 10 mm). Tracer particles are continuously introduced at the pipe inlet. Water flow passed through the pipe is injected into the channel through the two nozzles, and the jets are induced with two symmetrical cavity flows in the channel. Reynolds number based on the pipe inner diameter d and the average velocity in pipe U_0 (= 1.0 m / s) is 1×10^4. Measurements are made in a central vertical cross section (100 mm × 100 mm) of the channel illuminating by the 1 mm thick of the coded illumination.

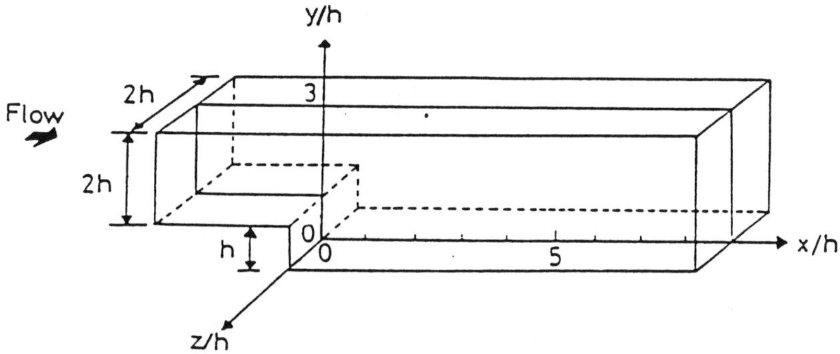

(a) Rectangular channel with a backward facing-step

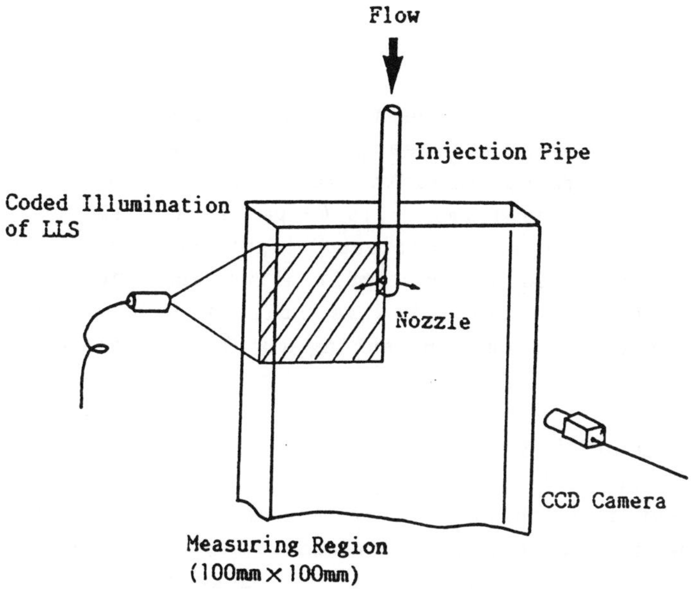

(b) Casting model

Fig. 5 Experimental apparatus

Table 1 Illumination Code

Backward Facing Step					
Code	T_1	T_2	T_3	T_4	dt
1 (l)	356H	890H	1406H	1931H	2H
2 (h)	50H	260H	280H	360H	2H
3 (h)	100H	260H	280H	360H	2H
Casting Model					
1 (l)	100H	356H	625H	881H	10H
2 (h)	100H	200H	400H	510H	10H
3 (h)	220H	270H	300H	330H	10H

1H = 63.56 μs

l : Low Speed Illumination Code

h : High Speed Illumination Code

The illumination codes used in this experiments are shown in Table 1.

In order to check the measurement accuracy of present technique, the velocity measured by PIV was compared with LDV data. The measurement volume of LDV has an elliptical shape, in which the diameter is 49 μm and the length of beam direction is 189 μm. Scattered light from the measuring point is collected by photomultiplier and the signal is processed by an universal wave analyzer. The sampling of bursting signal at each measuring point is performed during 1 minute.

RESULTS AND DISCUSSIONS

Measured velocity vectors are scattered at random in the flow field, and are rearranged by the interpolation method. The flow·field around the backward facing-step is divided into 46 × 25 non-uniform grid points, and velocity vector on the grid point is interpolated from several neighboring local velocity vectors. There are the number n of velocity vectors near a grid point. The interpolation area is defined at the vicinity of grid point as shown in Fig 6. From velocity vectors of $_nC_3$ combinations in the

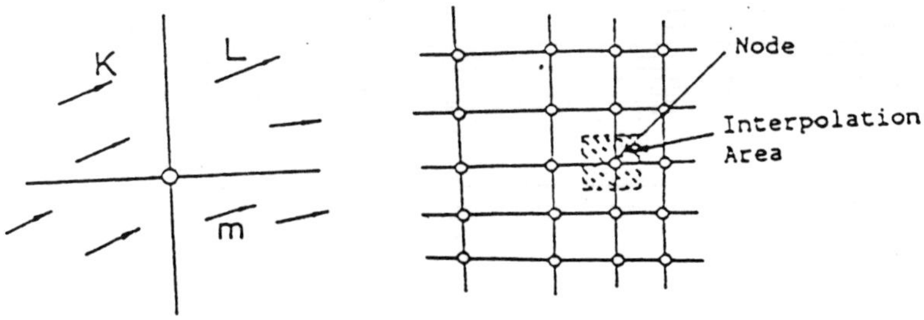

Fig. 6 Interpolation of mean velocity on the grid point from the measured velocity vectors

interpolation area, the grid point each velocity value can be calculated by the following equation.

$$\begin{bmatrix} 1 & dx_K & dy_K \\ 1 & dx_L & dy_L \\ 1 & dx_m & dy_m \end{bmatrix} \cdot \begin{bmatrix} U \\ \frac{du}{dx} \\ \frac{du}{dy} \end{bmatrix} = \begin{bmatrix} U_K \\ U_L \\ U_m \end{bmatrix}$$

The mean velocity value of grid point can be obtained by averaging all of the interpolated values.

In this experiment, 300 frames of images were processed at each 3h X 3h measuring domain, and 8,877 instantaneous velocity vectors were obtained. About seven to ten instantaneous velocity vectors existed in every interpolation area. The mean velocity vectors on the grid points are interpolated by these neighboring velocity vectors, and the resultant by PIV is shown in Fig. 7. To evaluate the present technique, the mean velocity profiles at the positions of x/h = 2.67 and 6.22 are compared with the LDV results. As shown in Fig.8, the present results agree well with the LDV results. It is suggested that the present technique has high accuracy and reliability.

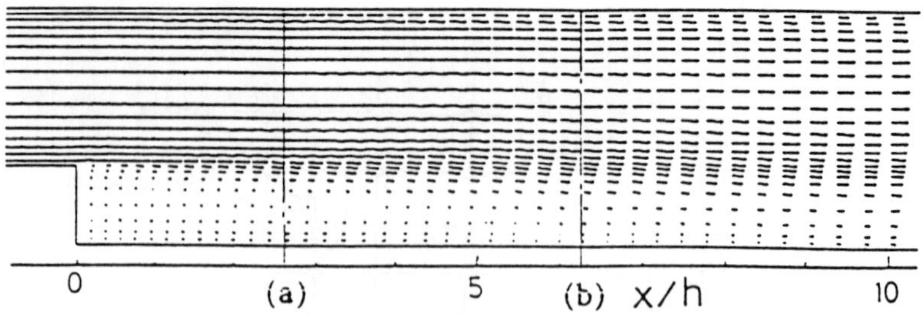

Fig.7 Flow over a backward facing-step

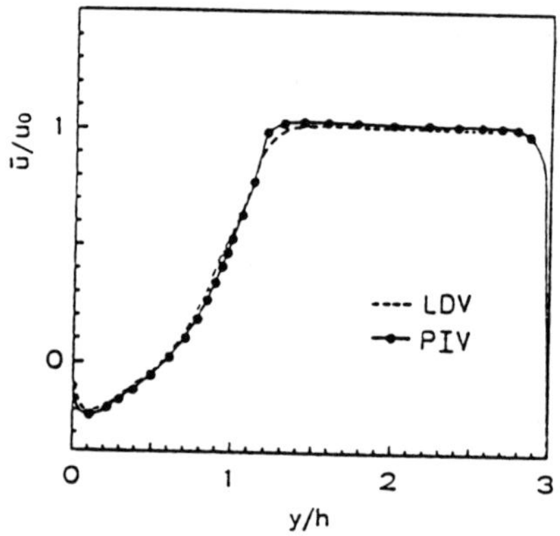

(a) x/h = 2.67

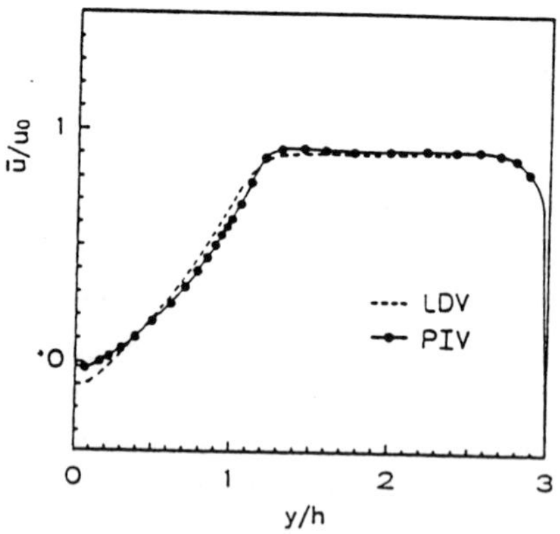

(b) x/h = 6.22

Fig.8 Comparison PIV and LDV results

Because of the applicability of the system is one of the important factor of the velocity measurement, we conformed the system function by the measurement of the complex flow field in casting model from this view point. A typical results of velocity distributions by using the three illumination codes are shown in Figs.9 (a), (b) and (c). In these figures, consecutive 30 frames (during 1 second) data were summed up, and the number of velocity vectors of each figure are 34, 526 and 45, respectively. From these figures, we can easily understand that this flow field consists mainly of the jet flow and the cavity flow induced by the jet. Final data are overlapped summation with a unified time base as shown in Fig.9(d). Figure 10(a) shows the interpolated velocity vectors from the Fig.9(d), and is also compared with the LDV result (Fig 10(b)). As shown in the figures, the profiles of averaged velocity distributions by PIV agree well with LDV results. At the velocity measurement and data reduction, it takes almost two days to obtain the averaging velocity on 11×11 grid points by using LDV, and in the case of the PIV measurement on 31×31 grid points we need only five minutes. It is clearly shown that the PIV has excellent performance for the full field velocity measurements.

In the present technique, the illumination code should be determined carefully according to the flow condition, and it is clear that a large number of the measured velocity vectors is required for an accurate prediction of the grid point velocity.

CONCLUSIONS

In conclusion, we have been presented an automatic image processing system for whole field flow measurement which consists of a coded illumination source, a high-speed image processing algorithm, and an image processing system.

(1) The ability and utility of this system is validated to the complex flow measurement over a backward facing-step in water channel. The mean velocity profiles measured by the present technique is in close agreement with LDV data.

(2) With good accuracy, the system performs a full flow field velocity mapping, rapidly and automatically.

It has been demonstrated that the technique is an important element in experimental research of complex turbulent flow fields.

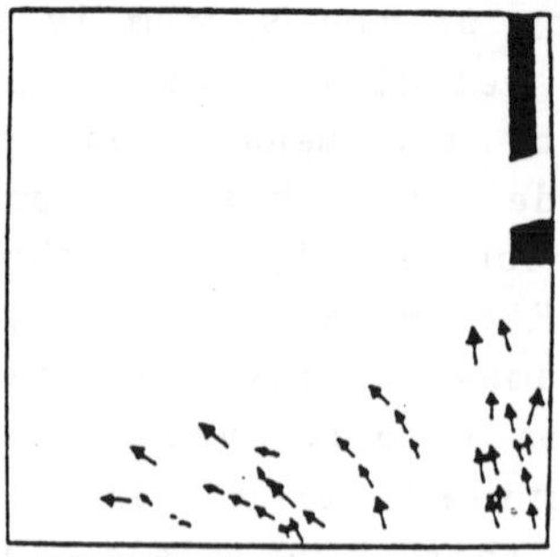

(a) Code 1

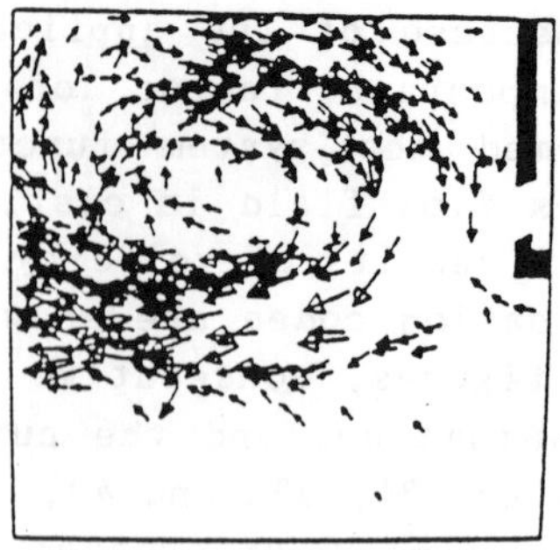

(b) Code 2

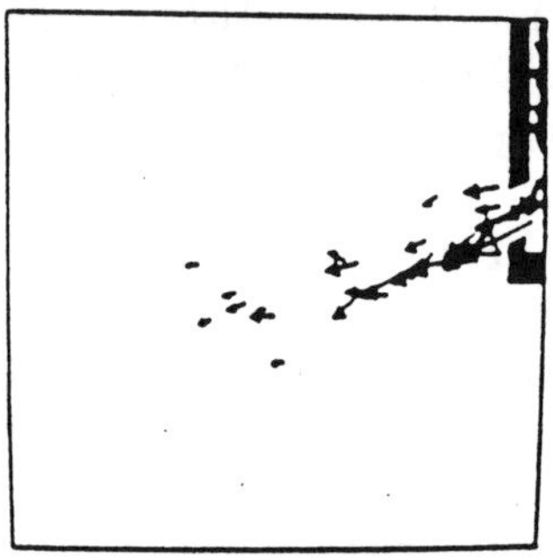

(c) Code 3

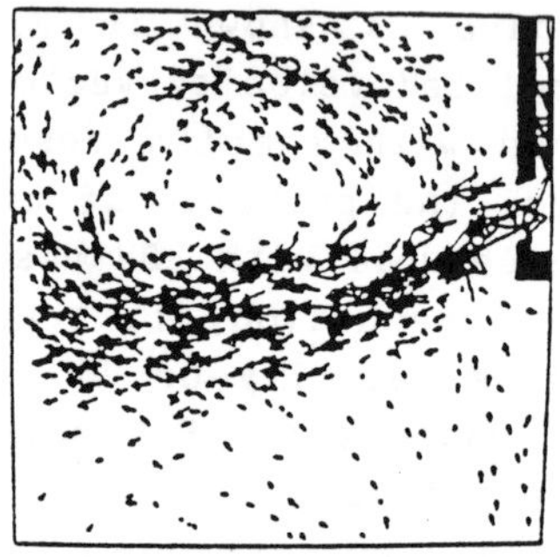

(d) Summed up velocity vectors

Fig. 9 Velocity Measurements in the casting model

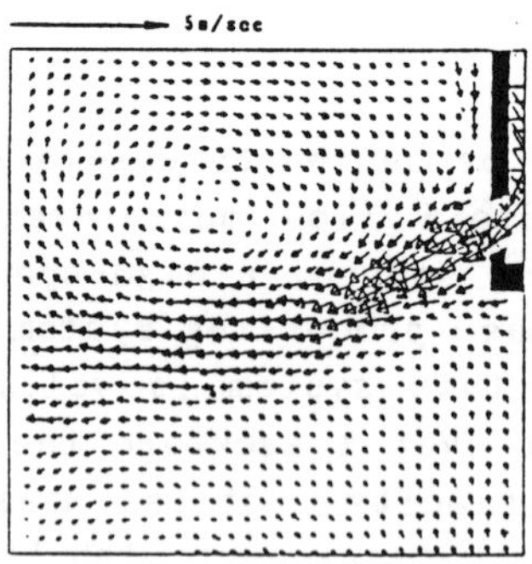

(a) Velocity Vectors on 33×33 Grid Points (PIV)

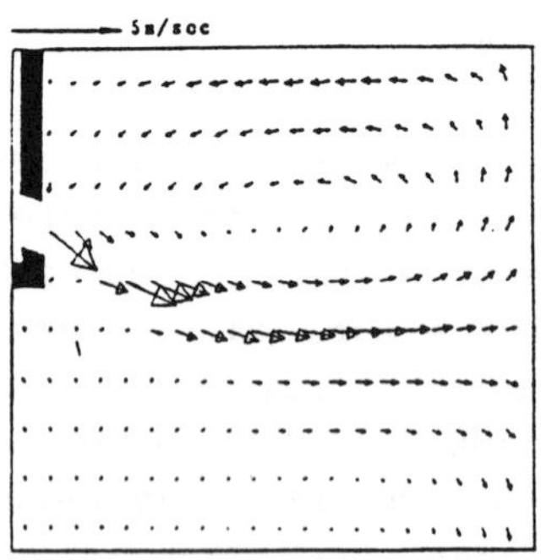

(b) Velocity vectors on 11×11 Grid Points (LDV)

Fig. 10 Comparison PIV and LDV results

REFERENCES

1.Kobayashi,T., Advances in computer-aided flow visualization, Flow visualization VI, Springer-Verlag, 25-37, 1992.

2.Hesselink,L., Digital image processing in flow visualization, Annu. Rev. Fluid Mech., 20, 421-485, 1988.

3.Adrian,R.J., Particle-imaging technique for experimental fluid mechanics, Annu. Rev. Fluid Mech. 23, 261-304, 1991.

4.Kobayashi,T., Saga,T. and Segawa,S., Multi-point velocity measurement for unsteady flow field by digital image processing, Flow Visualization V, Hemisphere, 197-202, 1990.

5.Kobayashi,T., Saga,T. and Sekimoto,K., Velocity measurement of three-dimensional flow around rotating parallel disks by digital image processing, ASME FED Vol.85, 29-36, 1989.

6.Kobayashi,T., Saga,T., Haeno,T. and Tsuda,N., Development of a real time velocity measurement system for high Reynolds fluid flow using a digital image processing design, ASME FED Vol.128, 9-14, 1991.

Visualization of marginal ice zone dynamics

M. Serrer,[a] S.B. Savage,[b] M. Sayed[a]
[a]*National Research Council, Ottawa, Ontario, Canada, K1A 0R6*
[b]*Department of Civil Engineering and Applied Mechanics, McGill University, Montreal, Quebec, Canada, H3A 2K6*

ABSTRACT

Ice floe dynamics in marginal ice zones are simulated using a discrete element model. The ice floes are modelled as random sized, inelastic circular discs floating on the water surface. Contact conditions are represented by a constant restitution and friction coefficients. The ice floes are driven by a uniform wind field. The simulations predict ice floe displacements, rotations, compaction and pressures. This paper focuses on animation and visualization of the simulation output.

INTRODUCTION

Marginal Ice Zones (MIZ) are broken ice fields which often form at the edge of the polar ice pack, next to open ocean. They form as a result of several processes that take place at the pack ice edge including the incidence of waves and increase of temperatures. A MIZ typically extends over a width of the order of 100 km, and consists of distinct floes with diameters varying from a few meters to several kilometres. Discussions of the morphology and behaviour of MIZ were given by Wadhams [1] and [2], and Leppäranta and Hibler [3]. Models of MIZ dynamics have been developed using a viscous-plastic rheology by Leppäranta and Hibler [4], and a "cavitating fluid" analogy by Flato and Hibler [5]. The discrete nature of the ice floes has also prompted the use of continuum models of flowing granular materials to treat MIZ (e.g. Shen et al. [6] and Lu et al. [7]).

The discrete element approach is used here to simulate MIZ dynamics. This approach has been used to model a variety of problems in soil and rock mechanics, flow of granular materials, and ice-structure interaction. Some of recent papers on this subject were compiled by Mustoe [8]. The present study examines the response of an assembly of ice floes to wind drag, water drag and Coriolis force. Interaction forces between the floes are calculated from an inelastic contact model. The simulation predicts the spatial and temporal distributions of ice velocities, concentrations, trajectories and pressures. Animation of the numerical simulations played an essential role throughout this study. It provided an effective means for debugging and verification of the model. Also with the very large amount of information generated by the simulations, animation was necessary for illustrating the salient trends of MIZ behaviour.

The present simulations are based on a model developed by Savage [9]. The following sections briefly discuss the discrete element model of the floes, the equations of motion of individual floes, and the boundary conditions. Presentation of a simulation run follows with emphasis on the animation approach and visualization of the results.

DISCRETE MODEL OF ICE FLOES

Contact model

The ice cover is modelled as an assembly of circular, random sized, inelastic discs. At the contact between two discs, linear loading and unloading springs are used to calculate the normal force as illustrated in Figure 1. The normal force per unit length is given by

$$F_n = k\,\delta \tag{1}$$

where k is the spring constant and δ is the displacement (taken as the radial dimension of the compressed part of the disc). During the loading phase, when discs are approaching each other, the spring constant is equal to k_1. This spring constant becomes equal to k_2 during the unloading phase, when discs are receding from each other. Energy dissipation occurs since the unloading spring constant k_2 is larger than the loading constant k_1. The restitution coefficient e would be equal to $\sqrt{k_1 / k_2}$. The tangential force F_S at the contact is calculated using a Coulomb friction assumption (see Figure 1). There are several other contact models that can account for energy dissipation by using dashpots along with springs. A few runs indicated that changing the contact model has negligible effects on the overall simulation results. The present model is used because of its simplicity, and because it gives a constant restitution coefficient for all disc sizes.

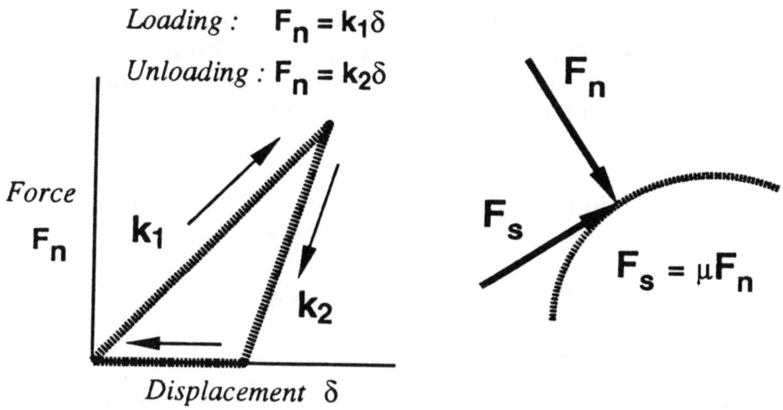

Figure 1 : Ice floe contact model

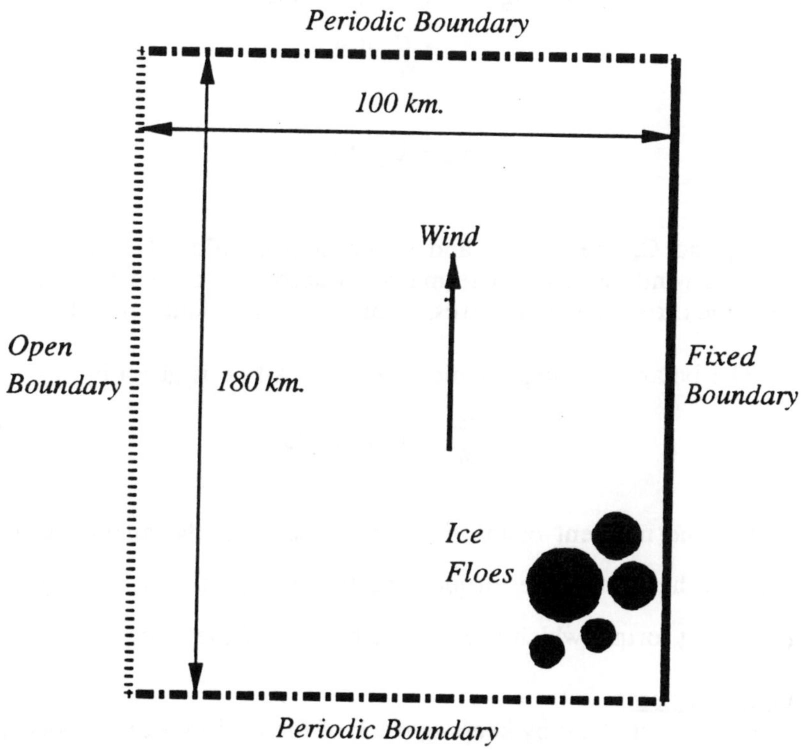

Figure 2 : Boundary conditions

Sayed and Daley [10] performed laboratory experiments to measure contact forces between colliding ice floes, and scaled the measurements to predict full scale conditions. Using their results and assuming that ice thickness is 1 m, contact parameters are chosen as: $k_1 = 2.5 \times 10^5$ kN/m and $\mu = 0.3$.

Equations of motion of an ice floe

The balance of linear momentum may be expressed as

$$m\frac{d\mathbf{u}}{dt} = -m\, f\, \mathbf{k} \times \mathbf{u} + A\tau_a - A\tau_w + \Sigma_{i=1}^{n} F_i \tag{2}$$

where m is the mass of the floe, **u** is the velocity, f is Coriolis parameter, **k** is a unit vector in the vertical direction, A is the surface area of the floe, τ_a is wind stress, τ_w is water stress and $\Sigma_{i=1}^{n} F_i$ is the vector sum of contact forces due to n contacts. The wind and water forces may be approximated by linear drag equations (Flato and Hibler [5]) as follows

$$\tau_a = C_a\, B\, u_a \tag{3}$$

$$\tau_w = C_w\, B\, u \tag{4}$$

where C_a and C_w are the air and water drag coefficients, and u_a is the geostrophic wind velocity. The matrix B accounts for the turning angle between the directions of the stresses and velocities (Flato and Hibler [5]).

The balance of angular momentum of a floe is given by

$$I\frac{d\omega}{dt} = \Sigma_{i=1}^{n} T_i - T_v \tag{5}$$

where I is the moment of inertia of the disc, ω is the angular velocity, $\Sigma_{i=1}^{n} T_i$ is the sum of the torques due to n contacts, and T_v is the water drag resisting torque which acts on the bottom of the floe.

Simulation scheme

Computations are done by keeping track of the positions and velocities of the discs. At each time step, the overlap between the discs and the contact forces are calculated. Wind, water and Coriolis forces are then added to the contact forces for each disc. Accelerations are calculated from the resultant forces. Velocities and displacements are next computed by numerically integrating the accelerations. The preceding scheme is repeated for each time step. In order to maintain numerical stability, the time step should be small enough such that any contact would extend over several steps (at least 50 steps). Computation time was greatly reduced by keeping a list of the neighbours of each disc, which was updated at intervals of several steps. This list was used in the calculation of contact forces.

BOUNDARY CONDITIONS AND SIMULATION PARAMETERS

We consider the case of geostrophic wind oriented parallel to a straight boundary of a solid ice cover. The ice floes would flow parallel to the solid boundary. This case is simulated by placing the discs within rectangular boundaries as shown in Figure 2. The wind velocity acts in a northley direction. The east boundary is fixed (representing the solid ice cover edge), and the west boundary is free. The north and south boundaries are periodic. A disc leaving one of the periodic boundaries is re-introduced at the opposite boundary, with its same velocity. The distance between the north and south boundaries is 180 km. The following parameter values are used in the simulation :

number of discs = 360
disc diameters = 6 km to 10 km
ice thickness = 1 m
ice density = 910 kg/m^3
restitution coefficient e = 0.5
geostrophic wind velocity = 10 m/s
Coriolis parameter f = 1.44×10^{-4} s^{-1}
air drag coefficient C_a = 0.0126 kg m^{-2} s^{-1}
water drag coefficient C_w = 0.652 kg m^{-2} s^{-1}
the turning angle between the drag force and the velocity = 22.5 ° (clockwise)
time step = 11 seconds
duration of the simulation = 40 days

ANIMATION

The animation program is a gl (Graphics library programming guide of Silicon Graphics Inc. [10]) based six degree of freedom geometry modeller. This modeller was designed to interactively display simulation results of multi-body dynamic systems. The user has full interactive control of all viewing and display functions as an animation is being displayed.

Overview of animation data

Input to the animator consists of:

1. An object definition file containing the geometry of the simulated system. The geometry is defined as a number of rigid bodies each with its own coordinate reference frame. An origin for each body is specified in the global reference frame. Each body can be composed of any number of cylinders, boxes, cones, spheres and polygons. The components are positioned relative to the body origin. Each component can have a unique colour/material property.

2. A position/orientation file containing one record for each simulation time step. Each record contains the spatial coordinates and orientations for each body in global coordinates.

3. A contact force file containing one record for each simulation time step. Each record contains the number of contacts followed by the coordinates of the contact point in global coordinates and the magnitude of each force.

4. A velocity file containing one record for each simulation time step. Each record contains the velocity components along the east-west and north-south directions for each body.

Although the current simulation is two-dimensional, future versions with full three dimensional force and velocity displays will be possible.

Graphical User Interface (GUI)

The user interface was created using the gl (reference [10]), and a GUI tool FORMS (Overmars [12]). Figure 3 shows a screen snapshot of the GUI. There are five basic functions available from the GUI:

1. Selection of data
   - geometry file.
   - animation file.
2. Playback of data file(s), rewind, single step, play,.. etc. On an IRIS 4D50/GT, an animation frame rate of two frames/second can be achieved displaying motion,forces and velocities of 360 particles.
3. Control of external single frame recording device. In this case a Panasonic Optical Disk Recorder.
4. Selection of the items to be displayed (e.g. bodies, velocities, forces).
5. Control of view which determines: position of eye, point of interest,field of view, clipping area.

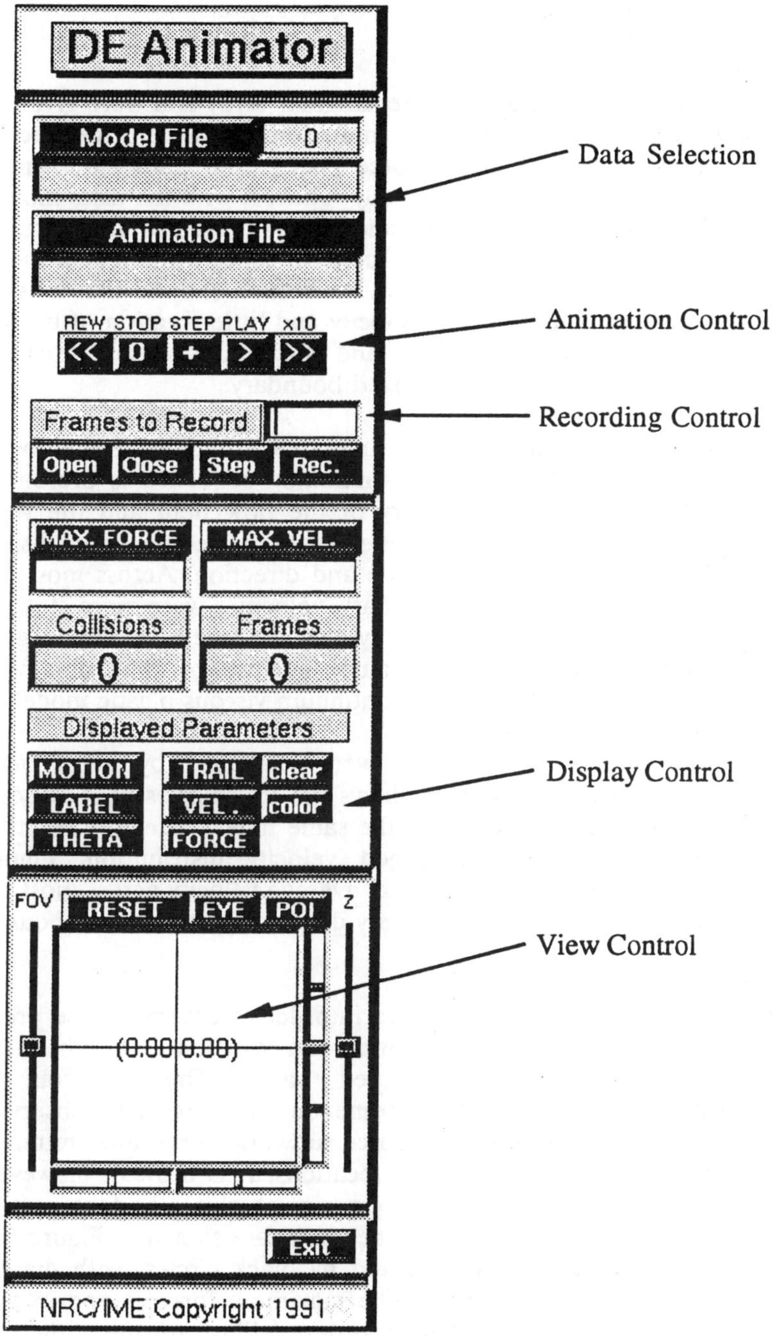

Figure 3 : Graphical User Interface

SIMULATION RESULTS

Velocities and trajectories

The discs were placed in a sparse rectangular array and were given small random velocities at the start of the simulation. As the velocities increased because of wind stress, Coriolis force packed the discs along the east boundary. A steady flow was eventually reached, with the discs moving in a North direction. The overall ice concentration (area of discs divided by the total area) is approximately 0.8. The local concentration near the fixed boundary has somewhat lower values. Animation showed that most discs move in one group with uniform velocity and little rotation of individual discs. Relative velocity between discs and fluctuations of each disc velocity occur over a narrow region at the fixed boundary.

Instantaneous velocities were animated as vectors originating from disc centres. One frame from this animation sequence is presented in Figure 4 which shows the geometry of the flow field and the velocity vectors of individual floes. The velocities near the fixed boundary displayed the largest fluctuations in magnitude and direction. Across most of the flow width away from the fixed boundary, velocities were more steady and uniform. The magnitude of the velocity near the free boundary (Figure 4) is approximately 0.2 m/s. This value is close to that obtained by Leppäranta and Hibler [4] using a continuum viscous-plastic model.

Displacements of the discs were examined by animating the trajectories of the discs. Figure 5 shows the trajectories of six discs that started from the south boundary at the same time. Trajectory lengths are representative of the time-averaged velocity distribution. Figure 5 corresponds, as expected, to a uniform average velocity across most of the width of the flow. The average velocity drops only at the fixed boundary.

Contact forces

Normal contact forces were animated in order to determine the pressure distributions across the MIZ. The animation revealed that contact forces follow networks of irregularly shaped "paths". This behaviour is in accordance with the "arching" phenomenon which is often observed in granular materials. The animated force networks were intermittent and always changed their geometry and locations. It is difficult, however, to quantify these geometries or the frequencies at which the force networks form. Examples of the contact force networks are shown in Figure 6. The normal contact forces are represented by black circles, with diameters proportional to their magnitudes. The maximum force is approximately 225 MN. Four cases are displayed in order to illustrate the variability of the geometry and magnitude of the force distributions.

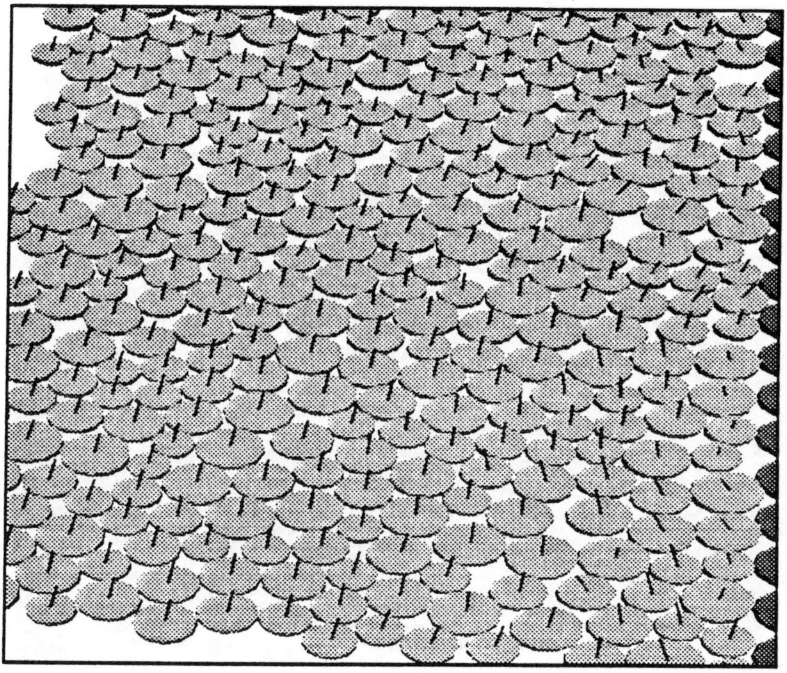

Figure 4 : Instantaneous velocity

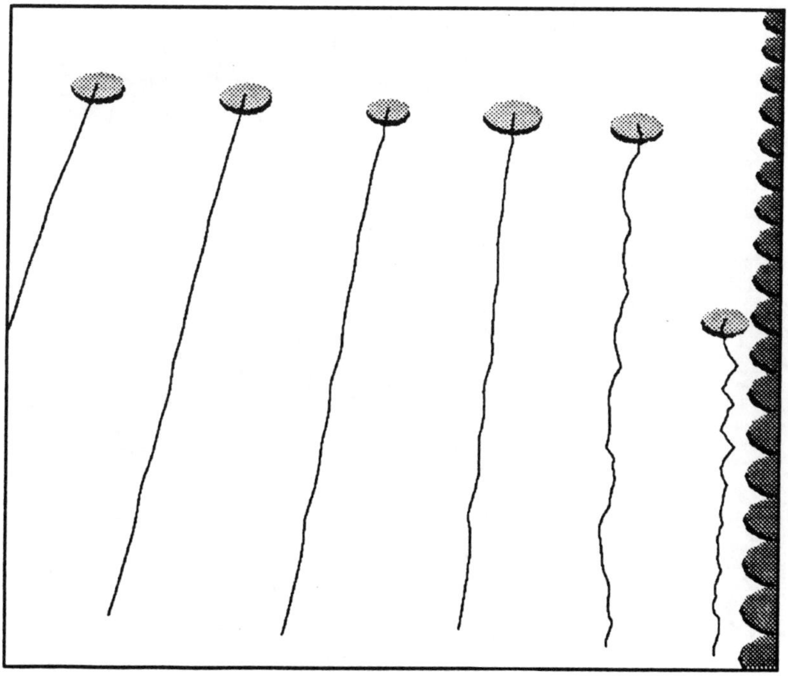

Figure 5 : Trajectories of six discs

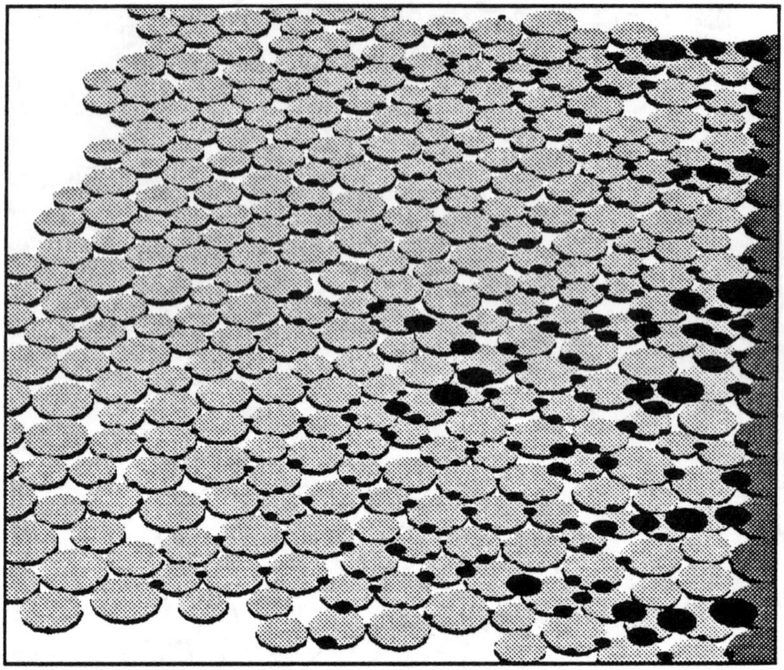

Figure 6a : Normal contact forces, simulation step 276010

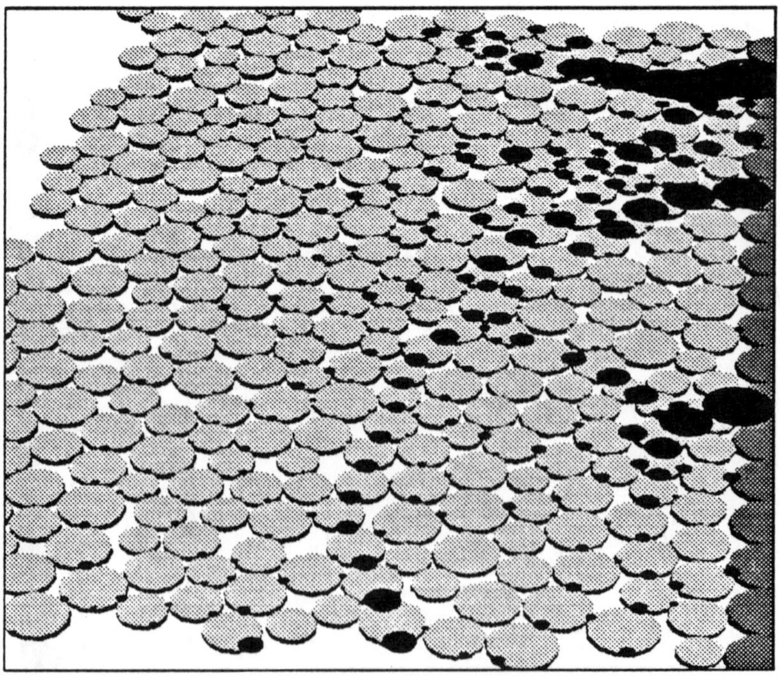

Figure 6b : Normal contact forces, simulation step 294290

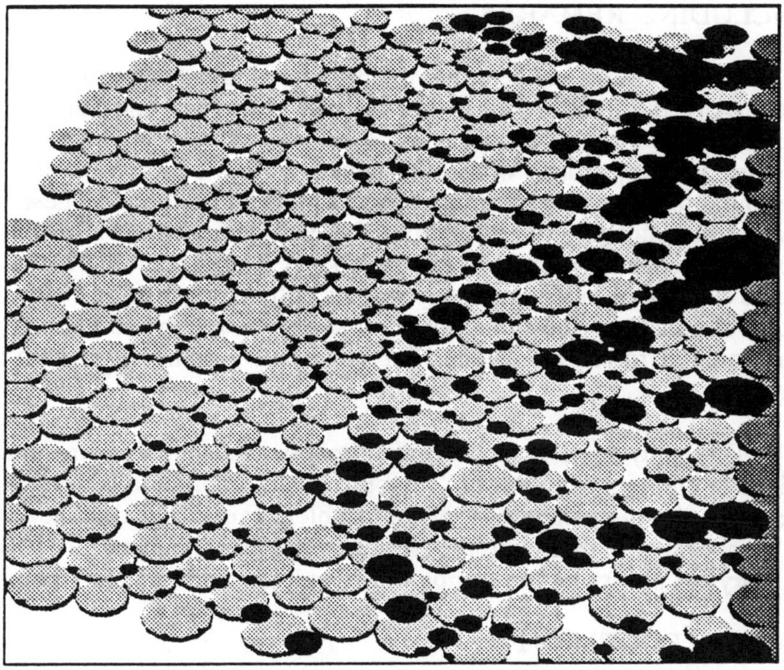

Figure 6c : Normal contact forces, simulation step 297120

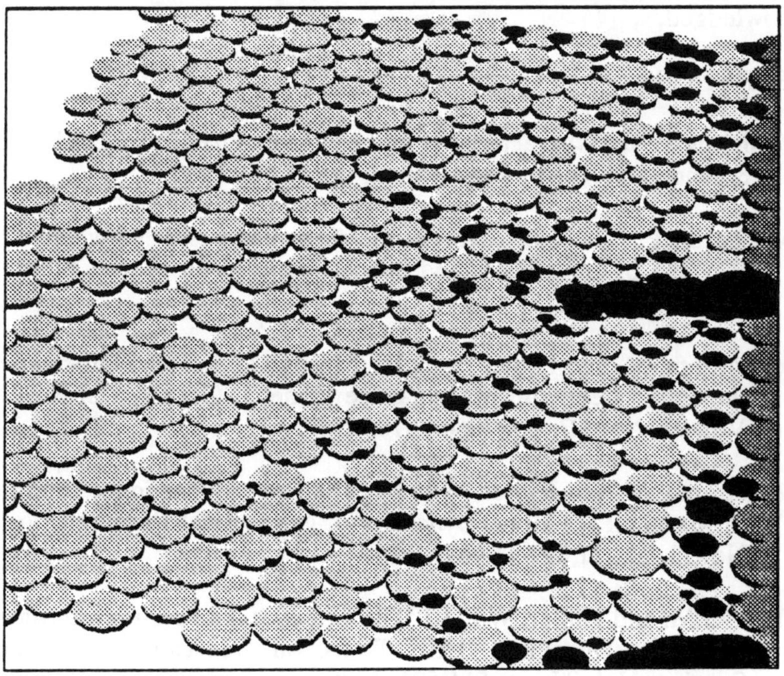

Figure 6d : Normal contact forces, simulation step 304000

CONCLUDING REMARKS

The preceding results indicate that the discrete element approach provides considerably more information about MIZ dynamics than the current continuum models. Fluctuations of local values of ice concentration, velocities and trajectories were determined. Forces between the ice floes were found to follow intermittent networks of irregular shapes. These force networks are radically different from the continuous steady distributions which are predicted by the continuum models. The present approach thus appears capable of modelling many aspects of field observations. However, further simulations involving more realistic boundary conditions and environmental forces are needed for quantitative comparison with field data.

This study has also shown that animation is essential for the analysis of discrete element simulations. The output of a simulation is usually very large. Also the discrete nature of the problem cause considerable spatial and temporal variability in the results. Therefore, animation become necessary for observing the trends which emerge from the simulation.

ACKNOWLEDGEMENT

The financial support of the National Energy Board of Canada is gratefully acknowledged.

REFERENCES

1. Wadhams, P. 'Ice Characteristics in the Seasonal Ice Zone' Cold Regions Science and Technology, Vol. 2, pp. 37-87, 1980.

2. Wadhams, P. 'The Ice Cover in the Greenland and Norwegian Seas' Reviews of Geophysics and Space Physics, Vol. 19, No. 3, pp. 345-393, August 1981.

3. Leppäranta, M. and Hibler, W.D.III 'Mesoscale Sea Ice Deformation in the East Greenland Marginal Ice Zone' J. Geophysical Research, Vol. 92, No. C7, pp. 7060-7070, 1987.

4. Leppäranta, M. and Hibler, W.D.III 'The Role of Plastic Ice Interaction in Marginal Ice Zone Dynamics' J. geophysical Research, Vol. 90, No. 11, pp. 11,899-11,909, 1985.

5. Flato, G.M. and Hibler, W.D.III 'The Effect of Ice Pressure on Marginal Ice Zone Dynamics' IEEE Trans. on Geoscience and Remote Sensing, Vol. 27, pp. 514-521, 1989.

6. Shen, H.H., Hibler, W.D.III and Leppäranta, M. 'On Applying Granular Flow Theory to a Deforming Broken Ice Field' Acta Mechanica, Vol. 63, pp. 143-160.

7. Lu, Q-M, Larsen, J. and Tryde, P. 'On the Role of Ice Interaction Due to Floe Collisions in Marginal Ice Zone Dynamics' J. Geophysical Research, Vol. 94, No. 14, pp. 14,525-14,537.

8. Mustoe, G.G.W., Special issue devoted to selected papers presented at the 1st US Conf. on Discrete Element Methods (DEM), held at Colorado School of Mines, Colorado, October 1989, Int. J. for Computer-Aided Engineering and Software, Vol. 9, No. 2, April 1992.

9. Savage, S.B. 'Marginal Ice Zone Dynamics Modelled by Computer Simulation Involving Floe Collisions' Report to the Institute for Mechanical Engineering, National Research Council Canada, Ottawa, Ontario, K1A 0R6, April 1992.

10. Sayed, M. and Daley, C. 'Experiments on Crushing Interaction Between Ice Floes' to appear in Proceedings of the 3rd Int. Offshore and Polar Engineering Conf. (ISOPE), Singapore, June 6-11, 1993.

11. Graphics Library Programming Guide, Silicon Graphics Inc., Mountain View, California, USA, 1991.

12. Overmars, M.H. 'Forms library, A graphical user interface toolkit for Silicon Graphics workstations' Department of Computer Science, Utrecht University, P.O. Box 80.089, 3508 TB Utrecht, the Netherlands, version 2.1, 1992.

Visualization software for injection molding

F. Trochu, D. Gour, M. Audet
Mechanical Engineering, École Polytechnique, C.P. 6079, Succ.'A', Montréal, Québec, Canada, H3C 3A7

ABSTRACT

Injection molding is a general manufacturing process used in various areas: metal casting, fabrication of plastic parts and composite materials, etc. The numerical simulation of mold filling can help the mold designer to properly position the injection ports and air vents so as to ensure an adequate filling of the mold. As finite elements are used increasingly to model the injection flow, the need for a specialized visualization software has arisen to simulate the real-time progression of the flow front and to display the temperature and pressure fields inside complex three-dimensional molds. The basic requirements of such an interface will be examined in the scope of finite element analysis. As an illustration of these concepts, the computer program Visuflot developed for resin transfer molding (RTM) is described. A video showing examples of real-time simulations will also be presented.

INTRODUCTION

A general software was developed during the summer of 1991 at Ecole Polytechnique[1] to visualize the flow front and pressure field calculated to simulate the filling of complex three-dimensional molds used in the manufacturing of composite materials. Although it was initially designed for a particular manufacturing process, namely resin transfer molding (RTM), its general features make it suitable for application to any type of injection molding simulation: thermoplastic flow analysis, metal casting, etc.

The purpose of this paper is to analyze the general software requirements of an injection molding simulation and to describe solutions devised to address these particular needs. An example of implementation is provided by program Visuflot, which was developed initially for resin transfer molding. Outputs from Visuflot permit to illustrate this approach, which remains general and can be applied to any kind of injection process.

The first part of the paper presents the programming philosophy of Visuflot and sums up the advantages of object oriented analysis for the development of software applications. The second part discusses the particular problems of injection molding, namely (1) the choice between a *fixed grid* or a *moving mesh* to follow the changing shape of the flow front as the fluid fills up the mold and (2) the methodology adopted to advance the flow front at each time step. Finally, as an illustration of these concepts, output results from Visuflot are presented and the basic requirements of a visualization software for injection molding are stated.

OBJECT ORIENTED DESIGN PHILOSOPHY

This work is part of a trend in software development made possible by the advent of object oriented programming[2]. Although it would be difficult to develop a general purpose program for injection molding because each process is governed by different mathematical equations, a great number of software modules remain identical, in particular the mesh generation and visualization modules.

Object oriented programming provides a general framework for software analysis[3]: the design of program modules is intimately connected with the specification of data structures, both being "encapsulated", so to say, in what is called a software object. Each object can be made completely independent from the rest of the system, the communication between the various modules being assured by specialized input-output functions. This permits to concentrate on the design of each object as a whole, without having to worry about its connections with other objects.

Object oriented programming has made possible the creation of specialized software modules general enough to be reutilized systematically. The software is specialized in the sense that it executes specific tasks. But at the same time, it is general enough so that the basic component modules can be combined easily to create new applications. An example of a successful object oriented software is provided by the OSF/Motif user interface and the X-Window toolkit, which have now become a standard for user interfaces on UNIX based workstations. The interface of program Visuflot is written with OSF/Motif.

A new generation of application software is based on object oriented programming. This approach helps to free end-users from a dependency upon huge "black boxes", i.e., the kind of program designed to incorporate all what the users might need. These programs have progressively grown to become "software monsters", relying heavily on vast hardware resources and highly sensitive to the computer system's environment. Everything must be planned through a complex design procedure by specialized software developers before any new feature can be added. This kind of program is not flexible and permits to accomplish only pre-programmed tasks.

On the contrary, the new generation of object oriented software gives end-users the opportunity to develop rapidly a specific application, using as building blocks the objects provided by the system's library. In this approach, however, the user must have a basic knowledge of programming, although he needs not be a specialist. It is sufficient for him to use the system's library as a tool for the purpose of his own application. The ACIS system[4] in geometric modelling is a good example of a successful software based on object oriented programming. The same kind of development is likely to occur in the field of finite element analysis, graphic applications, animation, etc.

COMPUTER SIMULATION OF INJECTION MOLDING

The main issue in the simulation of mold filling is the numerical treatment of the transient free surface or moving boundary. The resin entering inside the mold is constantly changing shape as it flows, so it is necessary to redefine the geometry of the saturated domain during the simulation. In the moving boundary approach, a new mesh is

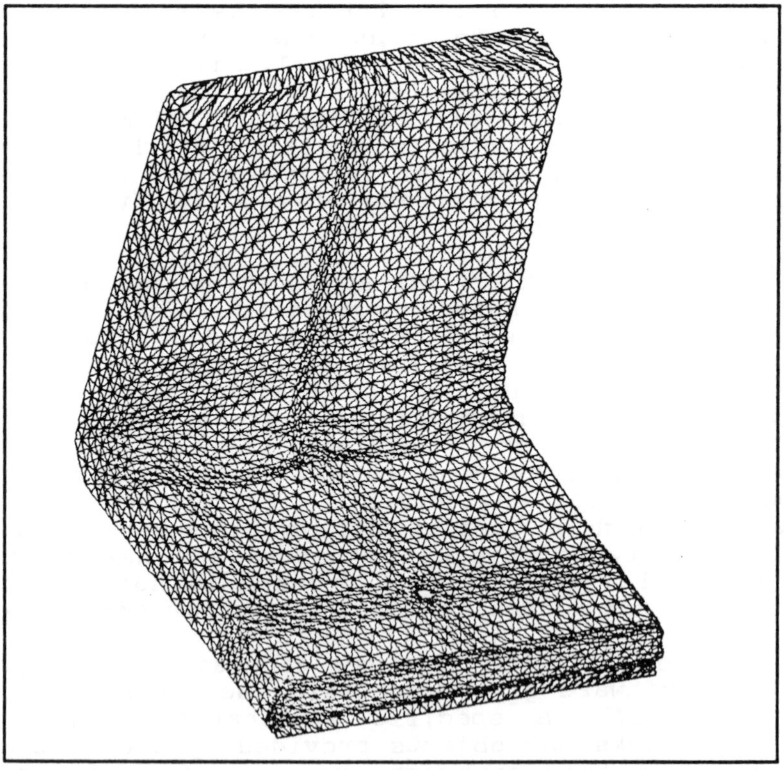

Figure 1. Example of fixed grid for a subway seat.

generated at each time step. The mesh generation could soon become the most tedious part of the simulation as experienced in [5] for example. Other problems occur when this approach is used. For example, it is difficult to divide the flow front inside molds with inserts or to merge the fronts coming from multiple injection ports. In order to overcome these limitations, it is easier to perform the simulation on a fixed grid (figure 1). But this requires the development of a special technique to follow the progression of the fluid flow on a fixed grid.

<u>Fill factors associated with the elements of the mesh</u>

The geometric domain representing the mold is decomposed into a finite number of solid elements as shown in figure 1 for a subway seat. In the fixed grid approach, a fill factor f, $0 \le f \le 1$, is associated to each element in order to track the progression of the flow front. These coefficients represent the level of fluid saturation inside an element (see figure 2). If f = 0, the element is empty; when f = 1, it is saturated. The saturation coefficients vary with time from 0 to 1 inside each element. At the end of the filling process, they become 1 for all the elements. Note that the fill factors employed by most investigators[6,7] are associated with the nodes rather than with the elements as in our model.

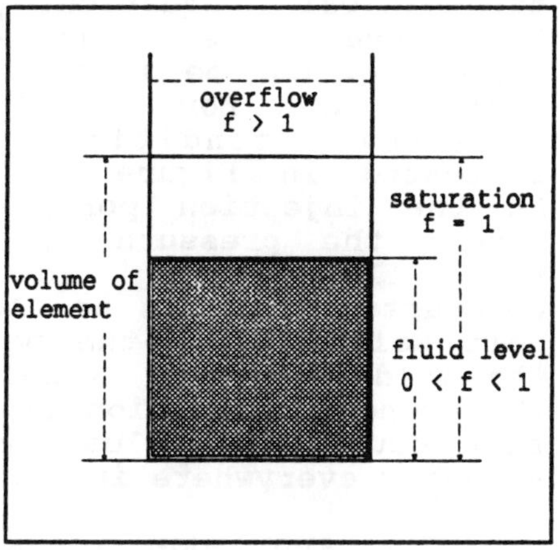

Figure 2. Fill factors associated with each volume element.

The fixed grid approach presents a computational advantage over moving grid techniques such as used in [5-8] because it is no longer necessary to reconstruct the mesh for each iteration. Moreover, the mesh generation can become rather complicated in the case of dividing or merging flow fronts. The drawback of the fixed grid, however, is a relative loss of accuracy in the shape of the resin front. Nevertheless, this problem can be overcome by increasing the density of the grid and by smoothing the front lines. The relative loss in accuracy arising from the mesh must be contemplated in the scope of the other approximations that come into play, namely the approximation error of the boundary value problem itself.

Calculation of the flow

The usual strategy to model an injection molding process is to consider the domain Ω filled up at a given time by the fluid (see figure 3). The viscous flow of a Newtonian fluid is described by the pressure field p and the velocity vector $\boldsymbol{V}$ of any fluid particle. The mathematical equations that govern the steady state flow can be solved inside Ω for the boundary conditions indicated in figure 3. At the injection port, either the pressure or the flow rate is constant. There is no flow through the impermeable mold walls, i.e., the normal velocity vanishes on the mold boundary. Finally, along the flow front, the boundary condition is $p = 0$. The solution of the flow equations yields the pressure and the fluid velocity everywhere in Ω.

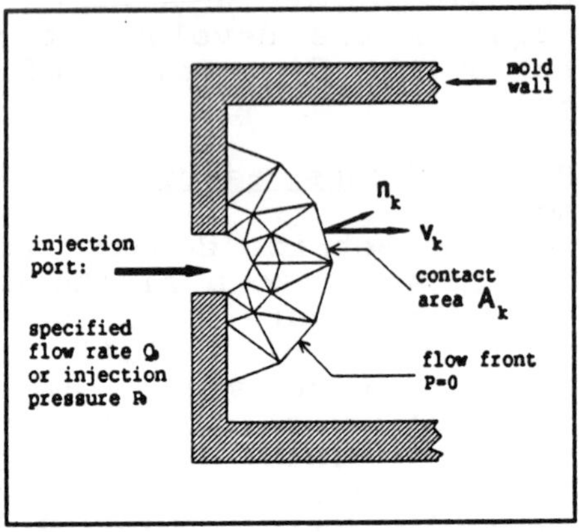

Figure 3. Solution of the flow problem in the fluid saturated part of the mold.

The knowledge of the pressure distribution inside each element permits to verify the integrity of the mold during the filling process. The fluid velocity along the flow front gives the fluid mass flowing through each element k of the front during time increment Δt (see figure 3)

$$R_k = \boldsymbol{V}_k \cdot \boldsymbol{n}_k \; A_k \; \Delta t \tag{1}$$

where $\boldsymbol{V}_k$ is the velocity vector, $\boldsymbol{n}_k$ the outward unit vector normal to the front profile on element k *and* A_k the lateral area of element k on the flow front.

It is important for the numerical scheme to conserve the fluid mass, because this is the basic variable that can be controlled during a filling process. This can be achieved by using non-conforming finite elements as presented in [1,9]. In order to respect also the balance of mass when moving the flow front at each time step, the following simple algorithm was devised.

Progression of the flow front

A new position of the flow front can be deduced at each time step from the previous one by updating the fill factors of the elements adjacent to Ω. The way the algorithm proceeds is indicated in figure 4 in the simple case of triangles. Two types of boundary elements are considered: elements with one incoming flow (case 1) and elements with two incoming flows (case 2). Note that this approach based on fluid mass conservation can be generalized to any kind of mold discretization (prismatic elements, tetrahedrons, etc.) and applied to any type of injection process.

For the algorithm to work correctly, it is necessary to allow a temporary overflow inside the elements adjacent to the flow front. Finally, after each iteration, a new saturated domain is created. The flow equations can be solved again in this new domain and the simulation is ready for a new iteration. Note that this algorithm is stable by itself when applied directly to simulate a filling process on a fixed grid. It permits to reproduce the experimental shape of the flow front for isotropic flows[10].

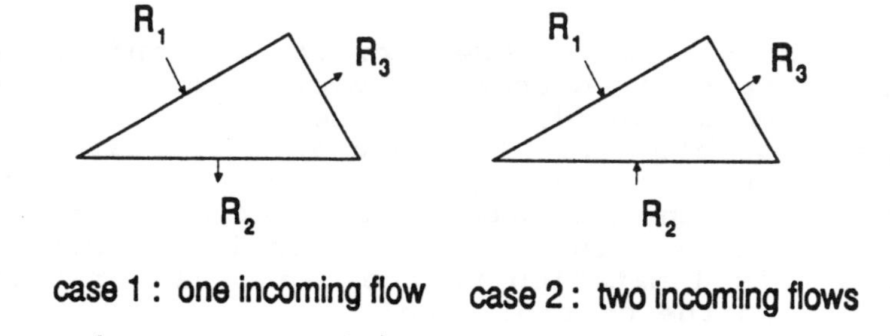

$$R_2 = \frac{A_2}{A_2 + A_3} R_1 \ , \quad R_3 = \frac{A_3}{A_2 + A_3} R_1 \qquad R_3 = R_1 + R_2$$

Figure 4. Evacuation of the overflow from triangular elements.

The choice of the time step Δt is important. It must not be too large for the accuracy of the steady state approximation to remain valid. It cannot be too small either if the flow front is to move forward. Actually, when the front becomes larger, its average velocity V decreases. The best strategy consists in increasing progessively Δt so that the average displacement of the front $\Delta x = V \Delta t$ remains approximately constant throughout the simulation.

REQUIREMENTS OF A VISUALIZATION SOFTWARE FOR INJECTION MOLDING

The calculation of the filling process is performed off-line by program Flot (finite element analysis). At each time step the pressure and velocity are computed in the saturated part of the mold and the position of the flow front is updated accordingly. For a typical three-dimensional mold with 5000 degrees of freedom, the computation may last from 45 minutes to one hour on a RISC/6000 workstation. The output of Flot is stored on a file, ready for interactive visualization by program Visuflot.

A visualization software for injection molding should be able to represent on the computer the real-time filling process. This task is not as easy as it might seem, because of the large number of volume elements that must be updated in real-time. This capability of program Visuflot will be illustrated by a video. Note that the graphic part of Visuflot is programmed in GL, the "Graphics Language" available either on Silicon Graphics or IBM workstations.

The simulation of the flow can be rendered at a requested speed in direct or reverse mode, i.e., for the filling or unfilling of the mold. It is not always possible due to hardware limitations to display the filling process with its actual velocity. Then the program automatically performs the filling at the maximum possible speed and signals this limitation to the user via a red light appearing on the control panel.

Successive positions of the flow front must also be recorded on a single image for subsequent analysis (see figure 5). Since the injection takes place usually in complex three-dimensional molds, it is important to position the object in space at the same time as the simulation is performed. This can be done with Visuflot via the control panel, a menu-driven interface allowing to position, scale and magnify the mold.

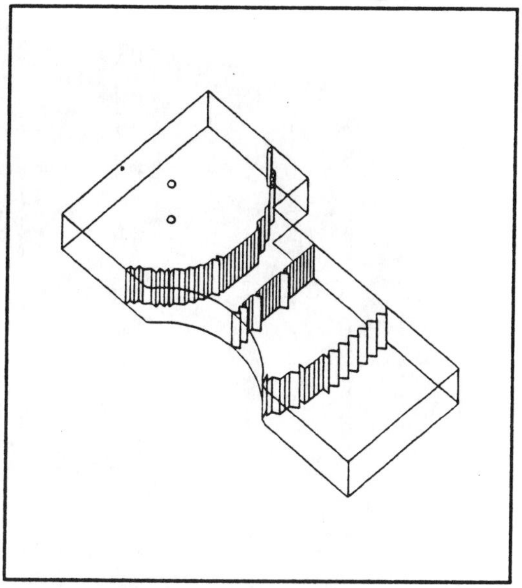

Figure 5. Example of successive positions of the flow front.

A sophisticated visualization program must also display the *saturated* and/or *non-saturated* flow fronts because this helps to verify and validate the numerical scheme. The saturated front is defined as the boundary between the fully saturated (f=1) and partly saturated volume elements (0≤f≤1). The non-saturated front is the boundary between the partly saturated (0≤f≤1) and empty elements (f=0). The real front lies in between the saturated and non-saturated flow fronts. By interpolating the fill factors of each element, it is possible to obtain an estimated position of the real front. Since all the surfaces representing the fronts are defined in a three-dimensional space, it is important to use shading to display the flow fronts. Examples of shaded fronts will be presented in the video.

Finally, another important feature is the ability to display the pressure field in the mold (see figure 6). This can be achieved at each time step either by a color gradation or by assigning different colors on a scale of pressure intervals. Note that if the pressure inside the mold exceeds a preset given value, either a bell is rang or the corresponding volume element is colored in red.

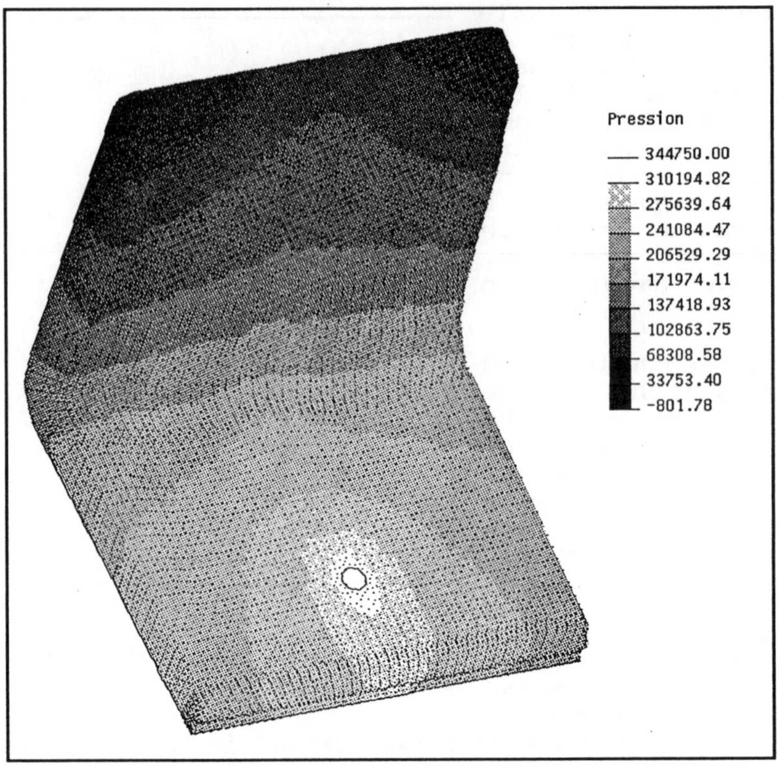

Figure 6. Example of pressure distribution in the mold of a subway seat.

CONCLUSION

The analysis is performed on a fixed grid and a fill factor, representing the fraction of each elementary volume occupied by the fluid, is associated with each element of the mold. This permits to simulate the case of dividing or merging fronts inside molds containing interior obstacles or with multiple injection ports. An important feature of a numerical scheme for the simulation of an injection process is the ability to conserve the fluid mass. This principle is respected by the algorithm proposed here to advance the flow front on a fixed grid, which is stable and can be applied to any type of injection process. A visualization software for injection molding must be able to simulate the real-time filling of complex three-dimensional molds. Successive positions of the flow front must be recorded also on the same image. The pressure field inside the mold must be displayed if requested. All these requirements are met by program Visuflot, which was developed to visualize the resin transfer molding process, but can be utilized also for any type of mold filling simulation.

ACKNOWLEDGEMENTS

The authors wish to thank Jean-François Boudreault and Dong-Ming Gao for their work on the subway seat. This work was supported by the National Research Council of Canada who is gratefully acknowledged.

REFERENCES

1. Trochu, F., Gauvin, R. and Zhang, Z. 'Simulation of Mold Filling in Resin Transfer Molding by Non-conforming Finite Elements' *Proceedings of the Conf. on Comp. Aid. Des. for Compos. Mat.* (CADCOMP 92), Newark, Delaware, May 1992.

2. Coad, P. and Yourdon, E. *Object-Oriented Analysis*, Yourdon Press Computing Series, Prentice Hall, 1990.

3. Hekmatpour, S. *C++ A guide for C programmers*, Prentice Hall, 1990.

4. *ACIS Training Guide*, Spatial Technology, Inc., Boulder, Colorado, 1992.

5. Trochu, F. and Gauvin, R. 'Limitations of a Boundary-Fitted Finite Difference Method for the Simulation of the Resin Transfer Molding Process' *Journal of Reinforced Plastics and Composites*, July 1991.

6. Bruschke, M.V. and Advani, S.G. 'A Finite Element/Control Volume Approach to Mold Filling in Anisotropic Porous Media' *Polymer Composites*, Vol. 11, No. 6, 1990.

7. Young, W.B., Han, K., Fang, L.h., Lee, L.J. and Liou, M.J. 'Flow Simulation in Molds with Preplaced Fiber Mats' *Polymer Composites*, Vol. 12, No. 1, pp. 30-38, 1991.

8. Coulter, J.P. and Güçeri, S.I. 'Resin Impregnation During Composites Manufacturing: Theory and Experimentation' *Composites Science and Technology*, 35, pp. 317-330, 1989.

9. Trochu, F., Gauvin, R. and Gao, D.-M. Numerical Analysis of the Resing Transfer Molding Process by Non-conforming Finite Elements, submitted to *Composites Science and Technology*, August 1992.

10. Zhang, Z. *Simulation par éléments finis du remplissage des moules pour le procédé de moulage par transfert de résine (RTM)*, Mémoire de maîtrise, Département de Génie mécanique, Ecole Polytechnique, Montréal, Canada, décembre 1992.

PTV measurement of a three-dimensional particle motion in a bubbling jet mixing water vessel

T. Uemura,[a] F. Yamamoto,[b] M. Iguchi,[d] Z. Morita,[d] T. Usui,[c] J. Ohta[b]

[a] *Kansai University*

[b] *Department of Mechanical Engineering, Fukui University, 3-9-1 Bunkyo, Fukui-shi 910, Japan*

[c] *Faculty of Engineering Science, Osaka University, 1-1 Machikaneyama, Toyonaka, Osaka 560, Japan*

[d] *Faculty of Engineering, Osaka University, 2-1 Yamada-oka, Suita, Osaka 565, Japan*

ABSTRACT

The present paper proposes a new stereo pair matching technique with correction of light refraction at water-air boundary using two video pictures and a new particle tracking algorithm for three-dimensional particle-tracking velocimetry (3D PTV) based on a cross-correlation method using two-consecutive-time-step binary pictures. The technique of 3D PTV is applied to measure three-dimensional velocity vectors of tracer particle motion in a cylindrical water vessel with a bottom blowing bubbling jet. The performance of stereo pair matching and the measurement results of the bubble-water two-phase flows are discussed, and as a result the proposed technique is shown to be useful in measuring the three-dimensional bubbling jet flow field.

INTRODUCTION

Particle-imaging velocimetry (PIV in abbreviation) using the techniques of flow visualization and digital image processing is a method for measuring fluid flow velocity indirectly by analyzing tracer particle motion at a time interval on the pictures. It has the advantages of whole field, contact-free and high speed measurement of instantaneous two- or three-dimensional (2D or 3D) flow, but it also has a disadvantage of low time resolution and relatively low spatial resolution, compared to the one-point measurement methods such as hot wire anemometer and LDV.

In the past ten-odd years, numerous papers on two-dimensional PIV have been published. One such method is the correlation method (Yano [1], Kimura and Takamori [2]) which computes the particle cloud displacement at a time interval and then the particle velocity vectors after finding the most similar distribution patterns of particle cloud density in two-consecutive pictures, using the cross-correlation coefficient to evaluate the pattern matching. Another is the so-called four-consecutive-time-step method (Kobayashi [3], Kaga and Yoshikawa [4], Kaga, Inoue and Yoshikawa [5], Chang, Watson and Tatterson [6]) which tracks each particle motion in four-consecutive pictures by finding the smoothest particle trajectory which gives the minimum value of deviation of the particle displacement and angle. The authors have proposed a new technique which is the so-called binary cross-correlation method (Yamamoto, Uemura, Koukawa, Itoh and Teranishi [7], Uemura, Yamamoto and Koukawa [8]). It also tracks the motion of each particle in two-consecutive pictures. The principle is based on the similarity of particle distribution patterns in the two-consecutive binarized pictures to identify each particle motion using the cross-correlation for the pattern matching. The technique has a high speed algorithm for the particle identification and takes a very short time to compute the velocity vectors. The authors have also proposed a double-check technique (Uemura, Yamamoto and Ohmi [9]) to replace some mismatched velocity vectors with the correct ones.

Recently the PIV has been applied to the measurement of turbulent flows, in which quantitative measurement of three-dimensional flow field is required. Three-dimensional PTV has been developed and applied to the unsteady flow measurement in rotating double circular cylinders to investigate the turbulent energy distribution of homogeneous turbulent flows in an agitated tank using the four-consecutive-time-step method (Nishino, Kasagi, Hirata and Sata [10]).

The present paper proposes a new stereo pair matching technique with correction of light refraction at water-air boundary using two or three video camera images and a new particle tracking algorithm for 3D PTV based on a cross-correlation method using two binary pictures. The technique of 3D PTV is applied to measure the three-dimensional velocity vectors of tracer particle motions in a cylindrical water vessel with a bottom blowing bubbling jet. The performance of stereo pair matching and the measurement results of the bubble-water two-phase flows are discussed, and as a result the proposed technique is shown to be useful in measuring the three-dimensional flow field.

MAIN THREE ROUTINES FOR 3D PTV

The 3D PTV consists of the following main three routines. (1) First, flow field is visualized by putting small solid particles as tracers into fluid flow, and pictures of instantaneous particle motion in a fluid flow are taken consecutively using two or three CCD type video cameras. (2) Second, three-dimensional X, Y and Z coordinates of each particle are decided by a stereo pair matching method using the two or three pictures taken at the same time by cameras. (3) Third, each particle is tracked or identified between the first and the second spaces at a time interval Δt.

The following two sections describe a new stereo pair matching method and a new three-dimensional particle tracking technique based on a cross-correlation method using binary imaging data, omitting the explanation of techniques for flow visualization.

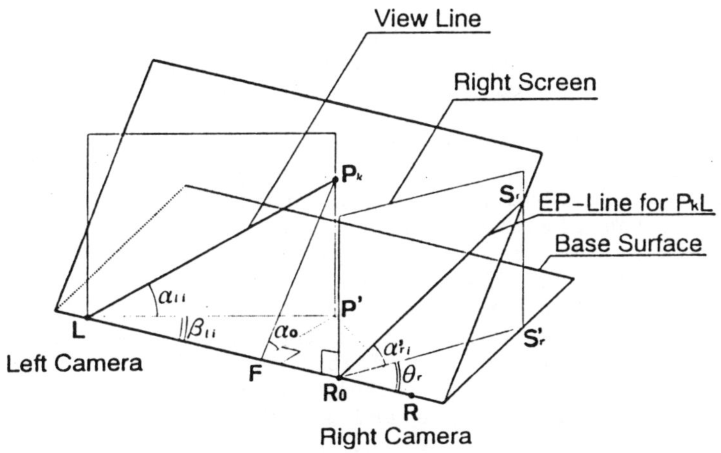

(a) Geometry of epipolar line for viewline, $P_k L$

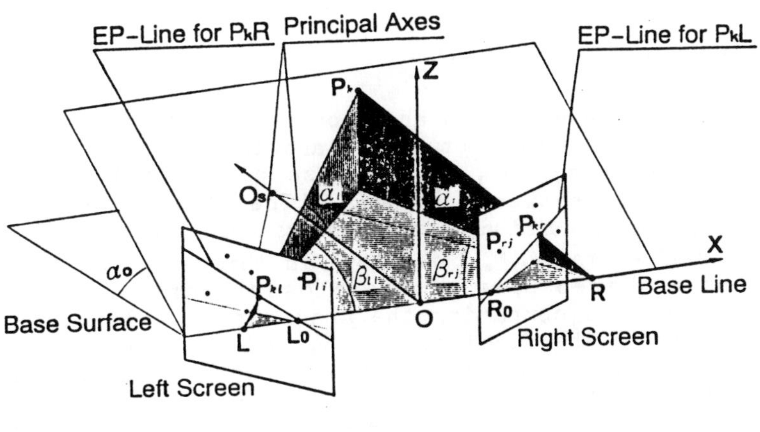

(b) Geometrical and camera parameters

Fig. 1 Illustration of stereo pair matching

Stereo Pair Matching Method

The visualized flow field is photographed by two video cameras, whose perspective screens are fixed in contact with a spherical surface and whose visual-axes intersect at the central point of the sphere, O_S. A pair of screens are aligned so that either one of visual-axes on each plane is contained in one common plane of a large circle, on which the viewlines are co-planar. Pictures are obtained with equal scale factor.

Geometrical and camera parameters are shown in Fig. 1. The viewpoints of left and right cameras are denoted by L and R, respectively. The length of the base line, LR, should be measured at a high accuracy because it influences the overall accuracy of the measurement results.

Suppose that a point, P_k, in Fig. 1 is a tracer particle in a three-dimensional space. For the left screen a line which connects a viewpoint, L, and an image point, P_{kl}, of a particle, P_k, is called a viewline. A line, RP_{kr}, is also called a viewline. A perspective projection of the line, P_kL, on the right screen is called an epipolar line (EP-line in abbreviation) on which the corresponding particle image should exist in the right screen. In reality, due to measurement uncertainty, the corresponding particle image may be found in neighborhood of the EP-line. One of some images that is nearest to the EP-line may be the most probable candidate for a stereo pair of the image, P_{li}. Then in the same manner, the second, third, and so on nearest candidates can be found.

A simpler method for calculating EP-line and the distance between a particle image and an EP-line might reduce the analysis time of the procedure. By noticing that every EP-line in one picture goes through a fixed point, that is, a perspective projection point, R_0, where base line LR is crossing. A simple expression giving a gradient of an EP-line is derived from geometrical relationships as follows:

$$\tan\alpha_{ri}' = \frac{\sin\theta_r}{\sin\beta_{li}}\tan\alpha_{li} \tag{1}$$

where α_{li} is the elevation angle of a left viewline, P_kL, β_{li} is the included angles between LR and the vertical projection of the viewlines, P_{kl}, on the base plane, and θ_r is the included angle between LR and R_0S_r'. The EP-line on the left screen can be calculated in the same manner. Since the gradient of an EP-line, $\tan\alpha_{ri}'$, is known in this step, it is easier to find a stereo pair of the closest particle image by calculating and comparing the gradient of a line, $\tan\alpha_{rj}$, which connects the fixed point, R_0, and a candidate particle image, P_{rj}, than by measuring the distance between EP-lines and candidate images.

The reconstruction of three-dimensional coordinates (X_k, Y_k, Z_k) of particles is illustrated in Fig. 2. A Cartesian right-handed coordinate system is defined, in which the origin, O, is taken at the middle point of the base line, LR, the X-

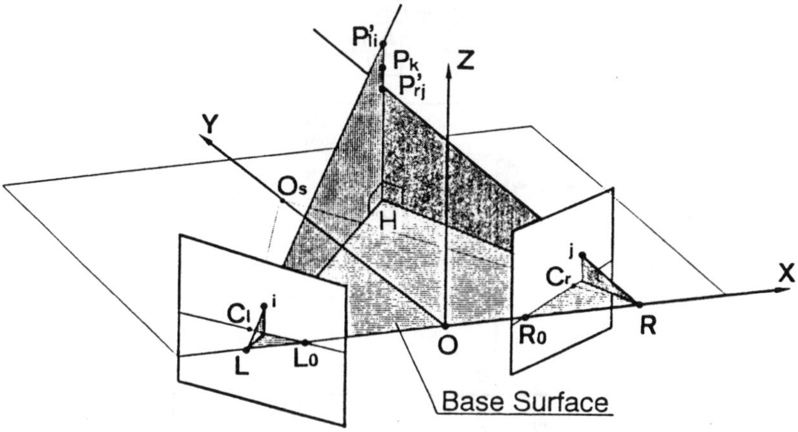

Fig. 2 Reconstruction of three-dimensional coordinates of a particle

axis along LR, the Y-axis on the base plane, and the Z-axis in the vertical direction. Since the two viewlines do not intersect in actual measurement, the particle, P_k, is assumed to be located at the point formed by the following three planes. Two vertical planes containing the two viewlines and an average of the two planes defined by the base line, LR, and the viewlines. Thus a three dimensional position of a particle, P_k, is decided by the following equations:

$$(X_k, Y_k, Z_k) = \left(\frac{LR\tan\beta_{rj}}{(\tan\beta_{li} + \tan\beta_{rj})} - \frac{LR}{2},\ \left(x_k + \frac{LR}{2}\right)\tan\beta_{li},\ y_k\tan\alpha_o \right) \tag{2}$$

here $\alpha_o = \frac{1}{2}(\alpha_{oli} + \alpha_{ori})$, $\alpha_{oli} = \angle P_{li}'LH$, $\alpha_{ori} = \angle P_{rj}'RH$

Correction of Light Refraction

Figure 3 shows an optical system of the arrangement of a camera and a measuring area in a water vessel. Suppose that a point, C, is a camera center, a point, P, is a particle in the water reservoir, and point, O, is a central point of the measuring area.

When a camera, whose view-axis, OC, is perpendicular to the wall of the water vessel, is installed at such a distance as the included angle, α, between the view-axis, OC, and a ray of light from a particle, $PP'C$, is small enough, the correction of light refraction can be made by the following simple technique.

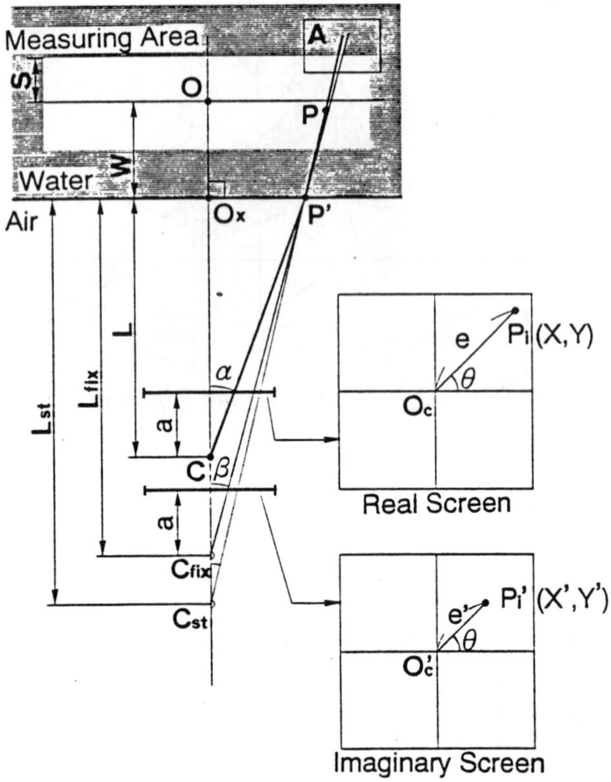

Fig. 3 Optical system for light refraction

An imaginary camera center is put at such a point, C_{fix}, as the following equation is satisfied;

$$L_{fix} = O_x C_{fix} = n L \tag{3}$$

where n (=1.333) is the index of light refraction, and a point, O_x, and the length L equals to $O_x C$ as shown in Fig. 3. Although a production of the line, PP', intersects at a point, C_{st}, on the view-axis, OC, in reality, for the simplification of calculation an imaginary camera may be placed at the distance, L_{fix}, instead of $O_x C_{st}$ which should be decided for each particle, from the wall of water vessel. The measurement error caused by this simplification is thought to be negligible. An image of a particle in the real screen is denoted by P_i on a real screen and the argument of the position vector, $\overrightarrow{O_c P_i}$, on the screen is denoted by θ, which is given by the following equation.

$$\theta = \tan^{-1}\left(\frac{Y}{X}\right) \tag{4}$$

Length of a position vector is denoted by $e(=\sqrt{X^2+Y^2})$. An imaginary point, P_i', corresponding to P_i, is taken on an imaginary screen so that the argument of the position vector $\overline{O_c'P_i'}$ is θ, and that the length of the vector is taken as $e'=e/n$ on the imaginary screen. The coordinates of the point P_i', X' and Y' are given by the following equation.

$$(X', Y') = (e'\cos\theta\ ,\ e'\sin\theta) \tag{5}$$

Using such imaginary coordinates of P_i' for both the left and the right cameras in the same way as mentioned above, the correction of light refraction can be made.

Particle Tracking Algorithm

Here it is assumed that the time interval between the two-consecutive pictures is short enough and fixed, and that flow velocities in the flow field do not change suddenly in time or space, that is, the similarity of flow patterns is preserved.

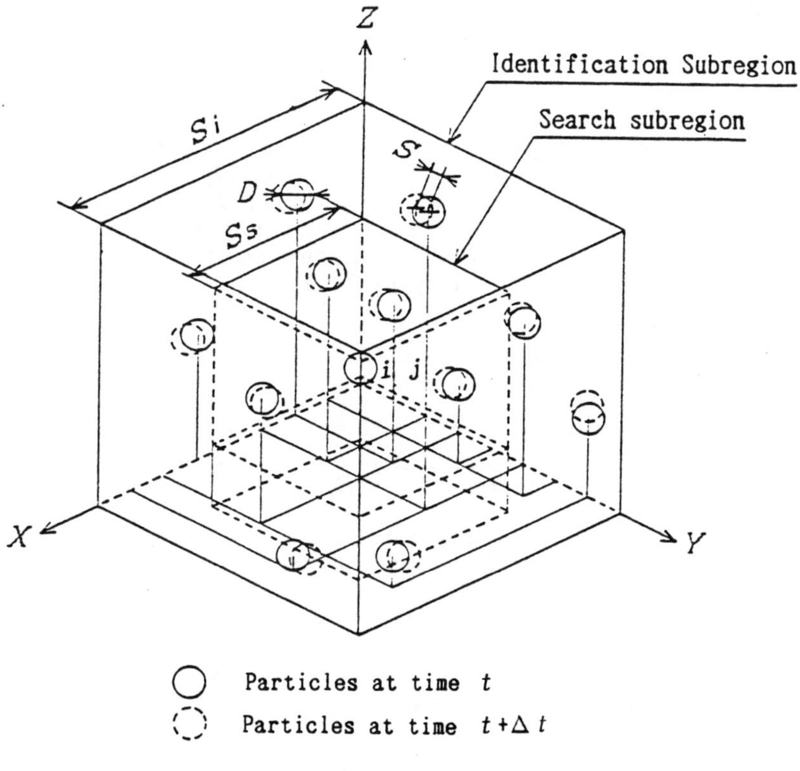

Fig. 4 Particle distribution in cubic identification subregion I and J when a reference particle i is overlapped with a candidate particle j at their centers.

The present problem is how to track each particle in the first space at time t with itself in the second space at time $t+\Delta t$, in other words, how to identify each particle. The procedures are as follows (see Fig. 4):

1) A cubic search subregion with a side length S_s is set up in the second space, whose center has the same coordinates as the center of a reference particle i in the first space. The number of particles contained in the search subregion is denoted by n. All particles in the search subregion are taken as the candidate particles, j_m $(m=1,2,\cdots\cdots,n)$.

2) The origin of the coordinate system in the second space is successively moved to make the center of a candidate particle, j_m, have the same coordinates as those of the reference particle i.

3) One cubic identification subregion each with the same side length S_i, is set up in both the first and the second space, whose center has the same coordinates as the center of the particle i. The particle distribution patterns I and J are made up of the particles in the first and the second identification subregions, respectively. A candidate pattern J for the particle j_m is denoted by J_m.

4) After the pattern I is overlapped with the pattern J_m, the number of particles which are overlapped with each other in the overlapped patterns, N, is counted. Here N is called the number of overlapped particle pairs.

If any two particles in the overlapped patterns meet the following conditions given by Eqs.(6) and (7), they are taken for an overlapped particle pair. Here, for simplicity of computation, the actual spherical shape of tracer particles is replaced with a sphere with diameter D. Such a spherical particle is called an imaginary particle.

$$\begin{aligned} \Delta X &= | X_{il} - X_{jm} | \\ \Delta Y &= | Y_{il} - Y_{jm} | \\ \Delta Z &= | Z_{il} - Z_{jm} | \end{aligned} \tag{6}$$

$$S = \sqrt{(\Delta X^2 + \Delta Y^2 + \Delta Z^2)} < D \tag{7}$$

where, X_{il}, Y_{il}, and Z_{il} are coordinates of a particle with the number l in the pattern I except the central particle i in the first space; X_{jm}, Y_{jm}, and Z_{jm} are coordinates of a particle with the number m in the pattern J except the central particle j in the second space.

5) The value of the cross-correlation coefficient C_{ij} defined by the following Eq.(8) is computed for the two patterns I and J:

$$C_{ij} = \frac{\sum_{k=1}^{N} \left(1 - \frac{3}{2}\overline{S_k} + \frac{1}{2}\overline{S_k{}^3}\right)}{\sqrt{N_i N_j}} \tag{8}$$

where N_i is the number of particles in the pattern I except the central particle i, and N_j is the number of particles in the pattern J except the central particle j. N and subscript k are the number and the order number of overlapped pairs, excluding the pair of particles i and j, respectively. $\overline{S_k}=S_k/D$, and $0\leq\overline{S_k}\leq 1$.

The above Eq.(8) can be derived from the calculation of total overlapped volumes of overlapped imaginary particle pairs in the two patterns, total volumes of particles in the pattern I and total volumes of particles in the pattern J using binary image data, based on the mathematical definition of the cross-correlation coefficient, as in the following form.

$$C_{fg} = \frac{\int f(x,y,z)\cdot g(x,y,z)\,dV}{\sqrt{\int\{f(x,y,z)\}^2 dV\cdot\int\{g(x,y,z)\}^2 dV}} \tag{9}$$

Here it is assumed that the pattern distribution functions $f(x,y,z)$ and $g(x,y,z)$ take binary data, that is, values of unity inside the particles and zero outside the particles and that the diameter of the imaginary particle is constant D.

6) The steps from 2) to 5) are repeated until both the values of N and C_{ij} are computed for all the candidate patterns. Which candidate pattern has the maximum value of N, N_{max}, can be determined after investigating the numbers of overlapped particle pairs for all candidate patterns. If there is only one pattern that takes the maximum value of N, N_{max}, the candidate particle can be thought to be correctly identified. If more than one pattern takes N_{max}, the candidate particle around which the pattern takes the maximum value of C_{ij} is taken as the correctly identified one. If the maximum number N_{max} is zero, the reference particle i is excluded for the calculation of identification. The number of such excluded particles is denoted by N_{ex}.

EXPERIMENTAL RESULTS

Performance of Stereo Pair Matching

In order to investigate performance of the new stereo matching method, small spots with a diameter of 0.5 mm are distributed on a model plate, which can be moved along Y axis by a precise driving system, as shown in Fig. 5, and are photographed by two cameras which are fixed on a large circle with a diameter of 910 mm. Two cases where the model plate is put in air or in a water vessel are investigated preliminarily for discussing the effects of refraction correction on the accuracy of three-dimensional coordinate reconstruction. The X, Y, and Z coordinates of the spots are known at a high accuracy beforehand. Misalignments of the camera setting can be corrected and estimated statistically using confirmed stereo pairs accumulated in the stereo matching processes. The

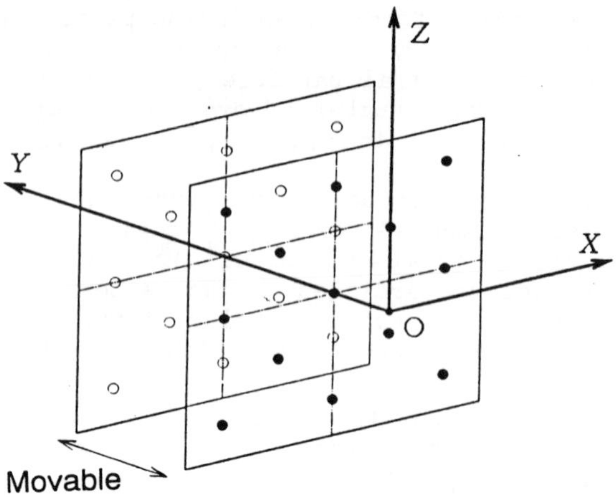

Fig. 5 Model plate for stereo pair matching performance

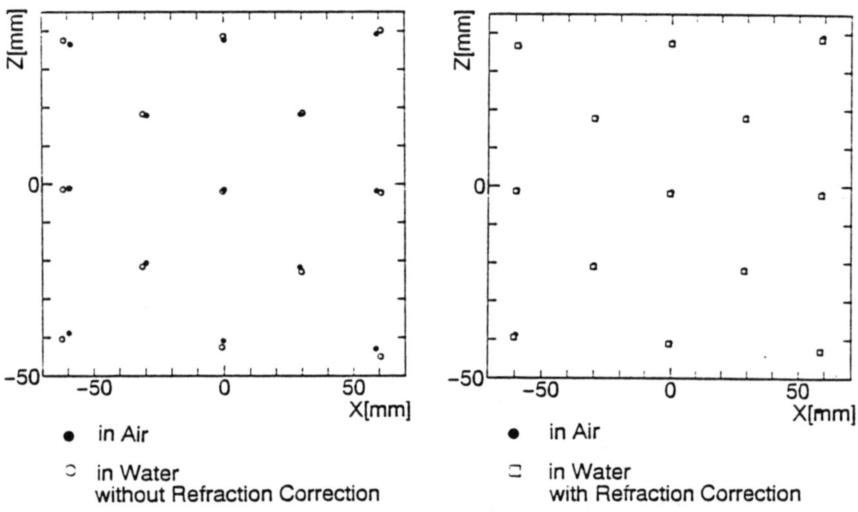

(a) without correction of light refraction

(b) with correction of light refraction

Fig. 6 Test results of stereo pair matching performance

coordinates of the spots are reconstructed by the stereo matching method as mentioned above.

Figure 6 (a) and (b) show the test results of stereo pair matching performance in the two cases without or with refraction correction. The average of deviations of the distances between the two test cases in air and in water, $\overline{\Delta S}$, are as follows;

$\overline{\Delta S}$ = 1.84 mm   when refraction correction was not made.
$\overline{\Delta S}$ = 0.51 mm   when refraction correction was made.

Besides the average value of Y coordinates was obtained as 909.96 mm and its standard deviation was 0.185 mm.

As a result, the satisfactory performance of three dimensional coordinate reconstruction was obtained by the present stereo matching method with the refraction correction.

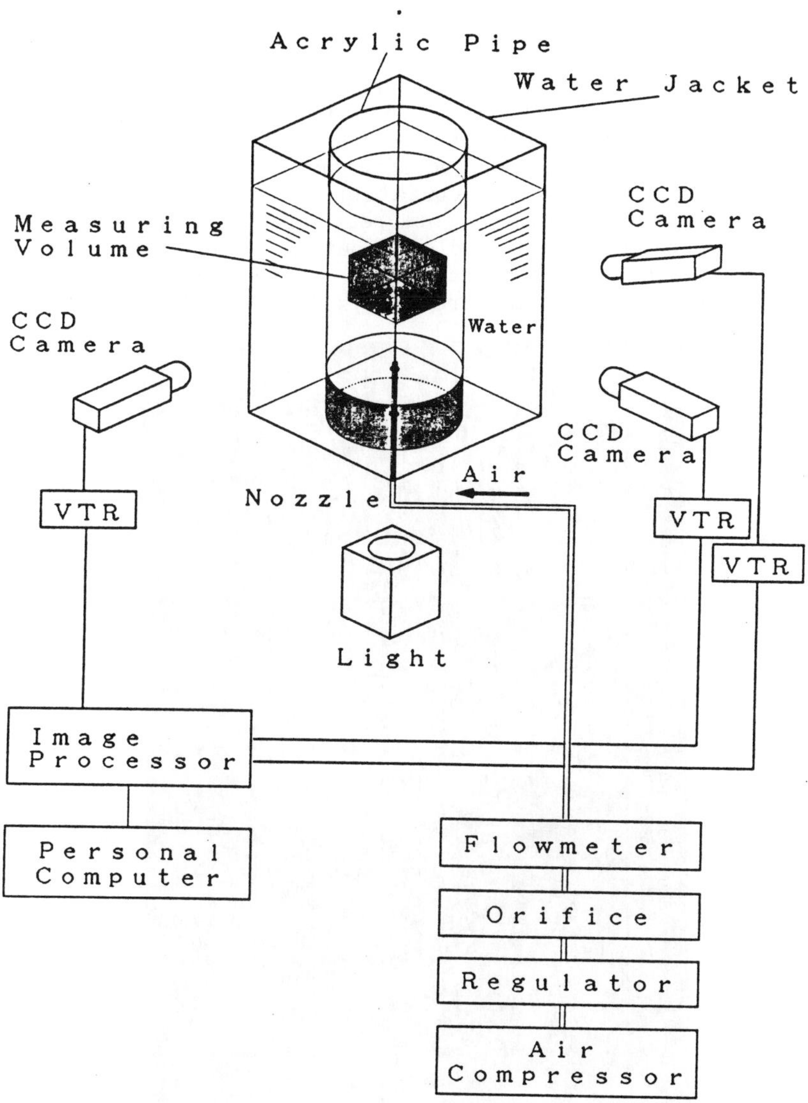

Fig. 7 Outline of experimental apparatus

Velocity Measurement by 3D PTV

The 3D PTV based on cross-correlation method using binary data was applied to gas-water two-phase flows of bottom blowing bubbling jet in a cylindrical water vessel.

Figure 7 shows schematically the outline of the experimental apparatus. The cylindrical vessel is made of a transparent acrylic pipe with a diameter of 200 mm and depth of 280mm and is filled with water. A water jacket made of transparent acrylic plates is installed to avoid the effect of refraction on images. Air bubbles are ejected vertically through a nozzle installed at the bottom of the cylindrical bath. The air flow rate is 1.4 cc/s. The measuring volume is a 100 mm cubic and is illuminated using a light system for a slide projector with a power of 650 W. Nylon-12 balls with a diameter of 0.9 mm and the specific gravity of 1.02 are put as tracers for the flow visualization in the bath. The water contains some salt to have the same specific gravity as the particles so that the particles float at neutral positions in still water. The motion of particles are photographed using two CCD type video cameras and are analyzed by the 3D PTV.

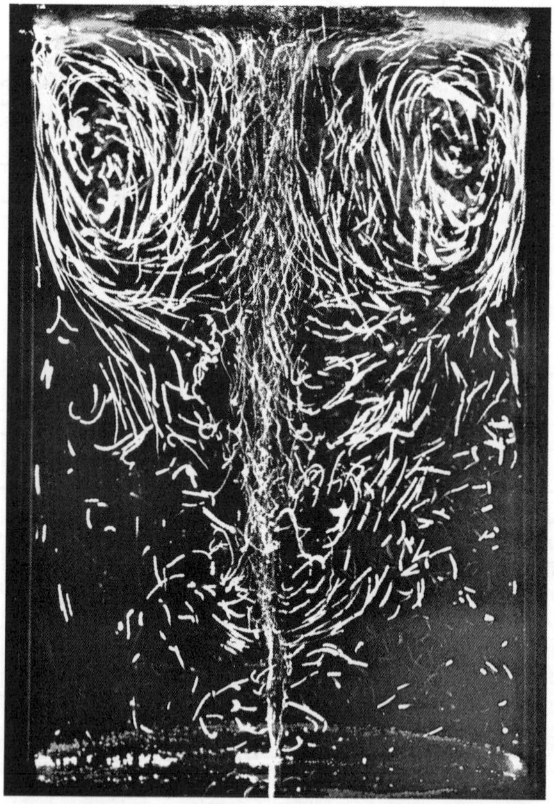

Fig. 8 Visualization of flow field in a bubbling jet mixing water vessel

Figure 8 shows a visualized whole field of gas-water two-phase flow in the cylindrical vessel. The pathlines of particle tracers were photographed using a camera and illumination through a slit type of light source with the exposure time of 0.5 s. Strong vortex ring is observed in the upper part of the vessel, and the air bubbles agitate and mix the water in the vessel. Note that such pathlines were not employed for measurement of velocity by the present technique.

Fig. 9 Bubble deformation and tracer particle motion

Figure 9 shows a picture of the flow field near the nozzle which is installed at the bottom of the vessel. It was taken by a camera which was set closely to the wall of the water jacket. The observation area was illuminated by a slit type of light with a thickness of about 10 mm. The initial form of a bubble is spherical at the exit of the nozzle. It soon deforms to a mushroom shape of hemisphere in the upper part and horizontally flat disc in the lower part. Then the mushroom shape deforms to a wholly flat disc and the bubble goes up along a zigzag trajectory. Each trajectory of a bubble is different from another. According to the observation by a high speed video picture, particles which are far from bubbles move at much lower velocity than the bubbles, but some particles which are very ahead near bubbles move at the nearly same high velocity as the bubbles.

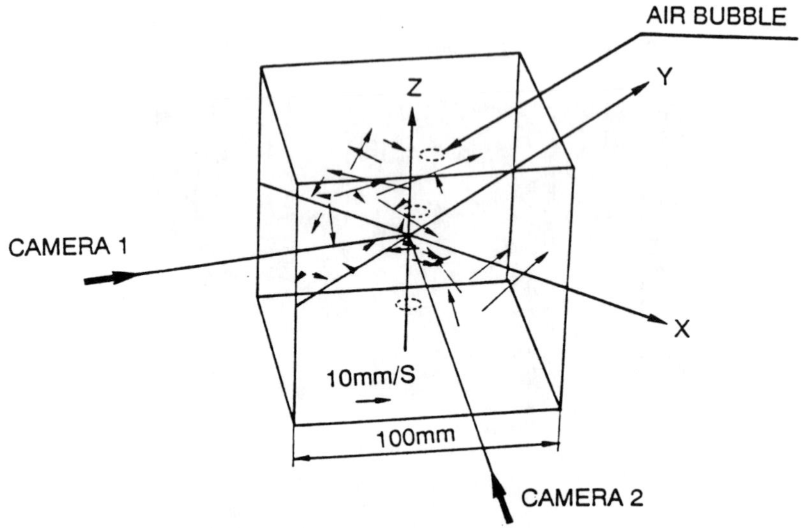

Fig. 10 Three-dimensional velocity vectors of tracer particles in a bubbling jet mixing water vessel

The authors [11, 12] have investigated the flow field using the 2D PTV by the binary cross-correlation method, and Hassan, Blanchat, Seeley and Canaan [13] measured the both components of a two-phase flow using a point-by-point digital cross-correlation analysis algorithm for the 2D PTV. Both of the studies described the velocity fluctuation of the water induced by the rising bubble motion. In present study, the three-dimensional velocity vector diagram was obtained as shown in Fig. 10. Although the number of the velocity vectors is not so large, it is observed that the water in front of bubbles moves upward at very high speed and that it enrolls following the bubble motion.

CONCLUSION

The present paper proposed a new stereo pair matching method with correction of light refraction at water-air boundary and a new three-dimensional cross-correlation method for particle tracking velocimetry using binary image data. The new 3D PTV were applied to measure velocity field of water-air two-phase flow of bubbling jet in a cylindrical vessel. As a result, the water motion around bubbles is explained clearly.

REFERENCES

(1) Yano,M., Velocity Measurement Using Correlation Concerning Digital Tracer Image, *J. Flow Visualization Soc. of Jpn.*, (in Japanese), Vol.3, No.10, pp.189-192, 1983.

(2) Kimura,I. and Takamori,T.,Image Processing of Flow Around a Circular Cylinder by Using Correlation Technique, *Proc. 4th Intl. Symp. on Flow Visualization*, Paris, Hemisphere Publishing Corp., pp.221-226, 1989.

(3) Kobayashi,T., Development of Flow Field Analyzing System using Digital Image Processing, (in Japanese), *Report of Grant 1986-C for Scientific Research by the Japan Ministry of Education*, ed.by Institute of Industrial Science, pp.63-66, 1988.

(4) Kaga,A. and Yoshikawa,A., Velocity Distribution Measurement through Digital Image Processing of Visualization Flow Image, *Building Systems, Room Air and Air Contaminant Distribution*, Edited by L.L.Christianson, ASHRAE, pp. 91-96, 1989.

(5) Kaga,A., Inoue,Y. and Yoshikawa,A., Velocity Distribution Measurement through Digital Image Processing of Visualized Flow Images, *Proc. of Second Intl. Conf., ROOMVENT'90*,pp.(22)1-15, 1990.

(6) Chang,T.P.K.,Watson,A.M.and Tatterson,G.B.,Image Processing of Tracer Particle Motions as Applied to Mixing and Turbulent Flow-1.The Technique, *Chem.Eng.Sc.*,Vol.40,No.2,pp.269-275,1985.

(7) Yamamoto,F., Uemura,T.,Koukawa,M.,Itoh,M. and Teranishi,A., Application of Flow Visualization and Digital Image Processing Techniques to Unsteady Viscous Diffusing Free Doublet Flow, *Proc. 2nd. Intl. Symp. FLUCOME'88*, Sheffield, pp.184-188, 1988.

(8) Uemura,T., Yamamoto,F. and Koukawa M., High Speed Algorithm for Particle Tracking Velocimetry Using Binary",*J.Visualization Soc. of Jpn.*, (in Japanese), Vol.10, No.38, pp.196-202., 1990.

(9) Uemura,T.,Yamamoto,F. and Ohmi,K.,A High Speed Algorithm of Image Analysis for Real-time Measurement of Two-dimensional Velocity Distribution, *"Flow Visualization", ASME FED-85*, pp.129-133, 1989.

(10) Nishino, Y., Kasagi, N., Hirata, M. and Sata, Y.,Three-dimensional Flow Velocimetry Based on Digital Image Processing, *Trans. Jpn. Soc. Mech. Eng.*, (in Japanese), Vol. 55, No.510, B, pp.404-412, 1989.

(11) Uemura,T., Iguchi,M., Yamamoto,F.,Morita,Z. and Mizukoshi, W., Video Image Analysis of Mixing Flow in a Cylindrical Vessel Agitated by a Bottom Blowing Bubbling Jet, *J. of the Iron and Steel Institute of Japan* (in Japanese),Vol.78,No.5,pp.738-744, 1992.

(12) Yamamoto, F., Uemura, T., Iguchi, M. and Morita, Z., PTV Measurement of Mixing Flow in a Bottom Blowing Bubbling Jet, *Proc. of the Intl., Conf., on Multiphase Flows '91-Tsukuba*, pp.251-254, 1991.

(13) Hassan,Y.A., Blanchat,C.H.,Seeley JR,C.H. and Canaan,R.E., Simultaneous Velocity Measurement of Both Components of a Two-Phase Flow Using Particle Image Velocimetry, *Int. J. Multiphase Flow*, Vol. 18, No.3, pp.371-395, 1992.

Visualization of reacting gas flows on non-structured grids

C. Best, D. Roebbelen
ISEI, JRC-ISPRA, Varese, Italy

Abstract

A software package **TURCOM** has been developed to visualize the results of 2-d fluid dynamics modelling on non-structured grids. This software is written in the PV-WAVE command language [1] and runs on most graphic workstations. Calculations are performed of chemically reacting gases within containers of variable geometry [2]. A database is used to store the results together with the grid geometry [3]. This database is accessed by TURCOM via mouse-driven menus and icons. An objective of this work has been to avoid interpolation onto regular grids. Therefore some new algorithms have been developed to present the results of model calculations graphically. Scalar fields can be displayed as iso-contours or as colour shaded iso-plots. Vector fields can be displayed as vector flows with or without a scalar colour index. Visualization of the time evolution of data is also possible.

Introduction

A research project for the "Numerical Simulation of Reactive Gas Flows" [4] is in progress at the JRC-Ispra. It's primary aim is to predict the development of explosions in confined or semi-confined environments. This involves simulating shock wave propagation and reflection from container walls. The calculations are performed on a non-structured grid in two dimensions. Complex geometries are defined by a mesh of node points and by triangular finite elements. Two examples of such mesh grids are shown in fig.1.

In general a 2-d non-structured grid can be defined as follows:

$$N_n, \underline{x}_i \qquad i = 1, N_n \tag{1}$$

$$N_\nabla, (C_{1j}, C_{2j}, C_{3j}) \qquad j = 1, N_\nabla \tag{2}$$

where:

N_n = no. of nodes, $\underline{x}_i$ = position vector of node i

N_∇ = no. of triangular elements

C_{1j}, C_{2j}, C_{3j} = node index for each vertex of ∇_i

From this definition the boundaries can be derived. This is done by identifying triangles with only two neighbouring triangles. The unconnected face forms part of a boundary. A boundary is defined as a set of connected faces. In this manner the boundaries can be identified as follows:

$$N_{Bdy}, \left[N_{fi}, \{f_{1ij}, f_{2ij}\}_{j = 1, N_{fi}} \right]_{i = 1, N_{Bdy}} \tag{3}$$

where

N_{Bdy} = no. of boundaries

N_{fi} = no. of faces for i th boundary

$\{f_{1ij}, f_{2ij}\}$ = end node indices for j th face in the i th boundary

Some further geometric properties of such triangular grids are found to be useful. These are indexed for each node and deduced by a sorting procedure.

They are:

L_i = 0: if node is internal

= m: if it forms part of boundary m

NP_i = number of triangles with this node as a vertex

$\{\nabla_{ji}\}_{j = 1, NP_i}$ = triangle index connecting this node

ordered anti-clockwise

This is illustrated in fig.2.

Figure 2 also defines what is known as the control volume for this node. The fluid calculations are discretised within these control volumes. This means that the resultant physical properties for each node point apply for all fluid within the control volume. These control volumes are therefore also important when visualizing the results of calculations. They are defined geometrically as the polygon formed by joining the mid-points of each triangle associated with a given node. The mid-point of each triangle is calculated from the intersection of lines bisecting each angle of the triangle.

Many runs of the code REAC are performed. The objective of the work described here has been to organise storage of results and geometry in a database [3] and to develop a sophisticated data visualization system.

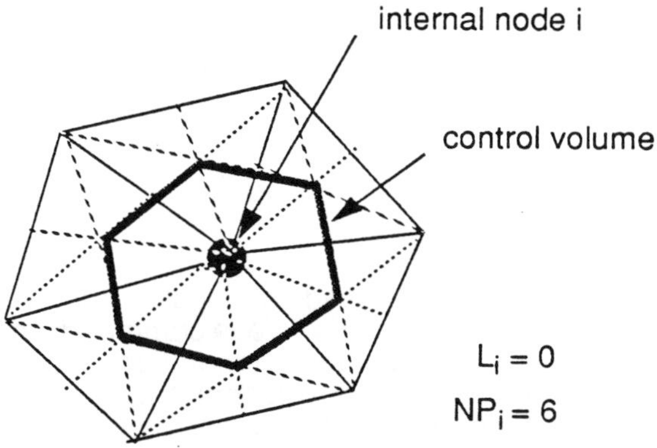

Figure 2: Control volume of a triangle node

EDF Database

The storage of results and geometries is handled by the **EDF** database, developed by one of the authors [3]. This database has previously been used to store experiment results from several experiments at Ispra. It has the advantage of a generalised format for vectorised data in 1 and 2 dimensions and of a standard Fortran/C interface library. A Fortran program was written to read the output files of REAC, calculate the grid geometry and store the results in the EDF database. The geometry code was provided by F.Ruel of IST, Ispra.

Each run of REAC produces one EDF file in the database. The database is distributed on the network and allows any workstation to access the results. At present data are written from an HP8000 and visualized on a Sun Sparcstation. The current situation is portrayed in fig.3.

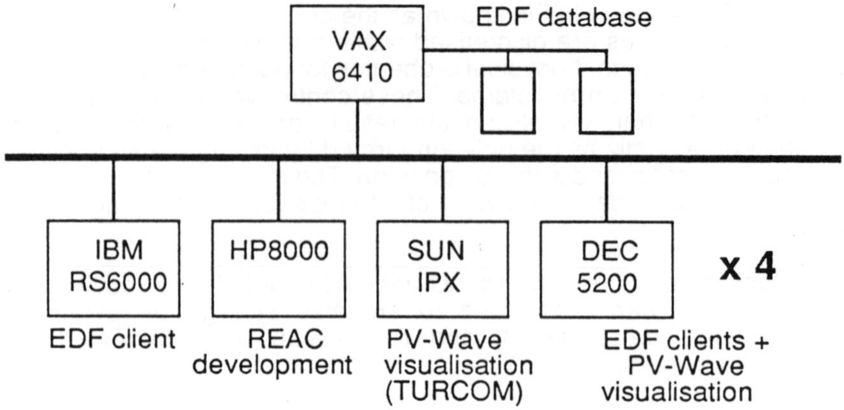

Figure 3: Network layout

TURCOM

TURCOM is the program used to visualize the results of each REAC calculation. It is written in the PV-Wave command language [1]. PV-Wave command language is a fourth generation program language which provides powerful graphics development tools. The software has been developed on a SUN sparcstation running OpenWindows, but will run on most graphic workstations.

The interface to the EDF database is made by a C program "edf-server" which communicates with TURCOM via remote procedure calls. This program sleeps until a request for information is made by TURCOM. As is shown in table 1, there are six levels of request:

Request	Action	Return Information
1	list of hosts	list of databases available on network
2	directory	list of REAC calculations
3	signal list	list of signals within a given REAC calculation
4	geometry information	return definition of non-structured grid
5	timing information	return timebase of calculation
6	signal data	return results for given signal

Table 1: Edf-server request

By means of the edf-server interface to the database TURCOM has developed a data retrieval interface, based on menus, which is dynamic. The user interface of TURCOM is realized either through mouse driven menus or through specially developed graphical icons. The menus use a modified form of the button interface found in the PV-Wave user library [1]. Fig.4a shows the main menu. The data retrieval button causes 3 levels of dynamical menus to appear corresponding to edf requests 1, 2, 3 and 6 (The geometry and timing information are retrieved automatically). Also shown in fig.4b-d are three examples of icons used for selecting contour heights, time slices and colours.

There are currently two methods to visualize scalar field quantities for a general grid geometry at a given time. These algorithms have been developed by the authors to avoid interpolation onto regular grids.

Contour Plot

A set of contour heights defined either by the user or by default (10 equal spaced heights) is used. Each triangle on the grid is considered. For each face of the triangle, the end node scalar field values are compared with each contour height, to find if there is an intersection. The intersection point is calculated by linear interpolation. Finally the contour element is drawn connecting the entry and exit point of each cell as shown in fig.5. This algorithm can be speeded up by following contours from one cell to its neighbour. A typical result for a pressure field is shown in fig.6. In such plots the boundary surfaces are always drawn in thick line-style.

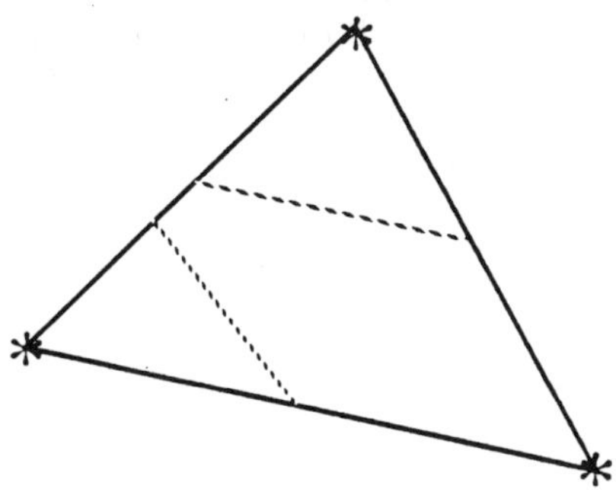

Figure 5: Contouring cells

Iso-Colour Plot

This method of visualizing the field uses a colour scale to quantify the field within each cell. Since the model calculations are performed inside the control volumes defined in figure 2, these are the cells that are used to colour code the field. A colour map of 211 colours is used (25 colours are reserved for special purposes). A colour scaling is either calculated from the range of the data or it can be user defined. The geometry of each cell is calculated from equations 1 and 2, whereupon it is drawn and filled with the assigned colour value. The grid boundary is always drawn, and as an option the grid geometry may be overlaid. A range of colour tables are available which are selectable by the user. These are either quantized in colour steps or may vary continuously. Fig.7 shows an output from the iso-colour algorithm.

Vector Field Plot

The velocity of the expanding gas is displayed as a vector field. For each node on the grid an arrow is displayed whose magnitude and direction is proportional to the velocity vector at that point. The PV-Wave user library procedure velovect [1] was modified to handle non-structured grids. The procedure has been further extended to allow a superposition of a scalar field. This is achieved by using a colour scale to colour each arrow according to the value of the scalar field at that point. This is useful to visualize for example the relationship between gas pressure and gas velocity. Fig.8a-b shows examples of both types of vector field plots. As options, the user is able to increase the size of the arrows and to define his/her own colour scaling.

Time Evolution of Single Nodes

The user may select up to 10 node points using the cursor from any of the 2-d displays. The values of the current scalar field are then plotted versus time at each of the selected nodes. This option is selected from the main menu. After each press of the mouse button the node whose distance is nearest to the cursor is selected. The selection of the mouse is terminated by "clicking" with the cursor outside the outermost boundary of the object. The algorithm used to determine if the cursor is outside the object is to calculate the number of intersections N_{int} made by a vertical line drawn from the current cursor position with the boundaries of the object.

$$N_{int} = \text{odd} \quad \text{- cursor inside object} \tag{4}$$

$$N_{int} = \text{even} \quad \text{- cursor outside object}$$

This is portrayed in fig.9. Once the list of selected nodes is terminated in this way, a time evolution plot at each node position is generated. An example is shown in fig.10.

User Customisation

TURCOM can be run on different workstations, X-terminals and PCs with X-emulation. To handle the different screen resolutions, printer and display characteristics, TURCOM has a utility menu to allow the user to customise the graphic output. Table 2 lists the options currently available.

Future Developments

Two main developments are foreseen in the near future. The first is to allow the user to select a linear slice and plot the scalar field value versus position along the slice at a given time. The second is to produce animations of data variation with time. This will be done both digitally (by storing frames in memory) and also to video tape.

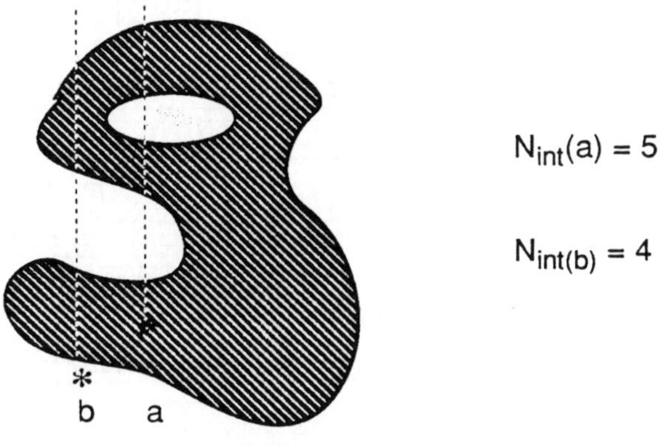

Figure 9: Number of intersections

Summary

A workstation based graphic program has been developed for visualizing the results of Simulating Reacting Gas Flows. New ways of visualizing scalar and vector fields on non-structured grids have been developed. A sophisticated user interface based on mouse driven menus and graphic icons has been written. Data management is handled by a distributed database. The software is in daily use at the JRC-Ispra, and further developments are underway.

The authors would like to acknowledge the advice and help given by G.Peter and F.Ruel of IST Institute, JRC-Ispra.

Text Input	Allows to label graphics displays
Colour Selection	Allows to select background, text and plot colours
Menu Scaling	Allows to magnify menus
Plot/Menu Text Scaling	Allows to magnify plot/menu text
Plot Geometry	Allows displays to be fixed geometry or calculated geometry
Hardcopy	Allows to select postscript (PS), colour PS, encapsulated PS, CGM etc. portrait or landscape mode

Table 2: TURCOM UTILITY options

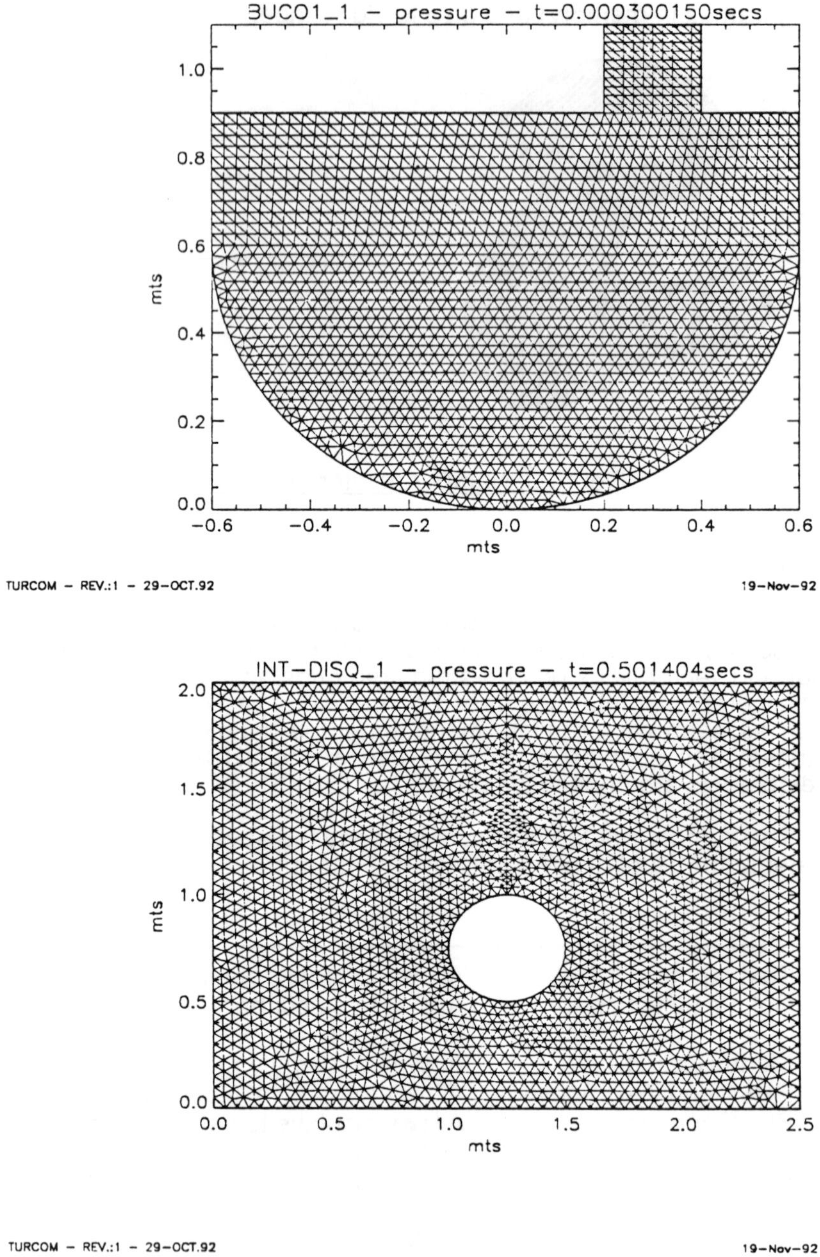

Figure 1: Examples of mesh grids

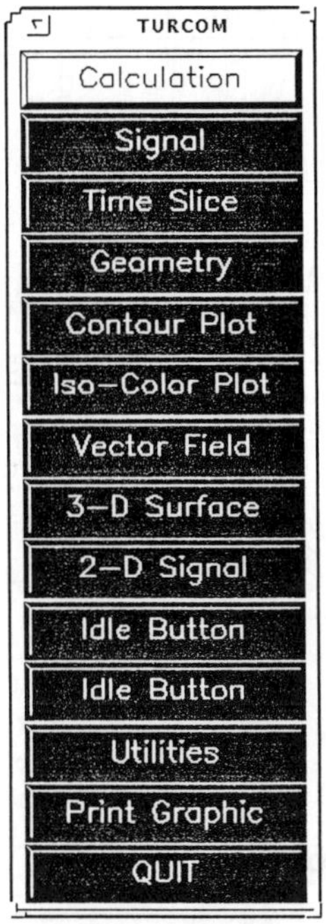

a. TURCOM main menu

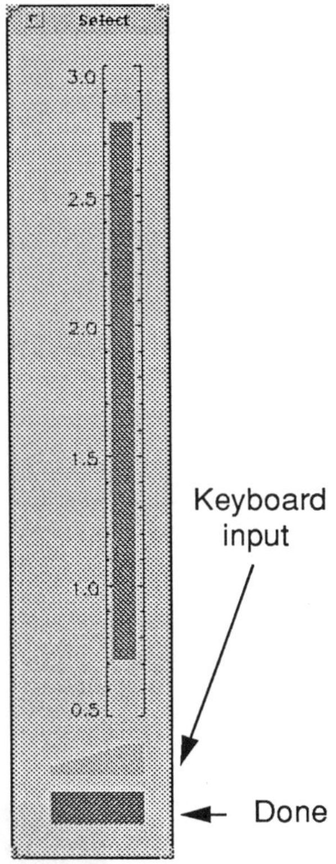

b. Input scale for contour heights

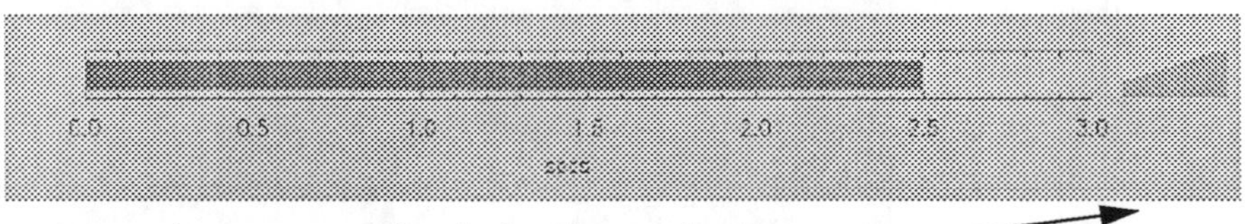

c. Scale for time selection

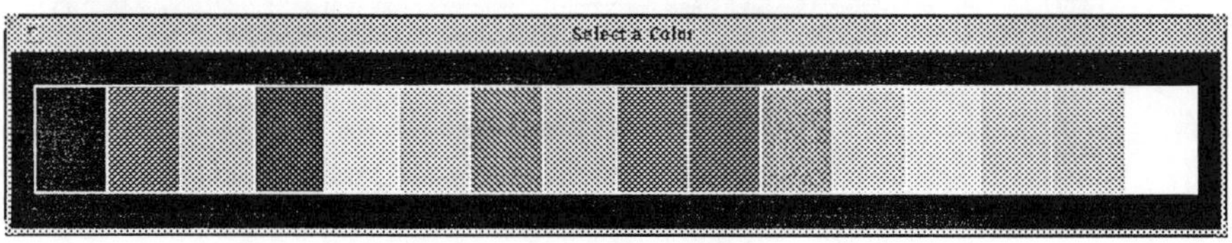

d. Colour scale for the selection of colours

Figure 4: Examples of TURCOM icons

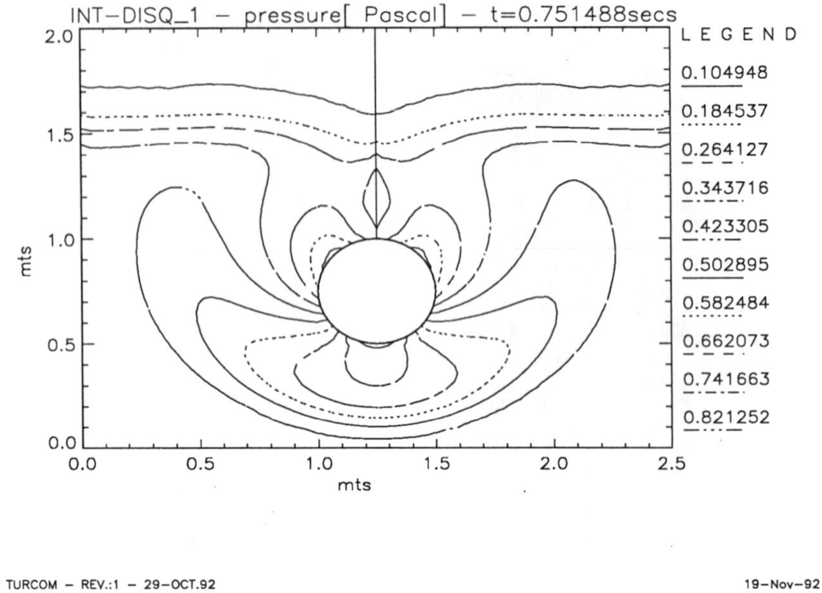

Figure 6: Contour plot

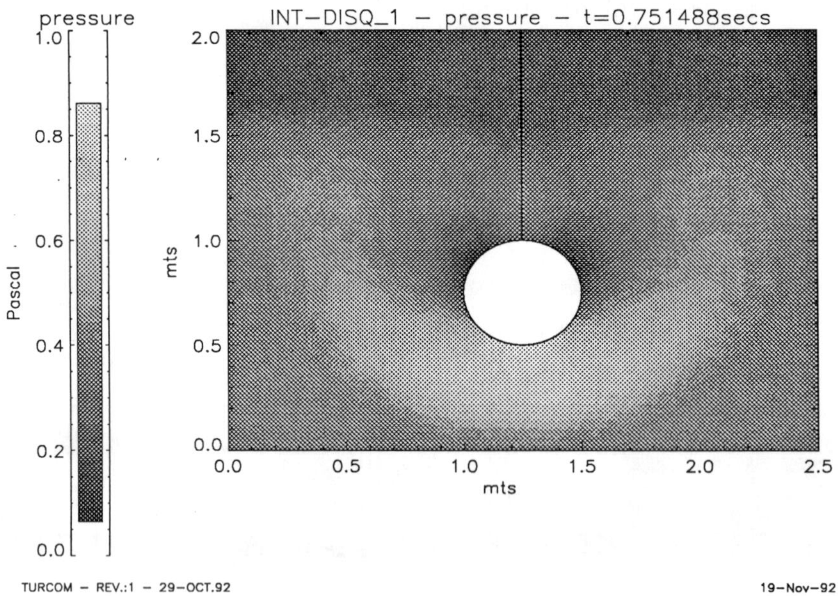

Figure 7: Iso-Colour plot

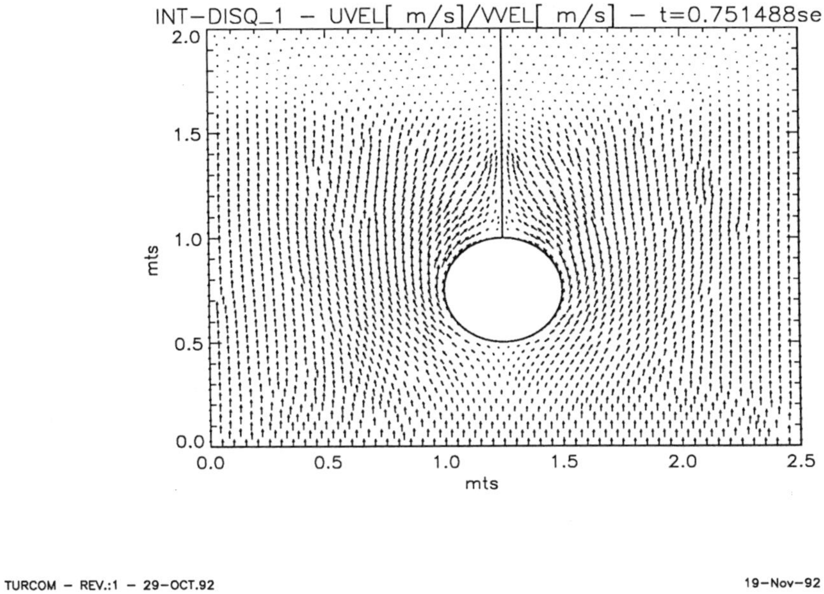

Figure 8a: Vector Field without scalar field

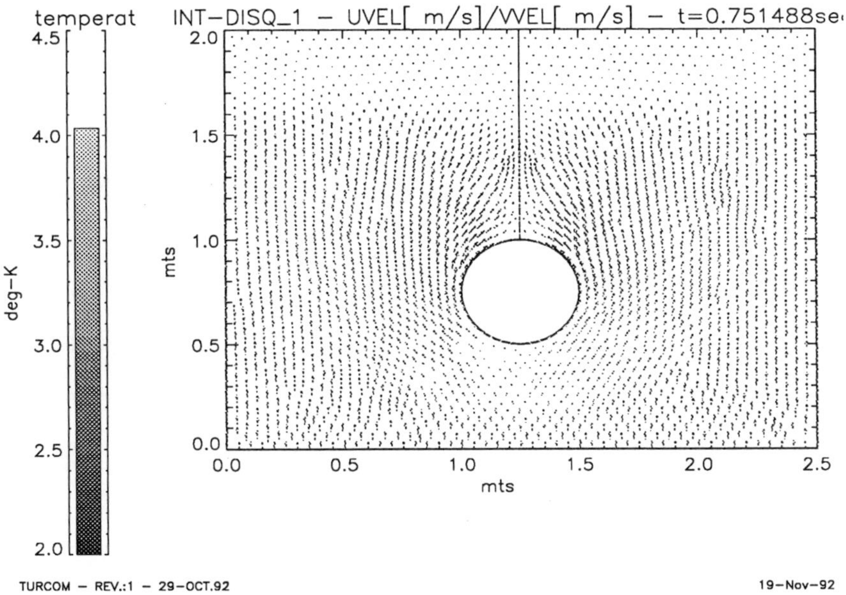

Figure 8b: Vector Field with scalar field

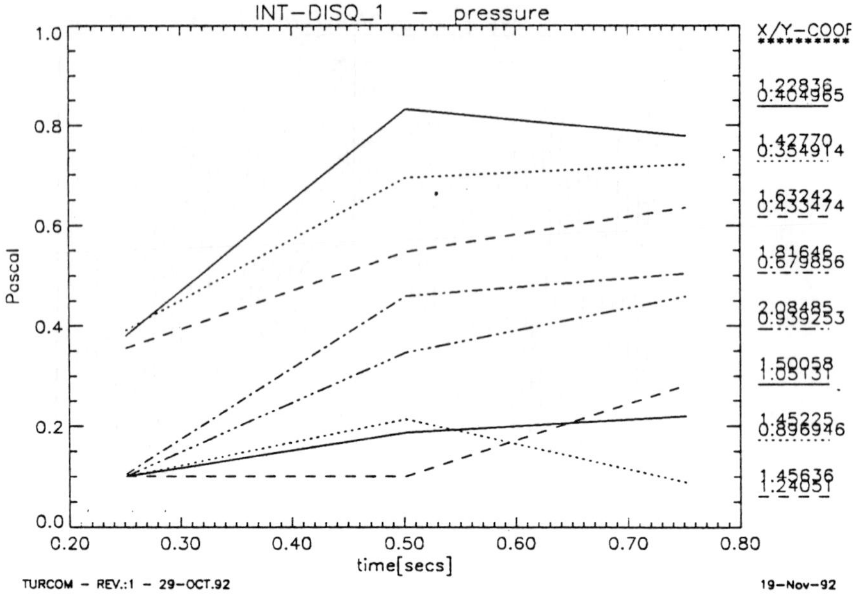

Figure 10: Time evolution of single nodes

References

1. Precision Visuals, Inc.,
PV-WAVE Technical Reference Manual

2. Ruel, F.,
Numerical Simulations of Reacting Gas Flows:
Assessment Calculations of Transonic Multidimensional Flows,
JRC Technical Note No. I.92.01

3. Best, C.H.,
Experiment Data Files:
A General System for Storing Data from Experiments,
JRC Technical Note No. I.90.96

4. Ruel.F.,
Numerical Simulation of Reacting Gas Flows:
Basic Models and Algorithms. Presentation of a Preliminary Code Version,
JRC Technical Note No. I.91.109

SECTION 6: ARCHITECTURE AND URBAN PLANNING

Invited Paper

The quest for the parrot on the shoulder: knowledge about emerging design solutions and its representation in a CAD system

B.R. Lawson

School of Architectural Studies, University of Sheffield, Sheffield S10 2TN, UK

In this paper I shall consider the way knowledge about a solution to a design problem emerges during the design process and the implications of this for its representation in any computer aided design system. In the context of this conference I shall concentrate mainly on the computer as a graphical device, whilst recognising that increasingly many researchers are interested in the potential it offers in applications which are only incidentally dependent on graphics. Consequently I shall therefore also be concentrating on knowledge about design solutions rather than problems although an essential part of the argument concerns their close inter-relationship. However, I believe this distinction can still usefully be made when discussing CAD which Galle has divided into what he calls CAMP or computer-aided modelling and presentation as opposed to CAPS or computer-aided problem-solving [1].

I shall explore this matter through an examination of that area of design I understand best which is architecture. However the issues raised here seem to me to be generic enough certainly to apply to other similar design fields, which would include industrial and product design, interior design, urban design and at least some areas of structural engineering. These fields all share in common a three-dimensional end product which will be lived in or used directly by people who are likely to be concerned about both its appearance and physical performance. I leave others to judge the extent to which these ideas are capable of application in other design fields.

These three-dimensional design fields share another feature in common. They are all areas where by now we might have expected to see much more real progress in computer aided design than has actually taken place in the professions themselves. After early enthusiasm for computer aided design in architecture for example, architectural practice has largely accepted the use of the computer in place of the drafting machine but not yet in favour of the sketch pad. Why should this be? There is no shortage of things computers can do which undeniably form important parts of the architectural design process. For example good reliable software exists so that computers can visualise three-dimensional forms, they can model energy flow, they can calculate structural forces and design structural members, they can model the effects of both artificial lighting and natural light whether in the form of sunlight or daylight, they can be used as cost-estimating tools, they can offer expert advice on a whole range of technical and legislative matters, and they can even plan albeit in a limited way. In fact they can do most of these things more reliably and in less time than most architects. Why then when the three-dimensional appearance, energy consumption, structural stability, lighting, capital cost and planning of buildings are so critical do architects fail so consistently to take advantage of these tools.

I have recently been conducting a study [2] in which I have interviewed a number of the world's leading architects about their design process, and I have found an almost unanimous view that we have yet to provide them with computer-aided tools which they are prepared to use in the design studio. As an example let us examine the views of just three; Michael Wilford, partner of James Stirling for over twenty years, the remarkable and highly acclaimed young Spanish architect/engineer Santiago Calatrava and Herman Hertzberger, one of the father figures of modern Dutch architecture. These three are all acknowledged as amongst the great designers of our time, and yet they could not be more different. Stirling and Wilford's work on museums, art galleries and university buildings is remarkable for its exploration of new territory and the development of urban form in the public domain. Calatrava has designed some of the most astonishingly sophisticated bridges and railway stations of our era, and Hertzberger is well known for his structuralist philosophy and humanistic attitude towards the users of his buildings. All three use computers as word-processors, and sometimes as drafting tools. Calatrava of course also uses them for engineering calculations, so there is no general antagonism to computers here. However none of these three great designers use or foresee using computers as design tools. Of course we can all say in these times of widespread criticism of the architectural profession that architects are just stupid and incompetent, but I prefer the alternative and rather more plausible theory that there may just be something wrong with the way we are designing CAD tools. In particular I am going to suggest to you that implicit in most such tools is a view of the design process which is based on several key fallacies.

It has long been recognised that the design process involves a peculiarly difficult combination of thinking skills [3]. The design of large systems such as buildings involves finding as well as solving problems in a context which has became known as 'wicked problem solving' [4]. Design problems are also notoriously 'knowledge rich' as opposed to 'knowledge lean'. By this is understood that such problems require considerable amounts of knowledge beyond that which is stated in the problem description to arrive at a satisfactory solution. Such knowledge comes in the form of expertise or the accumulated wisdom of the designer, and thus raises at least two important sets of issues for design research. These concern the way in which this expertise is acquired and the way in which it can be retrieved and applied to new design problems. Cross [5] has pointed out that knowledge itself is no guarantee of success but that 'a designerly way of knowing' is critical here. Lawson [6] has also shown how students of architecture may acquire knowledge about environmental science but remain unable successfully to apply this to design problems.

In fact there is now substantially more expertise in areas such as architectural science than is normally acquired by any one architect. For example in the Building Science Unit in my department at Sheffield University alone there have been numerous research students producing information and tools of use to architects, but their work fails to get into a format which makes it accessible to designers as opposed to research workers. There is a considerable difference between an analytical tool suitable for use by an environmental engineer and the kind of tool needed by an architect who is in the early stages of design [7]. The analytical tool thus tends to evaluate the proposed design in some way and if we are lucky it may check its performance against some declared criteria such as building regulations. This then is the first fallacy implicit in so much software design for use by designers like architects. It is based on the mistaken view that such design can proceed by a series of one dimensional analyses of successive attempts to form a solution. In fact there are two problems here. Firstly such software tends on the whole to be evaluative and analytical rather than goal directed or synthetically in its style, and secondly that each aspect of the design is treated in a discrete rather than integrated manner. Let us examine these two problems in turn.

The sequence of events in the design process is much less clear than was originally thought in the early days of design research which tended to favour 'analysis-synthesis-evaluation' type models in which it was assumed that problems must be analysed before solutions could be synthesised [8]. By contrast more recent empirical research has shown how designers tend to develop solutions in order to understand aspects of the problem, although this is still neglected in the references of many who write about CAD. The author in a series of laboratory experiments showed that by their fifth year of education architectural students had developed a cognitive style quite dissimilar to that adopted by post-graduate science students [9]. Whilst the science students preferred to analyse a design-like problem before attempting synthesis, the architects were more likely to develop solutions long before they had understood the problem. We might say that whereas their scientific peers seem to be problem-focused, design students appear to be solution-focused in their approach. Since we have already seen that such design problems are not capable of complete description, and since design is a prescriptive rather than descriptive activity this probably represents a reasonably sensible adaptation to the situation. Further evidence supporting this general cognitive strategy came from a study by Darke who conducted a series of interviews with award winning architects [10]. She found that when designing housing these architects frequently used some major form generating idea well before they could be expected to know what form would work best. This 'primary generator' acted as a sort of hypothesis enabling her architects to explore the inadequacies of their yet ill-formed solution and through this to learn more about the problem. In a yet unpublished study of a group of internationally famous architects the author has found evidence which while redefining the idea of the 'primary generator' strongly supports this solution-focused approach.

What appears to be the case is that much designer behaviour is based upon the general strategy of groping towards an understanding of both problem and solution together. Heuristic procedures are thus more likely to assist this process than purely analytical ones. For example, knowing with great precision just how much energy a proposed building will lose is much less help than knowing whether a change of orientation or double glazing would reduce consumption more.

All of this suggests that considerable attention must be paid to making information available in a form which is usable in a design process. Our failure to understand the significance of this has dogged the teaching of scientific material in architectural education. Moreover the problem continues beyond education into practice where particular problems may only be encountered infrequently thus causing designers at best to be unable to solve the problem unaided and at worst to forget to consider the matter at all. In fact this criticism can equally well be levelled at tools intended for designers to use which do not even require computers. For example the author has shown that most schools of architecture in this country have taught students to use daylight protractors, and yet those same students report virtually no use of this tool in practice [6].

This should not be taken to be a criticism of the idea behind the tool or its importance in design, quite the reverse. Others have shown that during the design process the techniques used are highly likely to influence the problems found and therefore solved [11] [12]. It is therefore critical that tools are designed which are indeed usable in the design process. To do this we must understand more about the nature of the generative ideas used by designers during the early stages of the design process and the language through which these ideas are expressed.

A study of the primary generators used by designers in both Darke and Lawson's studies reveal that these ideas are almost always integrative in their nature. That is to say they do not deal exclusively or uniquely with a single dimension of the problem. They do not therefore represent attempts at sub-optimisation. Such designers do not habitually try to design a solution which optimises energy efficiency, or structural rationality, or planning organisation. Take for example the attitude of Richard Burton who developed the first ever RIBA policy on energy conservation in buildings. Even this enthusiastic campaigner was to remark that "energy in building has had a fanfare lately and maybe it will have to continue for some time, but soon I hope the subject will take its correct place among the 20 other major issues a designer of buildings has to consider". This concern to ensure balance in the design process is nicely echoed by Ian Ritchie who describes technology as a 'beautiful parrot sitting on his shoulder' who constantly overlooks the process but occasionally gets out of control and 'jumps down and shits all over the paper before we've had a chance to think properly and then you have to get hold of him and stick him back up there'. This image of the parrot sitting on the shoulder is I believe one which should drive software developers in their search for the right kind of relationship between designer and computer, hence the title of this paper.

Most of the descriptions used by these designers about their early generative ideas are couched in the language of solutions rather than problems. Often these descriptions also reveal a process of discovery which again seems solution rather than problem-focused. "He suddenly said 'it's a wall!', and I said 'yes it's not just a lot of little houses, it's a great wall two hundred metres long and three storeys high'. The problems which have already been discovered are then fitted into this idea. "...but we'll make it a bit higher to get in the plant and service circulation and then we'll punch the residential elements through the wall as a series of glazed bays which come through and stand on legs."

Thus we can see here a process which consists of ranging across a series of problem issues very quickly through focusing on and developing a single solution idea. Many of these designers talk of the need to work very rapidly and intensively at certain key stages in the process. One of the main reasons for this seems to be the need to consider a wide range of problem issues, or as one leading architect described it 'keep a lot of things in play'. A common image used by designers to describe this is that of juggling. Michael Wilford, for example talks of 'a juggler who's got say six balls in the air...an architect is similarly operating on at least six fronts simultaneously and if you take your eye off one of them and drop it you're in trouble." The concentration required for such a complex juggling act is so intensive that one cannot imagine it lasting for long, or as Richard MacCormac puts it, 'one couldn't juggle very slowly over a long period'.

So how can computer aided design tools fit into such a process. One of the key features of this kind of design lies in its multi-dimensional problems. Some aspects are relatively objective and numerical, while others are more subjective. However as we have seen tools already exist to aid the designer with many of these problems. However our difficulty is that we have seen how good designers at least need to work extremely quickly, and many tend to dismiss computer-aided design for this reason. The whole process is destroyed by having to break off and enter data into a computer before getting an answer. This corresponds with the experience of studio tutors in design schools. Students often need and want advice very quickly. If they have to wait more than a few hours before the tutor can find time their thoughts have moved on and they may again be in that recurring phase of the process when they do not want to discuss it.

If we are trying to develop computer aids for such a process then it is simply no good writing individual programs, which while they may be able to give answers to their own questions will not be used by designers. Many years ago John Lansdown pointed out that we could either have what he called 'ad-hoc' programs or 'integrated systems' [13]. By this he meant we either have a series of quite separate programs each with their own inputs and outputs or we have a single computer description of the proposed design which a whole series of routines can operate upon. Clearly the 'ad-hoc' programs provide far too great a delay to be satisfactory for the design process as we observe it to be practised.

It was hoped that integrated CAD systems would offer the prospect of embracing at least all technical knowledge into a design tool, and the author has himself devoted many years to developing such a system for architects [14]. However this has proved something of a mirage I think for several reasons. Firstly the expense of such systems and the skill required to operate them has so far prevented them from becoming commonly used, although architects now make regular use of drafting systems. The second reason for the failure of such systems to make the predicted impact is of a more technical nature and concerns the difficulties of enshrining knowledge which needs updating and challenging into software which can only be updated by system suppliers. There is hope that techniques such as object oriented programming and the use of declarative languages may alleviate some of these problems but little real progress has been made to date. Even were this progress to be made such integrated systems all have a recognisable style with many built in preconceptions that designers find too inflexible in use. Creative and sensitive designers simply do not want to give up so much control of their process to software developers. The system feels as if it suffocates the designer. The tail wags the dog!

An alternative scenario might be for designers to be able to buy individual packages which can draw their data from a common and agreed conventional description of the design. The popularity of the Apple Macintosh amongst such designers suggests this idea is appealing. What the Mac basically offers is a common interface which is firstly relatively intuitive but also so disciplined that it forces software developers to construct programs with sufficiently similar interfaces for the user to be able to operate programs from several suppliers without suffering 'manual overload'. Unfortunately even though much work has been done I think there is little hope of a satisfactory common description in the foreseeable future. Indeed such an idea is positively dangerous in my view until we understand much more about what is required. Ill-considered and restrictive standards can impose serious constraints on innovative development. We already have enough obstacles posed by the early development of standards for microcomputer operating systems and data exchange formats for two-dimensional graphics.

However this discussion is already in danger of becoming as unrealistic as are many computer design aids, for it implies that a database of known facts about a solution can really exist during the design process. The problem before us is less a matter of how evaluative routines could be fed from a common database, but rather how that database grows and interacts with the designer as he or she gradually comes to understand the solution, which until the final decisions are made remains a nebulous and ever changing mirage rather than any concrete reality. Personality studies of designers have shown them to be prepared to accept high levels of uncertainty and change and to be capable of looking at things in many different ways. These then are also the characteristics we need from any computer system to which we might be describing our design solution.

By contrast the characteristics of the computer have lead software developers to create programs offering ever increasing levels of precision, accuracy and realism. Unfortunately these programs require huge amounts of input and can really only be justified once the design is more or less finalised. Perhaps the most obvious example of this can be seen in the world of visualisation tools. We now have the most astonishing levels of realism available using sophisticating rendering techniques of ray-tracing or radiosity. However these algorithms can produce nothing without data governing not only the geometry of the design but also surface colours and textures to say nothing of light sources and characteristics. All this is very nice when you want to convince the client that your design is just what he wants, but quite useless while you are still at the doubting stage yourself. What designers need are tools which give a rough indication. There is no point in being precise about something which is as yet only a vague idea.

This analysis leads me to suggest that the area in which we need to do more work is the way in which the designer describes this vague, imprecise, nebulous, shifting idea to the computer. I do not refer here however to the minutiae of ergonomics, to the detail of windows, icons, pull-down menus and pointing devices. These are in my view only the superficialities of interface design. What I refer to here is the way in which the computer must be able to infer and accept ambiguity, to live with incomplete descriptions and perhaps fill in the gaps, and most importantly interpret the constantly changing language of the designer.

We must finally learn the lesson that if computer aids are to play a useful role in design, as I believe they can, they must not require of their users to learn a foreign language. One simply cannot expect a creative co-operation to emerge from a situation where the key participant is continually reducing to interrupting the train of thought and argument by having to look up words in a dictionary! If we are to develop suitable interfaces then we must listen to how designers describe their solutions. We should set as our goal the objective of a system which can at least understand as much about a design as would another designer. The immediacy, subtlety and often vagueness of these conversations are not an eccentricity, but an essential component of the creative process. Many designers I have interviewed make reference to a phenomenon well known anecdotally, and best illustrated in Richard MacCormac's words. "I cannot say anything until I've got a pencil or pen in my hand." This supports the delightful concept developed by Donald Schon who describes designers as 'having a conversation with the drawing' [15]. Over the years I have developed such a relationship with my word processor. I wrote this paper by 'conversing with my keyboard and screen'. I can now type faster than I can write, and the liberating feeling I get from knowing how easily I can edit somehow enables me to work in a less inhibited and more productive way. Even the slight pause required to print offers that few moments to get up and walk around which somehow facilitates the creative process. I have yet to find any computer aided design tool which even approaches this quality of creative interaction. We have hardly begun in this field.

Of course it is far easier to develop a word processing tool than a design aid. We do not expect our word processor to understand and evaluate the quality of the arguments in the text. That is just what we do expect from design tools, and we would like this even before the argument is cogently put. This is a tall order, but I believe we can already see some of the things that will be necessary if we are to achieve this, and I shall devote the remainder of this paper to illustrating this through one example.

Let us then finally get down to asking ourselves what we want from tools which we might genuinely call computer design aids? Firstly let me say that I see almost no point at all in designers using computer graphics systems except for quite specialised purposes. Most designers have had to learn to draw and sketch and have a great facility with the pencil which is currently a far more flexible and sensitive tool than any computer drafting system I have ever seen. Donald Schon has taught us to understand that design seems to progress through what he calls a conversation with the drawing. That is to say the designer sees the drawing and makes a change and sees the result. Most usually the designer can and does compare the result very directly with what went before. Sketches are made either by drawing over the original on the same piece of paper, laying another semi-transparent sheet over it or starting a new sheet. In all cases the original and new can be compared. How many computer drawing systems allow this absolutely essential but simple requirement without enormously tedious and complex instructions? However computer drawing systems are often superbly efficient, accurate and reliable ways of constructing complex geometry and transforming it. Santiago Calatrava tells us that "I like to sketch quickly by hand", but that "I admire very much the precision of the computer maybe if you want to draw an arc between two points and you want to see immediately, you could not do that by hand". [2] This again reminds us about this need for immediacy, only if the computer is more direct and immediate will it be preferred by designers.

If we are really to produce useful CAMP tools then we must be able to interact with them in a very immediate and flexible way. However any observation of architects working shows that they frequently load their design drawings with meaning. These meanings are usually taken to be self explanatory and understandable by other designers. These can be seen at two levels. Firstly the kind of representation offered by the drawing and its relation to other drawings, and secondly the relationship between the drawing and the building it gives information about.

At the first level we therefore expect a plan to be recognisable as a plan rather than a section, elevation or perspective. If one of my students starts drawing during a discussion they will almost never say what kind of projection they are using and indeed bits of plans, sections and perspectives may well all appear on the same sheet and I am expected to understand this. At the second level my student will expect me to understand all sorts of things about the building from these drawings. For example as a plan is drawn it would be reasonable and normal to expect that the viewer could distinguish indoors from outdoors.

This last point about implicit information can be further elaborated to show that often what must be inferred by the reader of drawings depends on the ability to use multiple modes of thought about buildings or other design objects. It has been argued that architects organise knowledge about their emerging building designs into a number of quite distinct and structurally different forms. We may see these as 'modes' of thought each having their own 'features'. For example a building may be seen as a collection of components such as walls, floors, roofs, windows, doors and so on. We may thus call this 'component mode', and identify each building element as a feature of that system of knowledge. However the same building may also be seen as a collection of envelopes such as spaces, courtyards and external skins. What we might call 'envelope mode' then represents a quite different system of knowledge about the building to that which we see in 'component mode'. Unfortunately the mapping between these and other modes is rather messy, and the relationships vary from feature to feature. A wall which we would be inclined to think of as a single component may , along its length separate several rooms, so we have a one to many mapping from component to space. A door by comparison can reasonably always be considered to connect two spaces, or a single space with the outdoors.

We can go on to consider other modes such as 'system mode' where a building is thought of in terms of features such as structure, cladding, circulation, services and so on, or perhaps 'strata mode' where the features would include floor levels, facades and sections. Of course, all these ways of organising knowledge about a building are mutually interdependent and architects in conversation about their solutions frequently flip from one mode to another in quick succession. Regrettably however few computer-aided modelling systems for use by architects recognise this and require a form of communication in only one mode and that of the software designer's devising. There are two reasons why this poses a serious obstacle to the successful development of such systems. One is to do with drawing or input and the other to do with evaluation or output.

Observation of an architect sketching will show that as the 'conversation with the drawing' takes place editing is not done all in one mode. For example a sketch may show some rooms in plan draw in envelope mode which is then altered through a wall being added in component mode which might well alter several rooms.

However it is also the case that evaluative procedures may require data in different modes. Simple thermal evaluations need to be able to distinguish indoors from outdoors and thus need envelope data. Structural calculations need to be able to recognise those components which are part of the structural load bearing system, daylighting assessments are performed on rooms, routines to check for satisfactory means of escape in case of fire need envelope and stratum based data and so on.

It therefore seems necessary to envisage systems which are able to make inferences based on input in one mode about the implications to data represented in another mode. We have already built one prototype system at Sheffield which allows the user to draw and edit either components like walls and windows or envelopes like external skins and rooms. The System known as DAILY automatically keeps both modal descriptions of the building up to date [16]. There are however two serious weaknesses in this system. Firstly that the knowledge needed to make the inference remains in the program rather than in a data knowledge base, and secondly that the user must still decide before drawing which mode he is using.

This however seems to me to be the area most in need of work if we are really to make progress towards the 'parrot sitting on the shoulder'. Better and faster routines for rendering or more sophisticated algorithms for assessing energy flow will make no difference to the already disappointingly low take up of computer-aided design in architecture and its allied design fields. We need intelligent modelling modules which can automatically produce multiple modal descriptions of designed objects before we can sensibly feed design appraisal tools. Then and only then will we know whether such a 'parrot' is sufficiently tame and knows enough words to be useful in the design studio.

References

1.Galle, P. 'Computer methods in architectural problem solving: critique and proposals' *The Journal of Architectural Planning and Research* Vol.6, pp.34-54, 1989

2.Lawson, B.R. *Design in Mind* Butterworth's Architecture, London, in press.

3.Lawson, B.R. *How Designers Think* Butterworth Architecture, London, 1990.

4.Rittel, H.W.J. & Webber, M.M. 'Dilemmas in a general theory of planning' *Policy Sciences* Vol.4, 1973

5.Cross, N. 'Designerly ways of knowing' *Design Studies* Vol.3, pp.221-227, 1982

6.Lawson, B.R. 'Science, legislation and architecture' in *Changing Design* ed. Evans, B.N., Powell, J.A. & Talbot, R.J. John Wiley, New York, 1982

7.Brown, A.G.P., Wood, D.J. & Brown, G.Z. *A sunlight design tool* ECAADE, Aarhus Denmark, 1989.

8.Markus, T.A. 'A doughnut model of the environment and its design' in *Design Participation* ed. Cross, N. Academy Editions, London, 1972

9.Lawson, B.R. 'Cognitive strategies in architectural design' in *Developments in Design Methodology* ed. Cross, N. pp.209-220 John Wiley, Chicester, 1984

10.Darke, J. 'The primary generator and the design process' in *New Directions in Environmental Design Research: procedings of EDRA 9* ed. Rogers, W.E. & Ittleson, W.H. pp.325-337 EDRA, Washington, 1978

11.Eastman, C.M. 'On the analysis of the intuitive design process' in *Emerging Methods in Environmental Design and Planning* ed. Moore, G.T. MIT Press, Cambridge Mass, 1970

12.Rowe, P.G. *Design Thinking* MIT Press, Cambridge Mass, 1987.

13.Lansdown, J. 'Computer-aided building design: the next steps' *RIBA Journal* Vol.76, pp.138-140, 1969

14.Lawson, B.R. *Gable: an integrated approach to interactive graphical techniques for modelling buildings* On-Line Publications, London, 1981.

15.Schon, D.A. *The Reflective Practitioner: How professionals think in action* Temple Smith, London, 1983.

16.Lawson, B.R. & Roberts, S. 'Modes and features: the organization of data in CAD supporting the early phases of design' *Design Studies* Vol.12, pp.102-108, 1991

Adding visual recognition to the capabilities of computer-aided design

A. Koutamanis, V. Mitossi

Faculty of Architecture, Delft University of Technology, Berlageweg 1, NL-2628 CR Delft, The Netherlands

ABSTRACT

The paper presents an approach to reducing the labour-intensiveness of inputting and manipulating visual design representations in the computer by means of automated recognition of his/her drawing input. Design systems which employ computer vision can make or suggest corrections to the representations created by the designer; make explicit design aspects which are implicit in the representation; allow and facilitate the parallel use of several design representations. The same forms of recognition can also be instrumental for the use of autonomous robots in manufacturing and construction and for the transfer of existing archives of analog drawings to the computer.

INTRODUCTION

Computer-aided design (CAD) is rapidly and irrevocably changing the practice of most design disciplines. The advantages of computerization —compactness of storage, ease of modification, superior simulation facilities, extensive online design and decision support— are numerous and significant enough for advocating a transition form analog mechanical systems to electronic ones. However, these advantages should not obscure the many problems that remain to be solved, problems which may impede or even frustrate the intensive, demanding designer user.

One such problem is the labour-intensive character of inputting and manipulating visual design representations in the computer. In order to be able to use a drawing for anything beyond mere visualization, the representation of a design should be impeccably specified and structured into groups and layers, as well as properly annotated with physical or other properties. This structure and annotations link the visual representation with other aspects of design thinking and facilitate its use with other software.

The resulting intellectual burden of the computing designer can be reduced by means of automated recognition of his/her drawing input. Design systems which employ computer vision can recognize relevant entities and their interrelationships in detailed, complex drawings or models, as well as in elliptical and abstract sketches. The output of the recognition process consists of networks of design entities which describe with rather high specificity one or more aspects of the design. These networks form the basis for correcting / restructuring the representation if and where necessary, and for augmenting / complementing the representation with annotations or additional primitives which are implicit in the original representation. Such modifications are in turn the foundation for any transformation of a representation, for example of the floor plan of a building into a bill of materials or into a representation of pedestrian circulation in the building. The ability to transform a design representation automatically can have far-reaching effects for the usability of a representation throughout the design process and even after that, in the framework of manufacturing and construction.

In the following sections architecture and AutoCAD are used as sources for examples which are common to most drafting / modeling systems for any design discipline. The proposed solutions and extensions to these systems are applicable not only to AutoCAD but also to practically every programmable drafting / modeling system. In several cases it might be necessary to employ routines written in a programming language other than the built-in one as external resources. Far from being disadvantageous for the effectiveness and efficiency of the system, such external resources are often superior to the built-in programming facilities in terms of performance.

RECOGNITION OF INTERNAL DOCUMENT STRUCTURE

One problem that has been commonly neglected in studies of computerization is the labour-intensive character of inputting and manipulating information. In order to be able to operate effectively and reliably the computer requires explicit statement and identification of relevant information. In a text a word is not a word if not separated from its context by spaces. A sentence should always end with a full stop, a question or an exclamation mark. Deviation from the standard practices causes only problems: most spelling checkers are unable to recognize accidentally merged words; grammar control is generally impossible without a clear demarkation of sentences.

In visual design representations the computer requires that the logical and physical structure of a computer document is even more impeccable than in a normal text. Each graphic element must be placed on the appropriate layer; it must be properly connected to other, related graphic elements; it must be annotated with significant attributes which are not covered by its visual appearance. What this structure attempts is (a) modularization of information in order to facilitate processing, and (b) identification of the denoted entities, such as the building elements in an architectural floor plan. Without this structure it is impossible to use the visual representation for anything beyond mere reproduction and dissemination. The ability, for example, to produce a bill of materials on the basis of a computerized floor plan is dependent upon the internal structure of the floor plan: its layer structure, the predefined symbols used (*blocks* in AutoCAD), their correspondence with the denoted physical objects and naturally the conformity of the graphic elements which make up the floor plan to these. If a door is drawn on the wrong layer or without making use of a predefined symbol, then it is highly unlikely that this particular door will appear in the bill of materials.

The rigidity and strictness of a computer-based visual design representations can be seen as having a positive influence on the designer as it increases awareness of the structure and semantic content of a representation [1], despite the constraints it places on the designer's creative flexibility [2]. In practical terms, however, rigidity and strictness in computer-based visual design representations create a heavy intellectual burden on the designer who must specify and structure these representations impeccably. This burden is currently often obscured because many designers operate computer systems indirectly —through specialized draftsmen— but even then it is evident in the many revisions a computer document must undergo in order to become widely usable and reliable.

Automated recognition can be instrumental in safeguarding the internal structure of a computer document. Even wrongly drawn or placed graphic elements contain sufficient clues for the computer to recognize their semantic content and restructure the drawing accordingly. Such clues would have been even more and stronger if the history of the representation (i.e., the time each graphic element was input, and when and how it was subsequently modified) was also stored in the computer. Unfortunately this would require not only enormous storage capacity but also extremely efficient and reliable search facilities which are unattainable in practice.

The first step in the recognition of a computer-based drawing is the distinction between different types of information. In a floor plan we can distinguish between graphic elements which denote:

1. Physical objects (building elements) such as doors, windows, walls or stairs which must be explicitly specified by the designer.
2. Physical objects such as furniture which are only indicative of the use and possibilities of the building.
3. Properties and attributes of the building elements given in a graphic manner, such as dimensioning and textures (hatching).
4. Verbal annotations of other attributes which can be apparent (text in the drawing) or hidden (as annotated properties of a symbol).

Recognition of graphic elements which denote entities of each of the above categories is based on (a) implementation clues, (b) intrinsic representational characteristics, and (c) contextual factors.

The implementation of a design representation, that is, the making of a drawing or model, is governed by conventions and practical considerations which have to do with both the inherent structure of the representation and the implementation media. In CAD there are different types of graphic primitives and structures for each type of information, such as special facilities for inputting dimensioning and the ability to use predefined symbols for denoting standardized elements. In addition, the layer structure of a CAD document is based on a modularization of information which reflects the different types of design information in the document. Dimensioning, hatching or furniture are preferably placed on separate layers which are distinct from the basic layer(s) where building elements stand.

In addition to these implementation clues the standardization in the appearance of symbols denoting building or other elements facilitates recognition of discrete instances, even when they were drawn in a non-standard fashion. For example, it is only reasonable to expect that an architect inputs doors in a floor plan by means of one or more predefined symbols (blocks) which correspond to the basic door types that are to be used in the building, as use of predefined symbols offers significant advantages in terms of ease of modification, compactness of storage, processing speed and further document usability. Similarly, dimensioning should be input by means of the special facilities which exist in practically every drafting system. Nevertheless, it is not unthinkable that a door or dimension line might be drawn by other means. Limited experience, absent-mindedness, technical problems (e.g. inability to load the library of predefined symbols), organizational faults and many other sorts of human and computer errors do occur in the often tense atmosphere of a design firm and can be expected to influence negatively the quality of computer-based design representations. Automated recognition can compensate for these errors by recognizing non-standard symbols and then transforming them accordingly. This can be achieved by checking the graphic elements in a drawing against a set of rules or prototypical templates: a door in a floor plan is commonly represented by a straight line and an arc connected to one of the line's endpoints. The width of the arc is less or equal to that of the line. An alternative representation substitutes the arc with a diagonal line but without further changes (Figure 1). Similarly, a dimension line has minimal width and distinctive terminations (arrows or short vertical / diagonal lines).

Implementation clues and intrinsic characteristics are reinforced by contextual factors which have to do with the logical structure of the design or drafting conventions. A door, for example, is to be found in a wall. It is therefore only logical to expect that the free endpoints of the graphic elements used to represent a door in a floor plan should coincide with endpoints of lines denoting walls. Dimension lines can also be recognized on the basis of their context, i.e., their close proximity to a text block and the coincidence of their endpoints with perpendicular extension lines or lines denoting the boundary of a building or other element (Figure 2).

The above clues and characteristics can be used for checking the structure of a computer document, correcting inconsistencies in the use of graphic primitives or layers, normalizing symbolizations, and annotating graphic symbols with the names of the denoted entities and their default or inferred attributes. It should be pointed out that not all entities are as doors, windows and dimension lines. Elements with variable or even arbitrary forms and structure, notably stairs and walls, require particular attention. A promising approach is the decomposition of such elements into well-defined parts, such as

```
Endpoint 1 of arc = Endpoint 1 of line
AND

X-coordinate of Endpoint 2 of arc =
X-coordinate of Endpoint 2 of line
AND
Y-coordinate of Endpoint 2 of arc =
Y-coordinate of Endpoint 2 of line + length of line

OR

Y-coordinate of Endpoint 2 of arc =
Y-coordinate of Endpoint 2 of line
AND
X-coordinate of Endpoint 2 of arc =
X-coordinate of Endpoint 2 of line + length of line
```

```
Endpoint 1 of line a = Endpoint 1 of line b
AND

X-coordinate of Endpoint 2 of line a =
X-coordinate of Endpoint 2 of line b
AND
Y-coordinate of Endpoint 2 of line a =
Y-coordinate of Endpoint 2 of line b + length of line b

OR

Y-coordinate of Endpoint 2 of line a =
Y-coordinate of Endpoint 2 of line b
AND
X-coordinate of Endpoint 2 of line a =
X-coordinate of Endpoint 2 of line b + length of line b
```

Figure 1. Rules for the recognition of door symbols in architectural floor plans

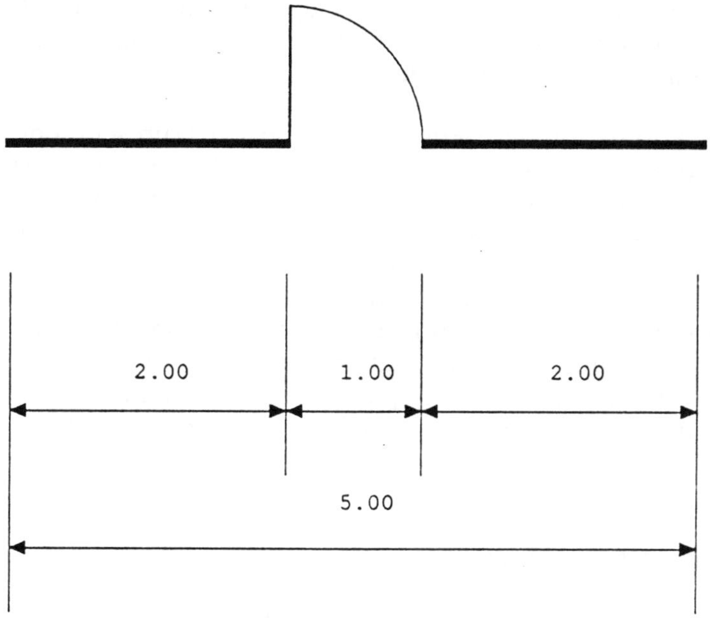

DOOR:

```
free endpoint of line = endpoint of wall
free endpoint of arc = enpoint of wall
```

DIMENSIONING:

```
distance of dimension line from text < [threshold]
endpoints of dimension line = points of extension lines

OR

endpoints of dimension line = points of building element
```

Figure 2. Contextual factors in the recognition of door symbols and dimensioning

straight wall segments, wall corners and wall junctions which facilitate recognition and description, as well as further utilization of the recognized elements and their structure.

Once the various graphic elements in a design representation have been recognized, it is possible to proceed to the relationships that exist between them, toward interpreting the representation as a coherent and comprehensive network of design entities —a network of building elements in architecture. The contextual factors which have already contributed to the recognition of building elements can be used now as expectations concerning the spatial relationships between a recognized element and its neighbours. Constraint propagation can be used to connect and develop these expectations into a semantic net which describes the structure of the represented design.

RECOGNITION OF IMPLICIT DESIGN ASPECTS

Correcting and restructuring a visual design representation is only part of the possibilities offered by automated recognition. Perhaps more significant is the ability to recognize design aspects which are implicit in the representation, such as the spatial articulation of a building in a floor plan. A floor plan is an explicit representation of the 'solids' of a building, its building elements, but also an implicit representation of its 'voids', of the spaces which are bounded by the building elements. The designer but also the lay person uses his/her cognitive capabilities to discern the duality of the representation and to shift seemingly effortlessly from the one to the other. Recognition and explicit representation of the spatial articulation of a design can be seen as endowing the computer with similar capabilities.

A space in a floor plan (or rather the closed loop of building elements which bounds it) can be recognized by means of edge-following [3; 4]. This is a straightforward approach which also builds on the network of building elements that has been already recognized. However, it is also computationally expensive, tenacious and quite blind to local problems: minor perturbations and small gaps may drastically reduce the performance of an edge-following space recognizer. An alternative approach which is based on the recognition of line drawings of three-dimensional scenes with trihedral objects [5] and also relates to aspects of human vision [6; 7; 8], is to concentrate on critical parts of a space, namely its corners. In this recognition of spaces in a floor plan is based on (Figure 3):

1. A typology of interior space corners. In rectangular floor plans there are eight types of such corners.
2. The connectivity of different corner types, which allows the development of expectations concerning the type and position of its neighbours (i.e., other corners of the same space). By propagating the connectivity constraints it is possible to identify the corners of each space and consequently its overall shape [9; 10].

The conceptual advantage of the above technique is that, while edge following deals with the elements that *bound* a space, this technique concentrates on the space itself, in very much the same way that illusory contour perception operates. The resulting practical advantages include direct recognition of the shape of a space and relative freedom from local perturbations and missing parts in the boundary of a space. The latter can be of particular significance for the recognition of drawings of buildings with semi-unified or flow spaces.

As with building elements, recognition of the spaces in a floor plan is followed by the development of a network of their spatial interrelationships which completely describes the spatial articulation of the represented design. This network complements that of building elements and can be linked to it, thus creating a dual network of the 'solids' and the 'voids' which comprise a building.

SUPPORTING THE DEVELOPMENT OF A DESIGN

Automated recognition of the relevant explicit and implicit elements and relationships in a visual design representation transforms the computer from a passive electronic equivalent of the drafting table into a knowledgeable and critical tool which can control and correct the content and structure of design representations without burdening the designer with unnecessary warnings and questions.

Recognition and explicit representation of the dual network of building elements and spaces in a floor plan allows for control and correction of the design representation also with respect to extrinsic domain knowledge, as it facilitates integration of formal and functional evaluation and analysis software. Such software operates mainly at the levels of building elements and spatial articulation and their interrelationships. The purely geometrical CAD representations are generally not compatible to this software without extensive grouping and

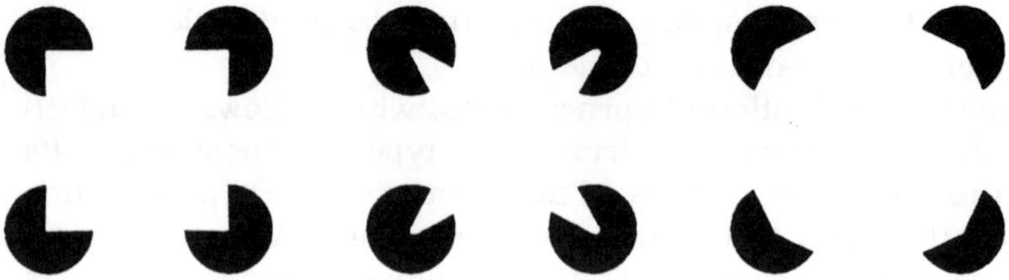

Illusory contours

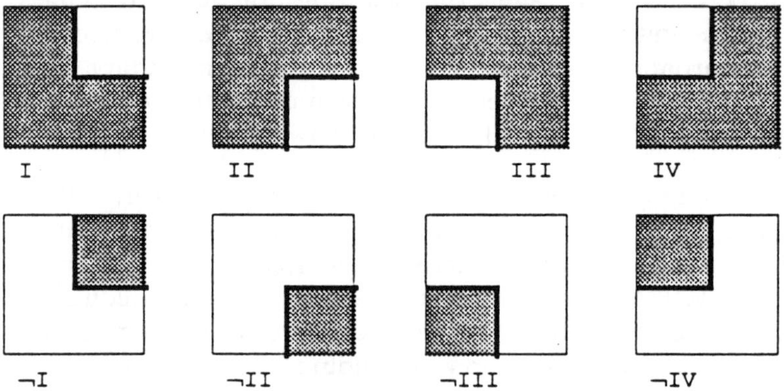

Space corner types in rectangular floor plans
black lines indicate the boundary of a space; white the interior of the space; gray other adjacent spaces

type	up	right	down	left
I	II•¬III	IV•¬III	none	none
II	none	III•¬IV	I•¬IV	none
III	none	none	IV•¬I	II•¬I
IV	III•¬II	none	none	I•¬II

Space corner type connectivity in rectangular floor plans

Figure 3. Space recognition

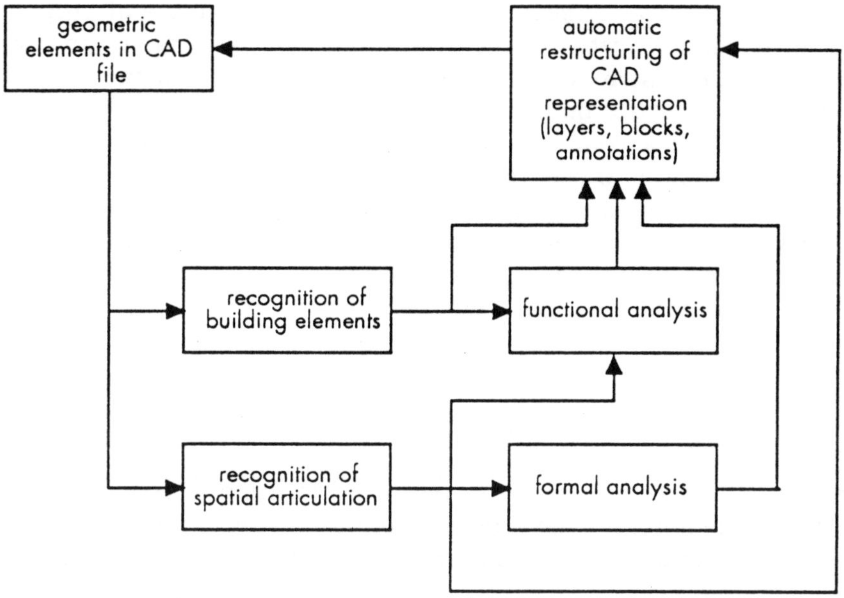

Figure 4. Automated recognition of computer-based visual architectural representations

annotation which identify the relevant entities, their qualities and interrelationships. Automated recognition can reduce this labour-intensive and time consuming task and furthermore offers the possibility of feedback from the analysis software to the CAD representation: conflicts identified in the formal and functional analyses can —via the networks of building elements and spaces— be pointed out to the designer in the CAD representation (Figure 4).

The computerization of architecture is characterized by a never-ending quest for the perfect holistic representation of designs and of design thinking, a representation which covers all possible aspects of the built environment and its manipulations in building design and construction. The results of this quest have been quite disappointing, not only because such holistic representations appear to be unattainable but also because the target of the quest has obscured several partial successes. Computer vision suggests an alternative approach which is more realistic and more relevant to our own cognitive processes, namely the parallel use of several representations, each catering specific design aspects, and the use of automated recognition for transforming one representation into another. In designing a building an architect might employ representations such as topological diagrams of spaces and their relationships (in the analysis of programmatic demands), orthographic projections like floor plans and sections (for the description of the spatial arrangement of the elements which comprise the building) and perspective drawings (for an impression of the appearance of the building). The computer's ability to recognize each of these representations offers a knowledge-based means of communicating design decisions between different representations and the corresponding computer systems, with a degree of explicitness and relevance, and at a variety of abstraction levels completely unattainable by data exchange schemes [11].

EXTENSIONS, I: ROBOTICS IN CONSTRUCTION

The use of the computer is not restricted to the production of documents for communicating a design solution. Direct use of computer-based representations for the computer-aided manufacturing (CAM) of building components is not commonplace today but does have its (early) applications, mainly in the form of data for the off-site production of standardized elements like doors and windows. Judging from other CAD/CAM processes where design and production are more closely organized, we can only expect a growth of such practices in the construction industry too. However, the nature of building construction necessitates more advanced solutions, such as the use of autonomous robots in building sites.

Autonomous robots offer the flexibility required in building construction but also present a number of yet unresolved problems, chiefly with respect to navigation. In an unconstrained, constantly changing environment such as the typical building site an autonomous mobile robot cannot rely on measuring rotations of step motors in order to determine its position. The intricate paths the robot must follow among the many obstacles in the site means that the accumulated errors of the step motor and the robot wheels become too gross too soon. Visual ability offers the necessary flexibility and reliability but not without further complications.

In a building site a robot requires an excellent sense of position and orientation in space because this space is not only the environment where it moves but also the object upon which the robot must operate (build). It also requires precise identification of the objects it encounters in the site, such as loose building components, components already in position and auxiliary structures like scaffolding. Moreover, the robot is asked to make very accurate estimations of the relationships which exist between these objects and the space of the site, e.g. in deciding where it should store temporarily the building components it unloads from a truck so that they do not impede in the current building activities and how the place of temporary storage should relate to their final position.

To perform such tasks efficiently and reliably the building robot requires accurate models of the site and its contents (other machinery and auxiliary structures), of building components, but also of the final and intermediate states of the building under construction. Without these it is impossible to reason where e.g. one common (i.e. one out of many identical or similar) building component should be placed in relation to another which is already in position. Architectural drawings and models, even if made by computer, indicate explicitly just two things: the initial state of the site and the final state, with the building completed, or rather with all building components in their final position. People working on site use their perceptual and cognitive abilities to read the drawings and discern or even guess what happens in between. Automated recognition of building elements and spaces in these drawings and models can offer similar abilities to the robot.

Recognition of building elements provides information not only on the final position of building components but also on their interrelationships. These represent the essential framework which determines the transformations of a site between the initial empty state and the final state of the completed building, as they also describe the various subsystems of a building (e.g. load bearing structure versus infill walls, openings or electrical networks) and hence indicate their temporal order of construction. The spatial articulation of the building under construction represents the final state of the building, the spatial model which verifies the completion of the project. Implicitly it offers the necessary information for understanding the incomplete building during construction, as it indicates the intentions behind the positioning of building components or auxiliary structures, as well as how the spatial characteristics of any intermediate state relate to the final state. These are essential for the navigation of autonomous robots in the site and offer valuable information concerning the forthcoming activities at any given part of the site and the incomplete building, thereby influencing decisions on e.g. the allocation of temporary storage areas.

EXTENSIONS, II: TRANSFER OF ANALOG DRAWINGS TO THE COMPUTER

Automated recognition of architectural drawings also applies to optically digitized (scanned) images [9; 10]. Computer vision allows the transfer of existing archives of drawings to the computer and, by developing them into structured design representations of building elements and spatial articulation, facilitates furthermore the utilization of these digitized drawings with CAD programs (Figure 5).

Recognition of digitized documents commences with geometric elements —lines and shapes formed by the pixels of the digitized image. This level amounts to a direct transformation of the digitized pixel-based image to a vector-based CAD document. Even though the computational tenacity of this level is such that recognition of geometric elements compares unfavourably to recognition at other levels, automated recognition of digitized drawings is a serious alternative to the even more tenacious and unreliable manual digitization with tracing devices like graphic tablets and mice —especially for large-scale digitization projects.

Recognition of geometric elements transforms a digitized drawing into a computer document which can be imported in a CAD system. The structure of the vectorized digitized drawing is obviously minimal and must be subsequently developed by means of the same recognition levels that apply to computer-based design representations. In orthographic projections such as floor plans and sections (but not elevations) this can be achieved by the recognition of the dual network of building elements and spaces which comprise the design. In perspective projections (perspectives, isometrics, axonometrics) and elevations recognition of spatial articulation of the design is replaced by recognition of its volumetric composition —which could be described as the aggregation of building elements and spaces. Recognition of volumetric composition in architectural drawings is directly based on the principles and algorithms which have been developed for the recognition of line drawings of three-dimensional scenes of polyhedra, including shading and shadows [5; 12; 13; 14].

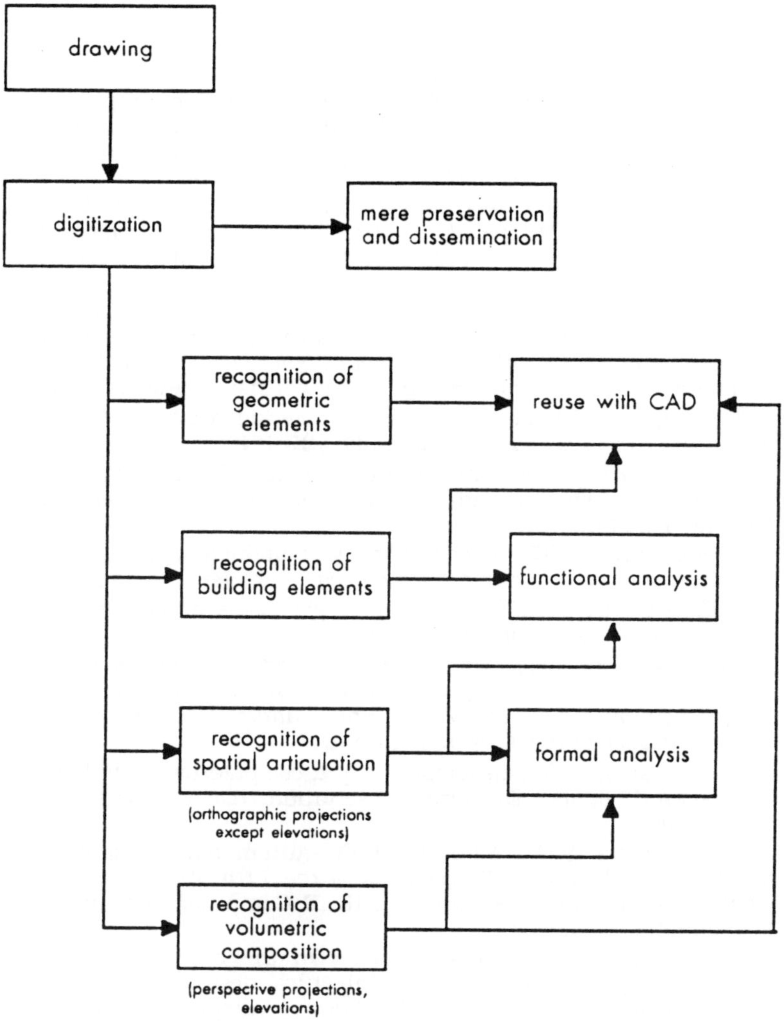

Figure 5. Transfer of analog architectural drawings to the computer by means of automated recognition

REFERENCES

[1] W.J. Mitchell, 1990. Introduction: A new agenda for computer-aided design. in M. McCullough, W.J. Mitchell & P. Purcell (eds) *The electronic design studio. Architectural knowledge and media in the computer era.* MIT Press; Cambridge, Massachusetts

[2] G. Stiny, 1990. What designers do that computers should. in M. McCullough, W.J. Mitchell & P. Purcell (eds) *The electronic design studio. Architectural knowledge and media in the computer era.* MIT Press; Cambridge, Massachusetts

[3] B.R. Lawson & J.P. Riley, 1982. ISAAC. A technique for the automatic interpretation of spaces from drawn building floor plans. in A. Pipes (ed) *CAD 82. 5th International Conference and Exhibition on Computers in Design Engineering.* Butterworths; Guildford, Surrey

[4] S.H. Joseph, 1989. Processing of engineering drawings for automatic input to CAD. *Pattern Recognition* **22** 1–11

[5] D. Waltz, 1975. Understanding line drawings of scenes with shadows. in P.H. Winston (ed) *The psychology of computer vision.* McGraw-Hill; New York

[6] I. Biederman, 1987. Recognition-by-components: A theory of human image understanding. *Psychological Review* **94** 115–147

[7] G. Kanizsa, 1974. Contours without gradients or cognitive contours? *Italian Journal of Psychology* **1** 93–112

[8] D. Walters, 1986. Selection and use of image features for segmentation of boundary images.*Proceedings IEEE Computer Society Conference on Computer Vision and Pattern Recognition.* IEEE Computer Society Press; Washington, D.C.

[9] A. Koutamanis, 1990. *Development of a computerized handbook of architectural plans.* Doctoral dissertation, Delft University of Technology

[10] A. Koutamanis & V. Mitossi, 1992. Automated recognition of architectural drawings. *Proceedings of the 11th IAPR International Conference on Pattern Recognition.* IEEE Computer Society Press; Los Alamitos, California

[11] A. Koutamanis & V. Mitossi, 1991. On the representation and analysis or dynamic aspects of architectural design. in F.H. Post & A.J.S. Hin (eds) *Proceedings of the Second Eurographics Workshop on Visualization in Scientific Computing.* Delft

[12] D. Huffman, 1971. Impossible objects as nonsense sentences. in B. Meltzer & D. Michie (eds) *Machine Intelligence 6.* Edinburgh University Press; Edinburgh

[13] M. Clowes, 1971. On seeing things. *Artificial Intelligence* **2** 79–116

[14] A. Guzmán, 1968. *Computer resolution of three-dimensional objects in a visual scene.* Ph.D. thesis, MIT

Reinventing the Danteum Project

E. Ng

The Arts Tower, School of Architectural Studies, Sheffield University, Sheffield, S10 2TN, UK

Abstract

This paper describes an experiment where architectural history and theories are re-interpreted using some of the latest visualisation and multimedia techniques. The Danteum Project by G.Terragni was used as a vehicle to conduct our investigation into some of the potentials offered by computer visualisation. On the one hand the project was meticulously modelled based on available plans and sections, on the other hand studies were carried out to understand the contextual and philosophical framework on which the project is based. The experiment then proceeds to combine the two levels of readings, physical and contextual, and went forward to 'Reinvent The Danteum Project'. The re-interpretation is an attempt to explore some of the meanings of the Project and to further one's appreciation of Terragni's notion of life and death. The re-interpretation relied heavily on a non-realistic and semantic approach to computer visualisation. This leads to a hypothesis on the brave new virtual world where, unlike physical modelling, computer visualisation can be used to represent the 'perceived world'.

Representation and Interpretation

The fundamental question of: 'What is Architecture?' can often spark off endless debates concerning its true nature and significance. However, before one can even enter the debate meaningfully, one has to ask some more fundamental questions: What Architecture? What exactly are we debating about? On what basis is the subject of our debate made known? Are we talking about the same thing anyway?

It is not the intention of this paper to enter into another philosophy discourse. However, a short diversion would help to illustrate the problem this paper is setting out to investigate, that is the notion of knowledge through representation and interpretation. Karl Popper in his *Objective Knowledge* outlined three fundamental mediums through which knowledge can be obtained.[1] They are, language, mathematics and artefacts. In his book he emphasised the importance of using correct and effective expression to maximise the integrity in the transmission of knowledge. For instance, an entity commonly known as 'circle' can be represented by the word 'circle' in English, or it can be represented mathematically by the equation $x^2+y^2=c$, or it can be represented by a picture 'o'. In all these three means of representation, the knowledge 'circle' is transmitted. However, it should be noted that although the knowledge is represented, it does not necessarily mean that it can be properly or effectively perceived. The perception of a piece of knowledge depends on the context from which the knowledge is 'needed to be perceived'. That is to say, the perceiver 'expects' the knowledge to match the context in which he or she is working. Take a typical art class as an example, the commands 'draw a circle', 'draw an $x^2+y^2=c$', or 'draw o' carry with them very different messages; and in the context of the action 'draw', the third command is most likely to yield the required action from the student.

The understanding of the nature of medium and representation has a profound effect on our ability to transmit and receive knowledge.[2] The realm in which architecture is operating means that the problem of medium and representation is a fundamental issue in teaching, research and practice. Architecture is art and science in one. It speaks both artistic and scientific languages. It relates to geometry and mathematics. It borrows ideas from poetry and literature. It embodies structure and environmental science. And it deals with people, culture and society. All these mean that in order to practice architecture, one needs to be familiar with all knowledge involved and understand all the media through which knowledge is expressed.

A Critique on Computer Visualisation

'In recent years, remarkable advances in computer visualisation of objects and physical phenomena have been made. Computer images can now represent real objects very accurately'.[3] However, although the claim signifies a notable achievement in the field of engineering, science and technology, it cannot be totally endorsed by the artists and architects alike. Computer associated research has always been criticised by artists and architects for being non-creative, too structural and hierarchical, lack of fussiness and ambiguity, and too precise and geometrical.[4] 'Tool without qualities' is how some architects come to describe CAD. The criticism is not without ground. In an evening reception held at the Mies Van der Rohe Pavilion, Gabriel Ferrater, Rector of Universitat Politecnica de Catalunya, challenged the participants of the ECADDE 92 conference that despite all the advances in CAD techniques, none of the many wonderful architectures that had been built in Barcelona was 'designed' on a computer. The challenge is not to be taken lightly, especially in the field of design and architecture. It reminds one of another challenge from Nikolaus Pevsner, 'A bicycle shed is a building, Lincoln Cathedral is a piece of architecture'. What exactly does it mean? And when one comes to visualise Lincoln Cathedral, what should one visualise? Is it enough to represent the structure as it is or what? What is the reality and meaning of the challenge? How can one make the distinction in method and in paradigm when one tries to model and visualise a bicycle shed and a cathedral?

The fundamental quest in today's urge for computer visualisation is realities, or how computer animation can mimic it better . But what is reality? Although many philosophers have put their thoughts to the subject, the enquiry has yet to escape the three realities Aristotle defined more than two thousand years ago. In his *Metaphysics*, Aristotle defined reality as: what it is, what it should be and what it is perceived to be.[5] He asserted that although reality could be made known through representation, physical representation was the least important in the world of perception. He further stated that reality was based on senses and not on physical objects, therefore, imagination and emotional impression had to be employed to meaningfully interpret the physical world.

In a similar vein, Sir Joshua Reynolds, first President of the Royal Academy, offered his most powerful criticism on the issue of realistic representation

> The Art which we profess has beauty for its object; this it is our business to discover and to express; but the beauty of which we are in quest is general and intellectual; it is an idea that subsists only in the mind; the sight never beheld it, nor has the hand expressed it; it is an idea residing in the breast of the artist, which he is always labouring to impart, and which he dies at last without imparting; but which he is yet so far able to communicate, as to raise the thoughts, and extend the views of the spectator.[6]

It is not difficult to understand the real meaning of his theory by looking at his painting 'Mrs Siddens and the Tragic Muse'. The painting is an example of sublime on canvas, not perhaps by means of a vast sublime scene but by means of the sublime mind it contained; the mind of the greatest actress of the age and her imagination is dramatised in the picture. One needs to have some ideas of art as well as personal knowledge of the sitter to be able to fully comprehend the picture. This notion of representation goes beyond what could be seen, as it also laid heavily on a triangular relationship between the painter, the subject to be painted and the viewer.

This notion of representing reality through interpretation is precisely what current computer visualisation is lacking. Unless the issue can be addressed and put into the agenda of further CAD research, it may be difficult to persuade artists and architects that the computer can be a 'tool with qualities'.

It is against this philosophical background that this paper is setting out to investigate the various possibilities and to offer some thoughts on the subject.

Re-inventing the Danteum Project

The Danteum Project was designed by G.Terragni in the 30s. Mussolini apparently approved the project, but it was never built because the war intervened. The project was Terragni's architectural interpretation of Dante's notion of life and death through Hell and Purgatory. It contains a richness of imagery comparable to the three 'canticas' of 'The Divine Comedy'. Despite the availability of some plans and sections, the true nature of the project is often mis-perceived as being an exercise in geometry and proportion. However, by carefully modelling and reconstructing its context, we are hoping to address some of its hidden dimensions vital to an appreciation of the project.

Thanks to generous support from the University, The Electric Design Atelier was set up in 1991 to conduct experiments to seek for this representation-interpretation interface. The grant allows us to purchase an Apple Macintosh based workstation complete with various input and output devices (see fig.1).

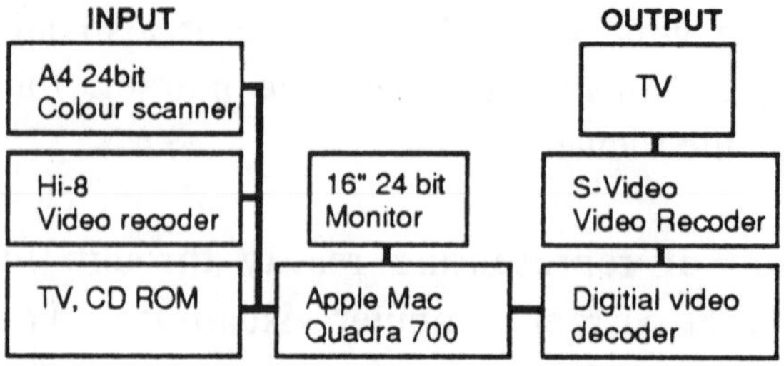

Fig. 1 System Set-up based on an Apple Macintosh workstation

Taking information from available plans and sections, a computer model was constructed using our in-house CAD software GABLE. The file was then transferred to the Macintosh through DXF and Kermit. It was opened in StrataVision, a rendering, visualisation and animation software, for further manipulation and image enhancement. Finally, a number of other softwares were also used to manipulate the images, to compose the visual and audio sequences and to group and present the final result.[7] It should be noted that all the hardware and software we used are straight forward off-the-shelf products. In addition, we are not the first to model The Danteum Project on a computer. Our contribution is based on the way we apply available technology to experiment on new ideas, and through it comment on the suitability of the tools for the task and offer our suggestions to improve them.

The physical modelling process of our experiment is straight forward and requires no further explanation. Our main task is to understand The Project, interpret it based on our understanding and represent it meaningfully. Lengthy research has gone into knowing the designer, his personal philosophy and the context in which his design works are based. In particular, we were interested in the way 'The Project' was originally drawn. We were also keen to see how this drawn representation came to reveal the designer's intention of the way his design should be perceived.

Although Terragni's original drawings contain much of his thoughts on number and geometry, they do not offer the same level of insight into the quality of the spaces and how they are related back to his interpretation of Dante's poems. To re-invent the missing picture, we embarked on a visualisation of the project based on Dante's The Divine Comedy. A number of techniques were used to represent the design non-realistic physically but realistic interpretatively. One of the rendering sequences is a walk through of The Danteum Project. The intention of the sequence is to explore the spatial quality of the spaces and their interlocking labyrinth. Context and content were overlaid to convey the overall atmosphere of each of the three main spaces. Dante's three canticas of The Divine Comedy were read during the walk through. There were no hard edges, no definitive geometry and no realistic texture mapping. The intention was to construct a sequence for feeling; hard physical facts were not included, instead, the sequence relied heavily on the viewer's ability to recall his knowledge of the project through association and interpretation.

Fig. 2 A view of The Danteum Project visualised non-realistically

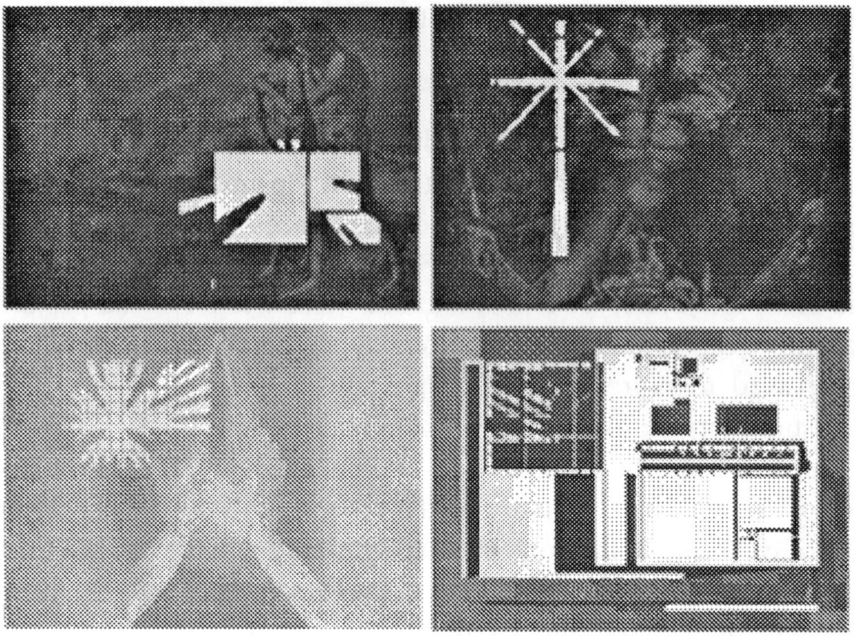

Fig. 3 A non-realistic walk through of The Danteum Project

Conclusion

In the foreword to his book *Space, Time and Architecture*, Sigfried Giedion actually attributed the sterility of modern architecture to the 'contemporary man's separation between thinking and feeling'. Recognising the problem, he went on to propose his solution in his *Mechanization Takes Command*. He was trying to show 'how this break between thinking and feeling came about', and 'how to bridge the gap between inner and outer reality by re-establishing the dynamic equilibrium that governs their relationship'. He concluded at the end of his book:

> The science and the arts are activities which, by exploring the unknown in the human mind, directly enlarge man's consciousness. Every scientist, every artist, is part of a long line of tradition. However, only a creative spirit can go

> forward, beyond the limits of that tradition, to explore what until then no one has known, no one has seen, no one has felt. By means of intuition, imagination, mystical impulse - what you will - he is able to open up new spheres of the unconscious. These spheres are distinguished from the outer world in that their essential development takes place directly, personally, without interference from any external power. They develop only in liberty, for no command can open the way to the unexplored.[8]

Today, computer visualisation is facing a similar dilemma. The admirable achievement in visualisation technology is not tested and used critically by artists and designers. Many designers either discard the whole idea of computer visualisation or accept it literally as it is. The situation reminds one of the time when photography was first invented. It took the artists many years to come to terms with the new technology and many more to develop a dialogue with the scientists. This paper offers a way to establish such a dialoque. This is not the only way, but the methodology offered is one which is based on content and context, the very things on which Art is based.At last, may I quote a passage from one of my previous papers to end the present discourse:

> Today, driven by the business world and market economy, technology is no longer a problem, humanising it is. The way we define technology is changing. The ability to go faster, better, cheaper and cleaner is no longer what the realm of technology is all about. Our ability to capture, to understand, to melt and to synthesize new technology into the 20th century culture and humanity is becoming a more fundamental issue in our practice and education. The digital revolution is coined by some as the Third Industrial Revolution. If we accept that its significance, for better or for worse, can be equated to what had been happening in the First Machine Age a century ago, we would have to accept that a new set of artistic, aesthetic, philosophic and social framework has to be formulated to guide us through this technological labyrinth.[9]

It is my own belief that only when the art side of CAD research can match the excellent standard set by the science side shall we begin to talk about a synergy of the two, and until this synergy can happen computer visualisation has very limited effects on the way things are designed, represented and interpreted.

Postscript

The student, now in his third year, who worked on The Danteum Project has now proceeded to apply the technique to his own design project.

---

[1] Karl Popper, Objective Knowledge, Oxford University Press, Revised edition, 1982.

[2] A very good paper on the subject of knowledge, medium and representation in architecture has been published recently in Design Studies. Although concerning themselves with a slightly different line of argument, the authors managed to delineate a hypothetical structure on knowledge and medium. See Philip Rhodes, Kevin McCartney & James Powell, 'Narrative structures within architectural videos', Design Studies, vol.13, no.4, October 1992, Butterworth-Heinemann, pp.355-378. For more information on creativity and representation, see Dalibor Vesley, 'Representation as a Vehicle of Creativity', Scroope, issue 2, Cambridge, 1990, p.11. Concerning structure of pictorial representation, see Ernst H Gombrich, Art and illusion: A Study in the Psychology of Pictorial Representation, Princeton, 1969. For information and method, see Hans-Georg Gadamer, Truth and Method, London, 1975, p.139.

[3] Taken from VIDEA 93 call for papers poster.

[4] See Francois Penz et al, 'Tools for Design: A Controlled Experiment Comparing Computer Work with Traditional Hand Drawings', Computer in Architecture, Longman, 1992.

[5] See W.D.Ross, Aristotle - Selections, Oxford University Press, London, 1927.

[6] Cited in Stephen Jones, The Eighteenth Century, Cambridge University Press, London, 1985, p.51.

[7] Hypercard, Adobe Premiere, Adobe Photoshop and MovieRecorder are used.

[8] S.Giedion, Space, Time and Architecture, 5th ed., 1982, p.876.

[9] Edward Ng, 'Towards the 4th Dimension', Conference Proceedings, ECADDE 1992 Conference, Barcelona, p.96.

Poiesis. Heuristic design and computer science. Interdisciplinar researches and experimentations about the relation between design and computer science

C. Talamo

Dipartimento di Programmazione, Progettazione e Produzione, Edilizia, Politecnico di Milano, Via Bonardi 3, 20133 Milano, Italia

ABSTRACT

The text reports the aims, the methodological structure and the results obtained with this interdisciplinary research that the Dipartimento di Programmazione, progettazione e produzione edilizia of the Politecnico of Milan has been developing for many years .

The goal of this research is to identify the opportunities of using the computer science in the creative stage of projects. In order to study in detail these subjects we can go along two ways : the theoretical one in which we try to achieve an explanation about the heuristic processes with the assistance of various experts and in the meantime to obtain a better comprehension of the relation existing between cognitive and creative processes and information systems : the experimental one in which we can consider this relation through the reinterpretation of the sketches with the computer.

The final aim is to outline a system of rules (or meta-rules) for the development of a program that we can use in the stage of heuristic design.

INTRODUCTION

The research "Poiesis. Heuristic design and computer science" deals with the relation between design creativity and computer science [1]. The purpose is to analyse and to study in detail the opportunities offered by the computer science to the designer in the starting creative stage of his work, that is the stage of heuristic design [2]. Researchers of different disciplines as neurologists, psychologists of

perception, semiotic specialists and computer science experts in artificial intelligence, have taken part in this study, in order to determine in an interdisciplinary way, the heuristic aspect of planning and to underline the potentialities of computer science. These contributions enable us to define some rules suitable for the development of a system we can use in the creative stage.
The main subject discussed in this research, is the beginning stage of the project where the designer must know the co-ordinates, the ties, the materials and the techniques to use, but also the shape and the "sign" of his work. The researching group has been dealing for many years with this subject through the analysis of heuristic sketches of architecture works, through the documented study of creative ways used by some designers and the surveys carried out in the analysis of the mental processes.

Our researching group has constituted an interdisciplinary group of researchers employed in various University departments that have organized a series of lectures based on different experiences, all involved in the subject of this research. We are expecting from neurologists and physiologists, precise information about the structure conformation of our brain in relation to its creative activities. The psycologists are giving us a useful aid to the comprehension of the images perception in order to understand the help that can be given by a computer. Bioengineers and experts in artificial intelligence can achieve the goal of creating a computer suitable to collaborate in the heuristic planning stage. The semiologist collaborate to the comprehension of designer's mental acts, in order to identify the constituent stages of project, in the same time he suggests the typical ways of the creation phase based on thinking and ponderated choices. At present, the aim of this analysis is to answer some questions related to the opportunities of :

1. establishing with the typical methods of the different disciplines if it is possible to identify and to separate some stages of the creative working;
2. finding the optimal modalities proper to create models of the mental acts that can be analysed and verified and used to formulate effective hypothesis;
3. setting data processing systems capable of assisting the designer in the heuristic phase on the basis of the typical methods of mental processes;
4. identifying the transfer manners from the sketch's stage to the computer through various devices (data pen,scanner,videocamera);
5. considering the impact that the use of uncommon instruments for the architects can have on the final project.

We think, this research can conduct to a better comprehension of mental processes and of working procedures used by designers and, in the same time, it can indicate new ways and tools for stimulating the creative abilities. The goal is the implementation of data processing systems capable to assist the designer from the planning stage to the creation of images [3].

WORKING METHODS

The aim of this research is to formulate theories related both to the possible effects, on our individual expressive abilities, by using a computer and, to define the specifications of a data processing project.

The analysis of these subjects is subdivided into three indipendent works that can interact together and that are charaterized by the different working methods according to the disciplines involved.

The first one, operating in a philosophical contest, involves two different kinds of experts: on one side the semiologists engaged to identify the paradigms of creative thought and the heuristic procedures used by men or by computers in the applications of artificial intelligence; and on the other hand the philosophers of science that analyze the fundamental elements of heuristics (project and scientific), in order to define thinking process (in its logical and methodological structure) considering the scientific discoveries and the innovations. This passage, established on a continuous comparison with the models and procedures of cognitive sciences and of artificial intelligence, supplies new cues and new interpretative structures for the definition of epistemological models and of rules proper to data processing systems devoted to the design.

The second one deals with the perception and cognitive sciences. We discuss the problem of creativity (considering the theory of mental models) and we identify some aspects of the thinking (creation and choice of options, for example) involved in the creative process.

Finally, the aim of the third one is to specify the questions emerged, to verify them and to legitimate them according to the usual designing methodologies. On the basis of this research we are formalizing some working hypothesis for an interactive and stimulating use of computer and in the same time we are establishing meta-rules for a data processing system capable to assist the designer in the creative stage.

EXPERIMENTATIONS

In order to check and enrich the work assumptions that have been formulated and to translate into practice the requests resulting from the different work groups, two experimentations have been carried out. The aim is to evaluate, from different points of view, the potentialities of information tools in comparison with designer inventive methods.

Both experiences start from a deep analysis of the sketches and the graphic documentation produced during the implementation of projects. The aim is the search for reading keys for the identification of the paths, the ways of representation and composition used for the development of a design idea.

The experimentation first lies in the use of information tools for this research and therefore in the implementation of innovative methods of analysis in comparison with conventional observation or redesign.

Secondly it lies in checking the compatibility of the structuring logic of cad systems and the procedures followed by a designer.

The first experimentation [4] is about the rereading of some architecture works through the computer; the aim is to make clear and to go through the creative processes lying between the first sketches and the last version of the project.

Three different methods for the interpretation of the heuristic design have been set up.

The first is about the possibility of getting some hints for the understanding of the creative path and the logic of composition of a project from the deep structure of the sketches.

Here is the procedure followed:

- capture by scanner of the heuristic projects
- vectorialization of the image
- importation of the image into a cad program for subsequent manipulation.
- cleaning of the image in order to highlight only the structure made of grids of reference, axes of symmetry, rotation angles and other useful clues,
- successive disaggregation of the references found in order to run backwards through the morphologic genesis of the project.

In this way, after processing in digital form a sequence of sketches represented in the plan it is possible, using manipulation procedures, to disaggregate the objects and to read the main axes of the project and the sequence generating them. It is possible, in the same way, to search for the representation schemes of the different meshes (structural, typological, installation and composition) for the connections existing among these and to check their definition degree already in the heuristic stage.

Figure 1 shows the result of the experimentation carried out on some of Mario Botta's projects. In order to make the operations easier, Botta's works have been chosen because they show in a constant and uniform way the design poetics of the author even though in a range of different architectural solutions; moreover the number of one family-houses allows to make useful comparisons. In order to check the rightness of the experimentation, the same method of analysis has been applied to a Frank O.Gehry's work, chosen for the deeply different approach.

Three Botta's projects for one family-houses have been taken into account. They deeply differ from each other as far as their morphology is concerned: the house in Manno, shaped as a triangle, the house in Stabio circularly shaped, the house in Breganzona shaped as a square. By applying the sequence of operations described before, it has been possible to run back through the creation path followed by the architect. It has also been possible to discover the importance of some initial references which are constantly found in different projects, such as the inscription inside a square area, the existence of central axes of symmetry - manipulated in different ways - inside which some architectural elements are placed.

The same experimentation, carried out in Gehry's Sirmai-Peterson House in Tousard Oaks, has shown the compatibility of the first layout and the subsequent definition of the project through a series of manipulations on the sketch.

The same procedure of disaggregation analysis is applied to the reading of the views: it is then possible to check by successive superpositions, the evolution of the initial idea when different variables, such as the material used, the construction techniques and rules are taken into account.

The analysis of all the sketches prepared by Botta for the house in Breganzona and the superposition of the main lines has made it possible to identify the idea originating the façade, unchanged in the different solutions proposed (fig.2).

Another method of analysis goes back to some studies on the graphs, considered as a form of representation of the conceptual structure of the images [5].

In this case the aim of the experimentation is to highlight the design logic followed in the typologic definition of the intervention.

Methods of digitalization implying the rereading of both the sketches and the definite project have been studied. The possibility of translating the project choices into connection schemes (among spaces, functional areas, architectural elements, expressive constant elements of a designer) has also been taken into account.

The results are connection schemes allowing to reread the project in its essential conceptual elements.

Using some information procedures it is possible to compare several schemes and to obtain a series of interpretation solutions about the interaction of different classes of design elements.

The experimentation applied to Botta's houses has shown permanence of topological relationships among the different internal spaces categories.

Finally, using Botta's designs, graphic data-base have been created: the objects stored are signs, symbols, architectural elements and all the constant expressive characteristics found in the work of an architect.

Once identified, these elements are processed in digital three-dimensional form and made parametric. It means that the basic element can be put in any project by changing the three dimensions in an independent or proportional way (fig. 3).

The program used for this experimentation allows to associate to the parametrization of the elements a series of algorithms where the variation rules are described. It is possible for example to describe the dimensional or morphological modification rule of a recurring kind of windows or doors, at width changing or with certain outline conditions. The images contained in the graphic data-base are then associated to the information contained in a relational database. It is then possible to interactively associate the images and the graphic symbols to technical data information, as for instance the materials and the techniques that are used or that can be used to build an architectural element.

The methods that have been described can be used in two different situations:

1. in case of existing architectures, they can be used for a critical reading and for a better understanding of the work: it is possible to analyse backward the sketches in order to recreate the genesis of the idea and the evolution of the poetic expression of the designer. This application is very interesting when the whole graphic documentation of major architectural works - from the initial sketches to the definite project and the photos of the work - is given;
2. in case of personal sketches, they can help to interpret and to disclose one's logic of composition and one's idea, in the designing stage and when the work is completed. In the designing stage the sketches previously made can be processed in digital form by traditional tools. Using the methods that have been described before, images can be disaggregated and decomposed, designs can be compared using automatic procedures and logic and inventive links can be looked for in the successive preparations of the sketches. The aim is to see one's project idea in heuristic stage and to run systematically through its evolution path. Once the work is completed, these methods can be used for better understanding of one's project poetics. Thus, comparing different projects personally made, searching for the constant elements, the signs, the symbols and the recurring schemes in the sketches it is possible to discover one's personal language of expression, one's unconscious references.

The result of this operation can be the creation of a personal data-base of symbols, images and quotations and also the contribution for a better understanding of one's creative universe.

The second experimentation [6] aims at comparing the procedures of project development and the characteristics of current instruments for computerized design.

The work starts from two surveys carried out at software houses both for the manufacture and the distribution of cad system.
The survey shows two main aspects: one is that project facilities tend to use cad programs for the definite project, but they continue to use traditional instruments in the heuristic stage. The other is a gap between the manufacturer, the distributor (charged of maintenance and training) and the user of cad systems. It can be argued from this lack of connection that the programs sold today do not cover all the needs arising from the use of programs in everyday designing practice.

Therefore the experimentation has been developed with two aims: on one hand to check if the structure of commands of the cad products commonly used is in accordance with the processes of the project development, from the heuristic stage to the definite formalization of the idea. On the other end to find a series of useful meta-rules for the development of a cad system specifically aimed at the heuristic project.

Two projects have been found: the Museum of decorative arts by Richard Meier in Frankfurt-on-Main (fig.4) and the Centre for bilogical studies of the Frankfurt university (fig.5).
Both projects have been analysed in the crucial moments of the definition process and some important logic-operating steps have been found:

1. context assumption,
2. schematization of major context elements,
3. choice of a reference, of an originating rule for the project development,
4. reference manipulation by combination-association schemes in the search of evocative hints,
5. sketch for a project,
6. successive manipulations of the sketch in the search of a formal coherence,
7. synthesis of the graphic preparations in a geometric scheme by dimensioning.

This process has been run through with the assistance of a cad program: the seven steps have been converted into program functions groups and a redesigning operation has been made (fig.6). Thanks to this it has been possible to check the flexibility of the functions in comparison with the project requirements and to identify their better use for the heuristic project. From this experimentation a requisition list has been set up in order to evaluate the performance of the different cad systems according to the typical needs at the ideas stage.

REFERENCES

1. The research "Poiesis. Heuristic design and computer science", directed by professor Guido Nardi, has raised funds from CNR (Centro Nazionale Ricerche).
   The described works have been developed inside the Laboratory of Informatic Technologies (LTI) of the Dipartimento di Programmazione, progettazione e produzione edilizia of the Politecnico of Milan.
2. About the theme of heuristic design we refer to: E. Biondi, *Modelli del capire e dell'inventare*, in M. Bertoldini (ed.), *La casa tra tecniche e sogno*, Angeli, Milano 1988, pp. 109-125; Gabetta, *L'euristica a partire da Leibniz*, "Correspondance", n.11 (1986-1987), pp. 1-22; B. Hamburger, *Dimensioni precise e approssimate*, "Lotus international" 87; E. Manzini (ed.), *L'architettura e la complessità del costruire*, Clup, Milano 1987; G.Nardi, *Dialogo sul pensiero inventivo in architettura*, in AA.VV., *Il pensiero inventivo* (eds. Boeri, Bonfantini, Ferraresi, Somalvico), Unicopli, Milano 1988, pp. 249-264; G. Nardi, *Le nuove radici antiche. Saggio sulla questione delle tecniche esecutive in architettura*, Angeli, Milano, 1987.
3. For a description of data processing system developed for the creative stages of design we refer to: P.N.Johnson-Laird, *The computer and the mind. An introduction to cognitive science*, William Collins Sons e Co., London, 1988. About the relationship between heuristic design and computer science we have published: C.Talamo, *Atti mentali e percorsi informatici*, in M.Bertoldini (ed.), *L'atto progettuale*, Cittàstudi, Milano, 1991; M.Bertodini, C.Talamo (eds.), *Poiesis. L'informatica nel progetto euristico*, Cittàstudi, Milano (in press).
4. This experimentation has been developed by Maria.A.Pievaioli with the assistance of Cinzia Talamo. The experimentation is described in M.A. Pievaioli, *Tre case unifamiliari di Mario Botta: rilettura sperimentale di alcuni disegni di massima con l'ausilio dell'elaboratore*, in AA.VV. *L'informatica per il progetto*, Cittàstudi, Milano, 1991.
5. The theory of graphs is described in L.March, P. Steadman, *The geometry of environment*, RIBA, London, 1971. Some hypothesis about the possibility to use the methods of rappresentation of the graphs for the analisys of projects have been expounded in: C.Talamo, *Progettazione e computer: ipotesi per un rapporto interattivo*, in Sinopie 4, January 1991.
6. This experimentation has been developed by Cristina Marini and Nadia Trenta with the assistance of Cinzia Talamo.

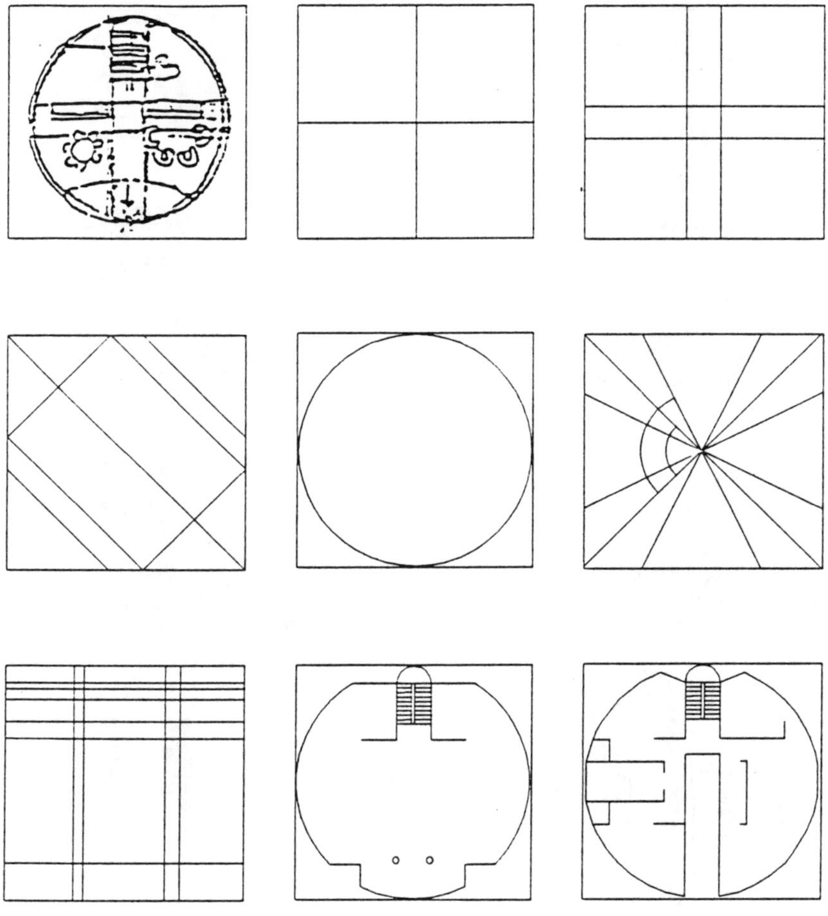

Fig. 1 - Mario Botta: project for one-family house in Switzerland
Disaggregation of the project for Medici House in Stabio (1980) in the search of the structure of composition
Summary table of the poetic path developed by Mario Botta, reproduced and synthetized with information techniques. The rereading of the initial sketch shows the shape characterizing the project and the design process developed.

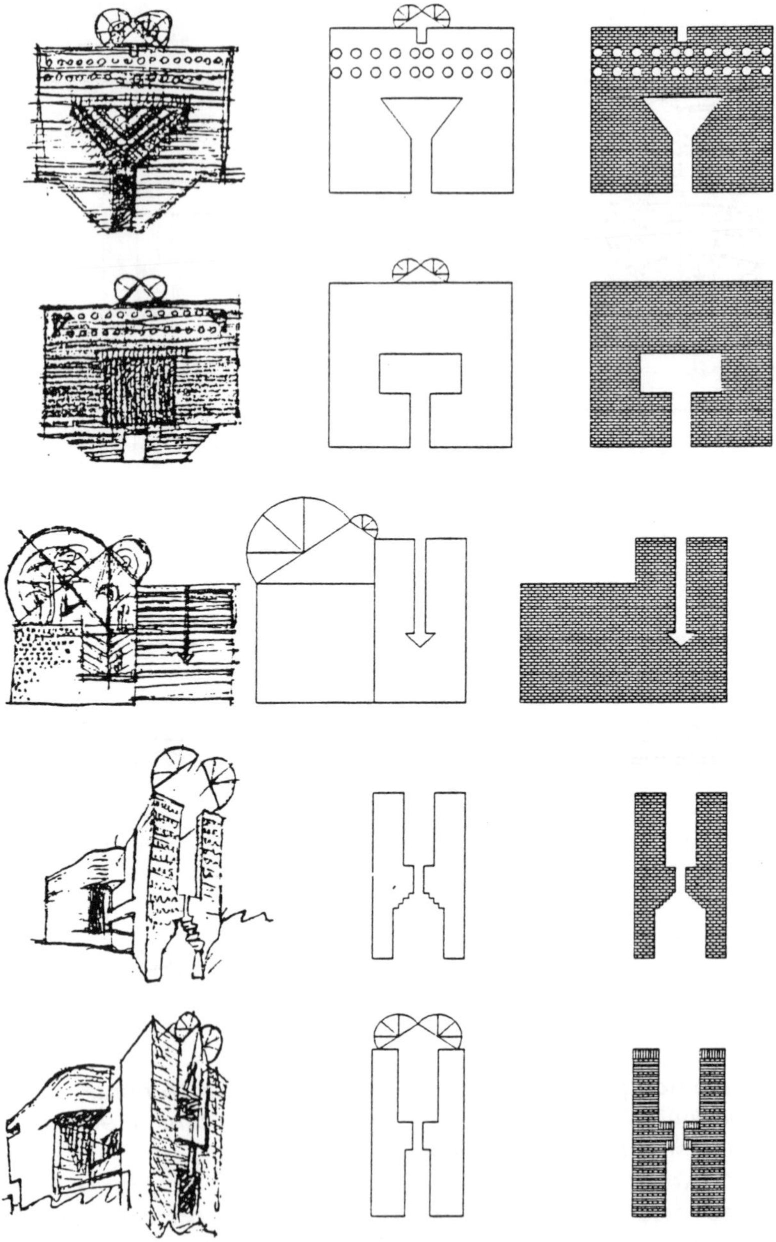

Fig. 2 - Rereading of the logic of composition of the façade of the Genini House in Breganzona (1984)

The images proposed are the result of the morphologic manipulation of the electronic graphic sign. Sketches have been obtained according to a cronological sequence and have been synthetized by the use of commands capable of manipulating the computerized sign.

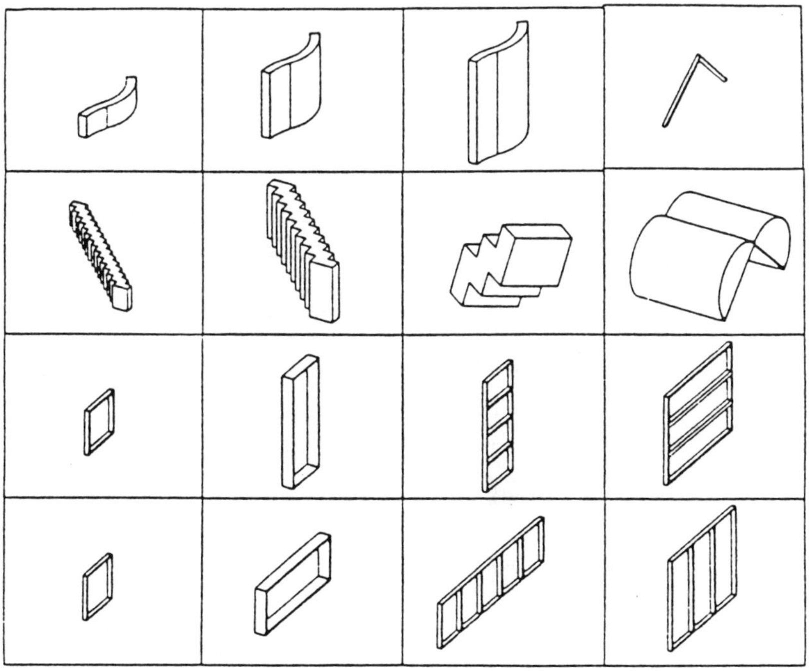

Fig. 3 - Creation of a database of architectural parametrized elements: second stage

The operation lies in the construction of parametrized macros of construction elements chosen and filed in a personalized submenu. In this way they can be put into other projects by changing the three dimensions.

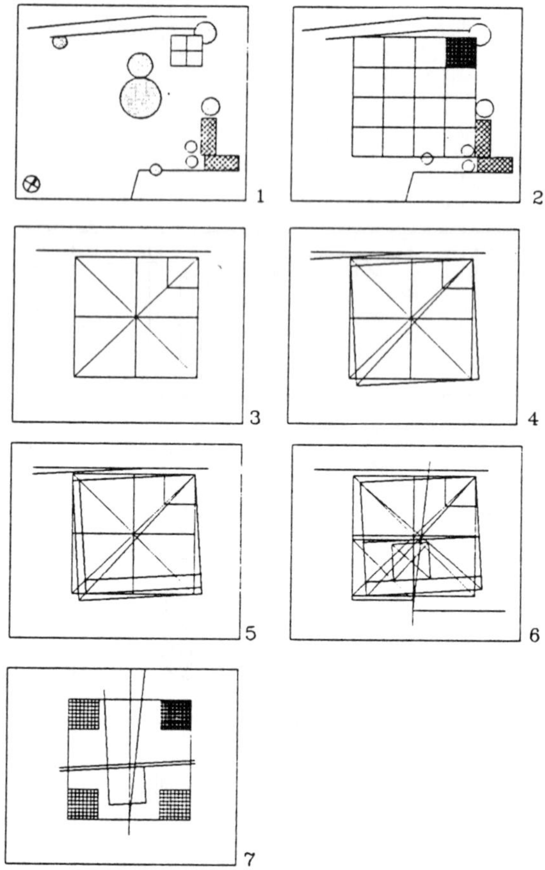

Fig. 4 - Formalization of the project stages through the redesign of the first project ideas by information tools.Richard Meier, Museum of decorative arts, Frankfurt-on-Main, Germany, 1979-1985.

1. **Context assumption**. Adoption of Villa Metzler's plan and façade and of the alignment with the river as generative elements of subsequent project proposals.

2. **Reference choice**. Definition of the physical boundaries of the intervention: inscription of the site into a square originating from the repetition of sixteen identical squares.

3. **Reference manipulation**. Through an anti-clockwise 3.5 degree rotation around the corner right above the square a second grid aligned with the course of the river is obtained.

5-6. **Reference manipulation**. Geometical studies on the two superimposed grids. The identification of basic lines of the settlement: definition of the buildings, the ways and the internal courtyards.

7. **Summary of project proposals**. The final geometric scheme of the whole settlement is obtained.

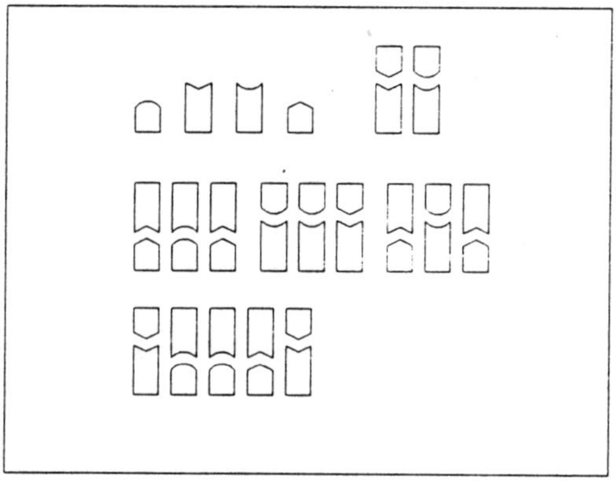

1

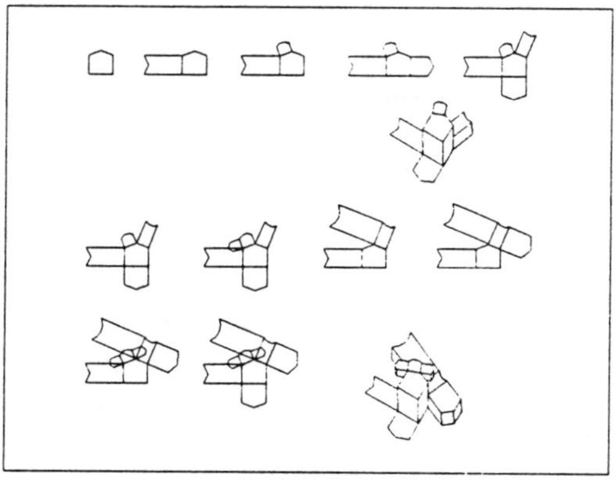

2

Fig. 5 - Formalization of the project stages through the redesign of the first project ideas by information tools.
Peter Eisemen, Centre for biological studies of the Frankfurt University, Germany, 1987.
1. The designer takes the four geometric figures used by biologists for the representation of DNA genetic process as basic elements for the formulation of first project proposals.
2. The five original pairs of images are manipulated, to develop the project, by using the same geometric principles governing the processes followed by DNA in the construction of proteins: **repetition - transcription - translation.**

STAGES	REQUIREMENTS
Context assumption and choise of a reference	- to import context images - to import images to be used as a reference - to vectorize the imported images - to draw over images - to manipute the images - to create a data-base of signs and images
Creation and manipulation of design drawings	- construction of the drawing with geometric primitive figures - manipulation of images - parametrization of images - contemporaneus bi and three-dimensional visualization
Check of the system	- to run graphically through the operations made - to reinstate previous system condition - to cancel operations - to correct mistakes - to change the application parameters - to know what the system is doing - to know, in a direct way, what can be done and how it can be done - to know the expected time for an operation - to personalize the interface

Fig. 6 - Summarizing scheme: basic software requirements for architectural design.

From the analysis of the design path we have observed that in the initial stage of the architectural design some important steps are recurring: context assumption, choice of a reference and reference manipulation.

After checking how these steps apply to the two projects, our aim is to translate them into a series of requirements for the development of a software program.

Experiences and rules in simulating architectural realities

S. Menoni, G. Vignati
Centro per la Grafica Informatizzata, Dipartimento di Programmazione, Progettazione e Produzione Edilizia, Politecnico di Milano, Via Bonardi 3, 20133 Milano, Italia

ABSTRACT

'The technical revolution has its mirror in language. A computer system gives us the opportunity to simulate reality not in a static way, as the classic perspective does, but in any visual and behavioural aspect.

We experience the environment, its dimensions, its lights, colors and walks. The graphic simulator draws plans, facades, sections, it makes us walk inside the building and in the city, and lets us compare alternative solutions' [1].

This quotation taken from Bruno Zevi's book 'Modern architectural language', lets us stress some aspects deeply related to the introduction of computer systems in design which have been experimented in two of the Laboratories of the Politecnico of Milan (Laboratorio di tecnologie Informatiche del DPPPE e Centro per la Grafica Informatizzata).

In a more specific way we chose two topics: the first which deals with the dynamical perception of space using a software written in the first Lab; the second through an existing program, TDI/Explore, tries to find out some rules in simulating indoors spaces and outside environments.

INTRODUCTION

The experiences presented below can be considered the result of research deriving from some needs that today architects feel in a deeper way than in the past. Perhaps these new needs are a consequence of the new dimensional scale the project deals with, and of the more complex problems it must solve. Perhaps they are an attempt to overcome a moment of crisis in theoretical and operative architecture, both in their relations to the past and to modernity. Last but not least they were caused by technology trying to find civil markets for techniques that have been developed in military research.

These techniques can be used for surface modelingand for visual simulations [2]; both ignored in the past, they have become the cross road of all the traditional ways of representing, that architects have been using for centuries: plans, sections, fronts, axonometrical and perspective views.

Those have been architects' tools for centuries but not always; in fact we must remember that they were born in a precise historical period and they were officially accepted in academical rules at the date of Le Monge treatise.

We wonder if today computer systems can modify architects' work, just as it was the case of traditional techniques of representation, ratifying an analytical and operative model at the time of Le Monge treatise. We are considering two major requirements that the new technology can satisfy or will be able to satisfy in a better way in the future (if this will be the direction of future research).

1) The first requirement is the opportunity to work on a model which is logical, numerical and iconic at the same time. On this model the architect can experience in a relatively short time what otherwise would require his whole life, through successes and failures. In fact the new technology, unlike the traditional drawing, gives the possibility to check chains of tests and errors associating 'from outside' the control on specific design parameters, such as lighting, colours, materials, details, to the visual simulation.

2) The second requirement concerns the possibility to obtain a model of our object or of our project which is not only a visual one, but can also control other aspects, especially the dynamic spatial experience.

The reliability of these controls is crucial to understand whether or not computer representations may have also an explorative value, if not a cognitive one.

Furthermore some obstacles are still to overcome fitting technical means to unskilled users, which however bring their own specific knowledge and skills.

We do not believe those obstacles are due to present computing limits, as producers too often say. Present hardware graphic oriented technology can trace vectors on the screen and cover pixels at increasing speed, using Phong or Gouraud shading algorithms at decreasing costs. In the very next future, parallel performed systems will be available to compute ray-tracing and radiosity images at acceptable frame rates. Today 100 Mips microprocessors, working in parallel systems, and graphic coprocessors of dozens of Mflops, are already available, as for example thePS of Intel.

RAM and VRAM memories needed in graphic applications can be reached at a speed of few nanoseconds. Decreasing costs are corresponding to these hardware betterments, so that larger sections of no-specialised users can afford sophisticated workstations.

But, to those increasing computing resources and to the rapid progress on hardware technology, it does not correspond a similar development in software. CAD Programs, in the widest meaning, for engineering, architecture, urban planning have been developed in different conditions, at times when they could rely on less sophisticated computing systems, on graphic libraries, languages, debuggers difficult to use, less effective and ductile than those which we were describing just now.

Besides, they were conceived in accordance to a different philosophy compared to the one emerging today: as every new technology does, computers for graphic applications tended to imitate traditional usage of drawing tools. First programs were automatic drawing oriented to obtain technical drafting rather than to support design process. Initial systems to produce territorial maps, for example, were conceived to get quicker thematic maps and not to put bases of a different cognitive model of territory.

We can argue that the gap between programs produced up today and specific requirements, made by specialised users, were the consequence of lack of cooperation between software developers and users. This is particularly evident in CAD and SGI applications, where graphic representation should be considered not only a technical paper, but also a cognitive model. Anyway, we consider this first software generation obsolete. For the last years specialists have been reviewing goals and solutions of their softwares.

A strategy to pursue is to get specialists of different fields and software developers working together in a more productive way than it was in the past. That is, not playing a prior defined role.

Before, it usually happened that the software developer asked for user's wishes, gave him a prototype solution to test, tried to improve his program... this sequence was repeated several times and the final one lead to a new release of the software.

We think that this approach should be left in favour of a greater involvement of users in software, creating a greater involvement of developers in architectural design problems; in other words we wish to improve an exchange of their roles.

The system is more than the sum of its parts and of its functions: we have to assume this to set new approaches, different design models and strategies, both in software creating and in architecture.

EXPERIENCES

First experience: dynamic and interactive simulation

The proposals discussed in the introduction are the background of the first experience, carried out in the DPPPE Informatic Lab, and which is the result of direct cooperation among architects, computer science engineers, psychologists of communication. The work is still in progress, trying to find out a potential way for the computer graphics systems to represent architectural and urban projects with respect to specific dynamic parameters, which cannot be studied using a static representation.

Our wish to represent in real time spatial perception of an urban site and of its changes when new projects will be realised, has some relations with virtual reality approach. Yet, our outlook is rather different. Our goal is to simulate interactively and in a dynamic way a walk through a designed environment, with the aim to identify strategic ways for a better quality architecture.

In the real world, architectures and especially their environment are perceived by users dynamically. The user is constantly moving and so are his eyes. All this is of special importance when perceptive tests involve not only the object (which is the contents) but the environment holding it as well (scenery) or when motion (as a path to be walked over) is an essential aspect of the project (roads, urban design, landscape design).

Traditional perspective ties up reality in a compulsory point of view, which is always subjective.

The temptation to overcome this constraint is not new. In the past some artists tried to represent a scene from several points simultaneously, so did Paolo Uccello in a fresco painted in the Florence Cathedral (1436), portraing sir John Hawkwood from two different points of view [3].

We consider important to let the point of view free to move in a project, measuring distances and relations among objects, shapes, geometries.

We are not able to achieve a realistic simulation yet, though we are trying to do so developing our software or linking it to an existing one, as for instance TDI/Explore.

The program is conceived in order to simulate alternative projects in the same urban site, represented at an urban scale (Photos 1 and 2), that is using simple shapes.

The alternative projects are represented at the scale of the building and the user can choose different levels in detailing them, even in interior design (Photos 3 and 4).

The structure of the program is shown below (Fig. 1); it consists on several files that we are going to describe.

In the first file the drawings (in Data Exchange Format) of the urban context and the files related to the alternative projects are described (Fig. 2).

We can also find the files associated to specific blocks (particularly important for interior simulation).

The numerical model of the project can be done in whatever graphic environment, provided that the final format is the DXF one, which is considered a standard *de facto* to transfer graphic information among different

CAD programs (and not only in the PC environment).

In the second file colours associated to different objects are defined by a code that assigns RGB components.

In the future this file will contain also information related to material aspects in order to reach more realistic effects.

In the third file light features used in the simulation are written (till now only Phong algorithms are used).

This is the general project of our software. Till now we have developed the heart ofit, consisting in objects movement and in lights features.

To get to the end of the whole project, we need a more powerful hardware, that can generate 10/12 thousand shaded poligons at second, gives the control of four/eight light sources, and that controls a double buffer for animation and the z-buffer algorithm for hidden surfaces.

Interaction and interactivity The model of interaction is perhaps the most difficult to set up, also from a theoretical point of view.

Major difficulties arise from the double meaning that the word interface assumes in our work. The first, most common one, is referred to 'how' to use program's functions.

We implemented a user interface, with a windows library OS-MOTIFS, but we still feel the need to create an interaction inside the representation itself, using virtual tools.

The second meaning refers to the kind of interactivity and relations to set out between users and virtual world, that is the level of involvement the person walking in the simulated environment must experience.

The involvement 'intensity' is particularly interesting when we introduce the system in didactic activities and in design laboratories. In this case the program itself becomes an interface to the project, a mean of evaluating and experiencing it. Interaction achieves a heuristic function, giving the user the possibility to choose which parameters to verify (moving from the environment to the object or from the object to the environment) and how to do so (step of the walking through, preview patchs, light intensity and position). To test this last kind of interface, some basic functions have been developed:

a) The gradual control of details which is possible to modify, depending on how much deeper the simulation must go.

In our simulation you can choose among alternative projects, identifying and changing some of their elements. Furthermore, it is possible to change objects position and orientation (Photos 5 and 6).

b) It is possible to assign a specific code to objects or to parts of the scenery in order to make them recognizable, when we wish to operate on them.

Second experience: realistic simulation of complex sceneries

The second experience has been carried out in the 'Centro per la Grafica Informatizzata' of the *Politecnico di Milano* where a Silicon GT/70 workstation and the sw TDI/Explore are available.

We tried to analyse potentials and limits of some sophisticated programs available in the market, realising tridimensional models and synthetic images.

The experience worked at different levels of representing architecture. The first one dealt with an indoor environment. This kind of rendering requires not only the iconic representation of a space, but also recreating its feature, its 'atmosphere', in order to check important design parameters, such as tactile feeling, warm or cold colors...

Programs' limits come out very quickly: the lighting model used is only local and it is very difficult to render the sense of a diffuse illumination. We did not succeded completely in verifying some tests on the relationship between artificial and natural light (though we could manipulate also some lighting indexes), and what we did in fact was to find some rules of translation from the real data to the realistic simulation.

The second level concearns the simulation of an urban area. Once we have a detailed complex model of the scenography, we can deal with very different projects. Till now we carried two experiences. We used it to test the visual impact of a sign panel that gives information about an urban parking and about places still available (Photos 7 to 9). The panel uses optic fibres and it has a logo that changes dynamically to indicate the density of stationed cars. Several tests have been done to decide which is the best shape for an urban area (with respect to the urban context); which are the best colors to use for the characters and for the logo to be seen during the day and during the night. The idea in fact was to put a driver on this virtual road and to check which was his time of reaction to the information received from the panel in order to eventually change his trajectory and reach the parking place.

This experiment is of particular interest, if we consider that generally speaking industries producing those panels are not able to make any tests to define better fonts, better colors, better rates of changing logos.

A second kind of experience used the urban area as a pure scenery in a video, which was the last part of a research trying to stress relations between cars and architectural/landscape background in car advertisment.

This work consisted on an assembly of images taken from different spots and synthetic images, which were a sort of critical interpretation of the advertisment. A trip from the mountains to the city was imagined, stressing the no-realistic views, shown in car advertisment, using the analogue frames and the real situation (traffic james) by the synthetic images.

In this work it was necessary to implement another level of representation, considering the landscape. This level will open for us the possibility to carry out visual impact studies of big projects which are not in urban areas. At the beginning of our work on TDI this was one of the most important goals to achieve, yet there are many difficulties, related to the simulation of vegetation, trees, countryside environment. A first attempt was done on a mountain road projected by two students in civil engineering (Photos 11 to 13). It was very interesting to check the difference between their numerical and accurate technical representation of the road, but was very poor regarding the relations to the natural environment, and the images produced by the computer using just their numerical data.

In another work still in progress, we try to analyse the visual impact of the new Malpensa Airportof Milan and the fragile landscape of the *brughiera* typical of the Padana countryside.

We are not so interested in flying perspectives as in the real views that people driving to and from the airport get from different places.

<u>Considerations on realism in virtual simulations</u> We would like to add some considerations concerning realism in visual simulations.

1) It is possible to set up a good computer model of a project, provided that it is well defined above a minimum level, under which a virtual simulation is no more consistent; and, furthermore, that we know what our simulating goals are.

2) If we recognise the simulation being a cognitive model, yet we must define at what conditions it is. For the architect, both technical and perspective play a historic role in communication with the customer and as a mental exercise in projecting. Referring to this context, we would like to discuss the iconic realism of synthetic images.

The sense of reality derives from some factors. The first is related to the possibility to see projects from different points of observation as it happens in real world.

It is also connected to the human attitude to complete lacking parts of a represented object when the essential structure of it is shown. According to some neurologists () 'the study of visual system is fundamentally a philosophical challenge. [...] Visual stimuli reaching the brain do not realise a permanent code of information; [...] the integration of visual images is a process in which perception and comprehension of the image are simultaneous.' [4]

Besides, this lacking parts and more refined details can be added working not directly on the image but at a logical level that produces the image itself.

There is a sort of promise in new technology, not completly realised yet, to codify at a logic level design parameters, providing an increasing precision, and to see them translated into images [5].

SOME CONCLUSIONS

Our approach to computer systems in architecture is more explorative than regulative. We did not move from the assumption of which are limits and possibilities of the device, trying to test it on applications to confirm designers' way of working.

We believed that such a new mean could influence architects' and planners' attitudes, satisfying never expressed needs, or needs never revealed in official histories. Some experiences make us distrust the naiveties common among pioneers of computer approach to architecture. It has been proved not so easy to translate design problems and processes in automatic systems, as Bruno Zevi thought twenty years ago. On the contrary, some betterments both in hardware and software did not always make users' life easier.

The 'manual' simulation, that is the simulation as a product of the operator's work trying to translate reality into technical devices is much more relevant than the automatic part of the whole process.

When we must solve a specific problem, in lighting representation, as well as in the material, in the 'true' colors, we meet with the limits of the hardware before and of the software later. Limits are to be overcome using more refined skills, alternative to those previously imagined by software developers.

The illusion of realism that can be obtained (and will be even in real time in the future) puts the question of how to use virtual representations in evaluating projects.

If we want to use this illusion in a decision making process, it is essential that other people can verify how the simulation had been carried out, to recognise in it a framework that can be taken down in parts and set up again several times. The simulation will be much more correct and honest if we are able to distinguish:

a) how much of the result depends on the computer itself, on its graphics potentialities (defining them better than we generally do speaking of color plans and resolution);

b) how much the results rely on the software, considered as an organised system of procedures, menus, instructions;

c) how much the results depend on operator's interpretation.

We still have a long way to define regulations in our work, in precising first hypothesis to be verified on the results. But we need all this if we wish to infer something from visual simulations to reality.

Finally, we would like to stress the importance of traditional drawing techniques, from free hand drawing to perspective geometry. Computers help us do perspectives and axonometries faster and better, but they do so relying on classic geometric concepts, which should still be considered the first 'education to see in the space' for an architect.

We here propose what had been said about multimedia learning programs. Technical tools should not free the designer from the troubles connected with expressing his ideas with signs. Yet, improvements obtained in precision and likelihood of models should prevent from vulgar errors and give the designer the possibility to compete with more complex and sophisticated projects.

REFERENCES

Some of the illustrated experiences are part of the general research 'Poiesis. Heuristic design and computer science' directed by professor Guido Nardi, and dealing with the relations between computer science and design.

In particular the first experience, about dynamic simulation was coordinated by Professor Claudio Molinari.

1. Bruno Zevi *Il linguaggio moderno dell'architettura* Einaudi, Torino, 1973.

2. Nadia and Daniel Thalman edited by, *New trends in animation and visualization* John Wiley & Sons, 1991.

3. Michael Kubovy *The psychology of Perspective and Renaissance* Cambridge University Press, 1986.

4. Semir Zeki *L'elaborazione dell'immagine visiva* in Le scienze, italian edition of Scientific American november 1992, p. 36.

5. Tomas Maldonado *Reale e virtuale* Feltrinelli, Milano, 1992.

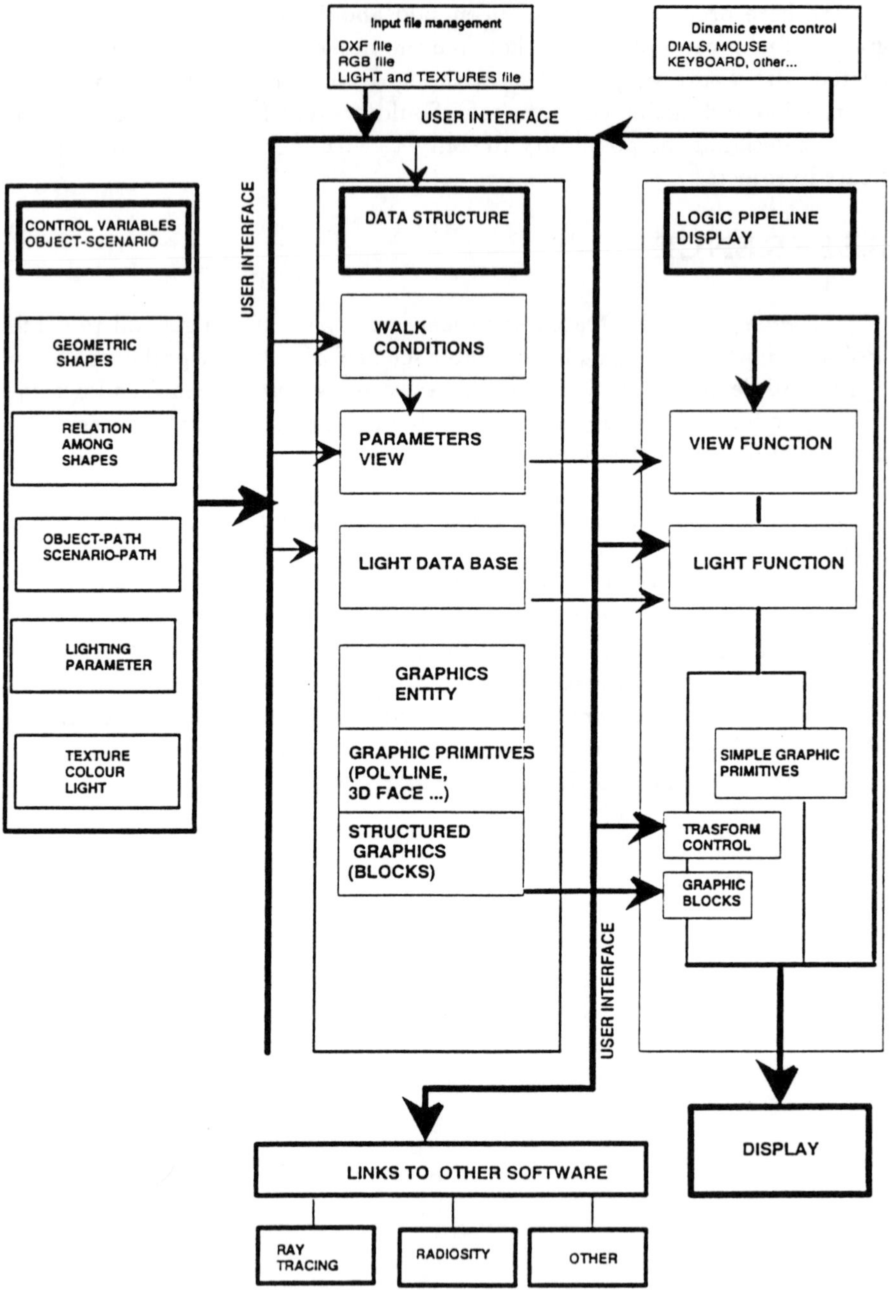

Input file management
DXF file
RGB file
LIGHT and TEXTURES file
Dinamic event control
DIALS, MOUSE
KEYBOARD, other...
USER INTERFACE
USER INTERFACE
CONTROL VARIABLES
OBJECT-SCENARIO
GEOMETRIC
SHAPES
RELATION
AMONG
SHAPES
OBJECT-PATH
SCENARIO-PATH
LIGHTING
PARAMETER
TEXTURE
COLOUR
LIGHT
DATA STRUCTURE
WALK
CONDITIONS
PARAMETERS
VIEW
LIGHT DATA BASE
GRAPHICS
ENTITY
GRAPHIC PRIMITIVES
(POLYLINE,
3D FACE ...)
STRUCTURED
GRAPHICS
(BLOCKS)
LOGIC PIPELINE
DISPLAY
VIEW FUNCTION
LIGHT FUNCTION
SIMPLE GRAPHIC
PRIMITIVES
TRASFORM
CONTROL
GRAPHIC
BLOCKS
USER INTERFACE
DISPLAY
LINKS TO OTHER SOFTWARE
RAY
TRACING
RADIOSITY
OTHER

Figure 1

DESCRITION FILE FOR SIMULATION

```
CATALOGS
Seat            #comment a
Table           #comment b
Armadi
BEGIN BASE
Urban_site              #comment
        LAYER on/off colour
        layer1  1       2
        layer2  1       3
        layer3  1       4
        layer4  1       6
        ENDLAYER
ENDBASE
BEGIN ALTERNATIVES
Alternative1            #comment1
        LAYER on/off colour
        layer1  1       5
        layer2  1       7
        layer3  1       20
        ENDLAYER
Alternative2            #comment2
        LAYER on/off colour
        layer1  1       5
        layer2  1       7
        ENDLAYER
        Alternative21
                LAYER on/off colour
                layer1  1       5
                layer2  1       7
                ENDLAYER
        Alternative22
EOF
```

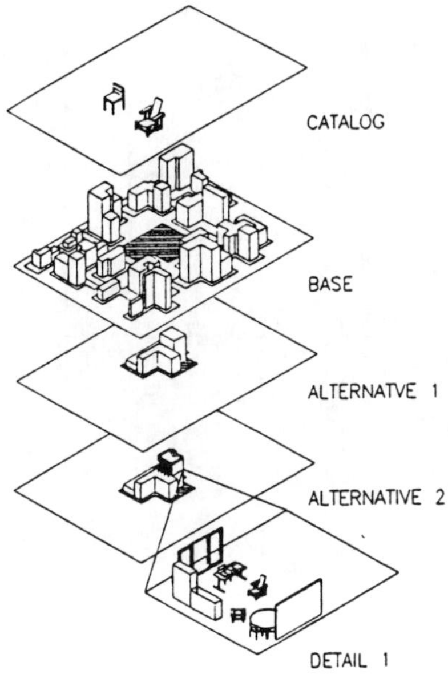

Figure 2

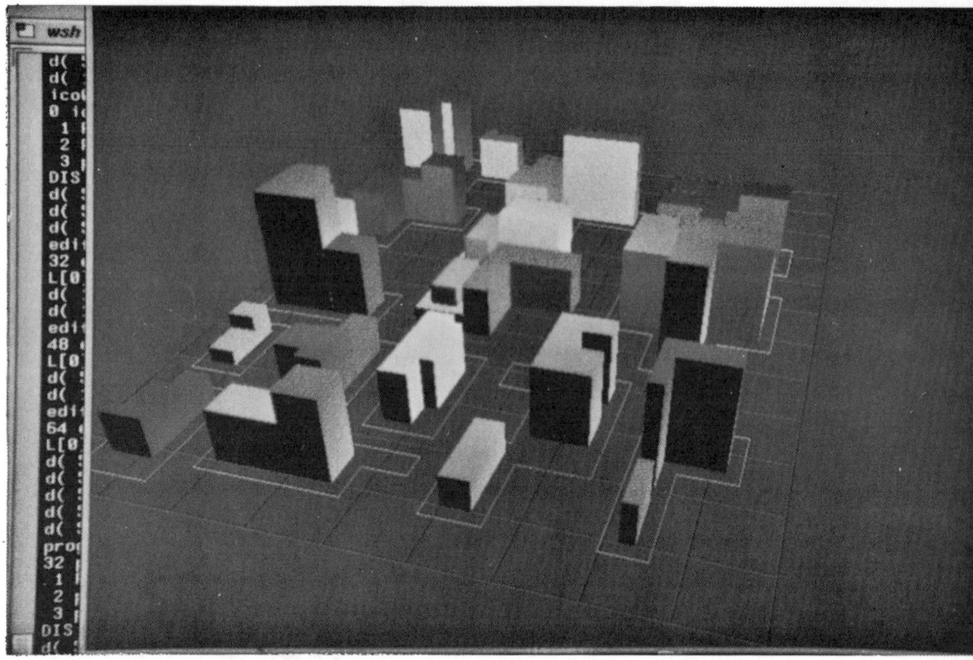

Photo 1

Photo 2

Photo 3

Photo 4

Photo 5

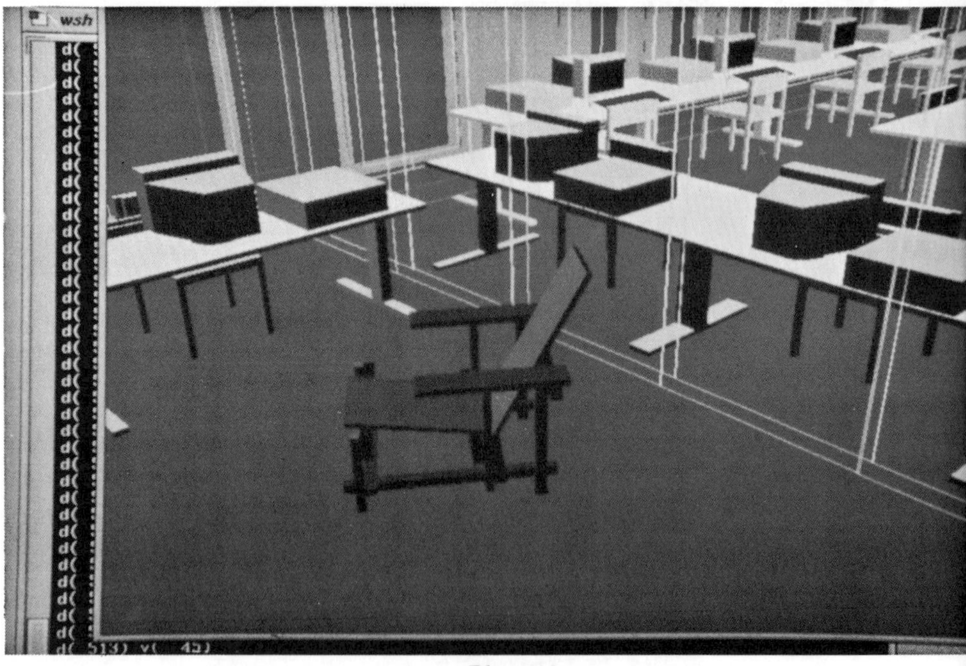

Photo 6

Photo 7

Photo 8

Photo 9

Photo 10

Photo 11

Photo 12

Photo 13

Studies and experiments for a laboratory of virtual construction elements

A. Campioli, C. Talamo

Dipartimento di Programmazione, Progettazione e Produzione Edilizia, Politecnico di Milano, Via Bonardi 3, 20133 Milano, Italia

ABSTRACT

The research [2] defines a methodology for the structure and organisation of information relevant to building construction elements and methods of assembly, by means of the computer. The construction elements become "virtually" represented in their geometric, technical and material characteristics with the objective of creating a support tool to both the design and the didactic stages in the production of professionals in the field of Architectural Design.

INTRODUCTION

The research proposes a prototype laboratory of the building elements which is implemented through the computer. This proposal comes from the necessary research into the systems, methods and tools by which the transmission , the communication of technical information and the verification of the design choices is carried out today, or will be in the near future, in the building sector.

The knowledge of materials and building methods has, up until now, been arrived at through the product manual, which constitutesa limiting element since the methods, elements and materials are always studied outwith the constructional context in which they are applied. The traditional manual, infact, comprises a virtual catalogue of possible solutions, already tested for combination according to the manufacturer's specification. If one considers that the result of a building is to be attributed in a particular way to the relationships which one establishes among the various elements, one can appreciate the negative aspect that this limitation assumes for the knowledge of building methods.

Furthermore, the present widening of the range of methods at one's disposal in the building sector makes the production of a physical laboratory of construction elements difficult: one would be dealing with only a few, hardly significant samples of an already wide range of products.

Furthermore, the group of components suffers from the same limitations as have been attributed to the manuals: presenting the components and elements without providing for their verification (in terms of language and performance) where they, within the building, form inter- relationships among themselves and the surrounding elements.

Many tools already exist today which allow construction elements to be treated as three dimensional geometric entities, as well as performing entities, and as icons to which are attributed a highly complex objective reality. One deals with Database, CAD systems, Rendering systems, the limitation of which is the lack of integration. In fact, the integration of these systems; which individually provide for the study of different aspects of one building method, of one component, of one material; without doubt constitutes a point at which to aim for the organisation of an information laboratory in which one is able to treat the elements in their geometric and performance complexity and in their aggregative potential combinations within the building system. However, an experiment starting with the intrinsic capability of each system seemed to be the absolutely essential departure point for devising a computer tool to train and assist with the design, enabling one to overcome the existing discontinuity amongst tools which were specifically conceived for each of the different phases and scales of the project.

This research has intended, therefore, to configurate a virtual locus in which to set up the construction elements and trace the information which is relevant to them: a laboratory in which; although not physically present; the building components, subsystems and construction systems can be studied in detail, regarding their dimensional, performance and linguistic characteristics.

The building elements installed in the laboratory are the result of a decomposition exercise carried out on buildings in their totality. Thus all the components from one building are present within the laboratory. This allows one to have all the detailed information on each individual component and simultaneously renders it possible to reconstruct the building in its totality, thus providing for comprehension of the relationships which one establishes among the various parts within the building system. With particular regard to the selected construction systems, it is possible to study the critical points of interface among the technical subsystems, between the technical subsystems and the components, and among the components themselves.

The scale which is applied for the definition of the elements to be installed in the laboratory is thus that of the building, adopting the following criteria for the selection of case studies:

- the buildings must be executed using dry construction methods, this allows for the recognition and interchangeability, within certain limits, of the components within the building system;
- the assembled components must be characterised by functional autonomy in respect of the other elements of the subsystem to which they belong (for example, in a cast in-situ concrete frame structure the beams and columns are inseparably correlated, while in a building composed of pre-fabricated elements every component -within its own limitations - has a geometric, static and functional individuality);
- every component must correspond to a small number of the other components;
- every component must be capable of being linked with organised information levels (geometric characteristics,response to stresses, performances).

In this way two categories of buildings can be established.

Those buildings in which the relationships amongst the subsystems or the components are inextricably linked are assigned to the first category.

In this case the writing of a graphic and informative repertoire permits:

- the creation of an informative and graphic catalogue which can replace the traditional manual, providing both performance-related information (relative to the class of the technical elements) and descriptive information (relative to the component under examination);
- the typological reconfiguration of the buildings, maintaining the unaltered characteristics of the components. In this case the facility of three-dimensional simulation allows immediate verification, although partial, of the design solution;
- the identification of compatible components for application within different building systems (even if the need is rare these days because of the structure of the building components market) ;
- the modification of some features of the components within the field of options predetermined by the informative system.

Those buildings in which the relationships of components are not specifically definitive are assigned to the second category. Within these building types the components constitute a selection from a wide range of possible and interchangeable alternatives. Their selection does not necessitate modifications to the other elements in the system. For buildings which are members of this second category, the writing of a graphic and informative repertoire provides for the same operations described in the first category and futhermore, the redesign of some components.

The laboratory of virtual construction elements pursues both a research and design experimentation purpose and a didactic purpose.

It is precisely to respond in an appropriate way to this duplicity of function that it was decided to organise the laboratory through the use of two tools with different characteristics. The first with a markedly informative vocation,, the second, on the other hand, concentrating on the projectual manipulation of the construction elements contained in the laboratory. In this way it was possible to cover the two stages which are fundamental to the design and didactics of the project: the acquisition of data and the regeneration of design.

The first tool is composed of a hypertext based on the personal computer through which one may access the information, the images and the drawings relative to the building, its subsystems , its components and its construction details. Organisation of the information in a hypertext means, in fact, the study of the characteristics of every single element, but at the same time, the study of their relationship - posssible or already existent- with all the other elements of the system.

The drawing up of an informative hypertext on the building methods at the disposal of the building sector today, more than establishing a tool which is useful for technical up-dating also offers the possibility of clarifying the relationships which one sets up among the various subsystems and components within the building, aims to highlight the compatability of the various elements, and to give an appreciation of the implications that a particular technical selection can have for the organisation of the whole building process.

The second tool is based on the integrated employment of the personal computer and the workstation, and aims at the maximum interaction of the informative system and the operator. To this end a system has been set up which, further than supplying a body of information, provides for the simulation of the elements of the laboratory and the simulation of the design interventions carried out on them.

Within this virtual interactive world

- the components and elements can be reconstructed in their reality, geometrically and in terms of performance, without occupying space: in this way it is possible to have information on a vast and ever-expandable range of construction elements;
- the components and elements can be assembled together to reproduce the building from which they have been extracted, or to determine other possible solutions; in this way it is possible to verify the relationships of an indivdual element within the system and experiment with it in other configurations;

- the components and elements can be manipulated in design for the research of solutions which may also be radically different to the original configurations.

All this is thanks to the possibility of relating the geometric representation of every element to a series of facts of varied nature in alphanumeric form .

Furthermore, in this locus of virtual elements, it must be possible to simulate design situations and verify every technical selection in the building system and its individual subsystems, regarding both the visual aspects and those of performance. The creation of the prototype of the laboratory of virtual construction elements was conducted through the following processes, which are common to both tools.

1. Collection of some designs whose features provide for simple characterisation of the subsystems, components and methods of assembly, with documentation also relevant to the construction details.
2. Analysis of the construction and decomposition according to the logic of the subsystems and components.
3. Graphic reinterpretation and digitising of the fundamental elements.
4. Definition of the linking structure of information to be associated with each construction element.
5. Determination of the process of correlation between the graphic images and the information structure.

Thus the construction of the informative system has been based on the features and specific potentials of the tools employed.

THE INFORMATIVE SYSTEM FOR INTERACTIVE SIMULATION

In the first phase of work, the requisites of the laboratory were analysed to determine the most suitable tool configurations for the construction of a system which could respond to some requirements with regard to the design and didactics of a project [3].

One of the first requirements was to make an informed design process possible, that is, one where every formal selection can be linked, in an immediate and interactive way, with an informative tool which can supply data about the technical quality of the construction elements that one intends to employ. It is intended to salvage the logical/inventive process with continual tracing of correspondence between the visual imagery and building methods; widening the opportunities for knowledge and the stimulation of ideas through reference to the data system. Such systems must be opportunely structured to reconstruct the matrix of relationships and reciprocal influences which are created at the point of design selections.

A computer tool must, therefore, offer two informative levels to guarantee these results: one graphic, in which it is possible to store and extract a group of signs and figures meaningful to the designer; the other alphanumeric with the possibility of organising different methods of accessing and linking information. It is fundamental that the connection between the two levels is direct, interactive and by multiple questioning. By "direct" connection it is meant that one may receive information based on the graphic representation of the construction elements, without the need to go through a special sequence of interrogation which could inhibit the creative processes.

The interaction between the informative levels represents the condition which is necessary to effect selections from among the design alternatives, in the sense that every proposal in the design phase must correspond to an immediate evaluation of the obtainable performances, both for the individual subsystem and the total building system.

One of the subsequent requirements is a realistic view of the construction elements with the possibility of working on them, becoming familiar with their material properties, redesigning them with the same degree of freedom afforded by their materials and methods of production. Furthermore it is important to be able to evaluate, at different scales, the impact that a design selection has on the environment which it creates and into which it is inserted. These opportunities are offered by those systems which allow the solid modelling of objects and visual simulation at the same time, and in which it is possible to graphically represent, with elevated levels of realism, the model and its contextual surroundings.

A third requirement is the recomposition of unitariety of the stage of design which can be followed through the coordination of those computers which provide for specialised modules. In this way the user designer verifies, during the progress of work, the design selections which have been made in respect of some parameters or specifications; checks the consequences of those choices in an interactive way on the basis of different categories of performance; verifies or executes correct dimensioning according to the rules of the different disciplines.

An analysis of the most diffuse systems existing in the field of computerised design was therefore conducted on the basis of these requirements, restricting the choice to those products which respond, in some way, to the requirements of a laboratory of construction elements. The following were identified:

- a C.A.D. system for the Personal Computer;
- a solid modeller;
- an integrated C.A.D. system for the workstation.

The objective is to arrive at the coordinated use of the three different tools, to employ in the system the full potential of each of them. The C.A.D. system for the personal computer was chosen particularly because of the greater potential that the tool is acquiring with every new version, the ease of use and its wide diffusion, which make it a particularly appropriate tool for didactic experiments.

Furthermore, the tool has the capability to combine with data bases and modules for rendering and three dimensional drawing.

The solid modeller, born as a support specifically thought for industrial design, turns out to be particularly useful for optimising the process of design and production of objects and components.The tool can draw and model any object, shaping the profiles and surfaces in an optimum way, as has been required by the designer and the processes of production; it can give the surfaces the look, the grain, the finish and the colours desired. Once the object has been created it is possible, with connection to a numerically controlled milling machine, to prepare an actual mould or true model. Besides this, it can produce wire frame drawings of the designed object on the plotter. The system however, even if it presents greater possibilities for the production of catalogues of objects, does not have the characteristics for complex development of architectural design projects as are typical of the C.A.D. system. Its use, for the purposes of this study, was therefore considered to be effective only if it is integrated with the C.A.D. system.

The integrated C.A.D. system, for a series of reasons which will be verified in the course of the research, seemed to be a tool which is sufficiently adequate to satisfy some of the stated objectives.

In short, its capabilities and limitations can be summarised as follows.

- The system was conceived and structured on the function of the design process and its decisional and operational phases. The organisation of the project derives from schemes which compel one firstly to consider the design of the process itself; to clearly define the various collaborators and the levels of their involvement. The conceptual and physical structures of the building object can be set up in the initial phase of work and these become the grid of reference for the total system, on which to base the project design and the coherent use of the calculation modules. This turns out to be an essential feature for the didactic requirements since it allows the experimentation in the process of product design and the acquisition of a comprehensive view of the components in play.
- The philosophy of the system assures an integration among the various special capabilities which can be introduced to the project: the modules which accompany it; with the opportunity of cross-verification; these offer the possibility to evaluate (during the design phase) which will be the effects of specific choices, which levels of performance can be achieved, and which interventions must be executed to ensure the integrity of the building organism.

  The wide picture of disciplinary fields covered (structural calculation, thermal balance, illuminotechnical appraisal, quantification of plant, client involvement, maintenance planning) allows a unified view of the various disciplines which come together in the design process.

The system is designed for connection to a data base which provides for the simple implementation and development of a vast and flexible archive of building components and the relevant information.

From the practical point of view two paths are followed.

In the first case the construction elements are drawn and stored in the C.A.D. system and are subsequently transferred to the integrated C.A.D. system.

In parallel, they are drawn with the same procedure for the solid modeller.

In the first case, since a detailed drawing of elements with complex geometries is not possible, these are three dimensionally represented in their mass, their general morphology and their details of connection to other elements. Coupled with these components are the tables of a data base which are structured so as to distinctly and exhaustively identify the properties of every individual construction element.

Once the construction elements are transferred to the integrated C.A.D. system the user is in the position to design architecture using the multiple possibilities offered by the system. One may redesign the elements, in part or in their junction with other subsystems, with natural reference to an ever concise representation of the component. One may analyse and study the construction elements both in terms of morphology and performance: in the first instance one can deconstruct and reconstruct the element with the aim of retracing the geometry and logic of construction; in the second case one may scan the tables of the data base in view of different parameters and, through the additional modules, one can execute a series of calculations and verifications, such as structural calculations, thermal balances and evaluation of lighting levels. One can finally assemble the elements to create a new architecture and evaluate, from different points of view, the overall visual effect of one's actual choices.

In the second case the same elements are drawn according to their actual geometry so as to achieve a model which is absolutely faithful to the original.

The construction elements can therefore be analysed in their true form through the use of legends of components which have been memorised by the modeller. In this case the component can be analysed, in respect of its visual qualities, observing it from different points of view and rotating it in space. One can direct the observations towards the individual detail; different solid colours and tones can be attributed to the surfaces, the position of different light sources can be determined, within the space; the object being to simulate the effects of shadow generated by the morphology of the parts. The individual components can be redesigned modelling the planes as one would plastic objects. In the presence of dry construction methods, it is possible to simulate, in a virtual space, the junctions of different components, as may occur on an actual building site.

The structure of the laboratory set up in the course of research comprises the following phases:

1. the elements are drawn through the C.A.D. system on the personal computer;
2. the elements are transferred to the integrated C.A.D. system;
3. the user, through the modules for evaluation of the integrated C.A.D. system, has a way in which to study the elements, to understand their morphological rules and performance characteristics;
4. the general directives are established for the case of redesign of the component, which is derived from the information emergent during the preceding phase;
5. through the solid modeller the element is further observed, for in-depth appreciation of the expressive features and to evaluate, in respect of these, the validity of the design selections effected in the preceding phase;
6. the component is remodelled according to the directives of phase four, or other specific indications which emerge in the process of observation;
7. one returns to the integrated C.A.D. tool, where the formulated modifications are translated into standard representations;
8. one finally assembles the elements to create an architecture, or the significant parts of same;
9. one verifies the visual result of the total building and of the individual components of which it is comprised.

The recourse to diverse systems offers a multiplicity of combined potentials.

In the first place the opportunity to experiment is created, both on levels of technology and methodology; communication among systems which were designed on different philosophies.

Furthermore, since it is a study of didactic scope, the use of two different systems of C.A.D. allows for different experiences, beginning with a singular archive of elements: for diffuse didactics (university courses) the C.A.D. system on personal computer provides for practice to be widened to a large number of students and the instruction to be relatively simplified due to the ease of use of the interface; for the didactics dealing with small groups of students or for the development of a design thesis, more sophisticated experimentation can be achieved,through the use of integrated C.A.D., having the opportunity to offer more profound training to the prospective operators.

The results of the two different levels of operation can also be exchanged, compared and developed in unison.

Furthermore, the implementation on the two systems opens the way for cascading development of successive experiments which make use of the implemented graphic and alphanumeric data. This was the case in the second experiment which will be illustrated and which was progressed using the documentation developed on the personal computer.

THE INFORMATIVE HYPERTEXTUAL SYSTEM

The second direction of experimentation, currently under development, aims to set up a computer tool of informative support for design education which provides for confrontation of the theme of the evolution and innovation of dry construction methods in the building sector through the consultation of graphic documentation and information.

The object is to structure a system which is able to articulate the knowledge necessary to the development of a project and, at the same time, the complexities of the connections which combine the information

The need to simulate, through the use of a personal computer,the complexity of the cognitive actions involved in the design process has produced the following structuring of the requirements essential for the determination of the features of the system:

- association of the information in a non-sequential manner;
- integration of heterogeneous information;
- conduction of interogation of varied nature;
- structuring of cognitive paths based on different design strategies;
- conduction of operations on the system through a *friendly* interface.

On the basis of these requirements the hypertextual tool [4] was created as an interface for a base of relative information, with the same scope for achieving the multimedia didactic environment.

The documentation in our possession, regarding works of architecture and construction systems, has been re-organised in five categories of information. Each category is an archive defined as follows:

- archive of architects;
- archive of architecture;
- archive of legislation;
- archive of images;
- archive of materials and methods of production.

The *Archive of architects* is represented by an index of architects and contains biographical and bibliographical information, as well as a list of completed works; the *Archive of architecture* indexes a series of buildings with a base of relative information - the materials used, the subsystems designed, the references to the images (of the building and construction details), the bibliography of each work; the *Archive of Legislation* contains the Italian and foreign legislation relative to materials and subsystems; the *Archive of Materials and Methods of Production* contains information on the characteristics of the construction materials and the manufacturing process; the *Archive of Images* contains a collection of images (drawings, photographs, films) regarding the included architecture, the subsystems, the components and the phases of construction.

The logical structure of the system, outlined in figure 4, is composed of *five archives* (or data bases), each one representing an autonomous entity; the information contained in the different archives can be correlated according to the information paths established by the user; the correlation takes place by means of the *database management system* which regulates the entry of information into the system; the selection of informative path and the interpretation of the results of the association of information occurs through a *multimedia hypertextual interface.*

The objective with which the interface has been structured is the simulation of the means by which information becomes linked during the design process, subject to a network of junctions which differ time and again. To achieve this goal, an operation was carried out to determine all the possible correlations of information contained in the archives. These correlations have become "keys" which allow the connection of archives by means of the data base management system.

The cognitive maps which can be configured in this way by the user, not being predetermined by a previous path, vary with the type of approach chosen: from the needs of further investigation which may arise in the course of enquiry to the different objectives (design, didactics, consultation).

The interface is based on two types of interaction: the lists of information accessible through the keys and the images accessible through touch-sensitive areas on the screen; allowing the deepening of knowledge of the represented elements.

One of the possible paths, for example, starts with an enquiry on all architecture which contains the presence of steel and brick; the response is a list of buildings on the basis of which one may, radially, connect a series of information: see the image of the building; become familiar with the subsystems and components which it comprises; have information about the architect. Every one of these possibilities opens the way to a further route which can be varied at any point in accordance with a circularity of more accesses; for example, asking to see the image of a building , one can choose the drawing of a typical plan, selecting a part of this, the drawings of construction details will appear, expanding the enquiry one can obtain information on the materials, and from this one can visualise an architecture composed of the same materials, or further still, other works executed by the same architect, and so on.

What must be emphasised is the horizontality of the interconnection of information and the extreme freedom of the combinatorial game with the knowledge that one is acquiring, value is attributed to exploration as a mechanism for autogeneration of semantic grids.

The system provides the user with two types of use: on one hand, self-teaching by methods which, surpassing the traditional consultation of technical manuals and publications in the sector, provide an ordered view of the tight connections between technological culture and poetic design; on the other hand, a support for the design activity as concerns the assumption of references and the widening of knowledge of precise thematics, technologies and architectonics.

The multimediality of the system is obtained by the engagement of instrumental apparatus which is composed of: scanner, telecamera, video recorder, tape recorder and optical disc.

The use of this equipment offers a vast potential for didactic and popular use. It is possible to combine the static images (photographs and drawings) with the information extracted from the data bank to produce a film which allows observation of the studied architecture by the interogation of the archives, and evaluation of the impact on its contextual environment.

It is possible, for unrealised projects, to attend to possible sequences of animation with sophisticated systems of simulation.

In the same way one can observe results of simulations of the behaviour of the materials and the built systems (for example, the static behaviour of the structure or the thermal and hygrometric behaviour of the envelope).

Yet, possibly the greatest value of this multimedial application is that offers the opportunity to widen the knowledge relative to the operational aspects through the visualisation of the sequences of the works and the phases of mounting the building systems and components.

By furnishing the possibility of linking information - on architecture, its author , its adopted materials, on its incumbant legislation - the visualisation of the operational phases of construction on site- the sequence of works, the procedure for installation of components, the construction of particular details - one considers the ability to contribute towards a more unitary vision of the interdisciplinary complexity attached to the design activity, and towards the resolution of the dichotomy between the technical act and architectonic culture.

REFERENCES

1. Andrea Campioli, Gabriele Gotti, Laura Pedrotti and Çinzia Talamo collaborated in the research *Studies and Experimentation for a Virtual Laboratory of Construction Elements*. Andrea Campioli edited the first chapter, while Cinzia Talamo edited the second and third.
2. This research enters the general research *Poíesis. Heuristic design and computer science*, directed by professor Guido Nardi, dealing with the relations between computer science and design. The described works have been developed inside the Laboratory of Informatic Technologies (LTI) of the Dipartimento di Programmazione, Progettazione e Produzione Edilizia of the Politecnico di Milano.
3. About the theme of computer aided design we refer to: Praderio, G. *Progettazione architettonica con il calcolatore*, Be-Ma, Milano, 1987; Jakendoff, R. *Consciousness and the computational mind*, Mit Press, Cambridge, Mass., 1987; Johnson Laird, P.N. *The computer and the mind*, Collins, London, 1987; Aa. Vv. *Elaborazione elettronica nel progetto di architettura*, Muzzio, Padova, 1990.
4. About the theme of hypertextual applications we refer to: Various Authors *Multimedia. Technology and Applications*, Ellis Horwood, 1991; Serra, L., Chua, T.S., Teh ,W.S. «A model for integrating Multimedia Information Around 3D Graphics Hierarchies», *The Visual Computer*, July 1991

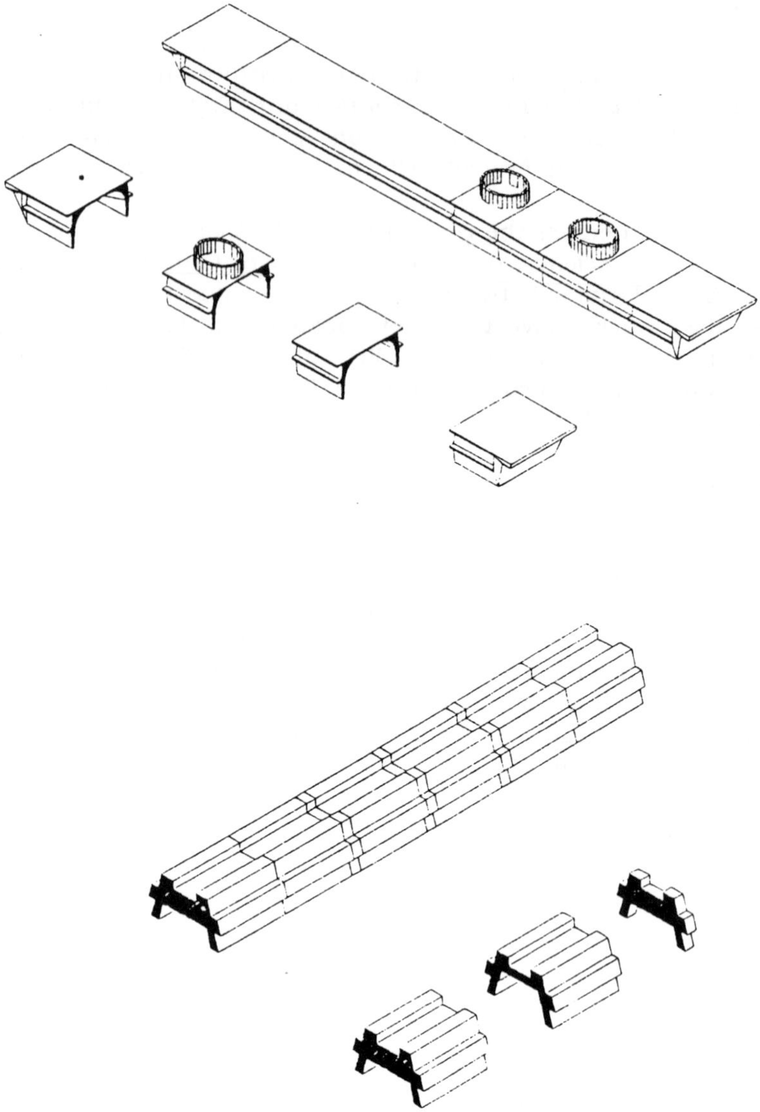

Fig. 1- Angelo Mangiarotti construction system in reinforced concrete, 1969. Digitising of the elements of the construction system by personal computer. Exploded axonometric of the covering panel (above). Exploded axonometric of the beam element (below).

The construction elements were drawn in segments of identical section to allow for, in the course of redesign, the parameterization of the different parts.

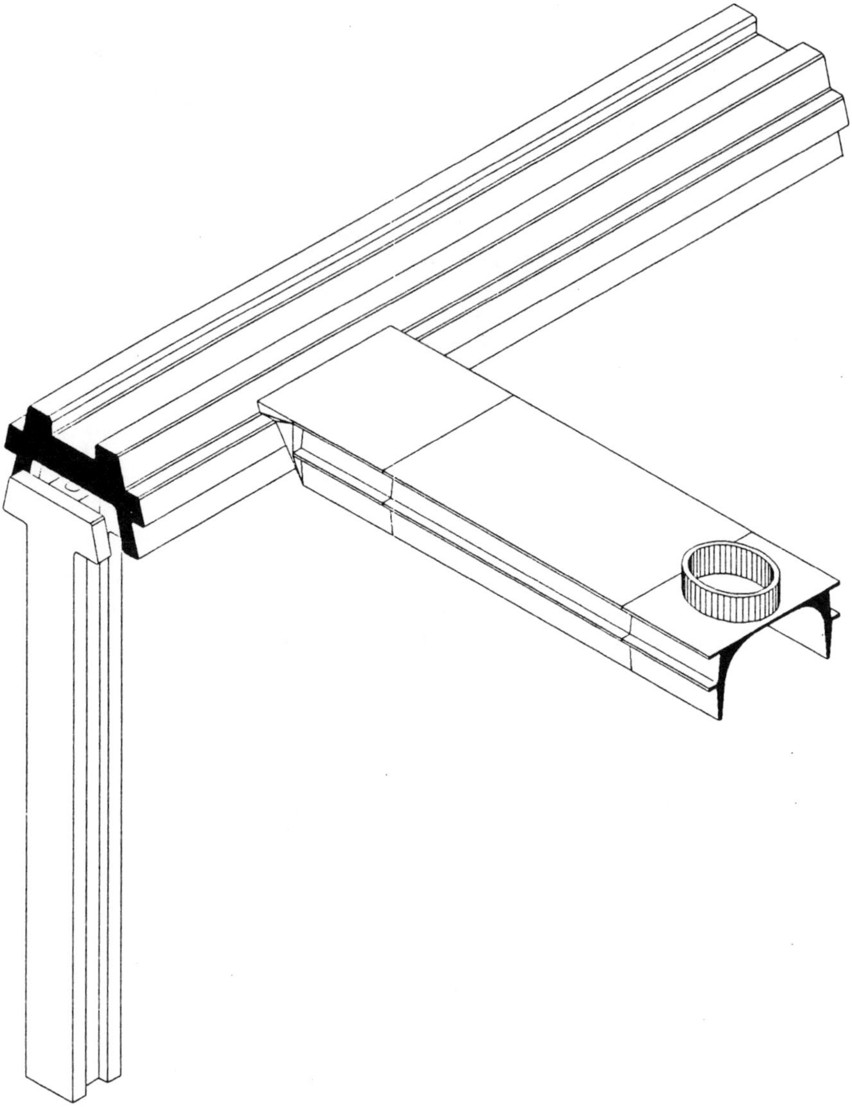

Fig. 2 - Assembly of the elements of the construction system by integrated C.A.D..

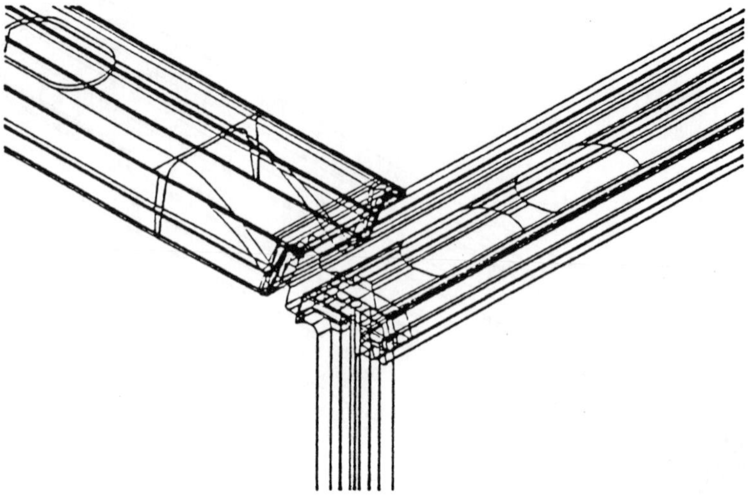

Fig. 3 - Three dimensionalmodelling of the components of the construction system by the solid modeller. Wire frame rapresentation.

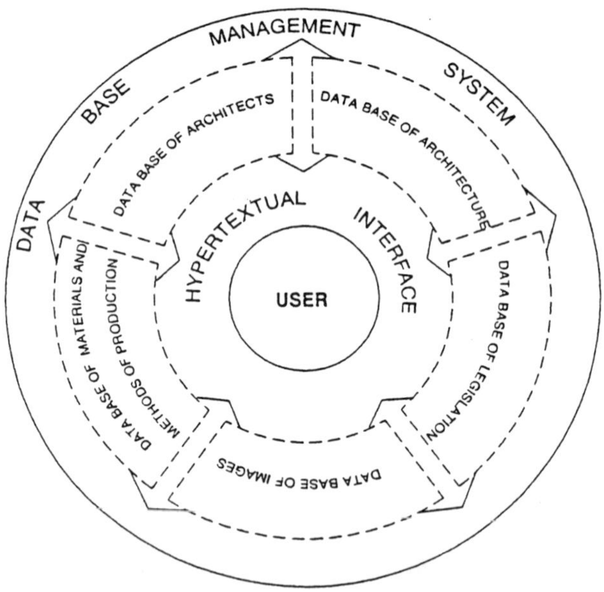

Fig. 4 - Logical scheme of the hypertextual system.

Computer visualisation in planning control: some case studies and issues

A.C. Hall

Division of Housing, Rual Development and Planning, Anglia Polytechnic University, Chelmsford, UK

ABSTRACT

Computer visualisation has a currently unrealised potential for assisting the control of urban design by planning authorities through its ability to provide and amend realistic pictures of proposed development and so enable lay groups, especially local politicians, and the public at large to view directly the impact of proposed development and to participate more effectively in negotiations concerning it. The author has been conducting research to test this assertion for some years, most recently via a 15 month project which involved taking examples of development being considered by selected British planning authorities, visualising them using micro-based systems and then monitoring and assessing the use made of the images produced. The principal system used was Autodesk 3D Studio on a PC platform, although Intergraph software and Apple Macintosh hardware were also employed. Hard copy was obtained by screen photography and supplemented by animated walk-throughs. The issues studied ranged from the impact of a large scale shopping proposal to the colour of a small apartment block, and included examples of both advance preparation for a design brief and reactive action by the controlling authority. The results confirmed the importance of the technique for design control and tests involving the design of domestic extensions revealed how questions of objectivity and level of realism could be resolved by reference to the position of the negotiating parties.

INTRODUCTION

The importance of computer visualisation for planning practice lies in its potential for the improvement of the quality of decision making by virtue of its ability to avoid misunderstandings in the negotiation of the outward form of development. One of the greatest stumbling blocks in the negotiation of the design of new development, be it between developers and their architects, applicants and planning authorities or planners and the public, is the difficulty in visualising proposed schemes in such a way that their three-dimensional visual impact on the user can be appreciated and discussed. Architects' plans and elevations are difficult for the lay person to interpret; models are expensive to construct and update and the devices required to view them as from ground level are cumbersome if not actually impractical; artists' impressions are also expensive and can be highly subjective in their interpretation of the proposals. Information technology has the ability to overcome this obstacle through its ability to model buildings and landscapes, both existing and proposed, in precise perspective, and to reproduce the effects of light, shade and shadows impinging on different building materials. The capability of computers to carry out these tasks has been in existence for at least ten years but the high cost and lack of convenience of use of both the hardware and software has previously put it beyond the day to day reach of those who were not experienced computer users. However, computer visualisation has now become available in user friendly form at affordable prices and is, therefore, within the reach of planning practitioners. Unfortunately, in Britain there has been no rush by planning authorities to test this potential, presumably through lack of both resources and information, and between 1987 and 1990 only a few isolated (but noteworthy) examples were undertaken[1]. One was the full colour visualisation for a proposed market cross to be constructed in the central square of South Woodham Ferrers new town in Essex, carried out by the author during 1987 and 1988[2].

A full-scale investigation of the application of computer visualisation to planning practice, financed by the Leverhulme Trust was undertaken by the author from January 1991 to March 1992. Its objective was to test the usefulness of computer visualisation in the context of day to day planning. The idea was to take a number of examples of development proposals under consideration by local planning authorities, visualise them, feed the images so produced back into the decision making process and monitor the results.

COMPUTING FACILITIES

Commercially available software on micro-computer was used to model the selected development proposals. Hard copy was obtained in the form of colour prints and slides by photographing the monitor screen. The first software and hardware combination used was Integraph Microstation 3.5.1 on an Apple Macintosh IIcx platform with 13" colour monitor. The program provided surface modelling facilities with limited colour and shading effects. Microstation was a general design package and which, while both powerful and easy to use on its own terms, was neither a specialist visualisation nor a specialist architectural package. Perspective views could be obtained although, with this version of the program, eye level views were difficult and views from inside a proposed development scheme almost impossible. In April 1991 the software was changed to Autodesk 3D-Studio, which had just appeared on the market, as it offered very high quality lighting and materials effects and facilitated the easy animation of images. This program was not available for an Apple Macintosh environment and so the hardware was also changed to a substantially enhanced PC386 with a Super-VGA monitor.The Super-VGA monitor (256 colours at a resolution of 640x480 pixels) was used to display the working environment. The Targa video card was used to supply a high quality RGB signal at a higher resolution to the Electrohome 19" monitor which was used to display the final rendered images.

The 3D-Studio program was a specialised visualisation package. Three dimensional modelling was facilitated both by extrusion and by use of a small range of primitives. The powerful features of the program related to the rendering of surfaces. 3D-Studio possessed a materials library which could be extended by means of a powerful on-screen menu. When compared with some contemporary advanced CAD programs it was easy to use for both modelling and the generation of perspective views. Its limitations arose from the fact that it was not a purpose designed CAD package. 3D-Studio made the process of animation very straightforward, and images rendered to sufficient quality for display on the VGA monitor could be animated within the computer's memory and replayed on that monitor. The animation could be stored in a file and, using an ancillary program, AAPlay, short versions could be put on to a floppy disk and replayed by other users.

THE BERKHAMSTED EXAMPLE

For the first test the CAD program Integraph Microstation on the Apple Macintosh IIcx machine was employed. The subject was a planning application for a major shopping centre and multi-storey car park on back land behind Berkhamsted High Street. A computer based model of the proposals was constructed from the architects' drawings and some adjoining buildings were modelled, mostly in block form, from maps and site visit information. A large number of colour perspective views were generated, both aerial and ground level. The planning officers dealing with the case visited Chelmsford to view the model interactively and were thereby able to see for the first time the true appearance of the proposed development and to assess the relative bulk of the structures. The development was delayed by negotiations and, ultimately, the recession in the property market. At the time of writing, the design was being reconsidered by the architects. Their views on the visualisation of their initial scheme were, however, sought and they showed great interest and were supportive of the idea. They were already using CAD (in the form of t^2-Sonata on an Apollo platform) for design but not, normally, for visualisation. They preferred, for the latter purpose, manually rendered computer generated perspectives as they considered that this gave a more accurate representation of the appearance of building materials than could be obtained with the computer software available to them. However, no such visualisations had been prepared for the Berkhamsted scheme.

THE GUILDFORD EXAMPLE

The next test, and the first using 3D-Studio, was undertaken for Guildford Borough Council and concerned the proposed redevelopment of a Borough owned leisure centre in the town centre. The aim of the visualisation exercise was to provide assessment of visual impact of the proposal in order to facilitate:-

a the selection of an outline design by the officers

b appreciation by council members of the physical implications of the proposal

c the production of a three-dimensional design guide for inclusion in the Council's corporate brief to potential developers.

The first step in the exercise was to model the surrounding buildings, roads and other structures, leaving a vacant plot where the leisure centre stood. The alternative outline designs for the development were then modelled and inserted into the model of the surrounding buildings. Oblique aerial and ground level views were produced together with animated walk-throughs. Hard copy was sent to Guildford for study in the form of screen photographs and floppy disks containing the animated walk-throughs. The Guildford architects and planning officers also visited Chelmsford to work interactively with the facilities, exploring views and design alterations of their choice. Figures 1 - 4 show the development of the design which was used as a basis of a joint officers report to members. The matter was still proceeding at the time of writing.

THE EAST CAMBRIDGESHIRE EXAMPLE

Both of the tests described above concerned major new development. However, computer visualisation has just as much application to day to day planning control when the design of small scale proposals, such as domestic extensions, is being considered. An example of such a use was sought for the third test and a case under consideration by East Cambridgeshire District Council in the summer of 1991 was selected. The owner of the existing house wished to extend the upper floor above the garage. As this was forward of the building line the proposal was of concern to the residents of neighbouring houses and to members of the Planning Committee. Using 3D-Studio, the proposed development was visualised from the plans submitted with the application, The house on one side was similar in design to the applicant's house and, therefore, it was convenient to model it explicitly. Views were obtained for five different directions specified by the planning officers, with and without the proposed extension, and also for winter and summer seasons as screening by vegetation greatly affected the view of the property. Plan views were also generated. Cast shadows were included in the visualisation as overshadowing was a significant part of the issue. Colour prints of the screen images were turned into OHP transparencies and were shown to the Development Control Sub-Committee when it considered the application. The application was refused by the Committee. After the meeting, a group of members expressed the view that the visualisation had enabled members who had not been on the site visit to participate fully in the discussion whereas normally they would follow the views of those who had without formulating a position of their own. The applicants and their architects did not attend the Sub-Committee Meeting and so a copy of the images were sent to them by the research team for their comments. They replied by letter raising a number of objections and criticisms particularly on the overshadowing issue, where they maintained a more realistic result could be obtained by means of a protractor and calculations. They also queried the *objective* status of the visualisation on the basis that judgement had to be used on what items are included or excluded, and maintained that architects' perspective drawings in colour could give a more objective impression, in appropriate circumstances. They believed that computer visualisation had a role in the presentation of projects but not any *special* or *independent* role in the control process. The architects appealed to the Department of the Environment against the refusal of planning permission. The Council forwarded copies of the visualisation to the Department for illustrative purposes to show the likely effect of the proposals. The inspector dismissed the appeal. Analysis of the arguments of the parties revealed two points of debate relevant to the study:

a was *overshadowing* a subjective reaction to the dominance of a building (the Councils usage) or a consequence of the precise incidence of the shadows cast by it (the appellants' usage)?

b did the computer visualisation give a realistic impression of either interpretation?

The position of the sun in relation to the structures in the computer model was checked and found to be realistic. The visualisation did not, in fact, show shadows falling on the neighbours' windows for most of the year and, somewhat ironically, it could be said to support the appellants' position. The Council, however, saw it as supporting their own interpretation of overshadowing. Their interpretation of overshadowing was ultimately that upheld by the inspector.

THE DANBURY EXAMPLE

The example described above did not test the full power of the computer visualisation technique as, although hard copy was circulated to the participants, they did not come together at any stage to negotiate using the computer interactively. An opportunity to observe interactive use in a routine development control context arose from an example provided by Chelmsford Borough Council. At the end of 1991 the Council had received a planning application for an extension to a detached house in Danbury, Essex. The adjoining house on the east side already had a modest extension. The planning officer handling the case was concerned at the bulk and appearance of the proposal and had suggested in informal negotiations with the applicants' agent, a builder, a repeat of the A-frame roof shape or, less enthusiastically, an extension similar to that on the adjoining house. At the beginning of February 1992 the issue was referred to the research project by the Council so that informal negotiations could take place, using computer visualisation, with a view to the householders submitting a revised application acceptable to the planning authority. Visualisation of the existing house, the adjoining properties, the proposed extension and the planning officer's alternative was undertaken in the days leading up to the negotiating session. Highly realistic effects (as used in the East Cambridgeshire Example) were not considered necessary for the purpose envisaged and the adjacent houses were modelled with colour but without surface detail. In late February 1992, one of the applicants, her agent and a planning officer (not the original case officer who had been prevented by illness from attending) assembled around the computer screen at the research unit for a 1 hour 40 minute session. The basic model of the house and neighbouring properties (Figure 5) was displayed on the large monitor and was agreed by all the parties to be realistic for the purpose intended. Oblique aerial (Figure 6) and eye level views of the applicants' proposal were examined and its merits debated, followed by the research team's understanding of the development control case officer's ideas (Figure 7). The discussion moved on to consider an extension of similar appearance to that next door (Figure 8). Unfortunately, examination of the plan (displayed simultaneously on the VGA Monitor) demonstrated that to achieve a room of sufficient depth would also have required a ground floor extension that would have been unacceptable to the applicants on grounds of loss of natural light and increased cost. The applicant then suggested that an A-frame extension at right angles to the existing roof would be acceptable to her because it would then have a blank end wall that would facilitate the arrangement of her furniture. She had brought with her a rough sketch of the arrangement and the idea was then visualised on the spot (Figure 9). This proposal met with some support all round. At the conclusion of the negotiating session, the participants were asked for their reactions to the use of the visualisation process. The applicant and her agent, the builder, both confirmed they had found it to be very helpful and the planning officer commented that it was a very useful procedure that had saved a lot of time. He commented further that it was especially useful for people, as in his case, who had not visited the site as it enabled them to participate fully in the discussions (c.f. the East Cambridgeshire example). He added that negotiations usually took place in the lobby of the planning office, away from the site in question, with ideas being sketched out in pencil on scraps of paper. Hard copy was subsequently produced by screen photography and the colour prints circulated to all the parties. A revised planning application based on the applicant's suggestion was subsequently submitted and approved by the Planning Subcommittee in April 1992. Selected visualisations were shown to the meeting and members were enthusiastic in their response to the use of the images.

THE WALTHAMSTOW EXAMPLE

The attitudes to the colour of new buildings by both users and neighbours can be an important factor in both sales and the obtaining of planning permission. This is illustrated by the 1990 example of the block of flats designed by Julyan Wickham Associates for Bisterne Avenue, Walthamstow[3]. Although the block was of modernist outline in a street of early 20th century houses, it did not excite the attention of the public until the striking post-modern colour scheme was applied to the exterior. The local residents objected strongly and tried, unsuccessfully, to halt the work via the planning process. This issue was used as an example within a limited study of lay reaction to building colour was undertaken as a separate exercise. Using the Integraph Microstation program on the Apple Macintosh platform, a three-dimensional model of the block was prepared and the colours varied to provide ten different combinations. Colour slides of the images were obtained and projected before a jury of students who had no connection with the built environment professions. The images were shown both in isolation and against the background of the adjoining houses. A questionnaire was administered to the jury to record their subjective reactions both to the colour schemes themselves and their relationships to their surroundings. The results showed the taste of the jury to be on the conservative side with stronger negative than positive views to the colours shown. A particular point of controversy with the new flats the use of a strong shade of blue, was confirmed. The investigation was, in effect, a pilot study as only a small jury, not necessarily representative of the population at large, had been used. There is clear need for further study of the reactions of the lay public to building colour, a project which computer visualisation technology can readily facilitate.

ASSESSMENT OF THE TESTS

The study took the form of a limited number of trials of the use of visualisation covering a range of likely planning control situations. An extensive investigation involving a large number of trials from which quantitative data could have been obtained was beyond the limits of the available resources. The results of the tests were bound, therefore, to be somewhat subjective, and have value more as pointers towards future developments rather than as providing the definitive and concluding word on the issues involved. Bearing this caveat in mind, the points that arise from a comparative assessment of the tests are as follows:

a All the participants were impressed with the quality of image that the current hardware and software could produce. The realism of the images depended on the human resources that could be employed and, in general, was found adjustable to the requirements of the task in hand. The only exception to this was the concern that the representation of the colour, pattern and texture of certain building materials was not as realistic as could be obtained by manual drawing. However, given the rate of progress of software development this is likely to be only a temporary problem.

b There was no problem producing images of development on small sites within the time scale required by the development control process. This was especially so for visualisations that did not require a great deal of surface detail and, moreover, such images could be produced "while you wait". On the other hand, visualisation of large scale development (such as the Berkhamsted shopping centre) was a very time consuming operation.

c The modelling of extensive background features was not found to be necessary when dealing with the issues affecting development on small sites. In all cases, the computer modelling of adjacent buildings sufficed and the overlaying of the images on to photographs of the background scene (by scanning or otherwise) was not found to be necessary.

d The question of objectivity (as raised by the architects in the East Cambridgeshire example) is an important one as decisions had to be made about the degree of detail to be included in relation to the issues at stake. The answer would appear to be that definition of objectivity is a matter for the judgement of the negotiators in relation to these issues. There is, therefore, a strong argument for obtaining the agreement of all parties to the visualisation of the existing development before proceeding to the negotiation. The benefits of this were illustrated by the success of the Danbury example in contrast to the negative reactions expressed by one of the parties in the East Cambridgeshire test. (The question of objectivity is discussed below at greater length).

e An important advantage of computer visualisation appeared to be its ability to bring the "site" into the office. The way that it could enable people without a detailed knowledge of the site to participate in discussions was frequently cited as a major innovation.

f There was direct evidence of the advantage to lay people of visualisation. The Danbury example, in particular, illustrated how it could enable an applicant to make a decisive suggestion that may not otherwise have come forward.

g There was also confirmation of the assistance that computer visualisation could give to professionals engaged in the design process by providing feedback on the appearance of their proposals.

THE OBJECTIVITY ISSUE

The issue of *objectivity*, referred to on several occasions above, is a central one and one that deserves further discussion. Although visualisation has an obvious use as a presentational device for advocating the merits of a proposal, enabling its initiators to advertise and clarify their position, it is suggested that this is neither its only, nor its most potentially valuable, role. It is asserted that a central component of the argument for the use of computer visualisation in planning control is that it should be seen to be independent of the positions of all the negotiating parties. It can, thereby, provide them with a common point of reference, a means for resolving disputes on points of fact and the basis for an independent test against which the participants can measure the strength of their arguments. However, this is not an assertion that meets with universal agreement and at least one respondent (see the East Cambridgeshire example, above) expressed grave doubts about its use in an independent capacity. It would seem unlikely, though, that the argument would lie with the extreme positions of total acceptance or rejection of the objective role. Even if the general assertion above were accepted there would still remain some important caveats to be debated.

The argument can be addressed most conveniently by considering one extreme on the scale, the artists impression. This is normally held to be the epitome of the subjective image: very much a device for advocacy and little use as independent evidence. Why is this so? The subjectivity arises where, in the course of producing the drawing, the artist, who is employed to represent the interests of developer, is confronted with choices which require judgement to exercised on the basis of opinion and interpretation. In attempting to break down the components of the subjectivity the following points of choice and decision can be identified:

a a single viewpoint (or a very limited number) is selected to show the proposal to advantage - unflattering views need not be made available

b perspectives, where not obtained geometrically, may be liable to distortion

c lighting, especially sunlight, and consequent shading and shadows may be exaggerated or underplayed for good effect and only ideal conditions may be shown

d the detail in which buildings can be shown can be selected for effect

e background buildings, parts of buildings, paving, street furniture and landscaping can be included or excluded to create the desired effect

f the appearance of materials to be used can be enhanced or played down as required.

The argument for the objectivity of computer visualisation rests primarily on the first three qualities:

a the viewpoint is infinitely variable and can be selected by any party

b perspectives are calculated to such accuracy that it is possible to fit a computer image on top of a photograph of the same site taken to the same visual parameters

c the lighting is variable in colour, intensity and direction and can be adjusted to positions requested by any party. Shadows can be calculated exactly (and can be checked for accuracy by comparison with those on existing buildings).

The argument for objectivity with regard to these three qualities relates to the independence of the observer. Assuming there is confidence in the methods of calculation used in the software, any desired image can be requested by any party and all will be obtained on an equivalent basis. For the other qualities listed for the artists impression, however, the matter becomes more complicated. It could be argued that the increasing ability of information technology to produce photo-realistic images could provide the solution. The test could be: "if the computer image is indistinguishable from a photograph for the existing situation then the modelled proposals must be completely objective". There is the immediate objection that photographs incorporate outside judgement and therefore, a level of subjectivity. There are also problems with the more general "realism equals objectivity" argument. Questions of level of detail and inclusion and exclusion of items will continue to exist for two important reasons. Firstly, the economics of the visualisation process imply that it is not always expedient to model to the maximum level of realism. For example, if cast shadows are the only issue then why spend time and money modelling the colour and texture of the brickwork? Secondly, negotiations concerning development proposals frequently occur at a stage where all the details have not been incorporated in the design, let alone agreed. To come to arbitrary decisions on, for example, such matters as choice of colours, building materials and landscaping merely in order to achieve total realism could confuse the deliberations by raising points prematurely or even go so far as to create a misleading impression. (It is not being suggested here that there would be a deliberately subjective input, merely that confusion could result). In practice, therefore, users will need to make decisions about what items should be included and the level of detail to which they should be modelled. The decisions thus made should relate to the purpose of the exercise and will include an element of subjective judgement. The *objective* quality will lie, as referred to in the test results above, in the way the images are used and in the basis upon which the negotiations are conducted. If all the parties agree, at the outset, that they have confidence in the ability of the hardware and software to produce a usefully realistic representation of the existing situation, then the subsequent visualisation of the proposals will command an objective status on the grounds that it has been deduced from a common position. (There is an analogy here with the view of scientific method post-Popper where the value of a theory stems from its ability to provide predictions by deduction which can be tested by experiment). The argument for the objectivity of computer visualisation rests, therefore, in part on the independence and validity of the algorithms employed within the working of the programs and in part on the way it is applied in the modelling and decision making process.

The arguments on objectivity and realism above and the results of the tests point to the solution of the more general problem in computer-aided design, how to determine the level of realism to be modelled. The concept of total realism is open to criticism as it is practically unachievable. Realism is a relative concept. In theory, finer and finer detail of the components of a building could be modelled given sufficient computer power and operator time. Every door handle, keyhole and grain in the wood could be shown. It has, however, been suggested above that the balance between the realism and economics is determined by the agreement in negotiation by the parties involved. This was readily obtained in the Danbury example and could point the way to a means of resolution in all CAD situations. The agreement between designer and client could be obtained prior to a discussion of the merits of the design by an interactive trial and error negotiation around the computer screen.

POSSIBLE LONG TERM DEVELOPMENTS

It is now conceivable that in a few years time computer visualisation facilities will be available to development control officers at their desks and will facilitate pre- and post-application discussions with all parties engaged in simultaneous negotiation, as in the Danbury example described above. As also mentioned above, planning applications may eventually be submitted on disk using a template supplied by the planning office. There is no reason why the disk could contain not only CAD files but also a variety of other media, such as visualised still images, digitised video recordings of actual and simulated development and an accompanying sound commentary. Technological innovation should also reduce the labour involved in entering the data into the computer for CAD and visualisation. Raster to vector conversion facilities are already available that can turn a scanned image into a digitised form. Their marketing in cheap and convenient form should ultimately mean that data held in the form of two-dimensional plans and elevations and, possibly, isometric and perspective drawings, could be read directly into the computer. However, the most impressive long term innovation is likely to be the use of virtual reality. Masks, gloves and earphones may not be necessary for planning control purposes but the advantages of seating a planning committee inside a simulator for a "drive around" a computer generated model of their town will be clear. Proposed buildings could not only be viewed from all angles but could be altered interactively. Moreover, individual members would be able to "explore" the town independently. Two dimensional virtual reality enables the user to move around the simulated environment, in vision only, on the monitor screen and is now available "off the shelf" though at a price. All these innovations will bring both the public and the participants in the process of negotiation ever closer to experiencing the true impact of proposed development.

CONCLUSIONS

Limited as they are, the examples studied have pointed the way both to the resolution of certain issues in visualisation generally and to its practical use in the planning office for design control. This has been achieved by making a central feature of negotiation between the parties to the development process. The ability of information technology to visualise development such that the images can be amended interactively facilitates its direct use in such negotiation. By removing potential obstacles to communication it helps to bring about agreement between the parties in situations where it may previously have seemed far from easy. The agreements between the parties will determine both the levels of realism modelled and the objectivity with which it is regarded. Moreover, the use of visualisation to assist negotiation of agreements will greatly improve the practice of design control.

REFERENCES

1. Hall, Tony, Computer Visualisation in Planning Offices - Some Examples of Recent Progress, *BURISA* Vol.99, pp 13-15, July 1991.

2. Hall, Tony, Computer Visualisation for Design and Control, *The Planner,* Vol.74, No. 10, pp. 21-25, October 1988.

3. See: Davey, P, Wickham in Walthamstow, *The Architectural Review*, Vol.188, No. 1124, pp. 59-64
Welsh, Colour Coding, *Building Design*, No. 1012, p. 2, 16 November 1990
Von Sternburg, *Building Design*, No. 1006, p. 1, 5 October 1990.

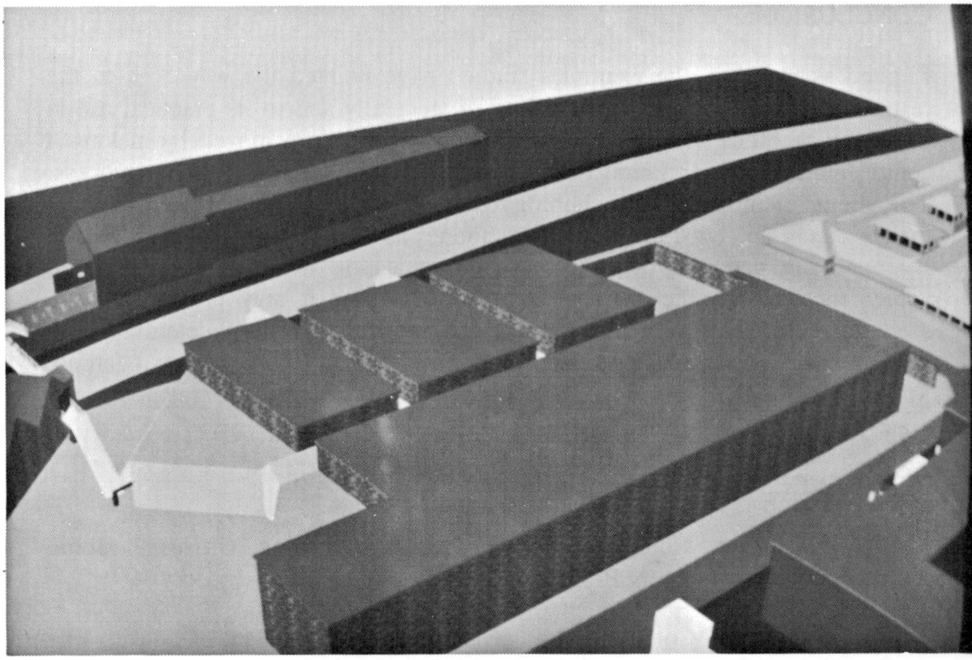

Figure 1 Guilford Example - first design - oblique aerial view

Figure 2 Guilford Example - second design - oblique aerial view

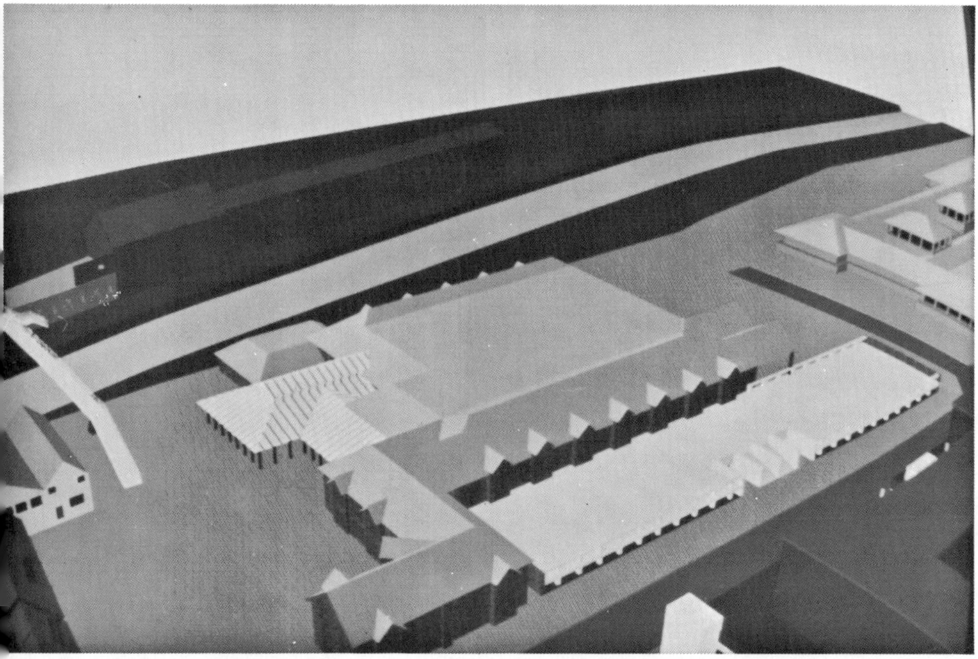

Figure 3 Guilford Example - third design - oblique aerial view

Figure 4 Guilford Example - fourth design - oblique aerial view

Figure 5 Danbury Example - before - oblique aerial view

Figure 6 Danbury Example - proposed extension - oblique aerial view

Figure 7 Danbury Example -planning officer's proposal- oblique aerial view

Figure 8 Danbury Example-extension similar to next-door's- oblique aerial view

Figure 9 Danbury Example - applicant's concept - oblique aerial view

Figure 10 Danbury Example - agreed proposal - oblique aerial view

Early attempts to combine the use of video and computer in architectural education

M. Linzey, A. Soutar
Department of Architecture, University of Auckland, New Zealand

SUMMARY

Computer and video belong to different codes of reference: while computer understanding derives from a mathematical, and in architectural education a CAD base, the time-based images collected on video belong to a filmic, literary foundation. Not only are their derivations very different, the people who choose to make special studies within each typically come from different educational routes.

The undergraduate course in architecture at Auckland University actively promotes use of both computer and video by students. However, student work which derives from both media together is more rare. The paper, supported by video clips, reports on undergraduate work performed between 1988 and 1992, illustrating early efforts to integrate computer and video, and comments on their relevance in an architectural education.

INTRODUCTION

The principal challenge to interactive multimedia, according to Kristina Hooper[1], is one of pedagogy and design. Pedagogy is a challenge because traditional teaching is already segregated into either computer or film, so multimedia tends to cut across traditional boundaries. And design because it is now up to designers themselves, and in particular students of design, to show the way to some extent, to use multimedia technology intelligently and demonstrate by its use just how to make the technology effectively contribute to creative work. Designers themselves have yet to demonstrate conclusively whether multimedia technology can generate a 'compelling experience' within the terms of reference and specific criteria of architectural design. Is interactive multimedia technology running the risk of developing in a user vacuum?

Knowing that interactive multimedia is on the horizon for educational institutions, the Auckland School has been at pains to encourage students to explore the interconnection between video and computers in a number of different ways. So far the response from students has been relatively slight. Three projects initiated and developed by students in the furtherance of specific architectural design projects are reported here in the form of short video clips. The videos were made without sophisticated post-production technology. The first tells a simple 'story' of a young woman imprisoned, her escape coming when she turns into a butterfly and flies through the cell bars—this metamorphosis was drawn in computer animation and simply edited into video. The second is an exploration of urban design at the level of street scape, with computer wireframes and negative video images playing over the geometry of architectural forms. Both these studies reveal something of the search for expression of an artist/architect no matter what the medium. The last item is a Hypercard^{TM} stack prepared for a video archive. Groups of students have been involved for four years on a community design project in the Cook Islands where successive cyclones had destroyed a formal palace of a ruling family. Each year a student team has flown to Rarotonga to work on rebuilding the palace. Some sixty hours of excellent video tapes have been accumulated, showing the re-building, also local customs, life on the island, talks by elders, etc. The Hypercard^{TM} medium is used to pull it all together, with a second-by-second shot list, and a cross reference to important key terms. It is a record in its own right of the historic project in words, animation and sound. Multimedia serves to provide continuity for student groups returning to the project from year to year.

CREATIVE PLURALISM IN THE EDUCATION OF ARCHITECTS

The Bachelor of Architecture degree course at Auckland University is one of only two courses in architecture presently available in New Zealand. For this reason the course has had to support a plurality of approaches and educational methods such as may be found only in the collective programmes of more than one school of architecture abroad.[2] Students are encouraged to explore and to integrate a wide range of techniques for design and communication rather than to develop only one method of achieving design. We are also very aware that changing vocational perceptions and opportunities in practice call for a wide range of choices and freedom to shape individual directions for study in the whole education of an architect.

A full range of graphical techniques from freehand drawing to computer draughting and video are taught in support of architectural design. It is recognised that different students have different natural aptitudes for techniques. It is also the case that different kinds of project are more or less suitable than others for different media. For example, when urban designers proposed a number of years ago to construct a monorail transportation system through the inner city area of Sydney in connection with the Darling Harbour redevelopment project, the medium of video was found to be a most appropriate vehicle to communicate to a wide audience a feeling for the anticipated impact of the scheme on the city environment. Many urban design projects of students in the School are presented to advantage using the broadcast qualities of video. Elsewhere in the architectural spectrum we have Peter Eisenman's radical design for the Guardiola House in the Bay of Cadiz, Spain, whose intricate geometry was intended to critically deconstruct the conventions of figure/ground, object/space through a controlled sequence of translations and rotations in three dimensions of a pure Platonic cube[3]. Students find the exploration of hidden relationships in this house is most edifying when it is explored in three dimensions with precision computer draughting techniques such as Microstation™.

The two technologies derive from different pedigrees — one from a mathematical and engineering world, the other from a filmic tradition. Video deals with a literal reality which is then interpreted by the medium. Computer is a contrived reality from beginning to end, the message is created entirely by the medium. In terms of the technology itself it is readily acknowledged that the two techniques are following convergent paths—towards interactive hypermedia and virtual reality—yet there may be deeper-seated psychological, educational or perhaps cultural reasons why this technical convergence may or may not be 'taken up' by designers. It may be the simple truth that one individual is better suited to expression in video than on a computer. Early indications of resistance may emerge in the relatively free choice yet critical design environment that the Auckland school offers in its design-based undergraduate course of study. Here the design studio may be imagined as a sort of 'free market' for creative visualisation and graphical communication techniques where the effectiveness of multimedia may be tested against a plurality of modes of presentation of graphical thought.

Video and computer classes in the Bachelor of Architecture course of study are limited by the commercial availability of affordable hardware and software. Computer work is supported on a network of mainly Apple Macintosh computers connected with a colour scanner, plotter, laser printers, and a colour printer. Classes offer training in the following computer graphics software products:

Claris CAD™
MacPerspective™
Microstation™
Photoshop™
Upfront™
Stratavision™

Other software products, including Autocad™, Hypercard™, Macromind Director™, are used for autonomous individual and group projects and for elective studies. All the available software is regularly and avidly explored by keener students. Practically all students in the Bachelor of Architecture course opt to study at least two semesters of computer classes in the first and second years of professional education. A fewer number of more senior students actively experiment with photorealistic techniques, walk-throughs, and so on in elective studies under the tutorship of post-graduate students and technical support staff.

Formal teaching in video technology is more severely limited by available resources. At present our workspace includes an S-VHS recording and editing suite, a simple sound mixer, and a Macintosh IIsi computer connected to the School network. A single semester course, *Audiovisual Techniques*, is offered in the third professional examination, but there is a class size limitation of 24 students on this class. As well as video, this course introduces skills and techniques in architectural photography, audio recording, and the critical analysis of these media. The examination is project based. As is the case with computer work, personal elective studies and additional projects are regularly undertaken by some students preparing and editing videos for and in conjunction with design studio presentations.

COMBINED USES OF VIDEO AND COMPUTER

Both computer and video techniques regularly find their way into presentations by students in the critical forum of the design studio, the heart of the architectural course of study. But there are fewer situations in which the combination of video and computer work has been successfully achieved in undergraduate design programmes.

The first example is a short video which was produced in 1990 by a team of three third year students, Belinda Clapperton, Anna Marie Chin, and Stuart Whitfield, for the *Audiovisual Techniques* course. It tells a simple 'story' of a prisoner whose 'escape' is achieved by turning

Fig. 1 A computer generated figure ...

into a butterfly and flying away. A young woman languishes in a cell. The angle of her body expresses depression and despondency. Her hand idly draws a figure in the dirt. Then the scene is suddenly transformed: an animated figure dances on a blank background. Slowly the figure cocoons itself in a black web (of depression?) from which, after a time a butterfly emerges and flies into the air. The butterfly departs through the prison bars. In a final shot we see the girl's clothes abandoned on the floor of the empty cell. This metamorphosis was drawn in Photoshop^{TM} and simply edited into the film video. Its justification in terms of visual design, beyond the mere technical exercise, is that the film-to-computer transformation is a metaphor for imaginative transformation in general. The video represents a vision of humanity trapped by architecture—a trauma with which many will readily identify—then finding release through the imagination. The project should be appraised in design terms—not in terms of glossy production values. As an adjunct to an architectural design, the design of an imprisoning place, it represents a thoughtful and creative use of available materials and technology. The switch from film to computer and back to film is of a part with the switch from monochrome to colour—a device to enliven the imaginative faculty of the viewer. These students explored the unique characteristics of the computer and video media and found ways within both to best express the simple idea of architectural confinement and release with an economy of means.

Fig. 2 ... metamorphoses into a butterfly.

The second example of mixed video and computer was produced by a third year student, Simon Fernyhough, to represent an aspect of his own design for an urban infill redevelopment. A number of simple wireframe and solid objects were manipulated on a scanned background image of an existing street scape. The dynamic sequence of computer images was captured on video output. The student used his own Omega home computer, on which he also composed his own soundtrack for the video clip. The dynamic and transformational possibilities of video were used in this case to comment metaphorically upon similar qualities (dynamism and transformation from old to new) in the architecture of the proposed facade. At the same time the lifting and transposition of pediment elements from the old street frontage, and the metamorphosis from elsewhere in the Pacific rim of a Japanese gate motif into the modernist grid of uptown Auckland were intended to communicate the dynamic merging of old and new, tradition and change—a tentative but thoughtful response to post-modern desires.

Fig. 3 The parapet above the old shops is transposed into the air.

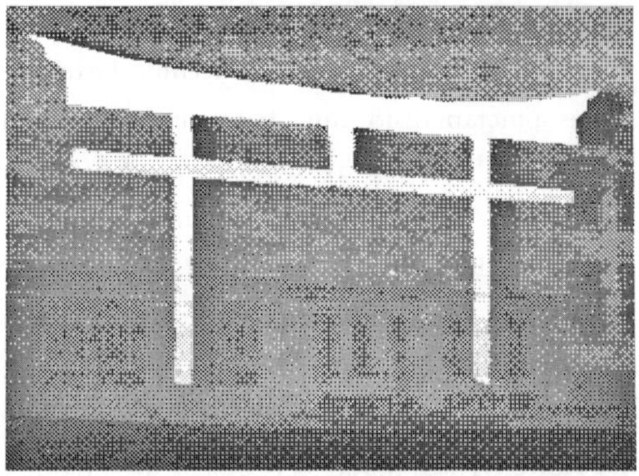

Fig. 4. A Japanese motif is introduced into the design.

Again the fact that the technology coincided with the task seems to be more important in this case in terms of successful architectural design than merely to demonstrate the student's versatility with technology for its own sake. It may be argued that these early assays into combined video and computer are irrelevant because their means of production is already out-of-date and superceded. But it is not the mode of production nor even the rapid strides that multimedia manufacturers have made in recent years that is significant here. When we are studying the learning process and the educational path of designers, sometimes the cruder and simpler technology, by placing a greater demand for problem solving on students, makes the problem solving that is going

Fig. 5. A computer rendered design is superimposed on the existing city scape.

on appear more transparent. Educators are more concerned with the means, and less with the ends of commercial design.

The third example of multimedia work to be discussed here is a Hypercard™ stack recording a longer term project undertaken by students to restore a ruined nineteenth century palace, Te Para-o-Tane, which stands on the main island, Rarotonga, of the Cook Islands group. This community based construction project has been an invaluable experience for students, providing not only practical construction and renovation experience with an antique method of coral-lime wall construction but contact with another Pacific culture closely related to New Zealand Maori, and insight into the values placed on architecture by another culture. Each year the participating students recorded their experiences in Rarotonga on video and with drawings, sketches and photographs. So far sixty five hours of video material has been recorded, much of which is excellent in technical quality and content. Although many students recorded the video material, measured the drawings, provided still photographs and sound tapes, in fact it was one of the authors [Anna Soutar] who is compiling and designing the multimedia stack as part of project work for her Master of Philosophy degree.

When authoring in multimedia one is struck at first by its non-linear, unstructured, apparently non-narrative quality. How does one design a resource not knowing how it will be used, in what sequence a user will skip from place to place within the web of information? What will be the 'learning experience' or the 'entertainment value' of the work, leaving aside its contribution to 'art'? But of course there is never any such thing as 'no narrative

Fig. 6. Te Para-o-Tane, an old palace at Rarotonga, is undergoing a programme of restoration.

trail.' We are indoctrinated by certain habits of discourse—by the linked flow and hidden hierarchical structuring of written text, or by the staged flow of film from one carefully framed shot to the next, the viewer carried along on a bed of music. All discourse, verbal and graphical, is a series of depictions and a sub-text of interconnections which subvert or reinforce the text with their own assertions, innuendo, collage. Integrated multimedia is no different in this regard except that the user retains control over the timing and ordering of the connections through the medium of the personal computer.

In designing the layout and transitions for the Rarotonga archive we also wanted to capture something of the soft yet complex tropical rhythms, the warm, unaffected grace of the island, the passionate intensity of wooden drums, the grace of the dance, and the symbolism of the hibiscus flower. And at the same time we had cause to wonder at the appropriateness of this novel electronic medium of infinite variety and interconnection to represent an ancient culture whose most respected attribute has been its ability to navigate by a traditional form of integrated multimedia—reading distance and direction in the rising and setting of the stars, the shapes of the waves, the seasons of the winds, in ocean currents, and by certain other strange appearances which will ever remain beyond the sensible comprehension of the West.

References:

1 Kristina Hooper, 'Interactive multimedia design 1988,' The multimedia Laboratory, Apple Computer, Inc., Technical Report #13, November 1988.
2 Anon., B. Arch 1993 Handbook, Department of Architecture, University of Auckland,1993.
3 P. Eisenman. Guardiola House, Santa Maria del Mar. in A. Papadakis, C. Cooke and A. Benjamin, eds., *Deconstruction Omnibus*, London, Academy Editions, 1989. p. 163.

SECTION 7: ENGINEERING APPLICATIONS

Satellite architecture – design of structures and mechanisms (visualization in praxis)

M. Hager

Austrian Aerospace Company, Vienna, Austria

Abstract

The new design methodology of recent years caused by the introduction of powerful computers i.e.:

* to assist and supplement the visualization process at the early stage of design
* to use solid model techniques to optimize the design configuration
* to establish kinematical movement models of the mechanisms
* to provide design interfaces for an FE model; is now an unavoidable tool in spacecraft engineering

It is normal systems procedure to divide a spacecraft system into two principal elements:

* the payload, and
* the bus

During the iterative process of design of the satellite bus structure by considering launch vehicle constraints and payload configuration drivers the initial design "grows" to the final design configuration synthesis phases. Computational geometry and 3D imaging provides an excellent opportunity for optimizing, modifying, adding and analyzing the structure using FE analysis.

The main task of this paper is to show through the representative examples, in terms of both structural and mechanical hardware, the satellite architecture as a new discipline and to highlight the use of visualization and intelligent design by aerospace applications.

Introduction

If one proceeds from the basic definition of architecture, design and engineering, the place the satellite occupies in any of these fields at the moment would be very difficult to define precisely. The satellite is certainly more of a place and a space where experiments and observations take place -- while it is changing position and taking up required trajectories -- than it is a means of transport such as a plane or a carrier rocket that has brought its payload -- satellite -- into a desired position. The development of technology brought us from the first Sputniks and Explorers to the present satellites, modern works of art in technology and the sojourn of the man aboard a space body (of a satellite type) gives a new quality to everything. This flying body becomes a **home** -- for the time being only for astronauts, but it is not hard to predict that in a short time, entire colonies, settlements will be evolving around the Earth. Perhaps only then will architecture become one of disciplines to be used when designing these objects or, to put it better, these objects will completely fit into architecture. Therefore, I think that today we can speak about a new discipline within architecture -- still in the cradle -- **architecture in outer space.**

The purpose of this article is to present its readers -- through examples from the practice -- one of the aspects of the application of modern designing methods: the visual representation of an object that is being created. As the design of the examples presented here begun in my firm two years ago and as we are still using the same system and program and present opportunities are greater, computers are more powerful and possess a larger capacity, and a virtual army of people deal with perfecting new programs, one must understand that some elements may already appear "old-fashioned" and outdated.

N. B. No programs will deliberately be mentioned here by their names, only by their capabilities.

Spacecraft Subsystems

The spacecraft system itself may be conveniently divided into two principal elements: the payload and the bus (support structure). The bus provides the payload with all the necessary resources for its functioning, it gives life to the payload and the mission target and provides certain shelter for the equipment. The breakdown into subsystems is shown on Figure 1.

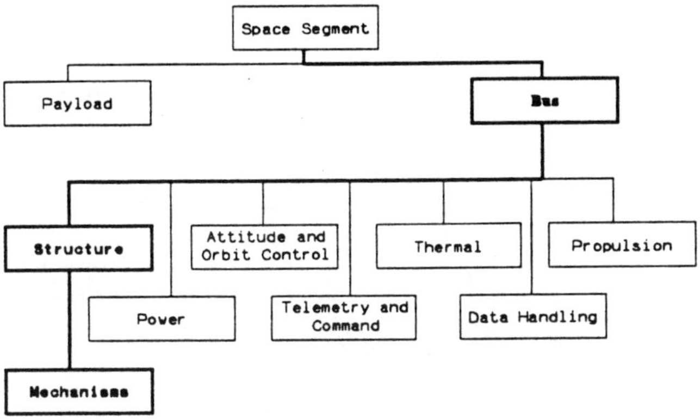

Figure 1: Spacecraft Subsystems

The subsystems interact strongly with each other and the design of any one of them affects and has resource implications on the other. The most significant characteristic of spacecraft system design is to identify the mission aspects and design elements that exert major influence on the type of satellite that may meet the specific mission requirements. The process is the identification of the "design drivers." In some cases, the drivers will affect major features of the spacecraft hardware. The varied mission requirements, combined with the need to minimize mass and therefore power, has thus led to a wide variety of individual design solutions.

The first step in structural design is to establish specification based on the mission requirements with as much detail as can be expected at the concept stage. A generalized list of requirements for structures is given as follows:

* Accommodation for the payload and spacecraft systems
* Withstand launch loads
* Stiffness
* Mounting/Interfaces
* Alignment
* Thermal and electrical paths
* Accessibility

Example 1 Design of the Gamma Ray Spacecraft

Figure 2 shows what the satellite consists of.

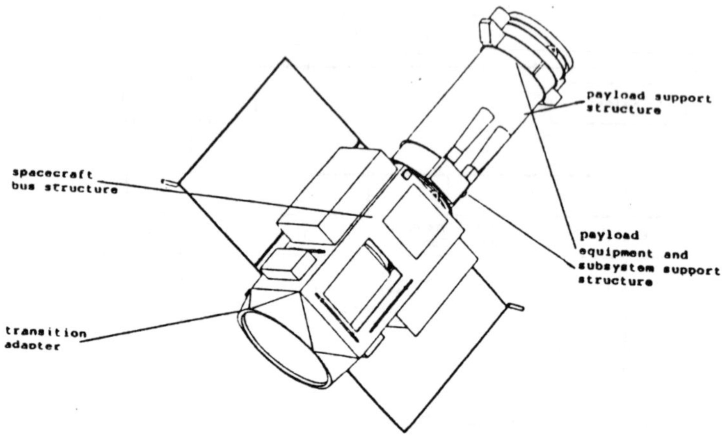

Figure 2: Gamma Ray Astronomy Satellite
Structure Definition & Description

Gamma Ray Astronomy with Spectroscopy and Positioning

Scientific objectives are as follows:

- imaging and accurate positioning (1 arc min) of celestial gamma ray sources

- fine spectroscopy of cosmic sources with high resolution

- study of active galaxies by

  - luminosity measurements
  - spectroscopy of line and continuum spectra

- investigation of point and extended sources within our galaxy

- observation of any transient events like gamma ray bursts or super novae

Payload Definition

The payload for the gamma ray astronomy with spectroscopy and positioning comprises:

- detector assembly
- coded aperture mask
- optical transient camera
- x-ray monitor
- data and power electronics

Brief description of step by step design:

Taking into consideration the basic requirements and defining design drivers, using the already existing types of the Ariane fairing, and permitted envelopes for double launch upper position (see Figure 3), sketches have become shapes of the outer permitted satellite contours. Of course, the entire geometry was done in cooperation with the firm that was responsible for designing the payload -- in the case of the gamma ray astronomy it is a telescope of very large dimensions that is isostastically fixed to the bus carrier. In order to economize when constructing the bus, the library of solids/buses was used to see which one of them meets the requirements best. The selected bus-structure (see Figure 4) was taken and the research directed towards coordinating the needs of inner space for mounting the other subsystems equipment. The finally created solid model started receiving inputs from various sides, from different firms which then -- based on the layout -- built and added their own elements. The thermal subsystem represented the greatest problem, but defining and calculating radiators was made easier by the possibility to manipulate with solid models in space and to simulate the positions of the satellite in relation to thermal sources.

The result of the entire analysis in this first stage of defining the spacecraft were 4 solutions that were proposed for further analyses. All these configurations could be very efficiently simulated with the help of solid models by very quickly changing the position of elements, mutual relations and subsystems. The fact that the mass budget and the center of gravity analysis were available to the designer engineer represented a particular advantage at any moment, so that entire trajectories of the C of G were made until the stage when the design was frozen and when it was established on the basis of trade-offs that the chosen configuration should be developed further.

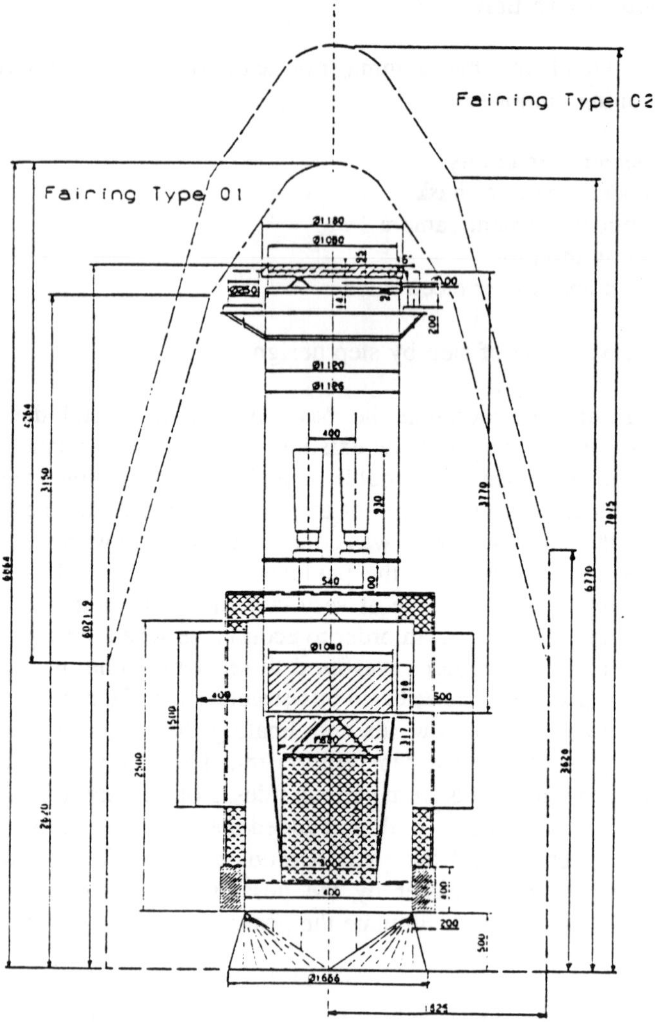

Figure 3: Satellite Structure in its Stowed Configuration within the Ariane Fairings

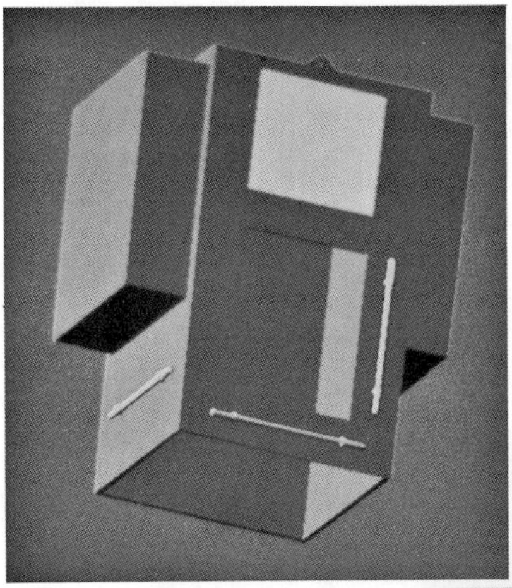

Figure 4:
Bus Structure; North and South Panel Compartments with Equipment and Instruments, Heat Shield Radiator, Magnetic Coils

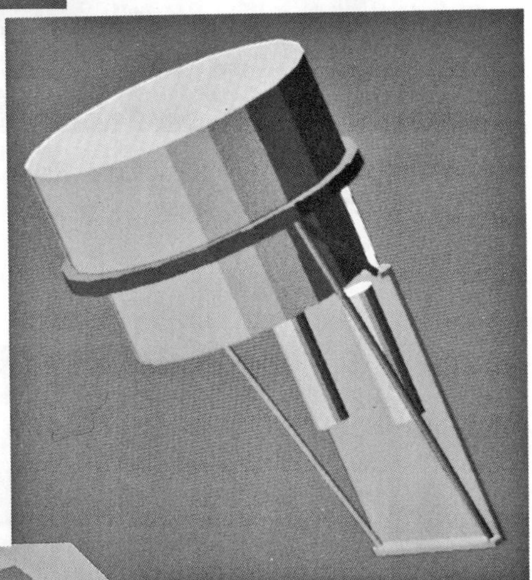

Figure 5:
Stirling Cooler Assembly Concept

Figure 6:
Transition Adapter -- provides the Geometrical Adaptation from the Square Spacecraft Bus Structure to the Circular Ariane Interface

Figure 7: Star Trackers, Gyro Package, Payload Telescope Tube Connection to the Bus

Figure 8: Mask of the Payload Isostatically Mounted to the Telescope Tube

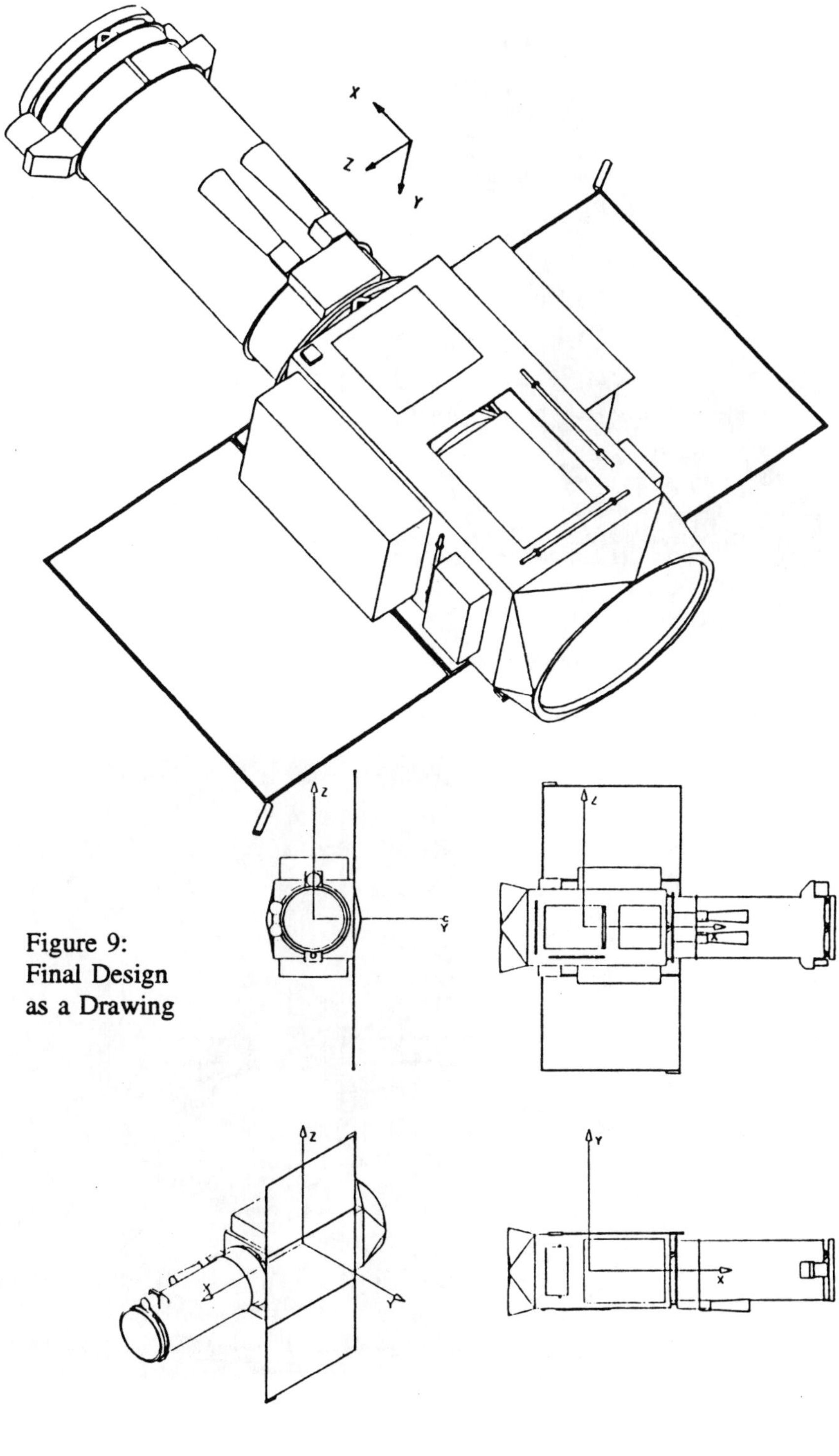

Figure 9:
Final Design
as a Drawing

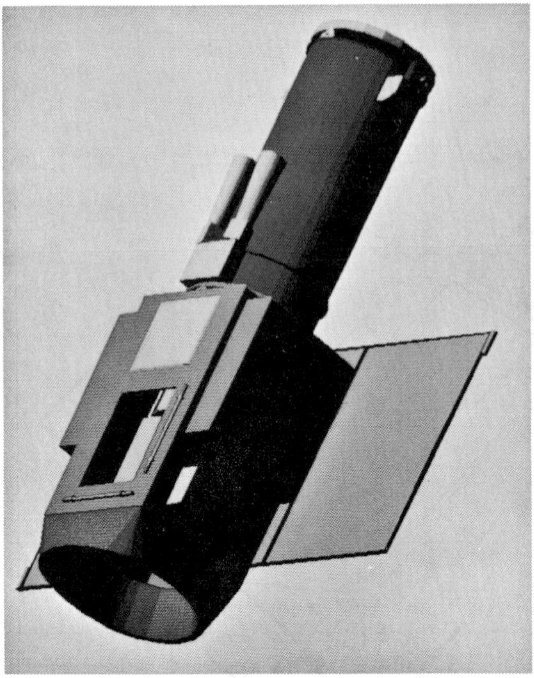

Figure 10: Final Design as a Solid Model

In this way, the first design, ready for further analyses and further optimization has been made. The finite element analysis has been made from solid models by directly applying already existing geometry, and dynamic analysis has shown how the mass and stiffness conditions are to be coordinated (see Figure 11).

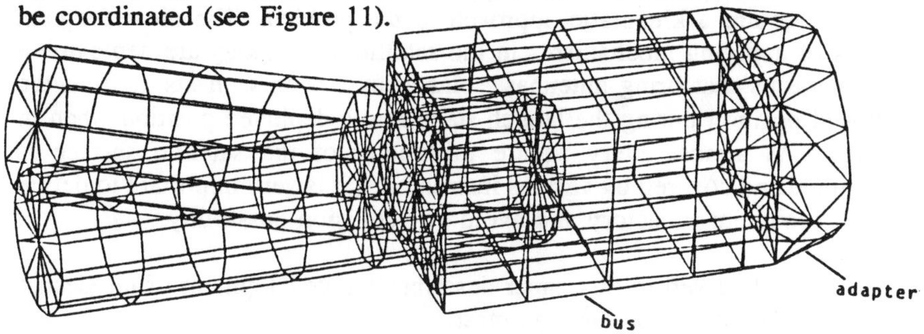

Figure 11: First Lateral Eigenfrequency = 15.9 Hz

Mechanisms meant for the deployment of the massive solar arrays have been taken in consideration and their basic kinematics has been performed by special programs that simulated boundary conditions and established movable joint-hinges. In such a way, the feasibility of design has been confirmed.

All these processes are iterative and inconceivable without computers in the cases of changes in geometry and positions of subsystems. The visual presentation in space -- using solid models technique and being able to see into the construction by "taking off" elements, even seeing and measuring movements of individual systems -- leads to an optimum in design and enables better cooperation within a project team and better organization of the design process.

The following figures show results of the above-mentioned process of designing the satellite structure.

Example 2 Design of Antenna Mechanism

Unlike the structure of the entire spacecraft, mechanisms -- in their nature and key assumptions of their application -- along with the solid model design require a kinematic approach to the problem. The most frequent cases of applying mechanisms in the satellite technology are deployment mechanisms that bring antennas, solar arrays, booms, masts ... from the stowed position during the launch stage into a deployed position when in orbit. As almost every square millimeter of the satellite surface is occupied and covered by various instruments, it is very important to define the deployment trajectory and avoid any collision possibility.

The example given in this paper describes the development of the design of the unfurlable mash antenna mast.

From its stowed position, in which the entire mash antenna is wrapped into a package of 790 mm diameter and 1,900 mm height, the antenna rotates to its final position and develops into more than 5-meter diameter. The entire process of different rotations has been analyzed for several different types of spacecraft buses and configurations and with different sequences of the antenna opening considering time and space (see Figures 12 ÷ 17).

The result of the final kinematic analysis was the obliquely positioned axis that made the movements from the start stowed position to the end position feasible.

The program that has solved kinematrical problems also gave inputs for the dynamic analysis of the entire movement.

Figure 12: Side Fixing Deployment of Mast Kinematics
Characteristic Angles
176.60437°, -245.4584°, 76.6831° (for Reflector)

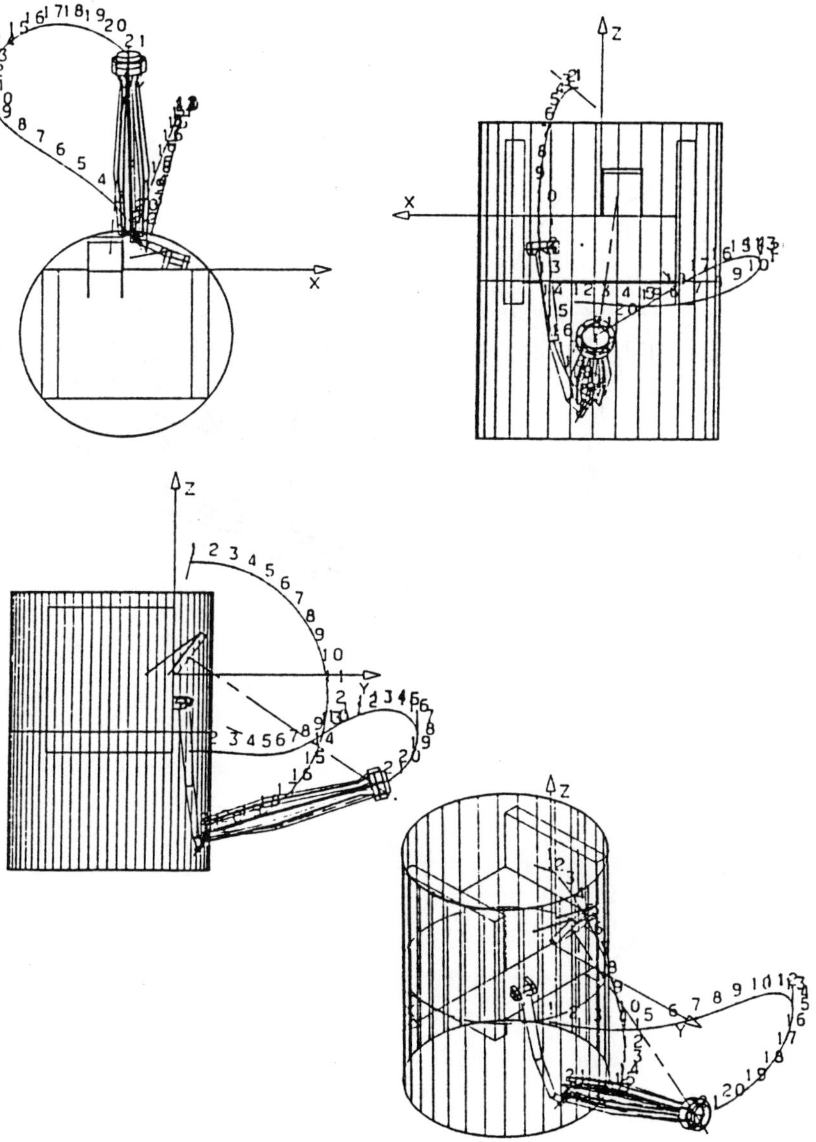

Figure 13: Satellite Bus, Mast and Deployed Antenna

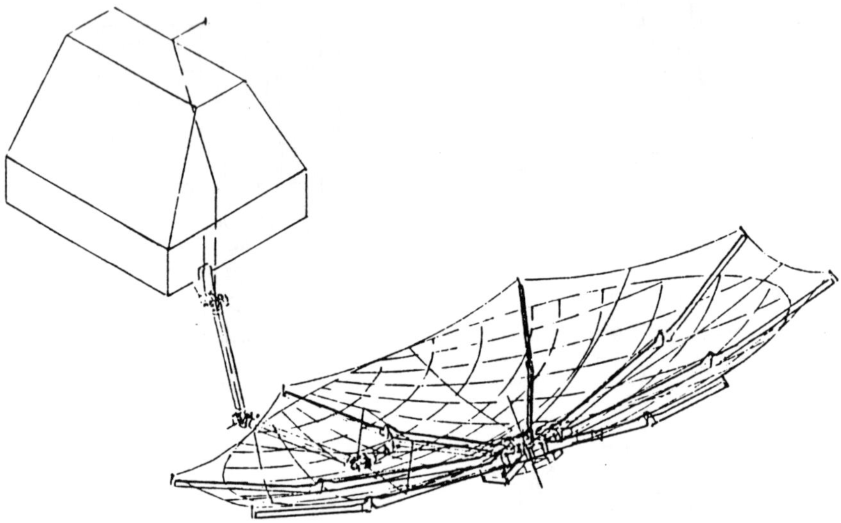

Figure 14: Sequences of Antenna Deployment

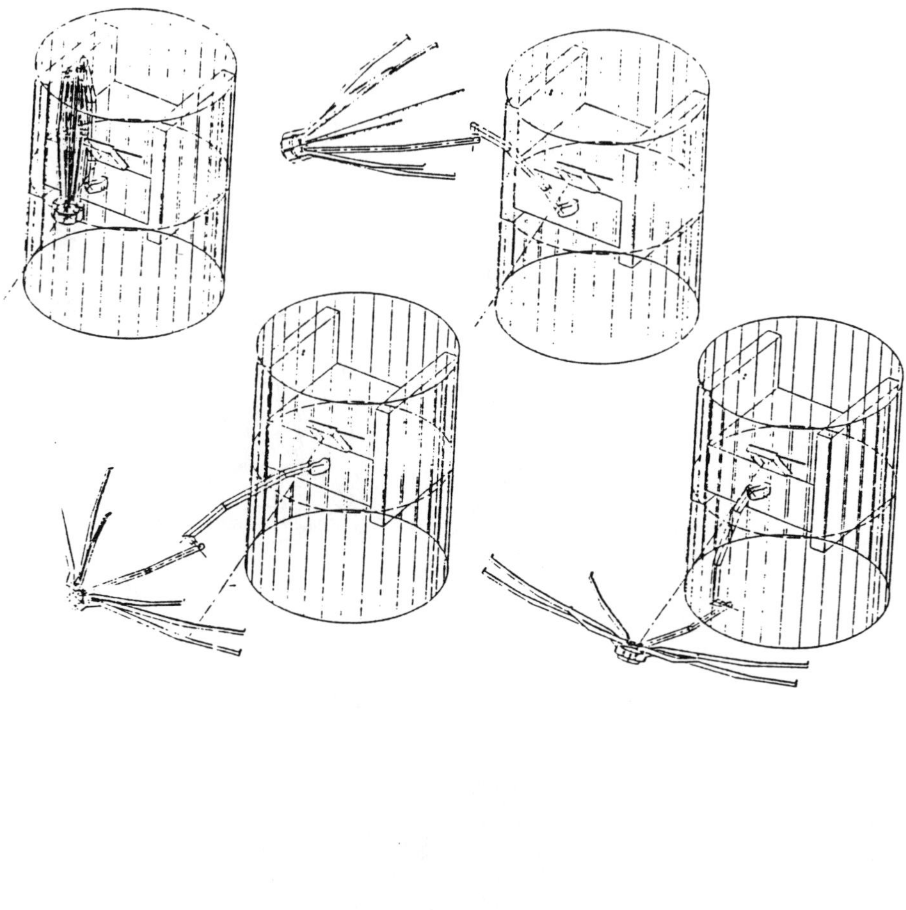

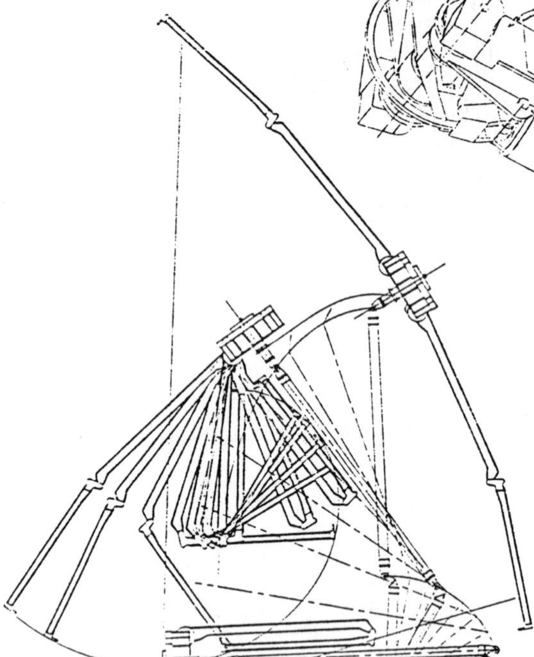

Figure 15: Sequences of Kinematics

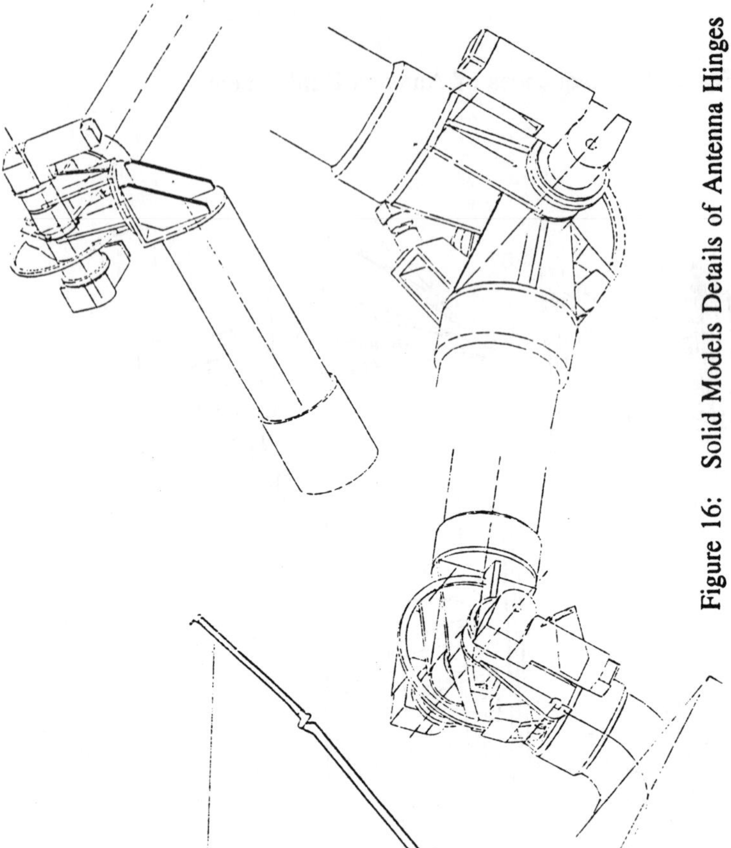

Figure 16: Solid Models Details of Antenna Hinges

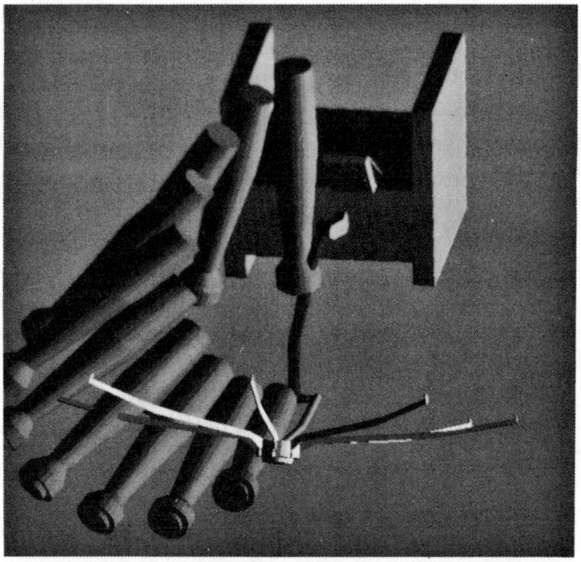

Figure 17: Antenna Deployment Visualization - Solid Model

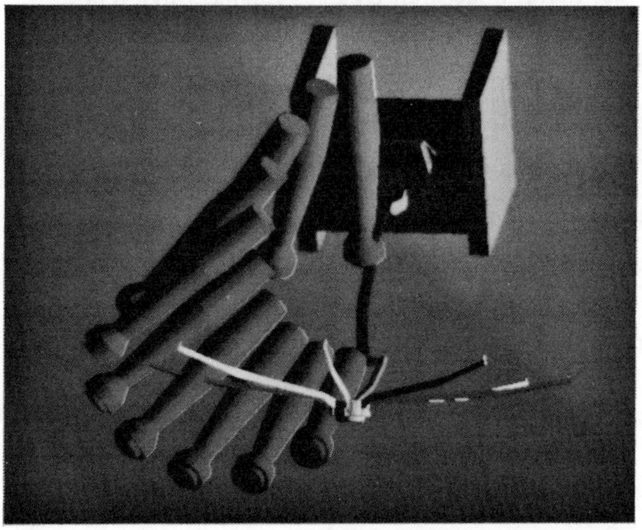

Conclusion

Both examples from the practice of designing satellites show that visualization is irreplaceable during the entire process of design. The processes of presenting the object of design and its development through 3D images contributes to better understanding of the problems, both in the architecture of satellites and engineering and designing of mechanisms.

Constant changes in design, the opportunity of directly analyzing structures on the basis of the 3D model and the kinematical presumptions of mechanism movements will continue to represent a challenge to new generations of engineers and scientists in the future as well. Both software and hardware programs are going to improve further and become more powerful and faster, albeit easier to use and uniformed so that they could be compatible throughout the world.

List of References

1. Fortescue, Stark: Spacecraft Systems Engineering, Wiley

2. Agrwal: Design of Geosynchronous Spacecraft, Prentice-Hall, N.Y.

3. Hallmann/Ley: Handbuch der Raumfahrttechnik, Hauser Verlag Vienna

4. Renner, Mauck, Balteas: Satellitentechnik, Springer Verlag

5. ESA/Dornier/MBB/ORS Projects: GRASP and UMA

6. Spacecraft Structures/ESA SP-321, 238, 289

7. W. v. Braun: Station im Weltraum (Across the Space Frontier), Fischer Verlag, Frankfurt

8. The Cambridge Encyclopedia of Space, Cambridge

A coupling approach between visualization tools and building energy analysis

J.M.S. Dionísio

ZGDV - Zentrum für Graphische Datenverarbeitung, Wilhelminenstr. 7, D-6100 Darmstadt, Germany

ABSTRACT

In the present paper an interactive simulation environment for integrated energy analysis in buildings is described. Supported by a database structure, main goals of STEPHANIE are the integration of calculation models and CAD systems, particularly related with thermal comfort analysis. Basic concepts and features that allow the calculation model to be built up from geometrical data with minimum additional user input are outlined. Special highlights are modularity and integration of the above mentioned environments using a common interactive Motif-based interface.

INTRODUCTION

The increasing concern about the quality of the indoor environmental conditions, the dual thermal comfort versus capital minimisation and future running costs lead researchers to develop a wide set of calculation tools. Traditionally numerical oriented, the input for these calculation programs was intensively created in alphanumeric form, manually typed, and only in a recent past graphical input interfaces start to be used by combining the applications with output from a CAD system.

Although integrated computer-aided building design (CABD) has been a subject of research for nearly twenty years [1], there is no argue if one says that existing simulation tools related with energy in buildings are not flexible enough. Therefore designers claim for flexible and interactive tools for designing and optimising. As mentioned, the basic theory was already quite well understood in the late 1970s [2] but due to limitations in computing technology, early prototype systems were rather cumbersome to use and imposed restrictions on possible design choices. Specially because of data management only recent advances in knowledge engineering and object oriented programming techniques, as well as increasingly powerful hardware, are providing the technology for developing full-scale integrated building design systems.

Our contribution deals with a CAD-based application [Fig. 1] for integrated architecture design and on-line calculations of energy in buildings willing to integrate and explore some graphics and database capabilities as well. The developed Motif-based interface allows quick and easy access to all relevant geometrical and non-geometrical data while maintaining on-line access and feed-back from all geometric elements in the design file within the CAD environment.

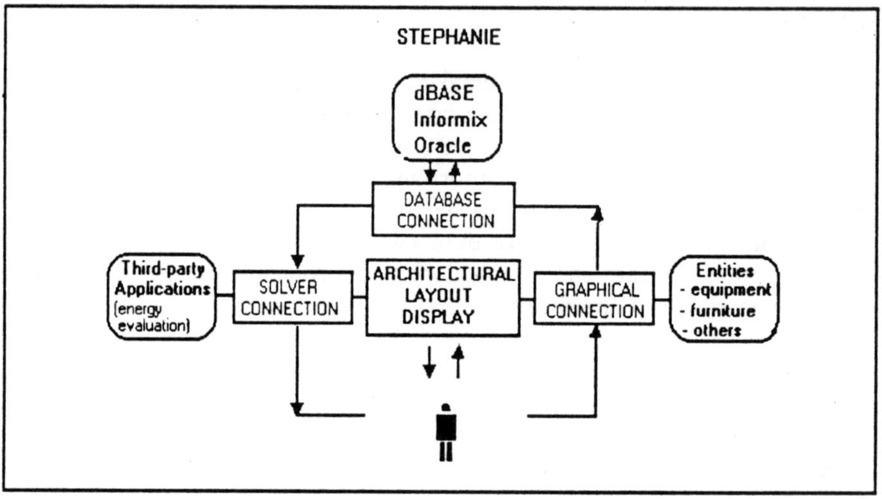

Fig. 1 General concepts of STEPHANIE

MAIN CONCEPTS AND SYSTEM FEATURES

At the present time a number of modelling systems is emerging which can address the range of cost and performance issues of importance in building design: from realistic visualizations, to a detailed appraisal of the indoor environmental conditions [3].

In this section we give a synopsis of our implementation that has been carried out so far. The approach that has been used as the basis for our analysis could be summarized in the three traditional levels, as follows:

1 - Conceptual design

2 - Preliminary design

3 - Detail design

In the conceptual design stage the building is treated as a whole, an entity. Topics related with the building and its environmental impact are discussed. Main lines of the architectural draft are also defined.

Follows the preliminary design stage and the definition of the building envelope, materials, surfaces and design variables such as design temperatures, occupancy, lighting and equipment types and schedules, etc. In each floor the space layout is known and it is time to the first test: the qualitative evaluation. Comparing with regulations imposed or other constrains that limit, e.g., *the thermal load/m2* in each surface and space (thermal zone), these values are evaluated and the surfaces coloured. This method alerts architects and engineers, in the early stages of design, about possible incorrect solutions without intensive calculations or detail [Fig. 2].

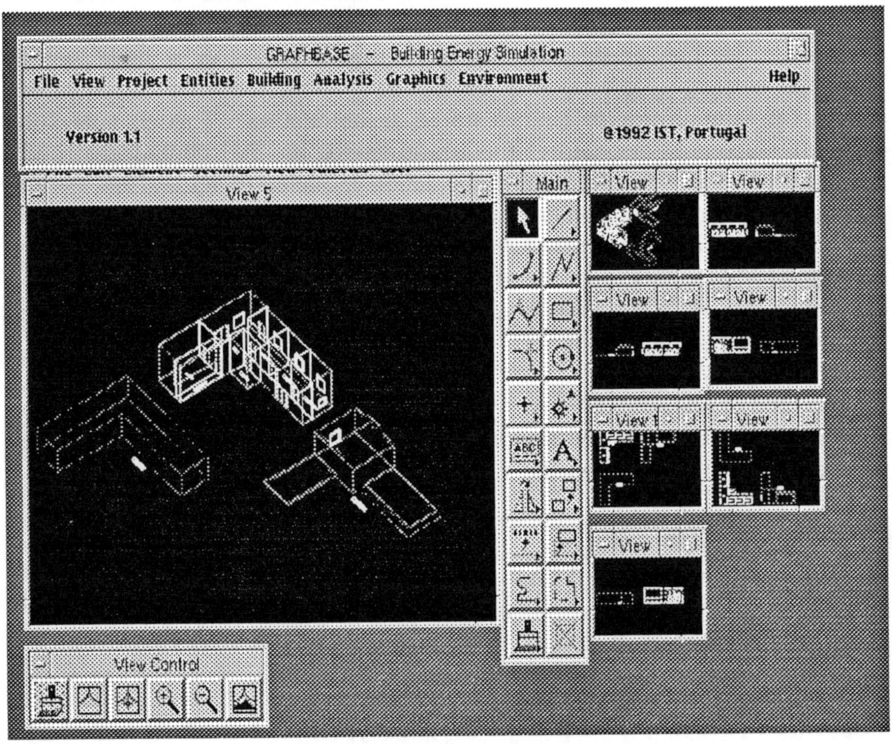

Fig.2 General overview

Once the project is checked and a possible compromise solution was found it is time for optimisation. That is to say the third and last stage, detail analysis. The quantitative analysis takes place and accurate calculations are performed. This application report hourly, daily, or monthly the building thermal loads and generate the information required to size and select the necessary equipment to maintain desired conditions within the building.

At the heart of the system lies a database structure that defines the model as a collection of class instances and their initialisation data [4]. This database structure is based on the active entity concept which we briefly describe as following:

Database core

As widely recognised DBMS support for CAD environments is faced with particular requirements that are usually not present in a traditional application.

The role of the database in the overall framework is to store predefined components in a catalogue format and to serve as a central repository of information that may be shared by calculation modules. As an example, several individual elements (objects) may be loaded within the architecture layout sharing the same prototype entities stored in the database. This eliminates the need to duplicate and store this information in the component itself. Furthermore, it is possible for the calculation modules to access this general information when required. For example, an architectural entity such as a wall can be queried about its perimeter, area, etc. while the database would be consulted for obtaining, e.g., thermal properties of each layer like specific heat or conductivity. Also stored in the database are relevant codes that specify each prototype component.

This kind of knowledge-based system incorporate recommendations derived from technical literature, and standard requirements and may be easily updated to incorporate specific knowledge of professional users.

Active entity concept

A design object such as a SPACE that has to be modelled in a CAD database typically is a complex, i.e., a structured one; each individual object (e.g., wall, door, window) consists of a hierarchy of other objects [5].

A wide set of prototype entities is already installed in the core. By means of instation, i.e., copying of versions of the prototypes, the overall set of elements (complex objects) may be linked with the database. Material properties, power supply of different equipment, etc., are being extended to support future analysis and reports. By reducing the complexity of symbols in the abstraction stage to a minimum, several libraries of graphical symbols are available as well as a large set of default values.

As mentioned in [6], CAD systems on the market offer very few possibilities for the development of appropriate dialogues and knowledge structures to represent the situations at each phase of product development properly. The task of a designer in a CAD system is to devise, on availability of some objects and operations on objects, new objects which fulfil desired functions. On the other hand, graphic interactive interfaces represent not only a suitable tool for the designer or draftsman, but also for all system users, as the development of graphics dialogues evolves, presenting them with easier and more comprehensible symbols for operation.

INTEGRATION: FROM CONCEPT TO PRODUCT

Follows a brief description of STEPHANIE graphical user interface and a few considerations about different data flows between the components. The user interface management system realises the dialogue between user and application modules [Fig. 3].

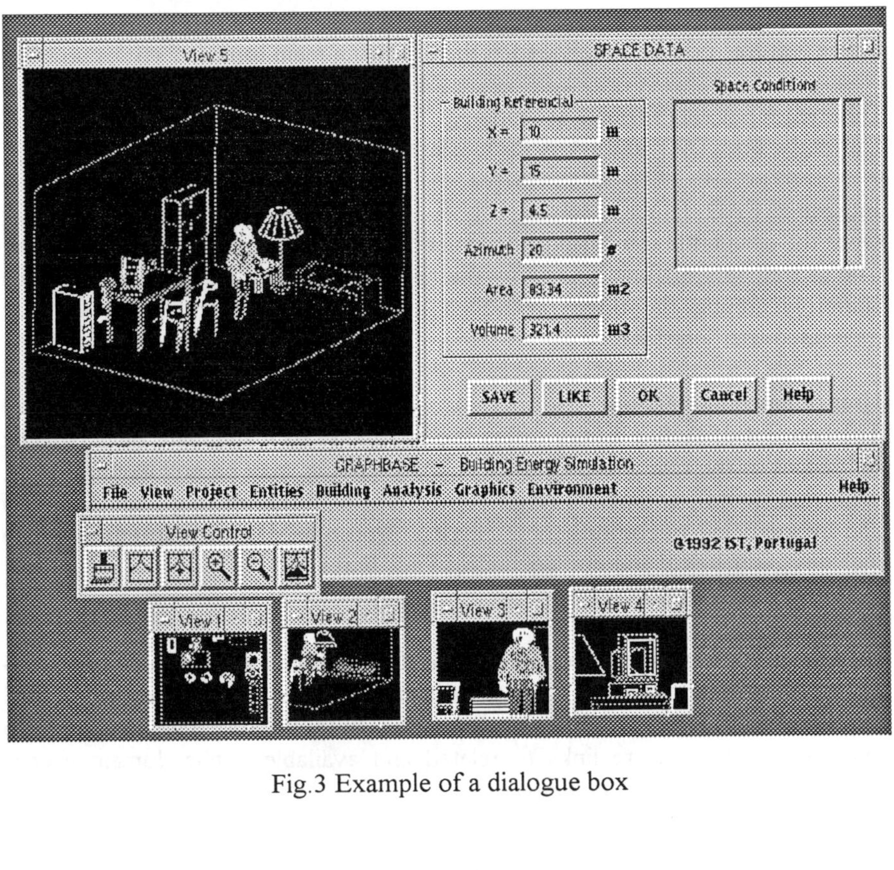

Fig.3 Example of a dialogue box

The major failure of CAD, especially in architecture, has been the lack of tools and procedures supporting the necessary abstraction level. One of the main problems is related with the way the user interacts with the machine although user interfaces have improved significantly over the last ten years. Multiple windows and pop-up menus are available to the draft system. However our concerns are still concentrated on data exchange. This data is being integrated, once entered, then usable in all drawings, specification, analysis and so forth.

The user navigates through the tree by selecting an item that offers additional menus and dialogue boxes where one can make a selection from a predefined set of alternatives at each specific dialogue set. Features and highlights include:

- dialogue boxes to access database tools
- interactive query about database attached architectural elements
- navigation through the building structure
- possibility of overall (building) or partial calculations (spaces or thermal zones)
- user direct manipulation of the in/outdoor design parameters

These features may be subdivided in the following main topics:

<u>Different input data formats acceptance</u>
Currently, both DXF and DGN formats are accepted

<u>Integrated CAD-database-solver analysis</u>
Once the design file is available all elements may be attached to the DBMS core. The overall sets of functions allow the needed user abstraction level to switch smoothly among the three mentioned main modules. Moreover it uses an integrated user interface.

<u>User oriented interaction</u>
The concept of Active Entity offers the possibility to manipulate objects using references of the real world. Therefore these entities may be architectural (e.g., walls and doors), equipment (e.g., computers) or furniture (e.g., sofas, tables) and by attaching operations they inherit a prototype database structure which includes geometrical and non geometrical attributes.

<u>Interactive and batch mode</u>
It includes a set of simple routines to perform quick energy calculations - *on-line mode* - and help designers to achieve easier and faster the initial requirements. However, different formatted output data files may be also generated aiming future links to related and available public domain energy calculations software - *batch mode*.

Input and output data visualization
Predefined time periods for calculations (e.g., hourly, daily, weekly) may generate large amounts of data. Occupancy, lighting and equipment schedules, outdoor hygrometric conditions may also be interesting to visualise graphically. Therefore its our purpose to take into account both sets of data and present the necessary display tools.

Several available libraries
Structured data sets like types of materials, surfaces, equipment not to mention geographical, and climatological data (e.g., locals, latitude, longitude, main design temperatures) can be loaded from available databases. Graphical entities are also available and can be loaded within the mentioned interface.

Portability and standards
The system is being developed in a C-shell under MDL. It does not depend on the system used although it uses different database software (dBASE, Oracle e Informix) depending on the operating system installed. Possible platforms are IBM PC or compatible, Macintosh and Unix Intergraph machines. In a near future SUN and HP alternatives are also expected.

CONCLUSIONS AND FUTURE WORK

We described an application which estimates the building energy consumption using databases obtained by simulation with third-party calculation programs.

At the current time, the mentioned prototype is operational within DOS and Unix operating systems and several refinements are planned or underway. Several additional user conceptualisations are being researched, with the focus on the more design oriented, and less technically motivated, users such as architects.

For the present it is assumed that integration allows higher proficiency and better final results. Graphical interface for input data and output visualization besides automation allows better decision-making. However, this may be called "blind" interactive simulation. Once the general concept of the use of knowledge-based systems has produced many expectations among professionals related to the construction industry, efforts are now devoted to the development of other simulation facilities as inference methods adding intelligence to design performance.

Conceptual frameworks for the use of graphic representation can be of major benefit in the architectural design process. By using objects, rather than simple graphical elements, besides high quality offers an effective abstraction level. Otherwise, an excessive amount of drawing would be necessary, possibly resulting in an overlay complex vocabulary or set of symbols thus detracting from the level of abstraction intended. In the design process it is often necessary in an early stage to consciously exclude certain complex representations to avoid being locked into premature decisions.

ACKNOWLEDGEMENTS

The author would like to thank Prof. J. L. Encarnação (GRIS/THD) for support and encouragement given, and JNICT for sponsoring this work. The author is also grateful to Prof. J. J. Delgado Domingos for his valuable contribution in the early stages of this work which was performed at IST, Lisbon, Portugal.

REFERENCES

1. Vanier, D. et. al: "Integrated Computer-aided Building Design Systems", ARECDAO 87, *Conf. Proc. ITEC*, Barcelona, Spain, pp. 34-37, 1987

2. Bijl, A. et al: "Integrated Systems EdCAAD Studies", *University of Edinburgh*, UK, 1979

3. Clarke, J. A., *Energy Simulation in Buildings*, Adam Hilger Ltd, Bristol and Boston, 1985

4. Dionísio, J. M. S., "Interactive Thermal Behaviour Simulation of Complex Buildings", *Proc. of COMPUGRAPHICS 92*, Lisbon, Portugal (ACM Portuguese Chapter), 1992

5. Dionísio, J. M. S. et al, "A CAD-based System for Air Conditioning in Buildings", *Proc. Workshop for Graphics Modelling and Visualization in Science and Technology*, Springer Verlag, 1992

6. Encarnação, J. L. et al., *Computer Aided Design*, Symbolic Computation Series, Springer Verlag, 1990

Visual manipulation of input probability curves in stochastic simulation models of construction materials handling systems

Z.B. Harun, G. Singh

Department of Civil Engineering, University of Leeds, Leeds, LS2 9JT, UK

ABSTRACT.

A probabilistic approach using the Monte Carlo Simulation Technique is proposed for modelling construction materials handling systems. The stochastic simulation models are clearer in nature and can be driven by probabilistic as well as deterministic input. The approach is able to simulate systems with up to seven cyclic phases and three service points. This paper highlights the benefits of the visual manipulation facility provided by this software. Other interesting features of this simulation software are also described. The features of the input/output facilities are finally presented and discussed to prove the usefulness of the system developed.

INTRODUCTION

Since 1976, work has been going on in the Dept. of Civil Engineering, Leeds University, to stochastically model the construction materials handling systems on the computers [4]. Work first started on a mainframe computer and then progressed to benefit from the increased capabilities of the micro computers. At present, the software has been converted, improved and made to run on the available microcomputers with 80386 and 80486 chips at 20 MHz and 66 MHz respectively. The authors believe that the combination of Monte Carlo simulation technique with microcomputers is the most suitable approach to appraising the various uncertainties [1] met in the different types of construction operations.

The software has been enhanced to run under the Microsoft's Windows environment which gives facilities such as multi-tasking and dynamic data exchange. At present, the UNIX operating system fully provides the said facilities. However, due to its higher price tag, this operating system is not well received by the computing community. The Microsoft's Windows is by far the most popular system and is used by over seven million users worldwide. It is therefore the most acceptable operating environment. The conversion (from mainframe to microcomputer and from earlier operating systems to Windows) has specifically been made to increase accessiblity of the software to the construction industry personnel.

As stated by Bournazos and Singh [4], materials handling operations in the construction industry involve the use of large and expensive equipments. It is imperative therefore that the material handling equipment should be selected and their operational pattern planned in such a way as to form production systems of minimum cost or maximum production for a given cost. This work has developed a microcomputer simulation package to achieve this result.

For the purpose of illustration, a ready-mixed concrete production and transportation in Malaysia has been chosen. Ready-mixed concrete companies have been formed in Malaysia since the seventies with RMC (Malaysia) plc. being the pioneer. Observations were made and data were collected at this company's main batching plant at Sungei Besi. The data were then run on the computer model for validation and optimization purposes.

MATERIALS HANDLING SYSTEMS

In all types of building work, construction materials such as concrete, bricks, steel, timber, etc. are handled either to form permanent stuctures or just temporary ones. At the foundation level, earth may be excavated to be replaced with concrete. Additionally, for finishings, fibre glass panels may be used for ceilings or cement rendering may be applied over brickwalls for aesthetic reasons. All these operations involve the handling of materials.

This also applies to all types of civil engineering constructions such as roadwork, bridge construction, dam construction, etc. In roadworks, the materials handled include earth material at the sub-grade level, sand for the sub-base, crushed rocks for the base course and asphaltic concrete for the wearing course. Similarly, in bridgework, relevant materials are handled from the start till completion of the whole bridge structure. The software developed in this work is aimed at simulating all types of operations encountered in the wide range mentioned above.

As a case-study, for this paper, a ready-mixed concrete operation is chosen. The operation involves a single concrete batcher at the loading point with single-type transit units (concrete trucks) and multiple unloading points. Figure 1 is an illustration of this system. The four main phases in this system are loading, delivery, unloading and return. In this system, queue formation only occurs at the loading point.

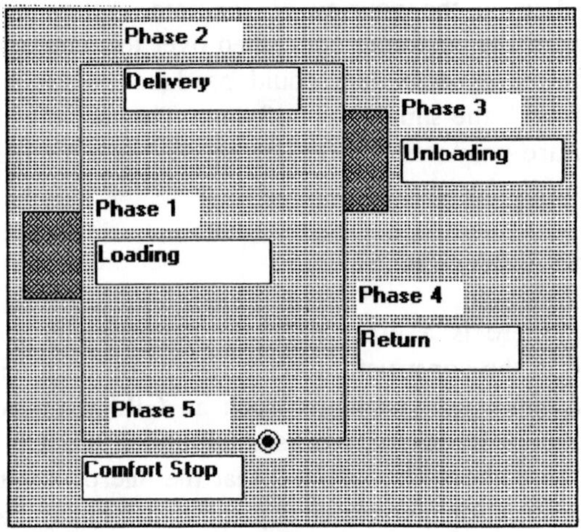

Figure 1. Layout of a Materials Handling System.

Two types of variables are encountered in this system. Namely, the decision variables and the time variables. The decision variables in this system include the capacities and numbers of batching plants and concrete trucks. The user may vary these variables in order to arrive at the optimum combination. The time variables are mainly the loading time, delivery time, unloading time and return time.

During the site observations at the RMC (Malaysia) plc main batching plant at Sungei Besi, certain points were noted. At the batching plant, the loading time was consistent at five minutes per cycle irrespective of whether the plant was preparing three cubic metres of concrete or five cubic metres. However, there was a large variation in the unloading time at the pouring sites. When pouring could take place as soon as a concrete truck arrived, unloading time was small. At times when there was rainfall or concrete formwork was not yet ready when a concrete truck arrived, then the unloading time became large. It was also observed that the truck drivers tend to take some time off for eating, toilet, etc. As such, another variable was introduced to cater for these times which is named comfort stop time. For the purpose of this paper, a fixed time was assumed.

The collection of data was made simpler by the fact that the concrete truck drivers were each given time sheets at the start of their each day's work. These sheets were filled with timings such as when the truck left the batching site, when the truck reached the pouring site, when concrete pouring was started and completed, and when the truck arrived back at the batching plant. These data are used by the Operation Manager to monitor the movement of the concrete trucks. The Manager is able to note from the time sheets if any of the customers delayed the pouring of concrete. Action could then be taken to advise or warn the customer concerned over this mishap. The authors strongly believe that these same data could produce greater benefits when used in conjunction with the simulation software. The manager would then be able to produce future production forecasts as well as optimize the daily production work.

SOFTWARE FEATURES

At present, there is a great need for user-friendly softwares with specific applications in the construction fields which run on a micro computer. There are general simulation packages such as GRASP which runs on the more powerful minicomputer or workstation under the UNIX operating system. However, the authors strongly feel that the microcomputer will be the choice of most construction personnel due to its increasing capabilities in terms of processing power and memory size and decreasing costs.

Another software which is worthy of mention is CYCLONE [5,6]. This software is able to simulate construction operations deterministically as well as stochastically. It was developed using the GPSS simulation language. A microcomputer version has also been developed, called microCYCLONE for educational purposes [3]. Due to the bulky nature of the original application software (which needs a mainframe computer to run), as well as the in-depth training required to familiarize the user with its concept, this software does not find wide appreciation amongst the construction personnel. This software tends to give the impression that the whole of the simulation procedure is more complicated than it actually is.

Earlier work by the authors [2,4] simulated cyclic systems consisting of four phases with two service points. Undoubtedly, many operations in the construction industry are of this type. However, the present work has been enhanced further so as to be able to simulate systems of three to seven phases with one to three service points. With the increased flexibility, the model is able to simulate most types of operations encountered in the construction industry.

This work has led to the development of a stochastic simulation package on an IBM-compatible microcomputer which can be used to produce construction forecasts and to optimize. To make the application of the model more viable in the construction industry, the model was developed to have the capability of simulating the major operations. These major operations may only constitute twenty percent of the total project work but at the same time they usually account for as much as eighty percent of the total project cost. Examples of such operations include major excavation operations, major concreting operations and major formwork constructions.

Through the use of the newly introduced Visual Basic programming language for Windows, visual manipulation facility is provided for the above system. The user of the system is able to change the shape of the probability curve for each time variables to suit his/her wish (based on past experiences) with the use of the mouse. The numerical values of the curve are updated automatically. Figure 2 is the screen print-out obtained via this facility. The user is allowed to manipulate/adjust the shape of the curves by clicking/dragging the corresponding scroll bars below the graph with the mouse.

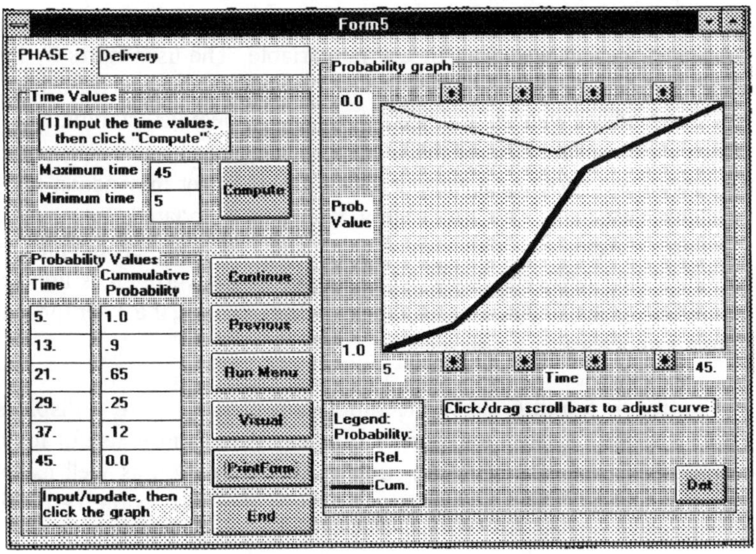

Figure 2. The Input Screen for the Delivery phase.

The probability values are obtained with much ease due to the visual display of the input system. There are cases where users, even with site experience, are not able to come up with probability values in numbers. With this facility, they are allowed to play around with the shape of the probability curves until they are satisfied. The probability values are worked out by the software and printed out. There may also be cases where with time, a user who has some experience with the probabilistic approach to site work, forgets the probability values for the different work activities/phases. However, he/she can still remember the shape of the probability curve. As such, the user may just reproduce the shapes of the curves while the software will work out the respective values. The authors take the view that in all instances where subjective judgement is required for input data, its visualisation and manipulation reduce the cognitive and emotive biases.

INPUT AND OUTPUT FACILITIES

For the purpose of illustrating the input and output procedures, the software is run with the data collected at the RMC (Malaysia) plc main batching plant. An earlier paper [7], discusses the results of the validation and optimization procedures for the simulation models. This section describes the input and output analysis carried out with the simulation package.

Sample screen displays during the input procedure are shown in Figures 2 and 3. The Figures show how the user can use the mouse to make his/her choice via clicking the buttons. Texts and numbers may be entered by clicking the respective textboxes and typing-in using the keyboard. The user may click one of the command buttons to go to different screen displays. Figure 2 is a sample of the input screen for a time variable. The user starts by inputting the maximum and minimum values for the variables. The software then calculates the intermediate values and asks for the respective cumulative probability values. The user may input these values if they are known. Otherwise, clicking the screen will produce a sample graph and the user may adjust the curve until he/she is satisfied. The probability values are updated automatically by the software. (Actually, the cumulative and relative probability curves are coloured blue and red respectively, while the background is light blue on the computer screen.) Once satisfied with the results, the user proceeds to the simulation phase.

Figures 4, 5 and 6 show the simulation phase. Figure 4 displays the input values for the decision and time variables. This form is useful for the user to review the input values before running the simulation as well as for updating the input values. Clicking the labels returns the user to the respective screens for updating the values. Figure 5 shows the actual running of the simulation. Clock times and waiting times are displayed in the respective boxes. This screen allows the user to observe the simulation run in compressed time. The waiting time values signify queue formations and plants being idle at each of the service points. The user should aim at reducing the waiting times in order to achieve optimization. Figure 6 displays the tabulated results of the simulation run. Average values are calculated for all the runs.

The user is expected to attempt at optimizing the system by varying the number of transit units (trucks). The results obtained are given in Table 1. This optimization procedure illustrates how easily sentivity analysis may be carried out. The user could carry out similar analysis by varying the capacities of the server units or transit units while observing the output results. At present, the optimization procedure may be partly automated by using the recorder or macro facility provided under the Windows environment. The user may use the macro to repeatedly run the simulation until the optimum decision variables are achieved.

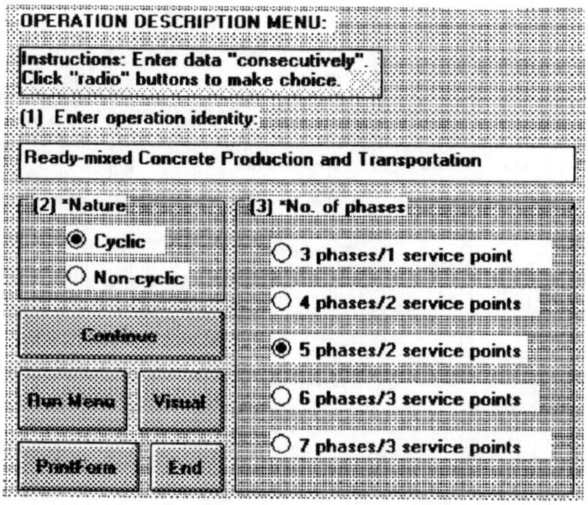

Figure 3a). The Operation Description Menu.

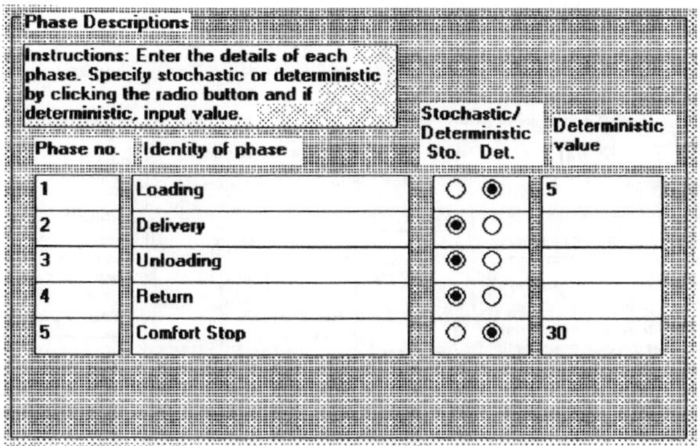

Figure 3b). Description of Phases.

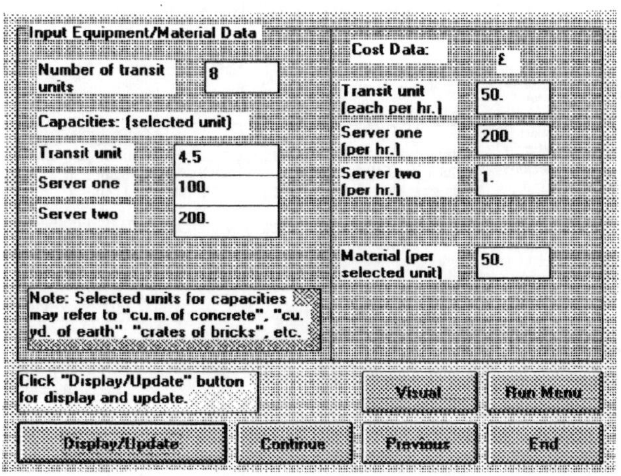

Figure 3c). The Input Screen for the Decision Variables.

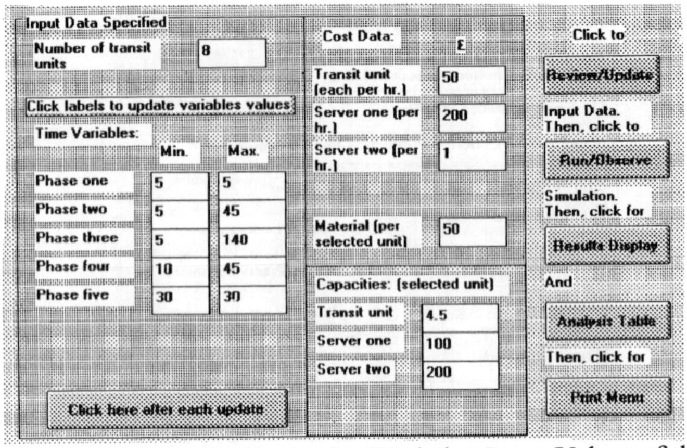

Figure 4. The Form for Reviewing/Updating Input Values of the Time Variables and Decision Variables.

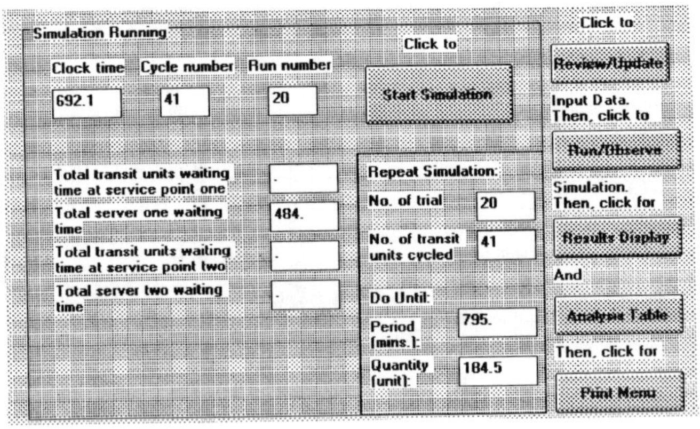

Figure 5. The Screen Output of the Simulation Run.

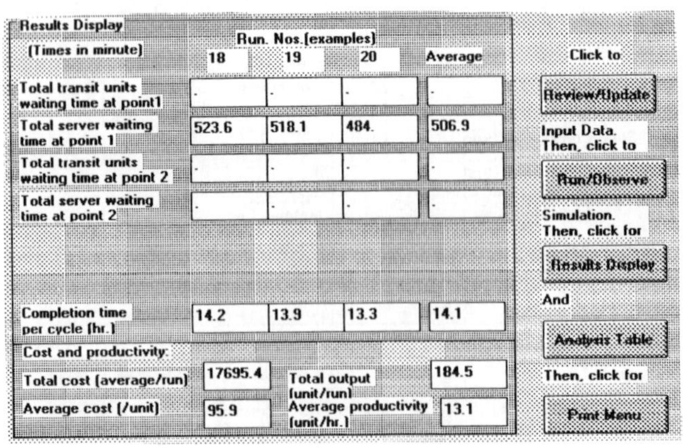

Figure 6. Tabulated Display of the Simulation Results.

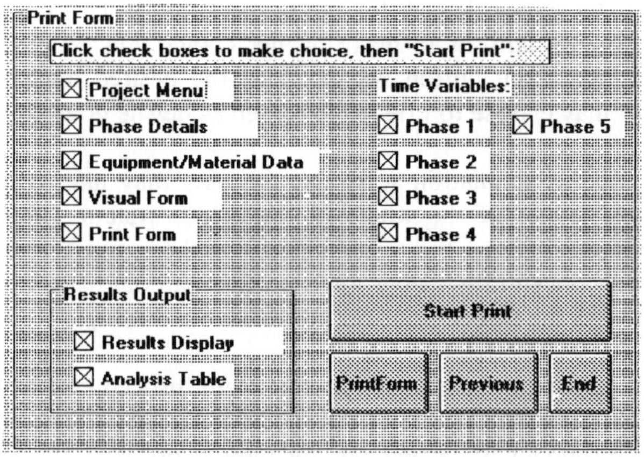

Figure 7. The Print Menu.

No. of Trucks	Av. Unit Cost (M$/cu. m.)	Av. Productivity (Cu. m. /hr.)
8	95.9	13.1
10	94.6	15.7
12	95.3	17.7
14	95.5	19.8
16	96.5	21.5
18	95.7	24.1
20	95.7	26.3
22	96.8	27.8
24	98.9	28.7
26	99.8	30.1

Table 1. Sentivity Analysis. The effects of varying number of trucks on average unit costs and average productivity.

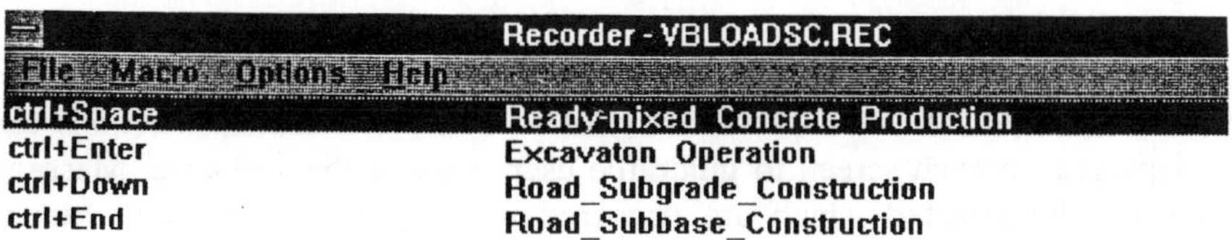

Figure 8. The saved files (Recorder) under different operation names.

Figure 7 shows the screen display for the preparation of hard copies. The user can click the option buttons to make his/her choice. Once the choice is made clicking the Start Print button will produce the desired copies. As an alternative, the user may choose to print each screen as he/she goes along.

In the running of this simulation software, the user may like to have the display of the visual layout of the whole of the actual system. Clicking the Visual button will instantly bring this visual display on the screen . This display can help the user to concentrate his/her mind on the material handling system chosen. Figure 1 shows the screen print-out of the layout display.

In running this simulation package, all the input procedures have been automated by the authors via the use of the recorder facility. Each run of the model is saved. Re-running is carried out by merely pressing the short-cut keys. In this manner, the tedium involved in re-running the model is reduced. Figure 8 shows an example list of recorded macros for the different construction operations.

DISCUSSION

This work attempts to utilize the different computer hardware and software facilities available in the market and tailoring them to support decision-making in the construction industry. The Windows environment has been identified to have great potential for applications in this field. Through the employment of this environment, great enhancements have been made to the graphical aspects of the software. Windows provides facilities such as multiple window displays and automation of input procedures via the use of the recorder.

From the start of the programming work using the Visual Basic programming language, it was noticed how easily the software could be developed. The structure of the whole software was initially prepared using the forms and toolbox without having to worry about the program codes. Due to the form structure, duplicating and updating of the software could be done easily. The debugging facility greatly eased the development of the finished software.

The event-driven approach to programming under the Visual Basic language has increased the interactive nature of the software. Instructions can be displayed on each screen to guide the user in using the software. Message boxes can be inserted which are triggered by wrong keypresses or mouse actions. These facilities actually increase the ease of use and robustness of the software.

The graphical nature of the user-interface under Windows has made the screen displays of the software more attractive. The colour pallete was used for adjusting the background and foreground colours of specific controls such as picture boxes, text boxes and label boxes. Psychologically, the friendly colours help the user to feel more at ease with the software. Instruction boxes and information boxes have been coloured differently to mark their respective significance.

Work is still going on to fully exploit the capabilities of the Windows environment. Different modules are being developed to integrate with the present software. The modules include a Help module, an Animation/Visualization module and a Database module. The authors intend to fully exploit the multi-tasking and dynamic-data exchange capabilities of the Windows environment for achieving the aforesaid integration.

CONCLUSIONS

The results of the work performed so far support the authors' view that the combination of Monte Carlo simulation and microcomputers has the greatest potential for delivering the most user-friendly and robust Decision Support Systems for optimisation of construction operations. The Windows environment, supported by Visual Basic has proved to be a powerful medium for creating the Decision Support System. The Seven-phases-with-three-service-points program created so far has verified the value of the approach taken by the authors. This approach allows sensitivity analysis to be performed very conveniently thereby leading to maximum production rate or minimum cost systems. The use of the program so far has proved the value of the visualisation and manipulation facility of the input probability distribution of the crucial variables.

References

1. Books:

1. Singh, G., Risk and Reliability Appraisal on Microcomputers, Chartwell-Bratt, London, 1987.

2. Harun, Z.B., "Microcomputer Simulation Models for Optimization of Materials Handling Systems in the Construction Industry"', M.Sc. Dissertation, Leeds University, 1987.

3. Halpin, D.W. and Riggs, L.S., Planning and Analysis of Construction Operations, Wiley, New York, 1992.

2. Papers in journal:

4. Bournazos, J.M., and Singh, G., "Materials Handling Systems - Production Forecast and Optimization by Simulation Model SIMAHAS", Highways and Road Construction Int., Oct., 1976, pp. 18-25.

5. Halpin, D.W., CYCLONE - Method for Modelling Job Processes, Jour. of the Constn. Division, ASCE, Vol. 103, Sept., 1977, pp. 489-499.

6. AbouRizk, S.M. and Halpin, D.W., Probabilistic Simulation Studies for Repetitive Construction Processes, Jour. of Constn. Eng. and Management, ASCE, Vol. 116, Dec., 1990, pp. 575-594.

3. Paper in Conference Proceeding:

6. Harun, Z., "Ready-Mixed Concrete Production Forecast and Optimization via Simulation Models on a Microcomputer", Proceed. Reg. Conf. Concrete Engineering and Technology, Kuala Lumpur, June, 1989.

Computer assisted design of a building site

J.C. Lafon,[a] S. Kezeu,[b], T. Soulier,[b] E. Christol,[c] F. Lambez[c]

[a]*Laboratoire I3S, C.N.R.S., URA 1376, Bat 4, Sophia Antipolis, 06560 Valbonne, France*

[b]*Société I3I, C.I.C.A., Sophia Antipolis 3, 06560 Valbonne, France*

[c]*Ecole d'Ingénieurs ESSI, Bat CERISI, B.P 145, 06903 Sophia Antipolis Cedex, France*

ABSTRACT

This paper describes the package CCAO that enables the graphical modelling, the visualization, and the computer assisted design of a building site. The design problem is to decide the location of the different objects needed for the construction with respect to a set of constraints. A hierarchical object oriented data base system is used which is well suited to the modelization of the different parts of a building site. A toolbox (written in C++) facilitates the creation of the hierarchy of classes.The visualization is made with the use of PHIGS+, and the graphical interface is developed in OSF/MOTIF with the use of the XfaceMaker2 interface generator.

INTRODUCTION

In the past few years , there were many works in Computer Assisted Design applied to architecture and construction.Some works are focused on the use of more realistic visualisation of buildings (image synthesis), to study for instance the future aspect of the building and its impact on the environment.Other works are concerned with the design phase of a building to extend the capability of traditional CAD system (Hall [11]). This is the case of researches on knowledge Based Computer assisted design in Architecture or the researches on the use of expert systems (Kimberley [12]).

Very few works have been made on CAD system for the building construction itself.This is the field of activity of the I3I society the objective of which is to offer to the building contractor a software tool to manage the whole process of the building construction. This paper presents the actual results of a two years joint research between this society, the ESSI school in Computer Science of Sophia Antipolis, and the I3S Research Laboratory in Computer Science. The whole system which is described here is named **CCAO** (in french Conception de Chantier Assistée par Ordinateur): Computer Assisted Design of a Building Site

I - MODELIZATION OF A BUILDING SITE

A building site is rather complex to model as it is composed of different parts:

a) the building land and its surroundings,

b) the block of flats to be constructed,

c) the different objects needed for the construction such as cranes, cement mixers, workers housings...

Some objects (for instance cranes) are moving objects, so we have also to model their animation. The position of objects can be changed according to the evolution of the construction and we have also to represent the different steps of the building construction itself. We have developed a graphical package to enable such a graphical representation. In doing it, we have first taken the decision that the modelling of the building itself was certainly already made (in the design office) by the use of a CAD system. This is not the case for the landscape and its surroundings and for the different objects needed for the construction of the building. This is the first aim of our system.

I.1 Landscape modelling

The problem here is to model the landscape on which the building site must be set up. Our approach is to start from the digital data of the terrain given by the geometer.We represent the land by the contour lines:

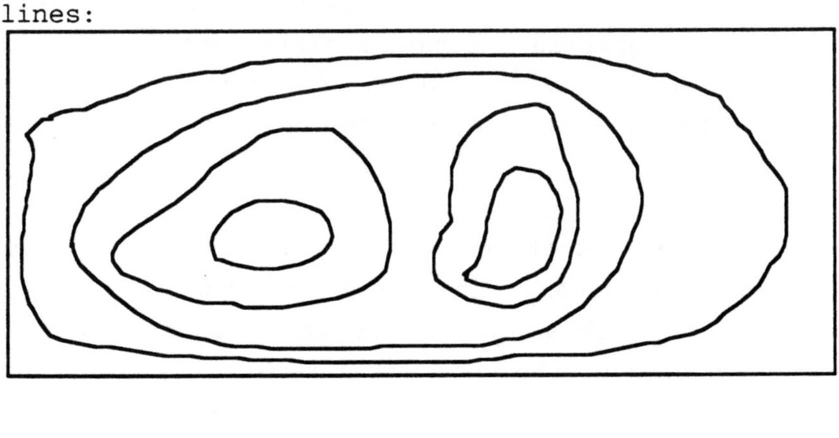

This enables us to have the possibility to visualize it in 3D with PHIGS. Realism is possible (as in Thiemann and All [14]) but it is not our first objective in this particular application.

I.2 Object Modelling

Object modelling is a broad field the evolution of which is described for instance in Eastman [7]. Here, we use a B-Rep model of polyedrons objects (other objects are approximated by including polyedrons).

Each object is defined in its local coordinate system. Obviously there is not a unique definition of such a local coordinate system for a given object. The default definitions of our modeler is therefore described to the user by the graphical interface.

In this local coordinate system, many parameters can be associated with each kind of object:

* variables representing the dimensions of the solid: for instance variables Sx, Sy, Sz for the length of a rectangle parallelepiped. Notice that it is only in the local coordinate system of an object that these parameters can have semantic meaning as the width, the height or the depth. In the real world, it depends of the position of the eyes (of the human observer),

* the name of its vertices and faces and some associated parameters: the local coordinates of each vertices, the normal vector of each plane faces, a variable designing the color of a face...,

For such polyedrons objects, that representation is completed with a derivative of the Brep model:

To each face of an object is linked:

* its normal vector (outward oriented),
* its algebraic equations (implicit and parametric),
* the list of parameters needed to define these equations,
* the list of the edges which defines its shape.

To each edge is associated:

* its direction,
* the two vertices which define that edge.

The geometry of each object is described with GPHIGS (a PHIGS+ implementation on the X Window system) the hierarchical data structure of which is well suited to the description of complex objects as cranes (with moving parts as the bottom on the rails, the tower, the jib, the wagon and the load itself).

Some of the surrounding objects are also important to be represented as they can represent obstacles for instance electric cables, trees, or existing buildings. As realism is not important here, it is adequate to model them as including polyedrons.

The different steps of the construction can be visualized in a wire frame representation or with a more realistic one with colours, textures and light sources with PHIGS +.

I.3 Animation of objects

Moving objects like cranes are considered as open kinematic chains of N rigid bodies called *links*.

The *joints* are the points of articulation between links and are numbered so that joint *i* connects links *i*-1 and *i*.

The motion is described in a Cartesian coordinate frame.The relative position and orientation between objects is represented by homogeneous transformations between coordinate systems attached to links. The geometric transformation is described by homogeneous transformation or homogeneous matrix multiplication.The compactness of the homogeneous transformation matrices is advantageous for expressing consecutive transformations. This is well adapted with the viewing pipeline of PHIGS (Hopgood and Duce [1]) which is used for our implementation.

II. THE COMPUTER ASSISTED DESIGN PROBLEM

The design problem is to decide where to put the different cranes, cement mixers, workers housings, roads..., on the building land to facilitate the erection of the building with all constraints being satisfied (no collisions between moving cranes for example).

New tools exist now to solve the placement problem of fixed objects on the construction site (see for instance Gleicher [10], Kondo [2], Nguyen and Lafon [13], Sohrt and Brüderlin [4], Verroust, Shonek and Roller[5]).

We have developed a graphical man-machine interface to enable the end user to specify the objects to be placed on the building land, to introduce the constraints to be satisfied and to interactively move these objects with the constraints being satisfied.

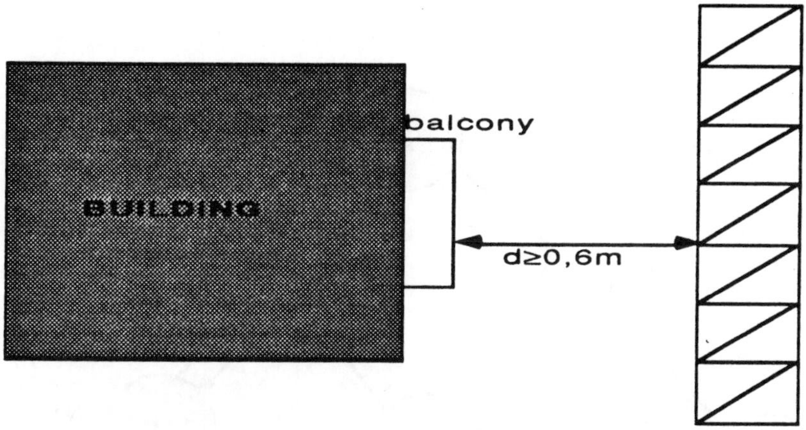

CONSTRAINT BETWEEN A CRANE AND A BUILDING

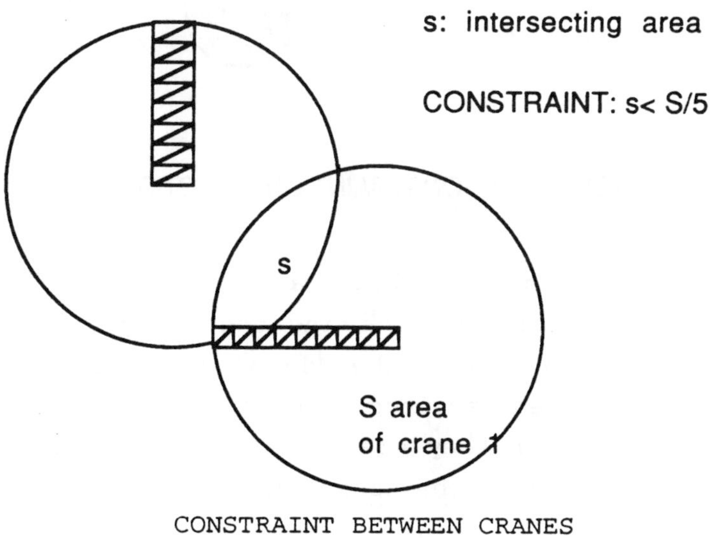

CONSTRAINT BETWEEN CRANES

The constraints involved will have to be resolved step by step during the placement of the objects, that is through the immediate resolving of the equation related to the creation of a constraint, and/or afterwards, while resolving simultaneously both the constraints and the equation system which they implied.

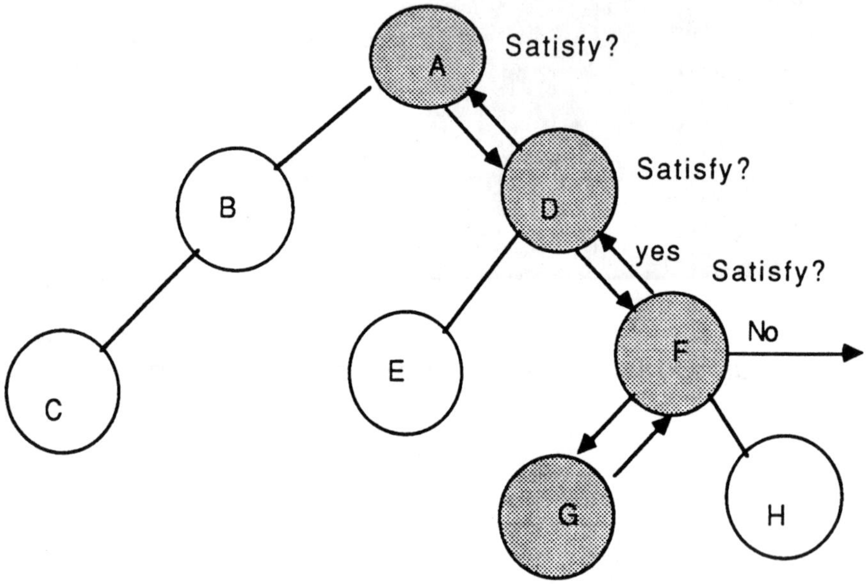

TEST OF A CONSTRAINT ON OBJECT G

The system will have also to transmit and maintain the constraints during the changes. That is solved in our system by the use of concepts existing in object-oriented languages: objects send messages to other objects when they are modified by the user.Each object then tests if the constraints are satisfied.

The efficiency and the reliability of a program written in an object oriented language clearly depend heavily on the inheritance mechanisms of this language. So, the hierarchies of classes given below have taken into account the possibility of the language used for the implementation (C++).

The main point here is the way constraints between objects are implemented: we have chosen to consider the different functions needed for the propagation and the solution of constraints as methods associated with each class of objects. For instance, a constraint of the type "distant from" may exist as different methods in the different classes of objects.

The inheritance mechanism is therefore very useful here to limit the duplication of methods.

III. AN OBJECT ORIENTED DATA BASE

The heart of our system is an object oriented data base of all the objects of the construction site. A hierarchy of classes is defined with associated methods.

The different classes of objects are the following:

```
Yard----------------------------------- Site
                                        Cranes
                                        Mixers
                                        Stacks
                                        Cabs
                                        Buildings
                                        Relief

                 ..........
```

```
Obj_to_build
                    Surrounding
                    Site--------------- Relief
                                        Surrounding
                                        Ways
                                        Obstacles
                                            Way
                                            School
                                            Dwelling
                                            Tree
                                            Cable
                                            ----

                    Storey--------------Walls
                                        Pillars
                                        Floor
                    ------------
Construction_resources
        Officering
                Yard_Chief
        Manual
                Worker
                Construction_Tools
                        Mobile_Tool
                        ----------
                        Crane
                                    Rail
                                    Tower
                                    Top_of_Tower
                                    Worker_Cab
                                    Carriage
```

Our approach can be generalized to the creation of a Knowledge Data-Base (see for instance the work of Formoso and Brandon [8]).

The following figure illustrates the hierarchy of classes created for this application:

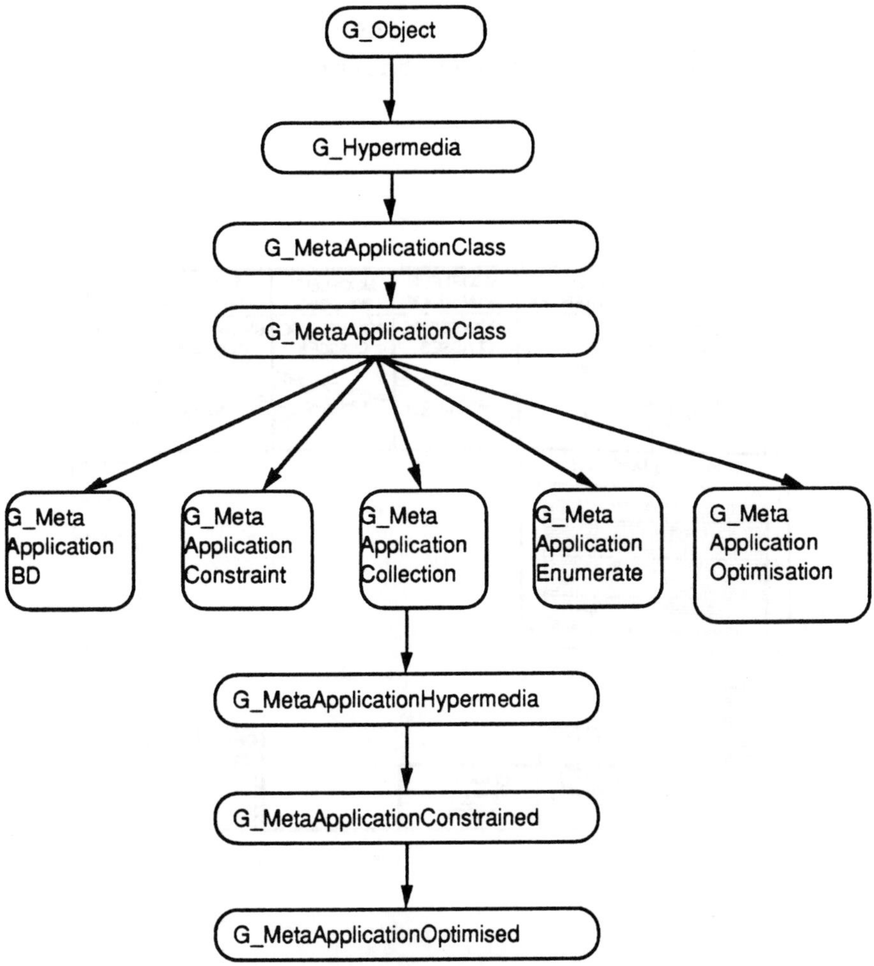

Different methods are associated to each class (the inheritance mechanism is used here): New Object, Delete Object, Undelete Object, Goto Father, Goto Object Name, Goto Root, Set-Get List Son, set-Get Owner, Set-Get First-Last Child, Set-Get Current Child,Set-Get Constraint, Test Constraint, Print...

IV Graphical Interface

The Graphical Interface is developed in OSF/Motif with the XfaceMaker2 interface generator. The following figure illustrates the architecture of this interface:

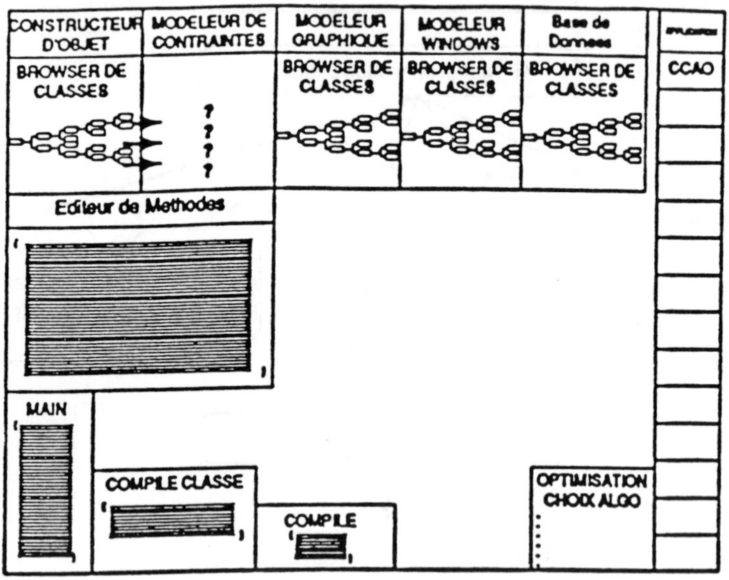

The user can interactively defined classes by only simple definitions as names, parameters, methods owing to an editor we have constructed to do this. He can include directly the C++ code in files whose name is the same as the one of his class and of suffix .h or .cxx.So, we have all the advantages of the Object-Oriented language without the disadvantages.

It is possible to pick one constraint with the hyper-media navigation mode and dynamically attached it to one specific object.

The following figure illustrates the graphical visualization of cranes with Phigs.

It is possible for the user to select different representation modes: wire frame, hidden Surface, colouring and shading.All these representations are implemented in Phigs +.

CONCLUSION

This Computer Assisted Design of a building Site is in fact a very big problem and many works remain to be made. In particular, we have not resolved in our project the planning problem of the construction itself. One classical approach to the problem of planning and scheduling of complex construction process is the use of simulation system (see Kano [9]) Some other interesting works exist on this subject (Kuczora [6], Raghunathan [3]). The knowledge-based approach, which is more general than our data-base approach is certainly also a promising area of research. We are convinced that their is also a great economic interest to develop an easy to use graphical interface to integrate and use different tools such as Graphical Modelling, Constraint solvers, Optimisation Methods, Branch and bound Methods...

REFERENCES

1. Book:

1. Hopgood, F.R.A, Duce, D.A. *"A primer for Phigs"* John Wiley § Sons, Chichester, 1991.

2. Paper in a journal:

2. Kondo, K. "Algebraic Methods for manipulation of dimensional relationships in geometric models", *Computer Assisted Design*, Vol 24, Nb 3, pp 141-147 March 1992.

3. Raghunathan, S., "A planning Aid: An Intelligent Modelling System for Planning Problems Based on Constraints Satisfaction", *IEEE Trans on Knowledge and Data Engineering*, Vol 4, Nb 4, pp 317-335, August 1992.

4. Sohrt,W., Brüderlin, B.D " Interaction with constraints in 3D modeling", *International Journal of Computational Geometry and Applications*, Vol 1, N° 4, pp. 405-425, 1991,.

5. Verroust A., Schonek F. Roller D. "A rule oriented method for parametrized computer assisted design", *Computer Assisted design*, vol 24, Nb 10, pp. 531-540, Octobre 1992.

3.Chapter in a Book:

6. Kuczora, P.W. "Planning and Project Managagement", in *Artificial Intelligence in Engineering* ed. Winstanley, G., pp 195-244, John Wiley § Sons, 1991.

4.Paper in Conference Proceedings:

7. Eastman, C.M. "Modeling of Buildings: Evolution and Concepts", *The Computer Integrated Future*, Proceedings CIB W78, 16-17 September 1991, Eindhoven , Lappeeranta Univ of Technoly ed, 1991.

8. Formoso, C.T, Brandon, P.S. " A Knowledge based Approach for Planning House Building Project", *Building in the future*, pp 40-51, Proceedings of the IAHS World Congress on Housing,8-12 October 1990, Sheinkman § Ural Ed, 1990.

9. Kano, N. "Simulation methodology in Construction Process", pp. 541-533, *Proceedings of the 1st Int. Conf. CAD and Robotics in Architecture and Constructions*, Marseille France, Hermes, June 86.

10. Gleicher, M. " Integrating constraints and direct manipulation", *Symposium on Interactive 3D Graphics*, pp. 171-174. ACM Press, 1992 .

11. Hall,R., Bussan, M., Georgiades, P., Greenberg, D.P. " A Testbed for Architectural Modelling", *Proceedings Eurographics'91*, pp 47- 58, Viena, Austria, 2-6 Sep 1991, North Holland, Amsterdam, 1991.

12. Kimberley, G.J., Hua, K., Donze, F., Smith, I.F.C. " Modelling Knowledge for Preliminary Bridge design", *Proceedings EuropIA- 90*, Liege, Belgium, 15-16 March 1990, pp 38- 46, Hermes, Paris, 1990.

13. Nguyen, C., Lafon, J.C. "Exact 3D Solid Modelling by Constraints" *COMPUGRAPHICS'92*, 14-18 December 92, Lisbon , Portugal, 1992.

14.Thiemann, R., Fisher, J., Haschek, G., Kneidl G. " Visualisation of Digital Terrain Data", pp 173, 196 *Proceedings Eurographics'89*, Hamburg, F.R.G, 4-8 September 1989,North Holland, Amsterdam, 1989 .

Visualization for decision making in construction planning

A. Retik

Department of Civil Engineering, University of Strathclyde, Glasgow G4 0NG, UK

ABSTRACT

This paper describes a new approach for planning and monitoring of construction projects. This approach is a result of two lines of research, which investigated the ways of and benefits from the integration of Computer Graphics and Artificial Intelligence tools within construction processes. The approach is based on visualisation of situations and/or solutions generated by a computerised system during the planning process. The implementation of the approach is outlined. The examples of visualisation are provided and discussed. The further development is proposed and conclusions are drawn.

INTRODUCTION

There is no doubt that one picture is worth a thousand words. For construction planning the question is not only how, but also (and mainly) what to visualise. Existing computerised design tools (i.e. CAD systems) are basically developed for designers to deal with a *product* design (i.e. drawing or 3D modelling of a building or structure). While the construction planning tasks (scheduling, time-cost forecasts, etc.) are a construction *process* design. Similar to the product design, the generation of the construction process is based on heuristic knowledge and its results are unique for every project. The difference is that these results (i.e. time schedule) are rather a sequence of events than physical objects and, therefore, it is more difficult to validate and evaluate them. Therefore, the visualisation may not only facilitate the validation and evaluation of the generated process, but also support further decision making (i.e. recourse allocation, progress monitoring, etc.), based on it.

Computer graphics development has proposed a variety of visualisation tools and techniques (see Brodie [1], Earnshaw and Wiseman [2]). There are already several different either explicit or implicit ways of integrating computer graphics with artificial intelligence tools in civil engineering. They can be viewed according to the aims of the usage. The obvious one is the graphical illustration of final results. In this case a decision to fire graphic procedures or take off predefined graphic solutions is made as a result of data elaboration or calculations. Typical examples are the RETWALL system (see Siller [3]), which allows the choice of kind of retaining wall, from 10 possible, and the drawing of it, and the GERT system (see Cornick and Bull [4]), which pulls out suggested solutions for gypsum wall connection details.

A different case is where a decision is made by the user as a result of visual examination of graphic pictures displayed by a computer. PLATFORM II (see Levitt et al. [5]) presents planned schedules in the form of Gantt charts to allow a user to do a solution validation. SABA (see Ligget [6]) displays suggested facility layouts. The system, that is described in Guena and Zreik [7], allows checking of the opening directions of doors in a given layout.

Realistic representation of the design alternatives or solutions becomes necessary where manipulation of the words and symbols carrying out graphical meanings is a vehicle for generation of a new design. Examples can be seen in Akin et al. [8], Bijl [9], Gero et al. [10], and many others. The development here is based on logic (e.g. PROLOG) or object-oriented languages (see Bourdeau [11], Keirouz et al. [12]). These allow us to build intelligent systems using logic representation of graphical and geometrical information. The synthesis of a new design (the system's output) is obtained as a collection of words and symbols or primitives (objects). The connection of the formers with graphical entities is a way for visual representation of the solution (see OPS/83 in Kano [13] for displaying simulation by Gantt chart or Oxman and Gero [14] for kitchen layout design.

Geometric and spatial reasoning is a relatively new and very powerful approach (see Woodwark [15]). This approach is based on defining the connections between the geometrical representation of design objects (location, dimensions, etc.), spatial relationships (composition, hierarchy, sequence, etc.) and their (the objects') functional roles. Therefore, visualisation and further manipulation of objects could be, for example, based on their geometry, but decisions may be made on functional attributes. The examples of these approaches are described in Fenves and Baker [16] for structural design and in

Woodbury and Oppenheim [17] and Giraud and Hanrot [18] for architectural design.

Two different patterns of visualisation for the construction planning are demonstrated below. Their further implementation is discussed later.

PLANNING AND MONITORING OF CONSTRUCTION PROJECTS

Interactive visualisation of the construction process

Perhaps the most potentially useful application of computer graphics in design is the possibility of a three-dimensional representation of projects or their particular details in the same manner as has been conventionally done with cardboard, wood or plastic physical models. Once such a model was created, it was possible for a designer or his client to observe its features from different angles, or even "walk through" its various parts.

In a similar manner the 3D graphics may be applied as an aid for visual simulation of a construction project progress. The planner, the contractor or the owner can observe the simulated project or "walk through it", when performed with different methods, or at various stages of its execution. The ability to show a real picture of the work progress, indicating deviation from the planned schedule, is very valuable for the senior managerial staff, especially those dealing with co-ordination of and resource allocation for several projects concurrently. To 'grasp' a visual status of several projects is not only easier but also quicker. The example on Fig. 1 shows how project status can be presented at various project stages. This approach is based on visualisation of basic activities of construction processes, as suggested in Retik et al.[19] for multi-storey residential buildings. Each of the activities has, as their attributes, the planned start, the duration, their graphical and geometrical representation and their location in the project space. Each activity is composed from a parametric 'library' of basic elements such as walls, columns, windows, etc., created for this purposes and as a result could be adapted to the desired dimensions and position.

This kind of representation can help not only the convenient visual validation and simulation of the project progress but also creates new possibilities, not available by conventional methods , for the planning process itself. During the monitoring of the construction progress, the deviation from the planned schedule may be shown by using a different colour. As a result, the planner may not only notice a delay in a particular activity but also see the actual place of this delay. Moreover, a knowledge-based expert system can be integrated within the 'managerial' part of the system, advising what actions may be taken to avoid such a situation.

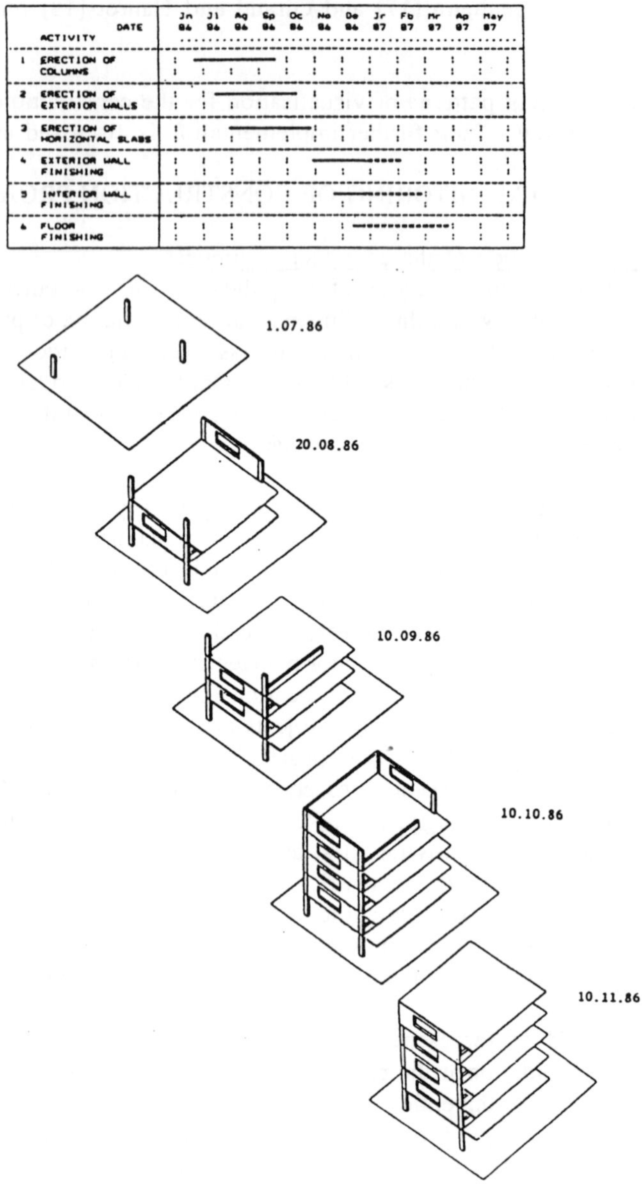

Figure 1. Construction Schedule Visualisation (from Retik et al. [19])

Graphic representation illustrated in Fig. 1, in conjunction with a regular bar chart, may be used in a similar manner in conjunction with a network, line of balance or any other technique. In such cases, it will present, in a visual form, those features which can assist the planner or site manager in a better perception of the construction process. For example, the system can present, in a graphical form, the activities on a critical path, or any other path (with network) or the progress of a specific work team (with lines of balance). Representation of such selected features can be requested by the planner from the managerial part of the system, appropriately programmed for this purpose.

Further development of the above approach may lead to the different realisation of the planning and monitoring tasks. The idea is to use dynamic visual status of all project activities during whole construction processes.For that, it is necessary to present the comprehensive picture of the project including new construction (structures, buildings, etc.), landscape development and site organisation with plant and facilities (see Fig. 2) and to schedule them. The visual status of construction work progress related to the site organisation is very important to the project managers. The possibility to check and compare visually several alternatives for plant and facilities location on the *site* and also to validate them in *time* may help to achieve better solutions.

Different levels of 'project picture' details are needed during the project implementation phases. For that, each building may be presented by work elements or segments which may be, for example, a structure erection or finishing works of a complete building floor section. Every segment should be associated with a group of activities. As a result, each segment may be later represented separately in more detail with its various composing activities or through the activities of each of its major trades, or responsibility groups - e.g. concreting, piping, etc., each of them monitored separately for its actual versus planned status (see Fig. 3).

One of the possibilities for achieving this multi-level planning is to present project data hierarchically. In this case, once the project is updated on the lowest level the result may be seen on all higher levels. To provide flexibility in defining hierarchy the 'hammock' approach may be used on every level, connecting a group of activities to one section or segment on the above level, consequently, a group of segments to a building, etc. (see Fig. 4). To avoid too much detail and make a picture understandable, every level can be connected to a *different* set of the parametric library of the graphical elements. For example, at the project level the following library elements may be necessary for generating the project picture: roads and approach ways; storage facilities, storage areas, etc.; cranes and rails; auxiliary structures; etc.

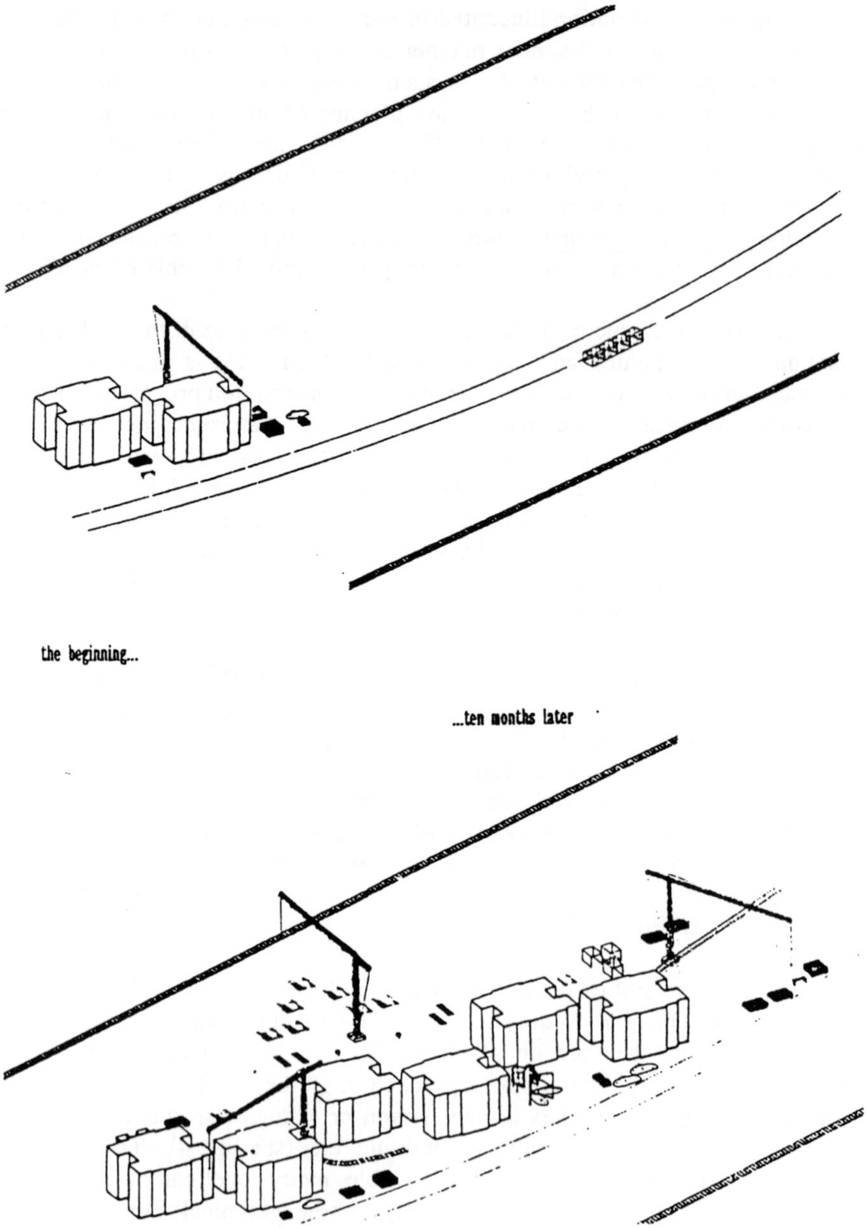

Figure 2. Example of Visual Monitoring of Construction Work Progress

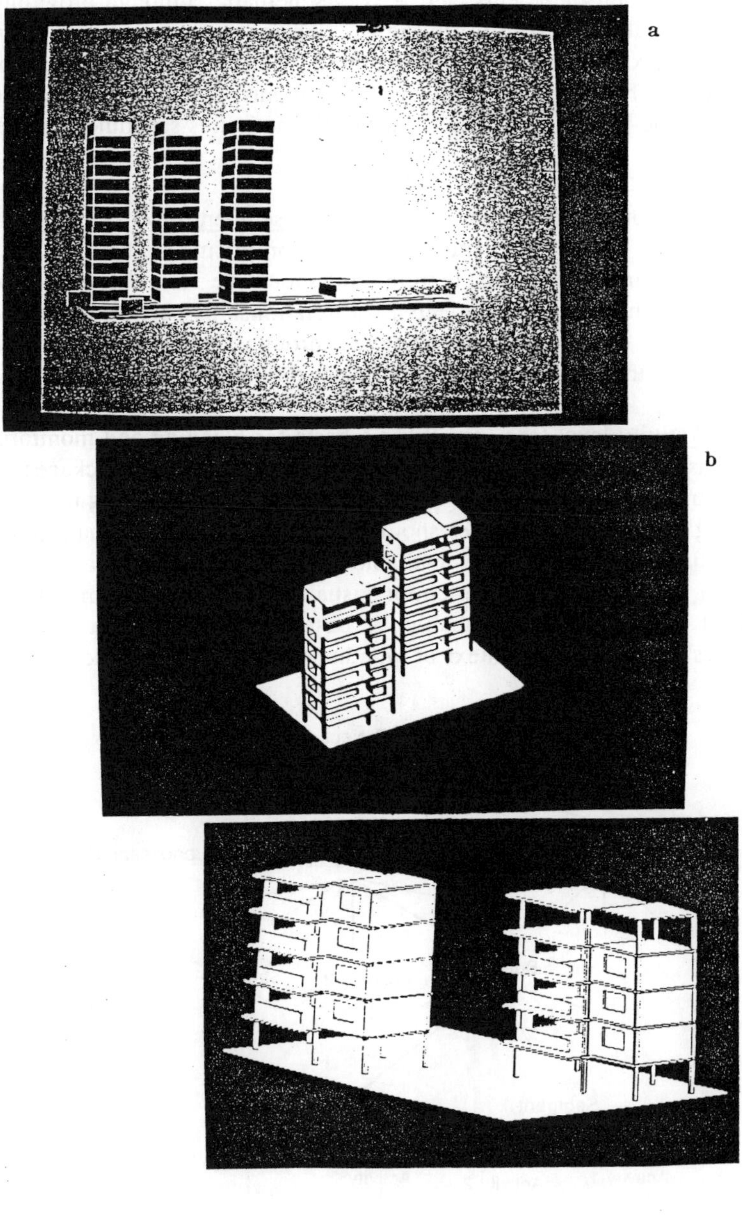

Figure 3. Different Levels of Project Visualisation:
(a) segment level
(b) activity level
(c) selected activities

The library elements can be very schematic, although providing enough realistic visual presentation (see Fig. 2). The adjustment of the elements to the project context may be done through the dimension parameters once the project is set up. The actual location of the elements during the project should be related to the project progress similar to the activities schedule (see Retik et al. [19]).

The buildings and structures on the segment level (see Fig. 3a) can be created from simple 3D parametric boxes or prisms. Their dimensions and visualisation images (colour, texture, etc) should be variables of the composing activities of their groups(segments). The relationships between the visual status and the work progress may be achieved through simple production rules or, in the case of the object-oriented approach, through message passing.

Activity level (see Fig. 3b) is a basis for planning and monitoring tasks. These tasks may be handled by existing project management packages, providing output to the upper level through the program interface. Setting up a project and breaking it down into activities should be primarily based on managerial criteria and actual needs (resources, available work gangs, work places, etc.). The decision which activities to monitor visually (some projects may have several hundred activities) will probably be based upon importance of the specific activities in the project context or whether they are subcontracted or not, etc.

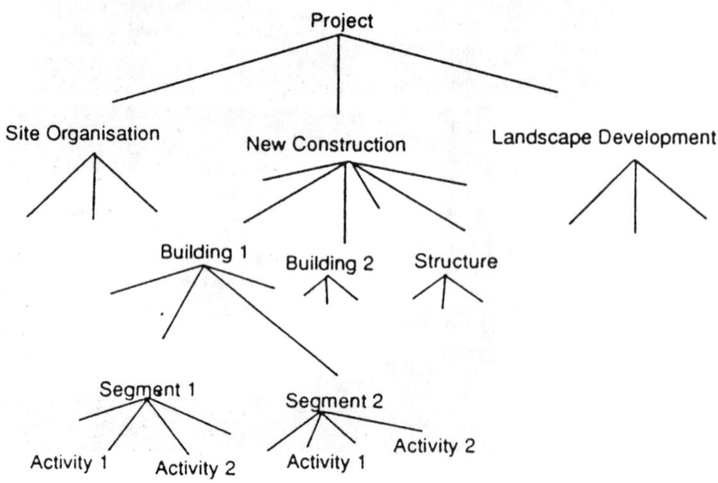

Figure 4. Hierarchical Representation of Project

Non interactive visualisation of the construction process

It is not always useful to visualise the construction process interactively. An example for such a situation is demonstrated further by the case of intelligent stochastic simulation of cost-time forecasting in Marston et al. [20]. In this case the *in situ* heuristic knowledge is integrated within the system for decision and action making during each cycle of the simulation process, if and when interference occurs in the work sequence.

The system prototype which is based on the above approach, consists of two main parts dealing with: (a) plan generation and (b) work process simulation. Plan generation involves (i) break down of work description into activities, (ii) establishing order and pace of the work as well as (iii) allocation of appropriate resources as materials, labour and tradesmen and plant and equipment. Once the time schedule is generated the simulation process may be started.

At this stage, the site production process is simulated by the Monte Carlo method. 'Actual' values of productivity rates are drawn randomly from probability distributions defined according to work context, load and domain. The resulting 'actual' duration of the involved activities is recorded. Progress is monitored to identify where departures from the plan cause interferences between activities or unacceptable delay, and management decisions are simulated to take corrective action where necessary. At the end of each simulation iteration, the 'actual' duration and cost is calculated, prior to a further simulation run being commenced. After completing a preset number of iterations (about 100), the final results are computed and reported.

Result representation, analysis and visualisation The primary outputs of the system - probabilistic estimates of total project cost and duration - can be presented both graphically and numerically. Various additional outputs may be produced for the purposes of explanation and validation including a graphical representation of the operational plan, reports on the decision responses to interferences and delays, and a facility to produce a trace of any simulation run.

One of the questions that appeared as a result of the simulation was how one can control decision making processes or know what kind of decisions were made and when. It is not realistic of course to visualise simulation processes interactively, showing every scenario a hundred times. Moreover, it is not necessary even since the decision will be made only when the interference has occurred. Therefore, only the conflict situations may be stored in the computer memory. Visualisation of the activities' interference may be achieved by graphic representation of a schedule such as line of balance graphs with the description of the decisions that were made. Evaluation of simulation results as well as their validity may be done later on. Because a large number of simulation cycles are necessary to achieve representative cost-time distributions, conflict situations will be stored in the computer memory together with the decisions taken. 'Post-mortem' analysis will then be possible (see Fig. 5).

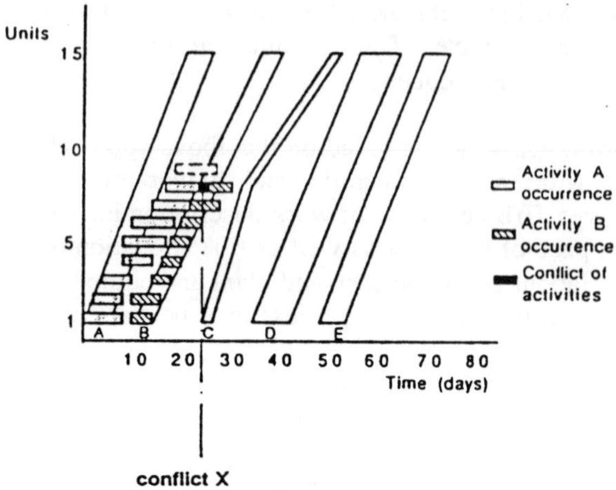

CONFLICT X:

DECISION: **increase the gang size of the activity A by one labour for a week**

IMPLICATION: **add to the 'extra-cost' slot the following value: 40 hours * (spec-lab-cost-p/h)**

Figure 5. Example of 'post-mortem' analysis

DISCUSSION OF APPLICATION ASPECTS

The application aspect of the interactive multi-level visualisation approach in the former section (see Figs 2,3) deserves further discussion.

The creation of a three-dimensional image for each element may be quite an involved task in a real life project consisting of hundreds of different work items. As noted before, such processes can be largely facilitated by the creation of a library of composing elements such as walls, windows, columns, slabs, etc. Nevertheless, even then the amount of work can be quite extensive, especially in large projects. The task may be quite simple however, if the project is anyway designed with the aid of 3D computer graphics - and has the desired image for the work items created in this process for different design purposes. In such a case, it would be enough to code each work element and the associated activity so that they could be properly referred to and recalled by the managerial part of the computerised system. It is not always a straight forward approach to relate a work element to an activity, but some abstraction/simplification may be done. On the other hand, the effort may be paid off in a complicated project where 'walk through' facilities will be necessary to explore project status on the activity level.

Some difficulties may arise with certain work elements which are invisible or not entirely compatible with the activities allocated for their execution. For example, activities such as formwork erection for concrete pouring, or excavating for foundations, would not be normally represented in a 3D building design. They can be added however, without difficulty as an auxiliary activity, coded in conjunction with the work piece they are to produce and represented by a different colour, shading, texture or annotation.

Another problem, especially at activity level, is the wealth of the graphical information contained even in a simple project - skeleton, fixtures, finishes, mechanical, electrical systems, etc., which can completely blur the image of the building under construction on a regular computer screen. A window in the screen may contain a symbolic representation of all or only special or subcontracted trades to be monitored acting as a checklist in the control process.

CONCLUSIONS

Visualisation of a building project may assist a schedule planner or user in a better perception of the construction work progress. The advantages of this representation may be more emphasised in large scale projects where the approach can be extended to monitoring not only the construction process itself, but also all auxiliary activities including on site plant and equipment. Here not only different locations of construction equipment and temporary facilities in space may be checked, but also be *traced in time*. In this case, especially in projects where heavy plant is used, delays and interferences may be prevented.

Monitoring of the work progress in large projects may follow whole work packages which can be subsequently separated into individual work items. It can also focus on individual work trades. The use of a graphical representation of the construction process is very much facilitated if the project is also designed with the aid of a 3-dimensional graphical system.

REFERENCES

1. Brodie, K.W., Carpenter, L.A., Earnshaw, R.A., Gallop, J.R., Hubbord, R.J., Mumford, A.M., Osland, C.D., Quarendon, P. (Eds). *Scientific Visualisation - Techniques and Applications*, Springer-Verlag, Berlin, 1991.

2. Earnshaw, R.A., Wiseman, N. *An Introductory Guide to Scientific Visualisation*, Springer-Verlag, Berlin, 1992.

3. Siller, T.J. 'Expert Systems in Geotechnical and Environmental Engineering', *Survey of the State-of-the-Art Expert/Knowledge Based Systems in Civil Engineering*, ed. Kim, S.S., pp. 101-120, Champaign, Ill., 1987.

4. Cornick, T., Bull, S.P. 'Expert System Integrated with CAD for Component Assembly', *Proceedings of Fourth International Symposium on Robotics and Artificial Intelligence in Building Construction*,pp. 657-671, Haifa, Israel, 1987.

5. Levitt, R.E., Kunz, J.C., Kartum, N.A. 'Using Artificial Intelligence Techniques for Automated Planning and Scheduling', *Proceedings of Fourth International Symposium on Robotics and Artificial Intelligence in Building Construction*, pp. 27-52, Haifa, Israel, 1987.

6. Ligget, R.S. 'A Designer-Automated Algorithm Partnership: An Interaction Graphic Approach to Facility Layout', *Proceedings of the SUNY Buffalo 3rd International Symposium on CAD*, pp. 87-102, Buffalo, NY, 1990.

7. Guena, F., Zreik, A. 'An Architect Assisted Architectural Design System', *Artificial Intelligence in Design*, ed. Gero, J.S., pp. 159-179, CML Publications, Southampton, UK, 1989.

8. Akin, A., Dave, B., Pithavadian, S. 'Heuristic Generation of Layouts (HeGeL): Based on a paradigm for Problem Structuring', *Artificial Design*, ed. Gero, J.S., pp. 413-444, Elsevier, Amsterdam, 1988.

9. Bijl, A. 'Human Knowledge: AI and CAD', *Proceedings of Fourth International Symposium on Robotics and Artificial Intelligence in Building Construction*, pp. 414-429, Haifa, Israel, 1987,.

10. Gero, J.S., Radford, A.D., Coyne, R., Akiner, V.T. '*Knowledge-Based Computer-Aided Architectural Design*', *Knowledge Engineering in Computer-Aided Design*, ed. Gero, J.S., pp 57-81, Amsterdam, North-Holland, 1985.

11. Bourdeau, M. 'Object-Oriented Approach Applied to Construction', *Proceedings of 5th International Symposium on Robotics in Construction,* pp. 609-618, Tokyo, Japan, 1988.

12. Kierouz, W.T., Rehak, D.R., Oppenheim, I.R. *Object-Oriented Programming for Computer-Aided Engineering*, Engineering Design Research Center, Carnegie Mellon University, Pittsburgh, PA, 1987.

13. Kano, N. 'A Knowledge-Based Simulation System for Construction Operation', *Proceedings of the Sixth International Symposium On Automation and Robotics in Construction,* pp. 499-506, San-Francisco, USA, 1989.

14. Oxman, R., Gero, J.S. 'The Integration of Evaluation and Generation in Knowledge-Based Design Systems', *Proceedings of the SUNY Buffalo 3rd International Symposium on CAD*, pp. 103-122, Buffalo, NY, 1990.

15. Woodwark, J. (Ed.) *Geometric Reasoning,* Clarendon Press, Oxford, UK, 1989.

16. Fenves, S.J., Baker, N.C. 'Grammar for Functional and Spatial Reasoning in Design', pp. 819-828, *Proceeding of Fifth Conference on Computing in Civil Engineering*, March 1988.

17. Woodbury, R.F., Oppenheim, I.J. 'Geometric Reasoning: Concepts and Demonstration', *Artificial Intelligence in Engineering: Robotics and Processes,* ed. Gero, J.S., pp. 61-76, Amsterdam, Elsevier, 1988.

18. Giraud, C., Hanrot, S. 'Elements for Spatial Reasoning in Construction', *Proceedings of 5th International Symposium on Robotics in Construction,* pp. 105-114, Tokyo, Japan, 1988.

19. Retik, A., Warszawski, A., Banai, A. 'The Use of Computer Graphics as a Scheduling Tool', *Building and Environment*, Vol.25, No.2, pp.133-142, 1990.

20. Marston, V. K., Skitmore, M., Alshawi, M., Retik, A., Koudounas, C. *The Use of Intelligent Simulation in Cost-Time Forecasts For Housing Rehabilitation Works*, University of Salford, UK, 1991.

[illegible] M. [illegible] Oriented Approach [illegible] Construction. [illegible] International Symposium [illegible] Robotics in Construction, pp. 529–[illegible] Japan, 1984

[illegible] W. [illegible] O. R. [illegible] Object Oriented [illegible] Engineering [illegible] Design, Research Center [illegible] PA, 1987

[illegible] N. [illegible] Knowledge-Based [illegible] Systems [illegible] Orientation, [illegible] Proceedings [illegible] and Robotics in Construction, pp. 499–506, San Francisco, USA, 1986

[illegible] The Information [illegible] Knowledge-Based Design [illegible] Engineering [illegible] [illegible] NY, 1986

[illegible]

[illegible]

[illegible] Environment [illegible] Construction [illegible] 1988

[illegible] Information in Design [illegible]

[illegible]

[illegible] M. Rezk [illegible] [illegible], 1991

EmDiSy - a personal computer program for emergency operating procedures graphical support

I. Šalamun, B. Mavko, A. Stritar

University of Ljubljana, Institute 'Jožef Stefan', Reactor Engineering Division, Ljubljana, Slovenia

Abstract

Use of computers in nuclear power plants can improve the emergency response during accidents. The operator can much faster get the information that he needs during an emergency. The idea of safety parameter display system (SPDS) was proposed to help the operator during accidents. The computerized emergency operating procedures (EOP) can be part of this idea.

EmDiSy is the computer code for the presentation of emergency operating procedures. It uses knowledge from the database file and presents it to the operator in different visualization forms. It starts with high priority procedure and then leads the operator through the procedures respective to the nuclear power plant condition. Every procedure is represented by six different windows. The first three windows describe the purpose of the procedure, the symptoms and entry condition of the system and the automatic actions at appointed condition. The second three windows represent a tree of steps through the active procedure, describe the action in active step and represent a 3D picture of the system or 2D picture of status trees. 3D pictures are drawn in the parallel projection and clearly describe the condition of the system and action that must be done. In a special window EmDiSy displays the connection between procedures.

EmDiSy clearly represents the operator action that must be done during an emergency in the nuclear power plant and points in the realistic picture the place of action. In the future EmDiSy could be extended with the expert system to give advices to the operator.

1. Introduction

Modern computer technology used in nuclear power plants can improve the emergency response during accidents. The operator can get the information that he needs during an emergency much faster. The idea of Safety Parameter Display System (SPDS) was proposed to help the operator during accidents. The purpose of the SPDS is to help control room personnel in evaluating the safety status of the plant. The primary function of the SPDS is to aid the operator in the rapid detection of abnormal operating conditions.

Emergency Operating Procedures (EOP) are plant procedures that direct operator actions, necessary to mitigate the consequences of transients and accidents that have caused plant parameters to exceed reactor protection system setpoints or engineered safety feature actuation setpoints, or other established limits. EOP can be used to bring the nuclear power plant under control without the use of other documents. The Computerized Emergency Operating Procedures (CEOP) can be considered as an upgrading of both systems.

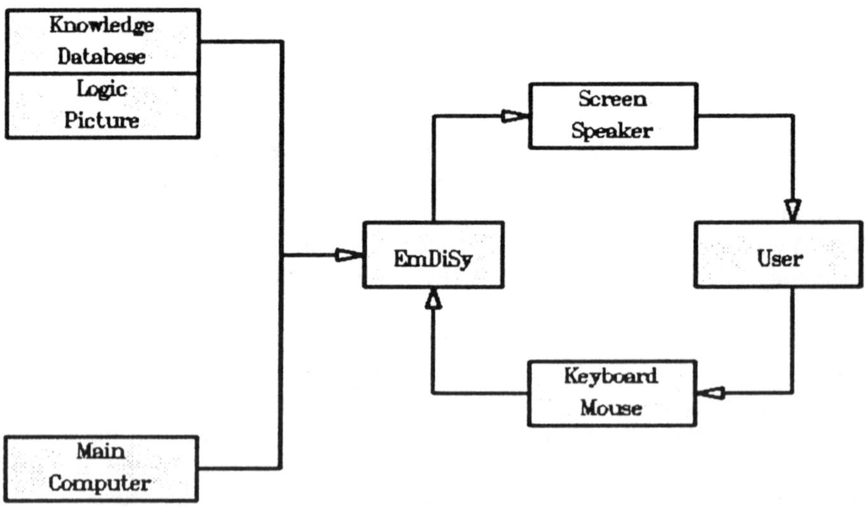

Figure 1: Scheme of EmDiSy computer code.

Each nuclear power plant has a single symptom-based emergency procedure which will be used by the operator to bring the plant under control after an emergency. The most fundamental requirement of an emergency procedure is that it must be usable. This means that the technical information is complete and accurate and that the information is presented so that the operator could find and comprehend it quickly. CEOP contents all these requests.

2. EmDiSy - computer code

Standard EOP is written in the paper form. When an abnormal event occurs, the licensed operator takes the EOP book and reads the action steps which make up the subsequent action. EmDiSy (Emergency Display System), which is being developed in our institute to support the operator during accidents in a nuclear power plant, is more than an electronic EOP book. It uses all recommendations for writing EOP and makes use of advantages of computer technology.

EmDiSy is written for personal computers and requests high graphically resolution (1024x768 pixel, 256 colours). The input units are keyboard and mouse, but the last one is not necessary. Data are alternated through network or serial ports. The database files are stored on the personal computer hard disk.

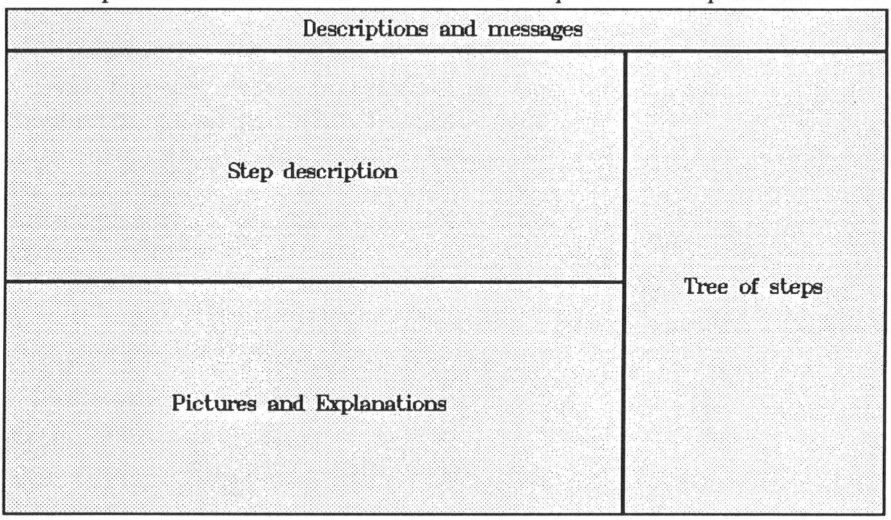

Figure 2: EmDiSy screen separation.

2.1. Database for EmDiSy

The base of the EmDiSy computer code is two groups of database files (Figure 1). In the first group are the knowledge database files, in the second files for graphical support. The knowledge database file contains all knowledge about the emergency procedures. In this file the purposes of the procedure, the symptoms and entry conditions of the system, the automatic actions at appointed conditions, the "foldout" for the procedure and the procedure rules are written. The components of the procedure rules are questions, instructions and pointers to the next step in the procedure or to the next procedure.

The CEOP has advantages in comparison with standard Emergency Operating Procedures (EOP) because the graphically designed computer code allows presentation of data on-line anywhere on the screen in any visible form. A file for graphical presentation contains picture saved as a bit map, name and position of data value. Pictures are drawn in 3D or 2D projection.

2.2. Graphical design

U.S. Nuclear Regulatory Commission recommends basic design for SPDS computer code [1,2,3] and EOP [4,5,6]. EmDiSy is function oriented CEOP which provides the operator guidance on how to verify the adequacy of critical safety functions and how to restore and maintain these functions when they are degraded. The operator does not need to diagnose an event to maintain a plant in a safe condition.

We have separated the graphical screen in four windows (Figure 2). Name and code of the procedure, date, time and icons are presented in the first window on the top of the screen. Next three windows (working windows) are reserved for questions, answers, data and pictures.

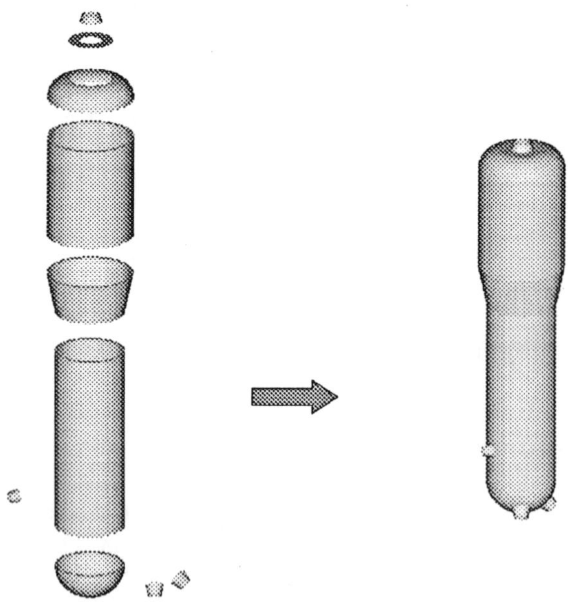

Figure 3: Steam Generator constructed with 3D base elements.

We have used all recommendations for writing standard EOP given by U.S. Nuclear Regulatory Commission. The EOP format helps to determine how accurately and quickly information can be located and acted upon by the operator. The Emergency Procedure is divided into four parts: purpose, symptoms, automatic actions, instructions and "foldout". Purpose section describes in few sentences the purpose of the procedure. Symptoms section has a list of possible initiating symptoms. Automatic actions section lists and describes the automatic system actions that should have occurred as a function of the engineered safety features. "Foldout" section contains the important data for the procedure and must be reachable at any time. All this sections are written in working window as set of sentences.

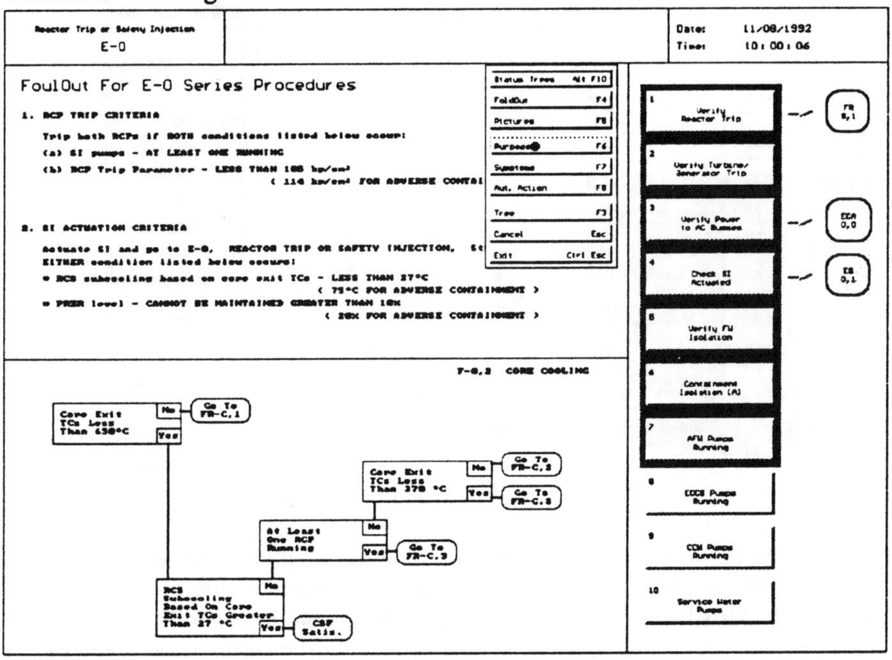

Figure 4: Graphical screen of EmDiSy computer code.

All action steps related to a particular task are grouped together (single procedure) and presented in the order in which they are to be performed. Each action step is limited to a single operation, it has a title and it is written in thirty words or less. The title of the instruction step is written in the first working window. The first working window contains ten titles and links between steps and other procedures. In this way the operator clearly knows the stepping order and which procedure can be reached from the appointed step.

The last two working windows contain instructions for the action in the selected step and a picture for description of the action. The data values are written in both windows. A two-column format presents procedural information for active step in the second working window. In this layout, operator actions are placed in the left column, notes and other supporting information are placed in the right column. Each action step is wholly contained on a single page, and it is not contained on the next page. Notes and explanations begin next to the action step to which they refer.

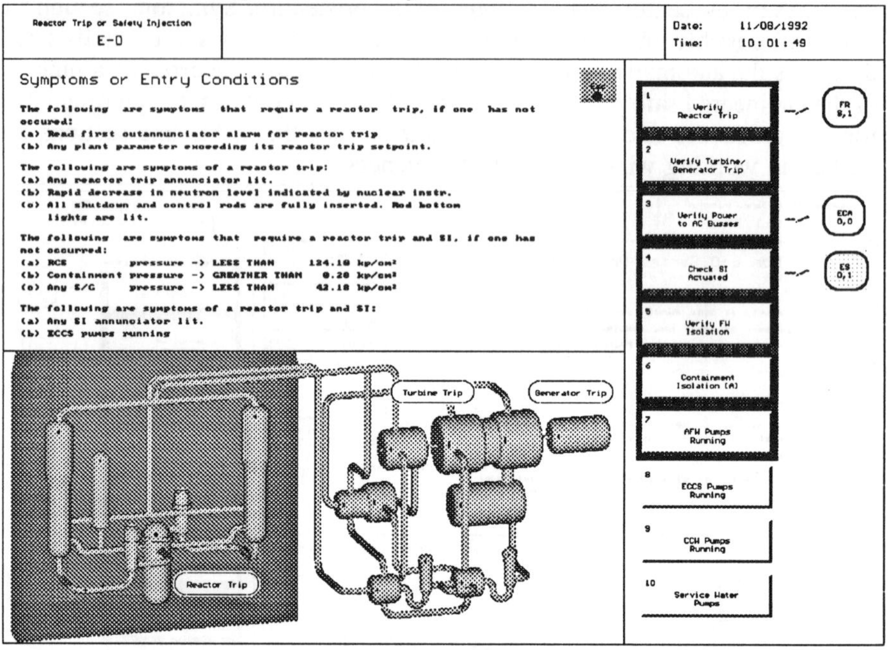

Figure 5: Systems picture presented by EmDiSy computer code.

Figures and tables are used in procedures to assist operators in making decisions and in locating information. Tables and pictures are explicitly and uniquely identified. Mostly the pictures are selected when the operator chooses the active step. Other pictures are very easy to find throughout menus, icons or keys. They contain only the relevant information needed to clarify or accomplish the purpose referenced in the text. Pictures are drawn in 3D or 2D projection. The pictures are prepared by any drawing software which can store it in some standard format (DXF, PCX, ...). The picture stored in the standard format is converted into internal graphic format which allows a fast manipulation. The last working window is also used for displaying purpose, symptoms, automatic actions and "foldout" section of active procedure.

For preparing 3D pictures we have developed a computer code for its creation. We have defined some 3D base elements and they have been used for constructing the whole picture (Figure 3). The computer code for calculating 3D pictures uses the input file which contents dimension of the picture, number of colours, observer position and list of base elements with dimension and coordinates. The output file contains the number of colours for each place in the 2D plane.

2.3. Using EmDiSy

EmDiSy always starts with the high priority procedure. It leads the operator through the procedure in accordance with his choices. The operator only has to compare two values and answer the question with yes or no. The code does not compare asked value with measured value and it does not take any action by itself. It is always waiting for the operator's instructions.

Particular data are reached through macro keys or menus (Figure4). The operator can switch between the pictures of the status tree, the pictures of the system and "foldout" data. One of those possibilities is always written on the screen. When the operator calls the editor to read data with the purpose, symptoms and automatic action, then picture or "foldout" is replaced with an editor. When the operator closes the editor, the old picture or "foldout" is written again and the code is returned to the place before this action (Figure 5).

3. Conclusion

Computer code EmDiSy is still being developed. Its purpose is to instruct operators and to mark all operator's actions. Our intention is to put together knowledge accumulated in EOP, data values displayed at an SPDS and some kind of expert system, which can analyze incoming data and point to errors in the system.

References

[1] U.S. Nuclear Regulatory Commission: Clarification Of TMI Action Plan Requirements; NUREG-0737; 1980

[2] U.S. Nuclear Regulatory Commission: Man-Machine Interface Issues in Nuclear Power Plants; NUREG/CR-5348, SAIC-89/1114; 1989

[3] U.S. Nuclear Regulatory Commission: Human Factors Acceptance Criteria for the Safety Parameter Display System; NUREG-0835; 1981

[4] U.S. Nuclear Regulatory Commission: Functional Criteria for Emergency Response Facilities; NUREG-0696; 1981

[5] U.S. Nuclear Regulatory Commission: Evaluation of Emergency Operating Procedures for Nuclear Power Plants; NUREG/CR-1875; 1981

[6] U.S. Nuclear Regulatory Commission: Guidelines for the Preparation of Emergency Operating Procedures; NUREG-0899; 1982

[7] Steven Harrington: Computer Graphics, A Programming Approach;McGraw-Hill; 1987

Decision support system for hydroelectric power plant design

D. Hajduković, B. Djordjević

Institute of Hydraulic Engineering, Faculty of Civil Engineering, University of Belgrade, P.O. Box 895, 11000 Belgrade, Yugoslavia

ABSTRACT

The decision support system for power plant design discussed in this paper is knowledge-based, and was conceived and formed in the Faculty of Civil Engineering in Belgrade. The system consists of a knowledge base, model base, decision generator and multi criteria optimization block, and takes into consideration all possible variations of power plant disposition. The method of multicriteria optimization (the iterative compromise ranking) is used to determine the optimal amongst possible alternatives, considering previously determined criteria. Three dimensional prototype drawings in relative scale form a visual database attached to the system.

INTRODUCTION

Possible disposition alternatives of a hydro power plant are numerous, both derivation disposition and those of every composing element, due to the great complexity of the task. It is solved only partially in the present time, for optimization methods are used for composing some elements of the power plant (turbine type, number of units , ...). Evidently, global disposition optimization is not possible on the manual design level, because designers and designing firms do not have sufficient knowledge and capacity to work on such a number of alternatives at the same time. It results that very alike dispositions are used in very different environments, sometimes in inappropriate conditions, because the designer is familiar with that particular solution.

Development of knowledge-based systems followed by development of computer systems in the last decade, has provided a tool which can be used as a decision support in global hydropower plant disposition design (Simonović [3]). The genesis of such a knowledge-based program package, developed in the Faculty of Civil Engineering in Belgrade, follows.

GENERAL IDEA

The general idea was to make a computer program capable of generating numerous disposition solutions for the same topographic and hydrographic situation, by modification of constructive elements and types of power houses and also disposition of composing elements (Djordjević and Hajduković [4]). The disposition variants are then ranked by means of iterative compromise ranking (ICOR - Opricović [2]) to provide the optimal solution concerning chosen criteria. In all the steps, expert knowledge is involved as is shown on a general diagram (see Figure 1. - Hajduković [5]).

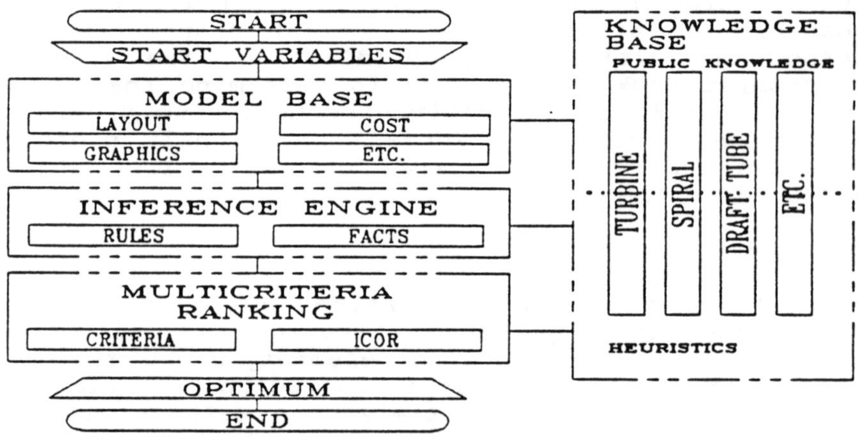

Figure 1. Scheme of hydropower plant design decision support system

<u>Knowledge base</u>
From Figure 1. one can see that the knowledge base provides expert knowledge for all the steps in

consultation with a computer. The knowledge in the knowledge base is generally divided into public knowledge (books, manuals, textbooks etc.), and heuristics (engineering experience and intuition), and formally divided into blocks containing data on composing elements (in other words equipment or constructions) of the hydropower plant. Each block contains at least one 3D drawing in relative scale to the corresponding element.

Relative scaled drawings are generated to exact specifications of their producers,or by using statistical data, in a way that enables their use in generating a 3D drawing of the entire hydro power plant which is under consideration. After a consultation of decision support systems, type and dimensions of all composing elements are calculated, and their prototype drawings are scaled to exact size and positioned (inserted) to form the global drawing. This drawing, which is a one on one model of the future object, is used for further examination, and, by its 3D nature, provides possibilities of any view, or cross-section of the object. This can be very useful, for numerical data are not often enough to support a decision as to whether the alternative would be taken into consideration, or discarded.

All the graphical support to the considered system is given by DESIGNCAD-3D graphics package, which showed to be very handy in forming 3D prototype drawings.

It is not possible to implement the mentioned methodology for all of the constructive elements of a power plant (Hajduković [6]). A basic presumption for the possibility of using relative scaled drawings is that constructive element dimensions in particular coordinate axes can be expressed by relation:

$$\mathbb{D}_{i,j} = \mathbb{C}_{i,j} \cdot \mathbb{D}_j$$

where

$\mathbb{D}_{i,j}$ is i-th dimension of power plant constructive element along axis j (eg. $\mathbb{D}_{2,1}$ is turbine outlet diameter in horizontal axis)

$\mathbb{C}_{i,j}$ is relative scale of an i-th dimension along axis j

$\mathbb{D}_j$ is referent dimension of a constructive element along axis j (eg. turbine diameter)

In other words, all dimensions of a constructive element of a power house must be a linear function of the single dimension of the same element. Standardization of the equipment in such a way is performed both by producers of the machinery, and also by the designers of power plants. On the other hand, there are some elements whose dimensions are not given in a linear way (e.g. turbines), but that problem can be solved by piecewise linearization of those relations. This means that, instead of one prototype drawing we should have a definite number of them. The blocks are:

Turbine The block "turbine" contains data on classes, systems and types of turbines; constraints in possible implementation; topographic as well as linear characteristics; extents of coefficients and parameters; dimensioning relations; cavitation characteristics etc. Prototype drawings in relative scale and in 3D of different types of turbines are generated, and, due to the problem of linearization discussed earlier, each of the turbine types has more then one drawing, for a different head condition. Figure 2. shows a Francis turbine for a design head between 100 and 150 meters.

Spiral This block contains data on: domain of implementation of different types of spirals as a function of installed power and head; included angles of concrete and steel spirals; construction and design criteria; enter velocity as a function of head etc. Spiral casing of the turbine mentioned above is also shown on the Figure 2.

Draft tube In this block there are data on: geometry of different types of draft tubes; special types of draft tubes used in non-standard conditions; types and installation data on draft tube gates; dimensioning procedures, etc. Draft tube for Kaplan turbine of design head between 70 and 80 meters is shown in Figure 2.

Valve Data on types and domain of implementation of emergency valves; ways of installation and manipulation with valves, standard dimensions and masses etc. A prototype drawing in 3D of each type of valve is part of the visual data-base.

Generator Containing data on: geometry, types and characteristics of generators; dimensioning procedures related to different types of generator cooling; design criteria for implementation of different types of generator suspension etc. Each type of generator, and suspension elements, has its prototype drawing.

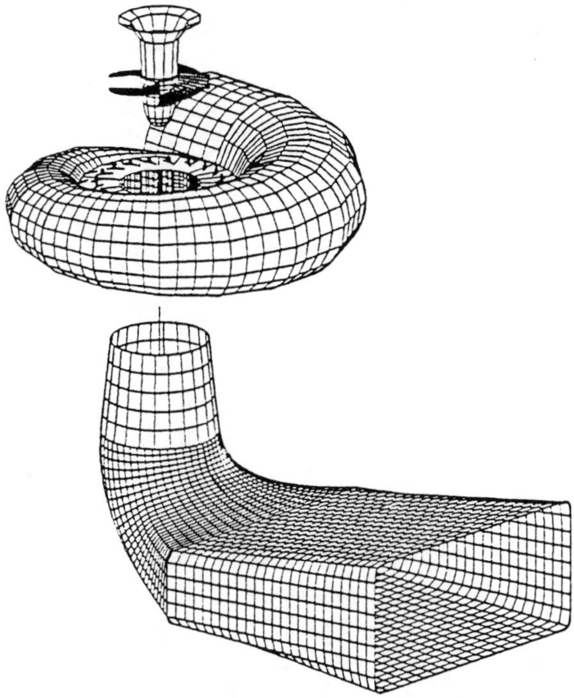

Figure 2. Kaplan turbine, spiral casing and draft tube for design head between 70 and 80 meters

Crane This block in the knowledge-base contains data on: types and domain of use of different types of main and auxiliary cranes etc. Each type has its prototype drawing.

Power house In this block there are data on: types and design criteria of power houses; specific forces for stress analyses; disposition of equipment due to manipulation zones of cranes etc. Constructive elements from this block are drawn interactively.

Expends This is a block with unit prices for excavation, construction costs and costs and installation of equipment etc.

Model base
Model base consists of mathematical models, defining disposition of equipment and disposition of power plant. Models are divided into blocks .

Lay-out With models for unit block dimensioning; dimensioning of necessary surfaces; disposition of control and maintenance rooms and unwatering galleries; etc.

Figure 3. Production section of power house in 3D

Elevation Models for determination of elevation of all equipment and constructive elements inside and outside hydropower plant eg. foundations, turbines, galleries, generators, floors, cranes.

Graphics Models based on software for interactive graphics (AutoCAD and DESIGNCAD-3D), are used for graphical presentation of every variant, which enables the designer to visually inspect the current variant of disposition. Figure 3. shows the formation of a 3D drawing of a power house. Generator shaft and floors of the power house are erased to make the drawing more clear. This drawing is performed by a decision support system developed in the Faculty of Civil Engineering in Belgrade for further research in this field.

Construction expends This model base block contains mathematical models for the calculation of excavation and concrete volumes, and also masses of reinforcement and equipment etc.

Inference engine
Inference engine infers new knowledge from the existing one using rule-based method of forward chaining. It is based on the formation approach (Hayes-Roth et al.[7]), in other words, it composes a variant of disposition of a hydropower plant from its composing elements. That variant of disposition is, together with the others, ranked by iterative compromise ranking in order to find an optimal one, considering the criteria selected.The VP-Expert system shell is used to couple its inference capabilities with knowledge and model bases discussed above.

Multi criteria optimization
Solution of the complex problem of disposition selection is unlikely to be found by consulting only one criterion. Also, criteria selection for optimization is a very complex problem, and it requires interaction with heuristic rules (Hajduković [5]), not only for selection, but also for the decision of what criteria weights and what ranking strategy should be used. The criteria are :

-minimization of total and/or annual expends, operative expends, energo-economic coefficients, environmental and social impacts etc.

-maximization of energy production, benefits, B/C ratio, security, operation and remount capabilities etc.

CONCLUSIONS

The methodological approach considered in this paper has many advantages in relation with conventional design methods due to computer speed and also to the accessibility to expert knowledge, and finally a much higher probability that optimal disposition is analyzed, because of numerous variants generated. This finally means a lower cost for the construction of such a hydropower plant.

Also, due to high user friendliness, interactive procedure and graphical capabilities, this software can be used for educational purposes, as the best way of expert knowledge transfer in the field of hydropower plant design.

Finally something must be pointed out about knowledge acquisition, which is the most delicate phase in knowledge-based software package generation. This is concerning a rule that the knowledge base has in such a package. Also, knowledge acquisition is a continuous process, for the hydropower plant equipment is continuously improved, as is the software package.

REFERENCES

1. Djordjević, B. Water Power Management: Hydropower Plants, Naučna Knjiga, Belgrade 1988. *

2. Opricović, S. Multi Criteria Optimization. Chapter 5, Iterative Compromise Ranking, pp. 50-70, Naučna Knjiga, Belgrade 1986. *

3. Simonović, S. Expert Systems for Water Resources Engineers, Lecture Notes, 1st International Course on Use of Expert Systems for Water Resources Engineers, Dubrovnik, Yugoslavia 1989.

4. Djordjević, B. and Hajduković, D. Use of Computers in Hydropower Plant Disposition Design, in SYM-OP-IS/88, pp. 489-492, Proceedings of the 10th Symposium of Operation Researchers, Brioni, Yugoslavia 1988. *

5. Hajduković, D. Hydropower Disposition Design Using Contemporaneous Methods of Cybernetics, Master degree thesis, Faculty of Civil Engineering, University of Belgrade, Yugoslavia 1989. *

6. Hajduković, D. Methodology for Hydropower Plants Graphical Simulation, in SYM-OP-IS/92, pp.483-486, Proceedings of the 14th Symposium of Operation Researchers, Belgrade, Yugoslavia 1992. *

7. Hayes-Roth, F.,Waterman, D.A. and Lenat D.B. (Ed.) Building Expert-Systems, Addison-Wesley, Reading, Massachusetts, 1983.

8. Begović, K. Constructions in Hydroenergetics: The Introduction to Equipment Selection, Naučna Knjiga, Sisak, Yugoslavia, 1986. *

9. Vora V. KBES and Interactive Graphics , Expert systems in Civil Engineering, pp. 219-223., C.Costem and M.Maher, ASCE, New York, 1986.

10. Mitchell, T., Steinburg, L. and Shulman, J. A Knowledge-Based Approach to Design, IEEE Trans. on Pattern Anal. and Machine Int., Vol.PAMI-7, No.5, pp.502-509, 1985.

* = in Serbian

Visualization of 3D scalar fields. Application to air pollution concentration fields

S. Ruiz,[a] L. Cremades[b]

[a]*ITEMA, Universitat Politècnica de Catalunya, Ctra. Nacional 150, km 14.5, Aptdo. 508, 08220 Terrassa, Spain*

[b]*Dpt. de Proyectos de Ingeniería, U.P.C., Diagonal, 647, Planta 10, 08028 Barcelona, Spain*

ABSTRACT

Air pollution distributions usually form finite volumes inside the 3D domain of simulation models. We have developed an application to calculate and visualize isosurfaces whose main restraint is that they must be closed (or at least almost closed) into the closed domain boundaries.

This application uses a 3D regular-spaced data mesh which defines the domain. This 3D mesh should be formed by a series of surfaces which should not be interlaced. These surfaces can have any shape.

The procedure for generating an isosurface has two steps: 1) the contour of the isosurface is computed in every plane. Every contour has to be closed. If open contours are created, they will be closed through the domain boundaries. 2) This contour is compared with the contours of the surfaces at the upper planes in order to assign contours to isosurfaces. So, the isosurface is generated as a whole series of curves in the space. To visualize it, we use the primitive B-spline surface of the standard graphics PHIGS.

This procedure allows us to get a good visualization of air pollution volumes in a relatively short computing time. Some tests and application examples are presented.

KEYWORDS

Visualization, isosurface, isocontour, scalar field, air pollution

INTRODUCTION

3D Finite Element Method (FEM) and 3D Finite Differences Method (FDM) used in simulation models are becoming more and more some of the most useful tools to engineers and scientist for studying and analyzing complex structures. The increasing computation power of the new workstations is allowing an increase in accuracy and resolution of the results obtained by FEM and FDM. Generally, after some variable time of computation, these results consist of a huge amount of values representing a scalar field in 3D.

The problem arises when one interprets these values. So far, the post-processors complementing the FEM and FDM's results usually force the users to imagine the volumetric distribution of the values through some 2D cuts in several planes. Visualizing this scalar field in 3D can be very helpful to engineers and scientists. To visualize an isosurface of stress values, for instance, can help the designers to optimize a part.

An isosurface is a surface comprising a volume where a variable has a certain constant value. Its visualization in 3D allows to detect, in a glance, some complex relations that otherwise would be very difficult to look at. Associated to the isosurface, there is the concept of isocontour. An isosurface is formed by an infinite number of isocontours, in particular horizontal isocontours.

We have developed a program to calculate and visualize isosurfaces in 3D, especially those which are closed into the domain boundaries of the scalar field or that can be closed through the boundaries. As a first application, the program has been applied to the visualization of tridimensional isosurfaces of air pollution concentration fields over topographic maps. These fields usually form finite volumes inside the 3D domain of the simulation models, then they are appropriate to be visualized with this program.

BACKGROUND

From the sixties, several researchers have focused on how to visualize scenes of the physical space through a computer output device, usually a CRT screen. Since then, several techniques have been developed to obtain this. We can divide them between those used to create 3D isocontours (wire models) and those used to generate 3D isosurfaces.

Among the first ones, we can mention, foremost, the method for tracing the isocontours directly (Sabin [1]). In this method, the isocontours are calculated from analytical equations of surfaces. It produces very good results, but can only be applied to singular cases.

Secondly, there is the method of interpolation in a grid. The isocontours are approximated by a series of straight segments which are interpolated bilinearly from the values at the mesh nodes (Sutcliffe [2]). An isocontour passes through one of the sides of a cell when one vertex of that side has a greater value than the isocontour and the other vertex has a lower value. This method can give rise to some ambiguities if, for instance, the case shown in Figure 1a happpens (cell nodes marked with "+" are nodes with a greater value than the chosen contour value, whereas nodes marked with "-" have a lower value). Both solutions shown in Figure 1 might be valid.

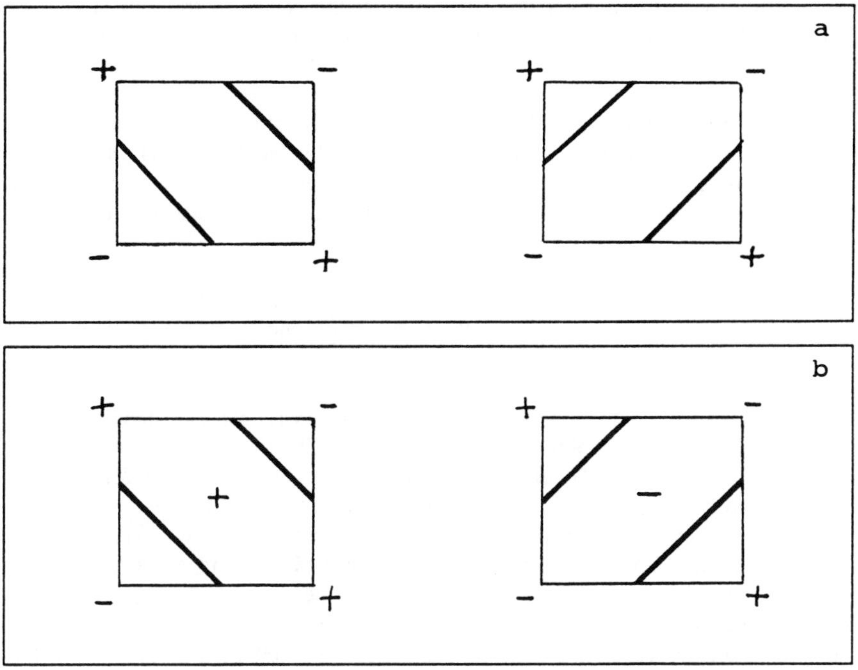

Figure 1. Method of interpolation in a grid cell: a) ambiguities, b) how to solve them.

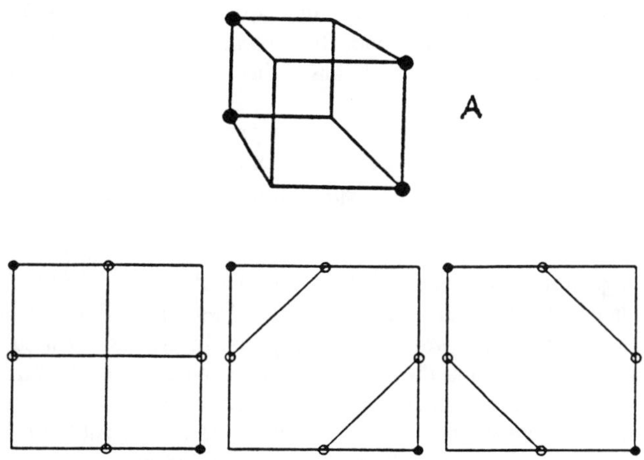

Figure 2. Marching cubes' algorithm. Ambiguities derived from the case A. Three possible interpretations (Collins [3]).

However, to choose the correct one, the value at the centre of the cell should be interpolated. In this case, we have two different solutions depending on the central value (Figure 1b).

The third method for tracing isocontours consists of following the contour, point after point, from two initial points (Sabin [1]). This algorithm has two main drawbacks: how to find the initial values and how to determine the ending condition in the case of a closed contour. It is more accurate than the previous one, but is also more complex.

As to the methods for generating isosurfaces, we can classify them in three large groups: methods of serial sections, methods based on cubic grids and methods of studying the volume itself (volume rendering).

The method of serial sections is useful when data are arranged as a stack of slices, such as, for instance, the data obtained from a CT scan. The first step is to identify the contours inside each slice. To do that, the methods above mentioned for creating isocontours can be used. Once the contours have been identified, those in adjacent slices are joined by a ribbon of triangles or higher-order surface patches. The problem with contour lines is that they do not contain sufficient information regarding the gradients associated with the 3D isosurface they describe (Collins [3]). So, the number of possible combinations to create the isosurface as a series of triangles can be very important. Nevertheless, some authors, such as Keppel [4] or Fuchs, Kedem and Uselton [5] have tried to find the optimum triangulation. An alternative approach taken by Sunguroff and Greenberg [6] represented sections of each contour by uniform B-splines. However, all of the above algorithms have problems in the case of bifurcations, i.e. where one contour in one slice divides into two or more contours in an adjacent slice. Nowadays, automatic methods are being researched to generate the complete isosurface in every case. Other applications, like TROTS (Veen and Peachey [7]) or ISOSRF (Wright and Humbrecht [8]), do not use any ribbon of patches to join the isocontours, but are based on removing-hidden-lines algorithms.

The methods based on cubic grids try to approximate the object volume as a set of voxels (Collins [3]). This technique, called "cuberille" technique, is usually applied in medical imaging, where accurate depiction of the object is more important than generating pleasing images.

Another algorithm based on data sets placed in a 3D regular grid is the "marching cubes". This is problably the algorithm which has been most widely implemented for the generation of isosurfaces whithin regular grids. In the case of a cubic grid each cube is considered in turn and the vertices of the cube are classified as being either "inside" or "outside" an isosurface based on the vertex value compared with a chosen isosurface value (Lorensen and Cline [9]). There are 256 possible ways in which a surface may intersect a cube, which are reduced to 15 if symmetry-related cases are ignored. The isosurface is finally approximated by a triangular mesh. The original algorithm has ambiguous cases, such as shown in Figure 2 (black nodes are outside the isosurface, whereas white nodes are intersection points of the isosurface): there are up to three possible solutions to visualize the isosurface.

Several authors, such as Wyvill, McPheeters and Wyvill [10] or Wilhelms and van Gelder [11], proposed some schemes to solve these ambiguities. In any case, other similar algorithms, the "marching tetrahedra", based upon tetrahedra instead of cubes, seem to work better. These algorithms create a tetrahedral grid from a cubic grid by dividing each cube into 5, 6 or 24 tetrahedra. This resolves the marching cubes ambiguity problem.

Finally, a number of techniques, collectively called volume rendering, have been developed to visualize the 3D structure of the interior of volumes, since it can be difficult to do by viewing individual slices or just isosurfaces. These techniques generally involve tracing an imaginary ray of light through the volume and applying a function, based on the data values encountered in the volume, to generate an intensity and opacity. These are then used to produce pixel values in the final 2D image (Collins [3]). These techniques produce very good results at the expense of important computation time.

To choose one of these methods to create isosurfaces is very dependent on the kind of application required and the graphic system (hardware and software) in which the application will run.

ALGORITHMS IMPLEMENTED

The goal of the application developed is to visualize 3D isosurfaces of air pollutant concentrations as fast as possible (interactively) from data sets arranged in a 3D Cartesian grid over a 3D topographic map. This grid should be regular in X and Y directions, but not necessarily regular in the Z direction. Taking into account these circumstances, we have chosen the method of serial sections to create the air pollution isosurfaces from isocontours, which in turn are created by following a modified method of interpolation in the rectangular grid. Isocontours are generated for each Z-dependent layer of the air pollution data set and afterwards the isosurfaces are created by joining iso-valued contours of the different layers. As we said before, the method of interpolation in a mesh is one of the least accurate methods to generate isocontours, but one of the fastest. To increase its accuracy, we can increase the resolution of the input data.

In the original algorithm all the grid cells were sequentially analyzed and then the segments were individually drawn. In our algorithm we try to find an isocontour by following the curve directly. So, once we have a point of the curve, we look for the next one and so we generate the curve point after point. Once a curve is completed, we look for other curves, but avoiding to analyse cell sides where a curve already passes through (only one isocontour segment can pass through a cell side). A curve then becomes a set of arranged cell sides, identified by three integer numbers (i,j,k), where k= 1 or 2. If k= 1, it is a horizontal side between the points (i,j) and (i,j+1); while if k= 2, it is a vertical side between (i,j) and (i+1,j). In this way, the algorithm works with integer numbers until the visualization step, in which the points where the "real" curves pass through the cell sides, are interpolated as real numbers. To visualize the isocontours, we make use of the non-uniform B-spline curve primitive of the graphic programming interface graPHIGS [12]. The graPHIGS is

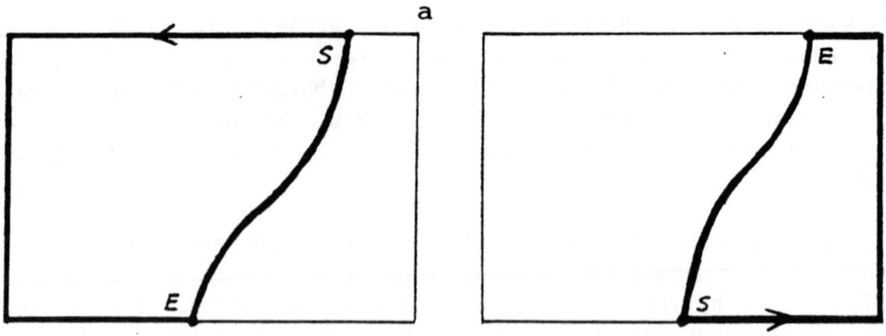

Figure 3. Two ways to close a contour through the domain boundaries (S stands for starting point of the contour and E for ending point).

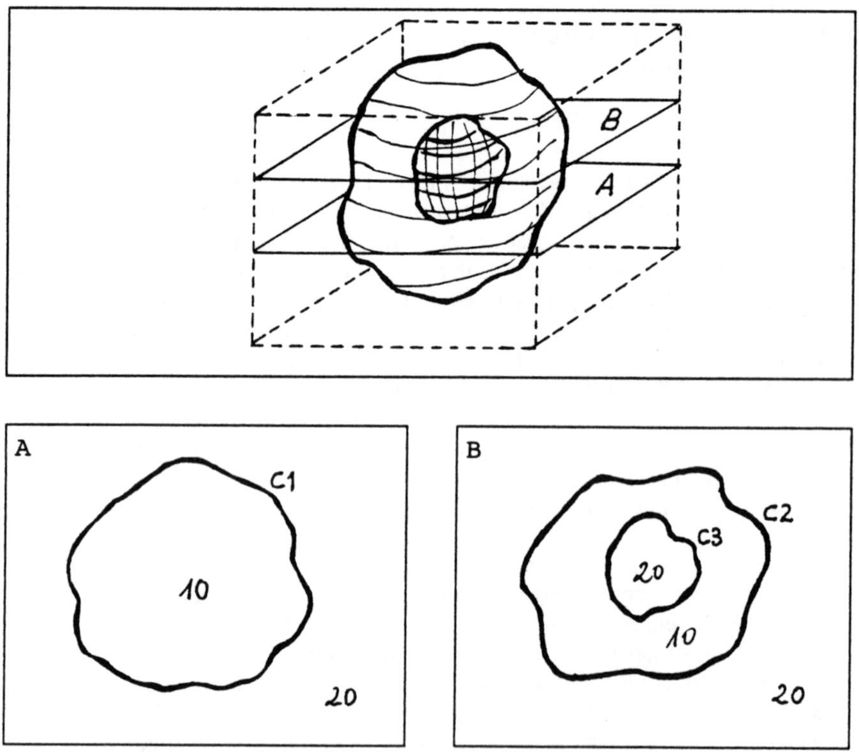

Figure 4. Presence of a bubble of 20-arbitrary units value inside a 10-a.u. isosurface (3D view and sections A and B). The value of the scalar field outside this isosurface is 20 a.u.

based on the ISO PHIGS (Programmer's Hierachical Interactive Graphics System) standard graphics.

This algorithm can be summarized as follows:

```
action: "searching isocontours in a layer"
   last = first
   while (not endsearch and not origin) do
      search origin from last
      if origin do
         last = origin
      endif
      while (not endcurves and origin) do
         search next
      endwhile
      generate curve
   endwhile
end_of_action
```

The application only deals with closed isocontours. That means that if a contour is open, it will be closed through the domain boundaries. As shown in Figure 3, there are two ways to close a contour through the domain boundaries. We have chosen the one allowing to close the contour through the shorter boundary pathway, in order to consider that most of the surface enclosed by that contour is inside the domain (b in Figure 3). However, the program is open for modifying this criterion easily. To close the contour, the boundaries are followed point after point counter-clockwise from the initial point (S in Figure 3) of the open contour to its last point (E in Figure 3). For each isocontour of a layer, the algorithm keeps the following information:

- number of points of the contour
- layer
- whether it is closed with boundary points
- size of the smaller rectangle enclosing the contour
- number of points enclosed by the contour
- map of the inner points
- value of the scalar field enclosed by the contour

At the end of this first step of the process, we have a set of layers with a series of closed contours defined in each layer. The next step is to create surfaces comprising volumes in which the scalar field has a given value, in our case, an air pollution concentration. An isosurface is created by relating an isocontour at a layer with contours of the same value at the upper layers (the algorithm works from bottom to top layer). The first condition to relate two contours is that their values at the inner scalar field should be equal. This overcomes the problem of the presence of bubbles (Figure 4). The second condition imposed by the algorithm is the need of continuity between isocontours at adjacent layers. Two isocontours at adjacent layers belong to the same isosurface if the intersection of the projected areas over a horizontal plane exceeds 70%. It is easy to modify this value or the criterion .

With this algorithm, an isosurface is by definition an arranged set of curves with indices according to increasing order of domain Z layers, with only one curve per layer. So, what happens in the case of bifurcations? The solution adopted is to consider a new isosurface (or more, depending on the number of bifurcations) from the bifurcation point up, in which the first contour is the contour where the bifurcation begins. In the case of unions, i.e. two or more isosurfaces converging on only one isosurface at the upper layer, the solution is analogous, but the last contour of the different isosurfaces corresponds to the first common contour. The volume resulting from this solution does not lack continuity.

This algorithm can be summarized as follows:

```
For each contour do
   if contour not related with contours at lower layers then
      create a new isosurface
   else
      if contour related with one only contour
         add contour to the isosurface
      else
         find the smaller surface among the surfaces related
         add contour to the smaller surface
         add contour to the other surfaces and close them
      endif
   endif
   if contour has more than one related contour at the upper
   layer then
      create new surfaces with contour
   endif
endfor
```

To visualize the isosurfaces, we use the non-uniform B-spline surface primitive of graPHIGS [12]. This primitive has an **n** x **m** matrix of real numbers as input data, where **n** and **m** are the number of control points for **u** and **v** directions, respectively. In our application, the **n** direction corresponds with the path of each contour in a layer, whereas the **v** direction corresponds with the Z direction. Then, **n** is equal to the maximum number of points of the isosurface contours and **m** the number of contours defining the isosurface. To apply this primitive, every contour should have its points arranged in the same way and its number of points should be equal to **n** (the algorithm inserts additional points by interpolation if necessary). For generating a surface interpolated by B-splines, the minimum number of data points in a direction must be greater than the surface order. The greater is the order, the smoother is the surface, but the longer is the process and more data are required. After some trials, we chose the order 2, since it guarantees continuity C1 and requires as few as three data per direction.

To visualize the topography, we also make use of this primitive from regular-spaced topographic data (terrain digital model). Some surface properties, such as derived from ambient and directional lighting effects, have been given to the surfaces by means of graPHIG's attribute subroutines.

Additional details about the program developed can be found in Ruiz [13].

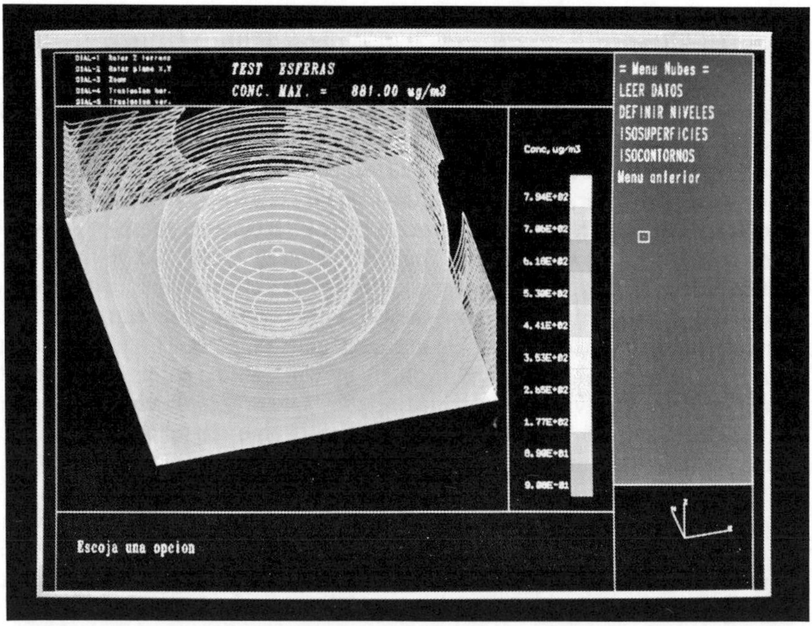

a

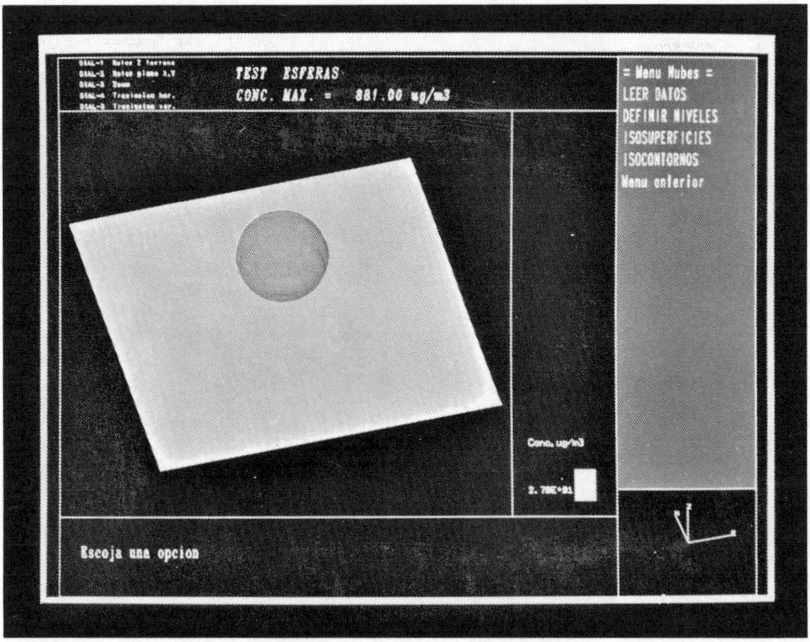

b

Photo 1. Test with a spheric field: a) isocontours; b) isosurface.

TESTS AND EXAMPLES

We have carried out some tests to evaluate the graphic abilities of the program to visualize isocontours and isosurfaces of scalar fields. Here are presented some of the pictures obtained for two of these tests. Pictures have been taken directly from a colour screen, but because of technical and economical reasons, are published in black and white. However, we think that their quality is enough to give a good idea about the visualization. The domain in both tests is a rectangular mesh of 40x40x15 grid points and the topography is flat. The first test deals with a scalar field where the isosurfaces are usually closed (spheres), such as for instance the field generated by an electrical charge. Photo 1a shows the isocontours for 10 concentric "spheres" of scalar values ranging from 0.9 to 794 arbitrary units (a.u.) in 88-a.u. increments. The interior of each "sphere" encloses a volume in which all the points have a smaller value than the isosurface value represented by the "sphere". It should be noted that the isocontours intersect the domain boundaries. Photo 1b shows the 27-a.u. isosurface wholly closed into the domain.

The other test deals with open surfaces. The scalar field has their values distributed in such a way that the isosurfaces consist of two parallel planes. We can see from Photo 2 how the planes are closed through the domain boundaries, transforming the isosurfaces into truncated pyramids. Actually, the values at the boundaries do not correspond to the isocontour value; only points inside the domain have the same value than that of the isocontour or isosurface.

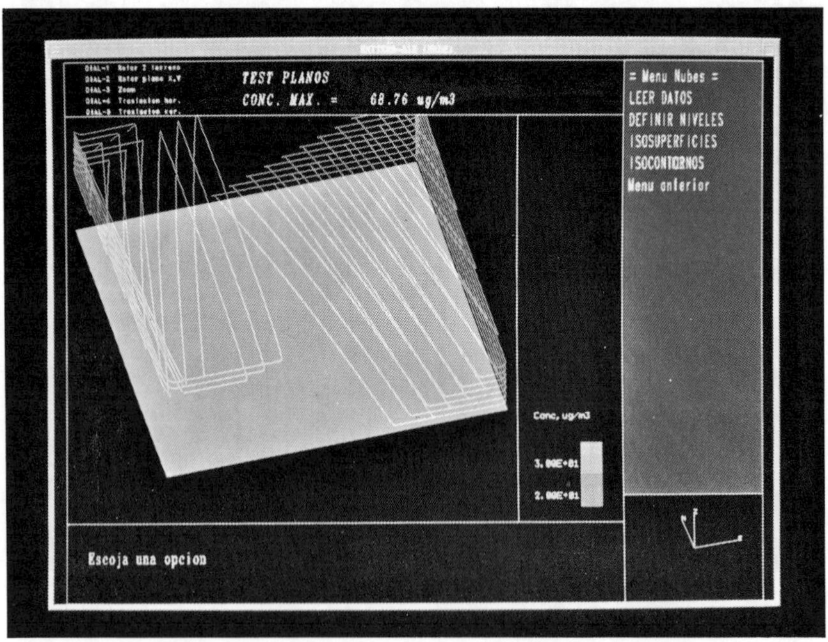

Photo 2. Test with open isocontours.

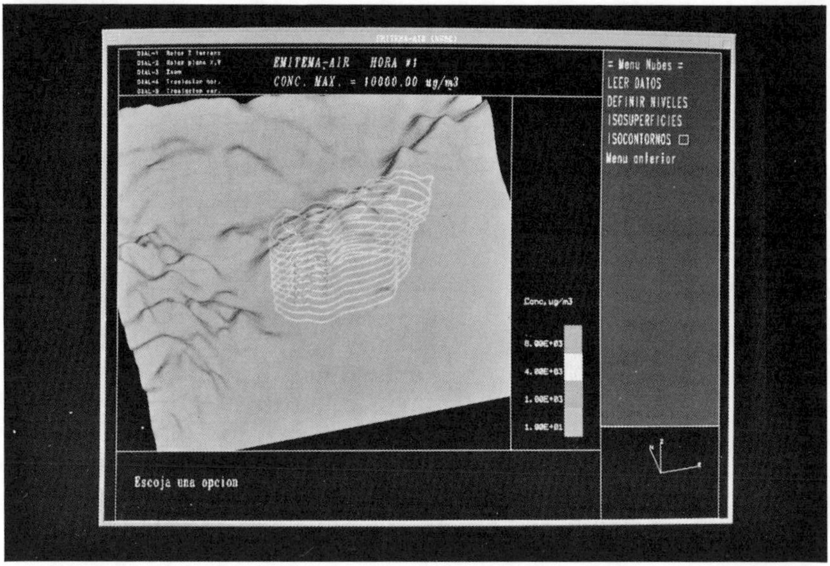

a

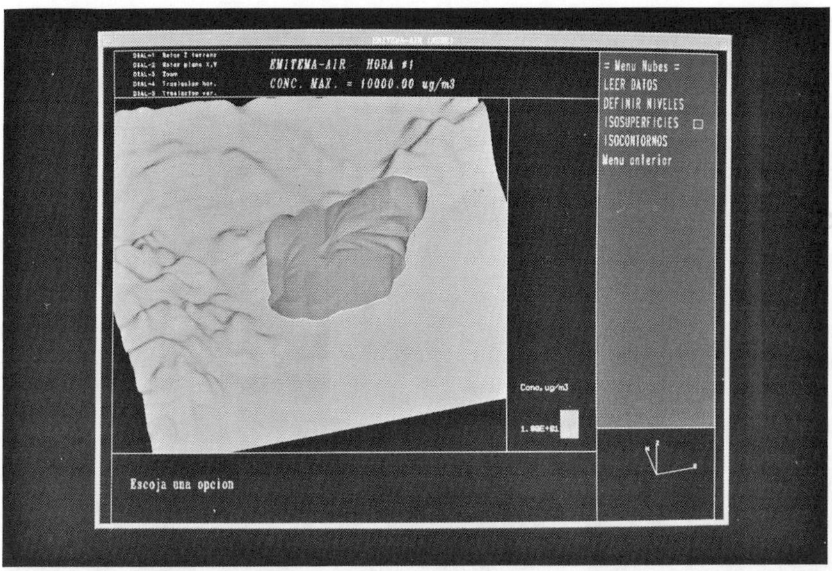

b

Photo 3. Air pollution concentration field from a power plant stack: a) isocontours of 10, 1000, 4000 and 8000 $\mu g/m^3$ SO_2; b) isosurface of 10 $\mu g/m^3$ SO_2.

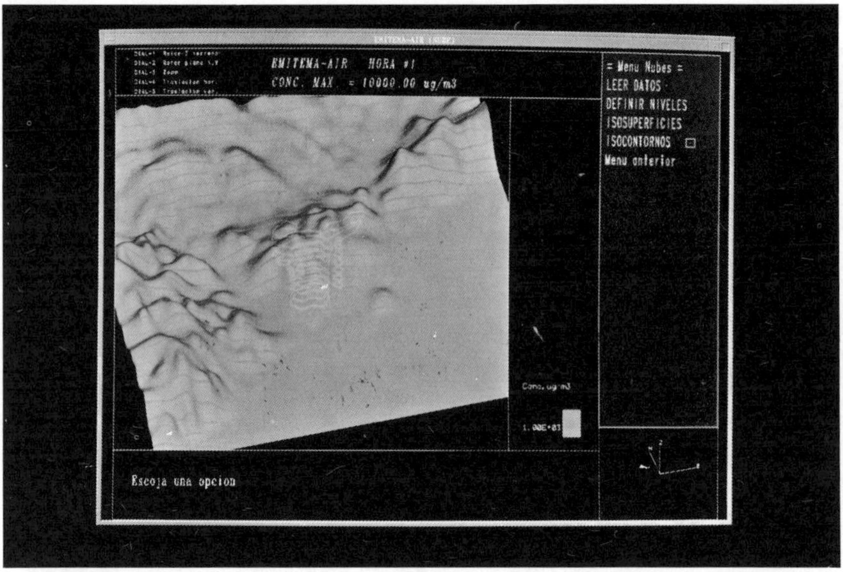

a

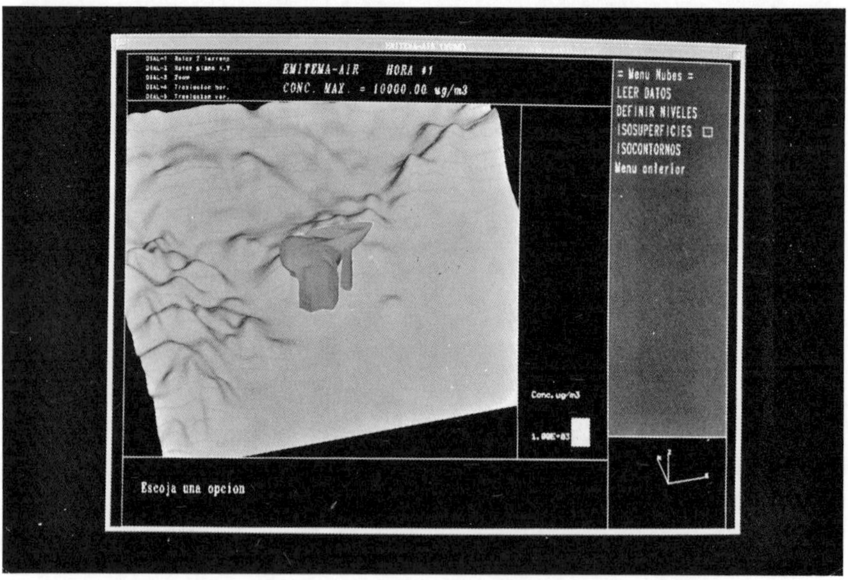

b

Photo 4. Air pollution concentration field from a power plant stack: a) isocontours of 1000 $\mu g/m^3$ SO_2; b) isosurface of 1000 $\mu g/m^3$ SO_2.

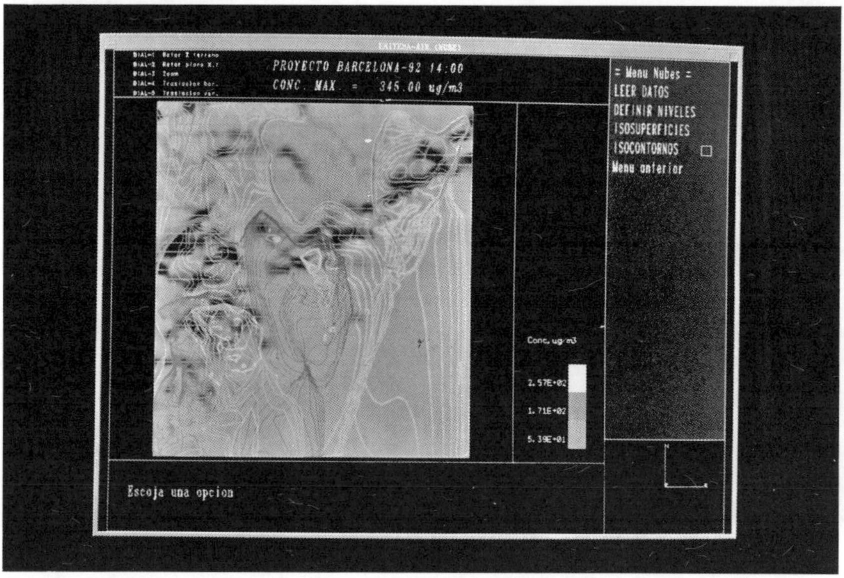

a

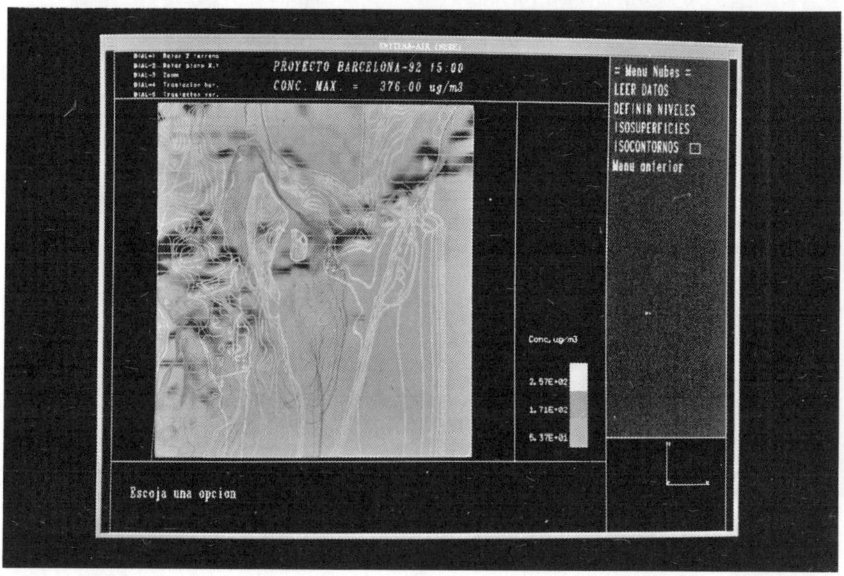

b

Photo 5. Ozone concentration field over the Barcelona area. Isocontours of 53.9, 171 and 257 $\mu g/m^3$ O_3: a) at 14:00 LST; b) at 15:00 LST (for a simulated summer day)

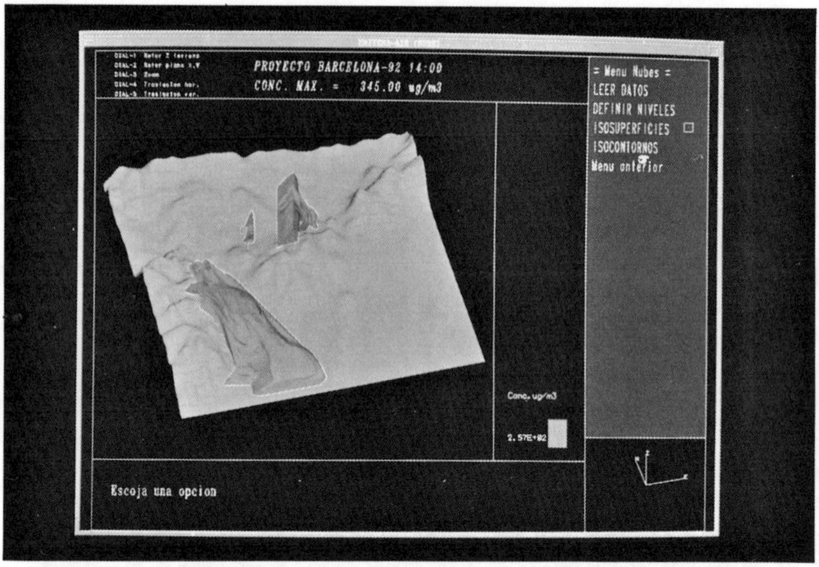

a

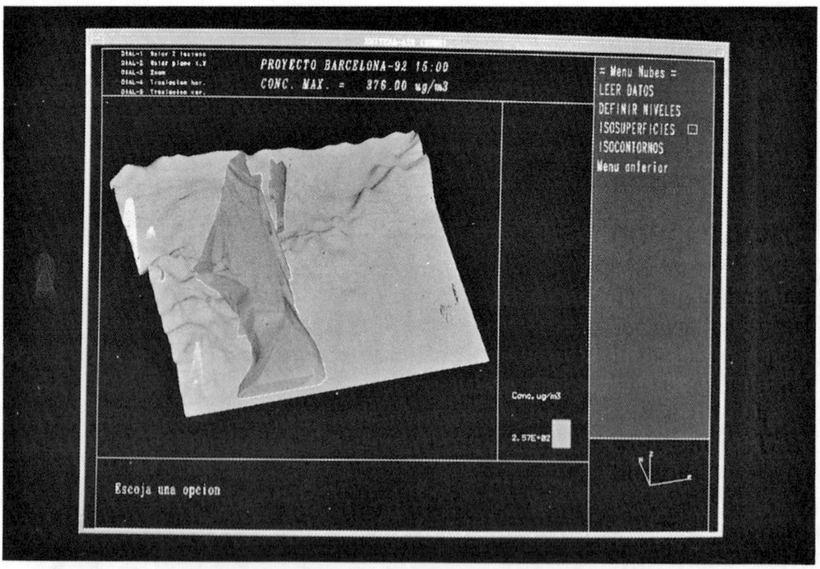

b

Photo 6. Ozone concentration field over the Barcelona area. Isosurfaces of 257 $\mu g/m^3$ O_3: a) at 14:00 LST; b) at 15:00 LST (for a simulated summer day)

Next, we present two application examples of the program by using air pollution concentration field results from dispersion models. The domain has 40x40x15 grid points as before. However, the Z layers are not horizontal but terrain-following coordinates. The topography map corresponds to the Barcelona geographical area which extends 39x39 km^2 (grid spacing is 1 km). The topographic data are given in meters above sea level.

In the first example, the cloud visualized in Photos 3 and 4 corresponds to the simulation results of a power plant plume after the first hour of operation by using the MATHEW/ADPIC model (Cremades et al. [14]). Air pollution values are expressed as $\mu g/m^3$ of SO_2. Isocontours of 10, 1000, 4000 and 8000 $\mu g/m^3$ SO_2 are visualized together in Photo 3a, from which 1000-$\mu g/m^3$ isocontours are separately represented in Photo 4a. Isosurfaces of 10 and 1000 $\mu g/m^3$ SO_2 are shown in Photos 3b and 4b. We can see from Photo 4b how two 1000-$\mu g/m^3$ isosurfaces have been formed from their isocontours.

The second example corresponds to some results of a simulation made by MARS model for the same topographic area as before (Flassak and Wortmann [15]). Air pollution values are expressed as $\mu g/m^3$ of ozone. 2D images of the isocontours corresponding to three ozone values (53.9, 171 and 257 $\mu g/m^3$) for two consecutive hours (14:00 and 15:00 LST) appear in Photo 5. The 257-$\mu g/m^3$ isosurface value is visualized in 3D in Photo 6 for these hours. At noon and in the afternoon on the Mediterranean coasts, a strong sea breeze develops, especially in summer. The surface concentration patterns of ozone between 13:00 and 15:00 LST show that high ozone concentrations are formed above the sea and are transported by the sea breeze into the Llobregat valley (located at the left-hand side in the pictures; Barcelona city is at the centre close to the shore). As a consequence surface ozone concentrations exceed the levels in Barcelona. This agrees with the visualization results obtained by the program.

CONCLUSIONS

The algorithms implemented (interpolation in a grid for tracing isocontours and serial sections for isosurfaces) are suitable for visualizing volumes closed into the domain. In the case of open volumes, the program closes them through the domain boundaries. These algorithms allow us to get a reasonable response time to work interactively. Visualizing an isosurface requires about 45 s on an IBM RISC/6000 mod. 550 workstation. This performance might be improved by using non-standard graphic libraries.

The program has been applied to visualize 3D air pollution concentration volumes, but is ready to represent any other variable from scalar fields.

The code was developed in FORTRAN. The executing program requires 13 Mb of disk space.

REFERENCES

1. Sabin, M.A. 'Contouring - A Review of Methods for Scattered Data', Mathematical Methods in Computer Graphics and Design, Ed. K.U. Brodlie Ac. Press, 1980.
2. Sutcliffe, D.C. 'Contouring over Rectangular and Skewed Rectangular Grids - An Introduction', Mathematical Methods in Computer Graphics and Design, Ed. K.U. Brodlie Ac. Press, 1980.
3. Collins, B. Data Visualization, Report UKSC 241, IBM, January 1992.
4. Keppel, E. Approximating Complex Surfaces by Triangulation of Contours, IBM Journal of Research and Development, Vol. 19 (1), pp. 2-11, January 1975.
5. Fuchs, H., Kedem, Z.M. and Uselton, S.P. Optimal Surface Reconstruction from Planar Contours, Communications of the ACM, Vol. 20 (10), pp. 693-702, October 1977.
6. Sunguroff, A. and Greenberg, D. Computer Generated Images for Medical Applications, Computer Graphics (Proceedings of SIGGRAPH '78), Vol. 12 (2), pp. 196-202, August 1978.
7. Veen, A. and Peachey, L.D. TROTS: A Computer Graphics System for Three-Dimensional Reconstruction from Serial Sections, Computers and Graphics, 2, pp. 135-150, 1977.
8. Wright, T. and Humbrecht, J. ISOSRF - An Algorithm for Plotting Iso-Valued Surfaces of a Function of Three Variables, Computer Graphics (Proceedings of SIGGRAPH '79), Vol. 13 (2), pp. 182-189, August 1979.
9. Lorensen, W. and Cline, H. Marching Cubes: A High Resolution 3D Surface Construction Algorithm, Computer Graphics (Proceedings of SIGGRAPH '87), Vol. 21 (4), pp. 163-169, July 1987.
10. Wyvill, B., McPheeters, C. and Wyvill, G. Data Structure for Soft Objects, Visual Computer, Vol. 2 (4), pp. 227-234, August 1986.
11. Wilhelms, J. and van Gelder, A. Topological Considerations in Isosurface Generation - Extended Abstract, Computer Graphics, Vol. 24 (5), pp. 79-86, November 1990.
12. The graPHIGS Programming Interface. Subroutine Reference, Version 2, Release 2.0, SC33-8194, 2nd Ed. IBM, November 1990.
13. Ruiz, S. Visualización de Isosuperficies en 3D. Aplicación a Volúmenes de Contaminantes Atmosféricos. Proyecto Fin de Carrera, UPC, 1993.
14. Cremades, L., Calbó, J., Sainz, M. and Baldasano, J.M. 'IMPACT': Instrumento Modular para la Previsión del Impacto Atmosférico de Centrales Térmicas. Memoria y Manual de Usuario, Dpt. Proyectos de Ingeniería, UPC, Barcelona, October 1989.
15. Flassak, T. and Wortmann, M. Barcelona '92 Project. Dispersion Simulations. Draft version 1.0, part 2, ITT, Universität Karlsruhe, December 1992.

From the point to the hyperbolic paraboloid: 3D CAD for teaching and profession

L.F. Meyer, H. Feliciángeli, C.F. Meyer

Department of Applied Engineering, School of Engineering, National University of Asunción, P.O. Box 765, Asunción, Paraguay

ABSTRACT

This paper presents a macro package that deals with geometric entities to be used either for teaching of Descriptive Geometry or as a professional tool mainly for engineers and architects who work with hyperbolic paraboloid structures, both at the design and at the construction phases. For all cases, visual representation may always be stereoscopic (3-D image with the help of bicolor filters), using an incorporated algorithm developed previously by one of the Authors.

The package (a library of macros designated AutoVEC), written in AutoLISP, internally uses vector analysis to handle geometric entities. The entire presentation is in AutoCAD Menu Format and allows students, and engineers or architects to directly visualize spatial drawings. Geometric data may be introduced either through the keyboard (as numerical coordinates) or by snapping in the graphics area of the screen.

The Computer Aided Teaching package may consider sets of points and functions of sets of points, giving linear geometric entities like lines, planes, or surfaces and solids as combinations of them. At any time, their projections onto the horizontal and vertical planes used in Descriptive Geometry are available.

For professional use, emphasis is given to hypar structures, in which case it is being necessary to give only the coordinates of four points, in order to obtain all the general parameters needed for design: skew axes X_0 Y_0 Z_0, vertex V, angle between Z_0 axis and original Z axis, angle between X_0 and Y_0, and warping coefficient k_0. Through the use of parametric coefficients α and β, previously developed by one of the Authors, it is possible to deal with each individual point on the surface to find useful information, such as the distance to a plane, the angle between straight generators, etc., for multiple applications. In addition to giving the components of external loads, the package also provides: total surface area, list of scaffolding and forms, and a materials estimate, like concrete and reinforcement bars, for use by builders.

INTRODUCTION

Three elements, each coming from each one of the Authors, have been combined to define the research program which originated this paper:
- Descriptive Geometry,
- 3-D Representation, and
- Hyperbolic Paraboloids.

They agreed that vector analysis and graphic computing should be their tools.

With these thoughts in mind, they set up as an objective the development of a modulated program that could either be used for educational purposes or for professional applications. So that it could really be useful, especially for students and architects, it should be able to provide a three-dimensional visualization of the geometric figures. Also, to make communications easy for untrained users, it should be possible to access through a menu format with the help of an I/O mouse.

VECTOR ANALYSIS

The parametric method developed to define an arbitrary hyperbolic paraboloid in space is vectorial in nature. Right from the start, the Authors agreed that the same approach should be used to define all other geometric entities in order to avoid a hybrid way of dealing with them, and also because it was deemed as the most appropriate.

Vectors, *i.e.*, straight lines with a specific length and direction, are defined and handled either as two ordered points or as a point and a line stemming from this point. Points and lines can be picked on the screen or defined as a list of three numbers (Cartesian coordinates or components).

AutoCAD

As is well known, AutoCAD is a Computer Aided Design user friendly program developed by Autodesk, Inc., with the objective of providing a drafting and design environment. In essence, AutoCAD can only draw straight lines and ellipses (circumferences are particular cases), surrounded by all the amenities of good draftmanship: labels, texts, dimensional lines, etc. With these entities, it is capable of drafting 3-D lines and surfaces and modeling solids. Needless to say, it can enlarge, erase, move, copy, change scales, draw lines with different widths, hatch figures, and view them from any direction whatsoever, among several other surprising capabilities. As contracts drawn up by lawyers read, "this is not an unabridged or limiting list".

What AutoCAD cannot do is draw other figures, unless you replace them by combinations of straight lines and ellipses, or deal with variables, or perform arithmetic operations, or plot arbitrary functions, or work with vectors.

AutoLISP

As is also known, AutoLISP is the language used to program for AutoCAD. It is a specific version of the Common LISP programming language that is prepared to operate from an AutoCAD drawing.

As we have seen, AutoCAD is very limited as to the geometric entities it can draw and manipulate, and is not prepared to deal with formulas or to plot them as functions, and in fact, cannot deal with the simplest of arithmetic operations or even treat numbers or variables. It is no more than the equivalent of a drawing pen, a very sophisticated one, indeed, but only that. The above mentioned operations, however, among some others, may be performed through the use of AutoLISP.

Programs in AutoLISP are but text files, that allow the user to introduce one or several programs of his own, which he can use later on as if they were AutoCAD commands. This means that the repertoir of AutoCAD commands can be augmented to any desired extent. It is also possible to modify original AutoCAD commands so that they will act in a different way than usual.

Last but not least, AutoLISP can access directly to the data base of an AutoCAD drawing, to either modify it or export information from it to an external data base.

Thence, the potentialities of AutoLISP are quite important.

AutoVEC

In this paper, the possibilities inherent in AutoLISP are used to define a set of instructions or commands prepared to deal with vectors. Such a library of macros becomes, in this way, a real programming language fit to work with vectors in a graphics environment; that of AutoCAD.

AutoVEC is a library of functions (close to seventy) oriented to define points, vectors and planes as well as handle some geometric transformations on them. AutoVEC combines perfectly the simplicity and clearness of the vectorial support of geometric properties with the flexibility and strength of lists: an outstanding data structure. This combination allows facing complex processes in a clear and simple way. AutoVEC could become an important tool for treatment of engineering problems in fields like structures, land surveying and design.

Through AutoVEC it is possible to, vectorially and in space, define or find: points and sets of points, straight lines and sets of straight lines planes, distance between two points, distance from a point to a straight line or a plane, and distance between two lines, vectors and operations with vectors, like all arithmetic operations with their components, plus scalar, vectorial and mixed products, moduli and unit vectors, angle between lines or planes, angle between a line and a plane, lines (or planes) perpendicular to planes (or lines), intersection between two straight lines, or between a straight line and a plane, or between two planes, orthogonal and general projections of a point on a straight line or a plane, decomposition of a vector in three given directions, and other operations and generations.

With it, it was quite easy to develop the pedagogical package and to redefine the parametric system of coordinates presented in 1964[15] and adapted to AutoCAD through AutoLISP in 1990[13]. Annex gives a list of all the macros developed so far.

STEREOSCOPIC IMAGE

It is a known fact that human sight (as well as that of animals in the upper echelons of the evolution ladder) is of a three-dimensional nature. Because a human being has two eyes on the same face, his mind has enough information to, automatically and involuntarily, compute distances, an operation which is the cornerstone of three-dimensional natural vision. Every point located in the field of vision is apprehended as being situated at a determined distance by itself or in relation to other points. However, as the distance between the two eyes is quite small, this property of the human mind is lost at a distance of about 25 meters, where it becomes impossible by this sole mechanism to distinguish, between two isolated objects, which one is closer or farther from the individual. In this situation, past experience regarding the visual appearance and real size of objects, or perspective of the view, as well as light and shadow of objects and their expected shape, or any other information will help the mind in its interpretation of visual reality. [It is interesting to note that not all people and societies developed these visualization aids up to the same stage; when they contemplate pictures or photographs, perspective and lighting are not as readily interpreted as it seems to be for us].

Methods of 3-D artificial vision are means to deceive the mind, to make it believe that points all located on the same plane (usually a piece of paper, or a movie or computer screen), perpendicular to the direction of the composite beam of vision, and thus at approximately the same distance from the eyes, are at different distances. We are then confronted with an optical illusion; a figure with three dimensions where there is only a drawing on a single plane.

This effect is achieved by the artifice of presenting two images (called anaglyphs) of the same object, as if they belonged one to each eye, displaced from one another the same distance than the central or polar projections of the object, taking each eye as a center or pole, would have on the plane where they are. See Photographs 1 to 5.

It is then necessary to make each eye see only one of the two images, the one that, intersecting with the other image from the other eye, will produce the desired virtual image of the object. Several devices have been used for this purpose, like polariscopes or bicolor filters, this last being the case in this

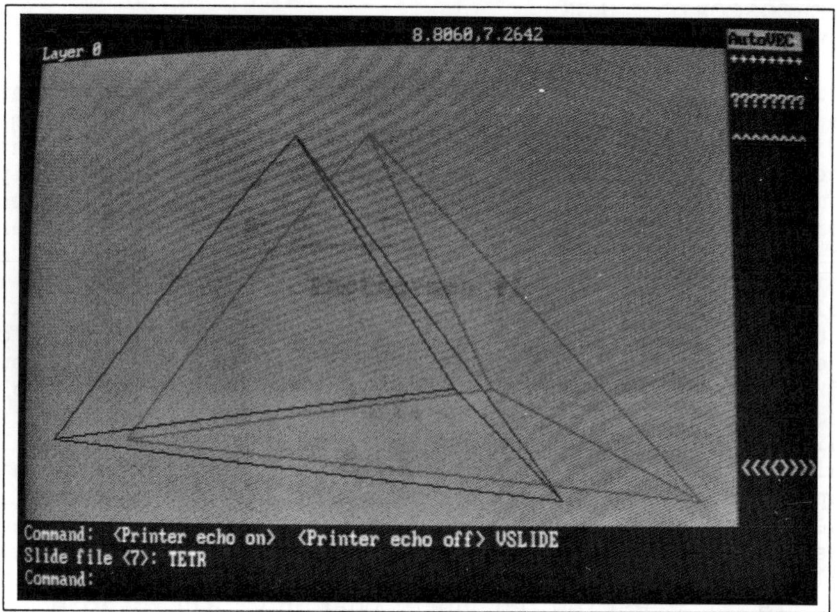

Photograph 1 - Stereoscopic Projection of a Tetrahedron

paper.

Unexpectedly, color selection was an operation of its own; very bright colors would not completely interfere with one another, or give a black or dark grey composition. For this reason, sepia and light blue (colors #10 and #12 in AutoCAD Version 11) were used after several trials excluded other colors from being used. These two colors are very close to being complementary, thus rendering dark grey lines which pop up readily.

3-D images, of course, can be formed either behind or in front of the screen. The quality obtained is quite good, even though some people, resisting the deception, perhaps, might require a few seconds to adjust their eye muscles when they start using the program. Some might require a certain training.

The algorithm developed by one of the Authors[9] some years ago, was incorporated into macros, and its presence is latent, available at any time on request. This approach is quite different from the one based directly on stereoscopic screens.

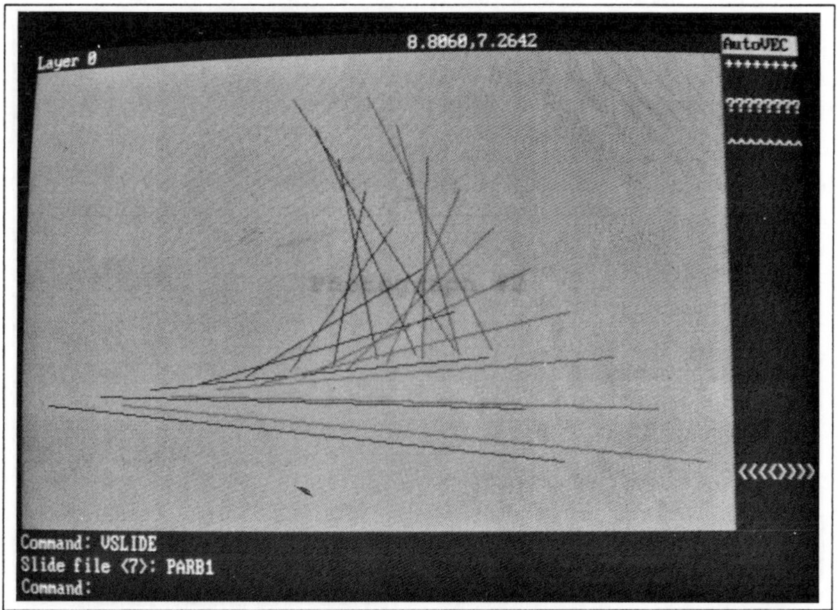

Photograph 2 - Stereoscopic Projection of a Hyperbolic Paraboloid

To efficiently deal with spatial figures, it is usually necessary to have developed a spatial imagination. It is a known fact that Descriptive Geometry helps develop spatial imagination[10]. With 3-D images, this spatial imagination is no longer necessary, but ironically, it helps its development even more[18].

COMPUTER AIDED TEACHING: THE PEDAGOGIC MODULE

The pedagogic module was planned to assist in the teaching of Descriptive Geometry, subject matter of the lectures of one of the Authors at the School of Engineering of the University. Because it is very easy to use, parts of it could also be very useful in technical schools at the secondary level or for adult education, in the teaching of drafting, mechanics, construction, carpentry, etc.

The module starts by presenting the two orthogonal planes, one horizontal and the other vertical, defining the four dihedral angles in space, as deemed originally by the French scientist Gaspar Monge, famous engineer and *revolutionaire*, in the so called Monge Projection Method[16], published in 1799. Any point in space, can then be projected onto any of the two planes by a line

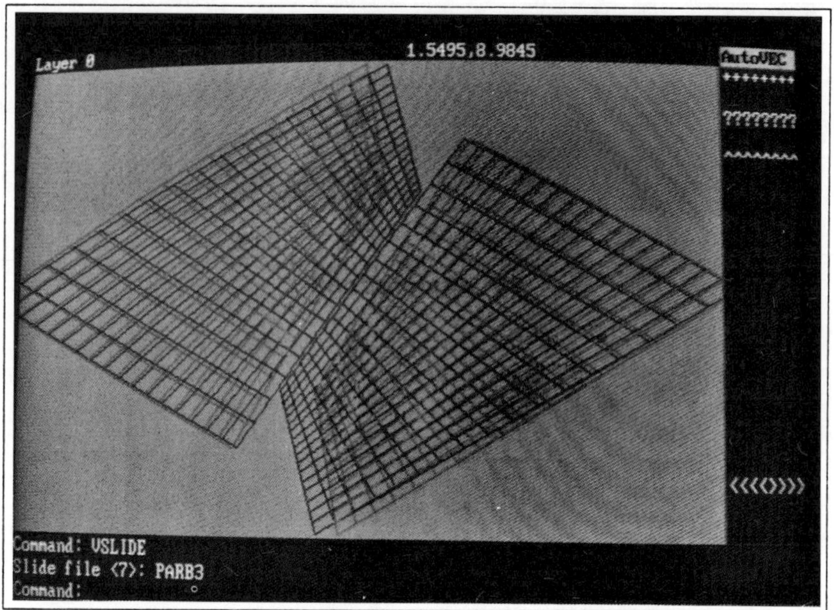

Photograph 3 - Stereoscopic Projection of two Symetrical Hyperbolic Paraboloids

perpendicular to it, and in a similar way onto the other. These projections, called orthogonal projections, constitute a particular case of parallel projections.

In an inverse way, it is possible to reconstruct the original point if its orthogonal projections are known.

To make things more convenient without losing information, it is possible to make both planes overlap, by revolving the horizontal plane 90°, as the vertical plane coincides with the screen. [In a book or writing piece of paper, it is usually the horizontal plane the one kept fixed]. This methodology is what Descriptive Geometry is about.

If there is more than one point, the same rules apply.

With the definition of two points, we can have a straight line; with a line and a point, or with two intersecting lines, we get a plane; two non-intersecting non-parallel lines, define a hyperbolic paraboloid. Combinations of these geometric entities will yield many more possibilities, all available

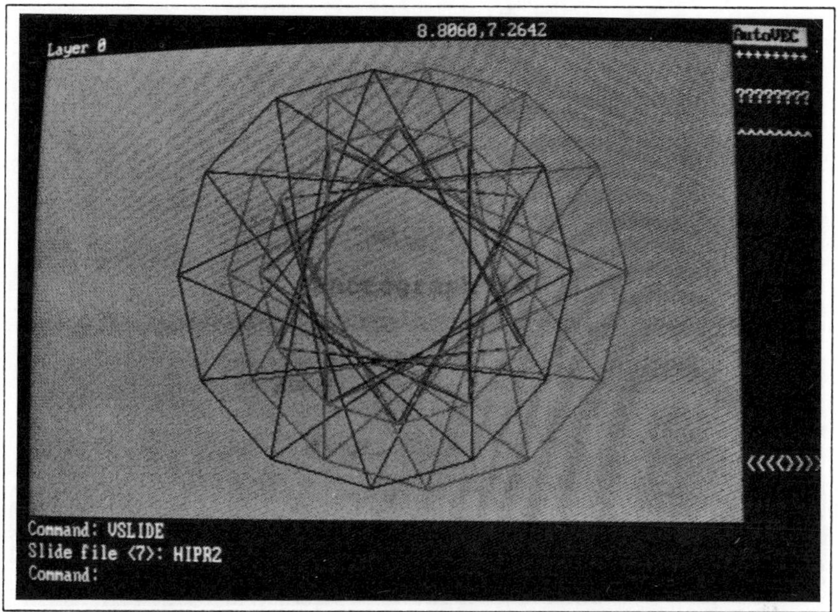

Photograph 4 - Stereoscopic Projection of an Hyperboloid of Revolution - Axial View

through the pedagogic package.

THE PROFESSIONAL MODULE

For convenience of use and description, the Professional Module is divided in two parts, the first one, much more developed than the other deals with hyperbolic paraboloids. The second is an open box, bound to be filled up with future developments. For the time being, only unsought applications have been found, not being reported here.

Hypar Structures
Hyperbolic paraboloids, usually shortened to hypars or even HP, are outstanding shapes to be used in structures made up of a number of different materials, but especially of reinforced concrete.

Geometric Considerations This surface may be considered to be generated by a parabola, called parabolic generator, travelling always parallel to a plane, called director plane, on another parabola of the other sign, called parabolic director, located on the plane containing the axis of the first parabola and perpendicular to the director plane. Interchanging the director and the generator parabolas, the same surface

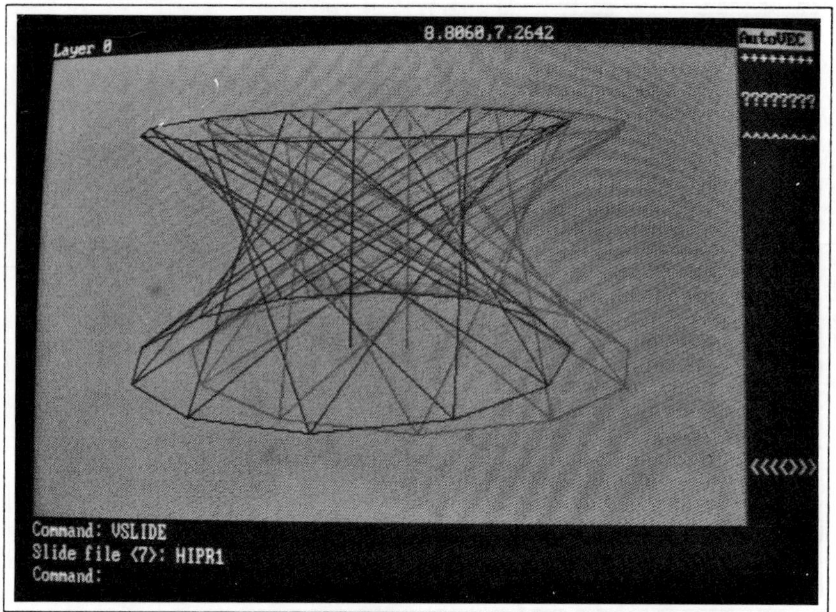

Photograph 5 - Stereoscopic Projection of an Hyperboloid of Revolution - Side View

is obtained. This way of generating the surface gives rise to the known canonical formula of the hyperbolic paraboloid:

$$z = \frac{x^2}{a^2} - \frac{y^2}{b^2}$$

Another way of generating the surface is to make a straight line, called straight generator, travel always parallel to a plane, called director plane, intersecting two non-coplanar straight lines, called straight directors. If two positions of the straight generator are taken as straight directors, and any of the directors is taken as a generator travelling parallel to the plane parallel to both original directors, the same shape is obtained. Surfaces generated by a travelling straight line are called ruled surfaces; as hypars comply with this condition in both directions, they are ruled surfaces *par excellence*. This way of considering the hypar gives rise to the also well-known formula

$$z = kxy$$

where k is the warping coefficient.

This way of generating the hyperbolic paraboloid is also the basis for the parametric definition of the surface presented for the first time in 1964 by one of the Authors, giving rise to parameters α and β, used in the derived formulas (a mathematical demonstration of this way of considering hypars, and the resulting principal characteristics are available upon request to the Authors).

For the canonical formula, axes of coordinates are defined at the vertex of the hyperbolic paraboloid, which is that point on the surface being at the same time the vertex of both parabolas, the director and the generator, mentioned in the first definition. Two axes are the tangents to both parabolas at the vertex and the third one is fixed by the intersection of the planes containing them. A classical Cartesian system of coordinates, with straight angles among the axes, is the result. It is, therefore, a tri-orthogonal system.

When the hypar is considered as a ruled surface, the coordinate system is defined by the two straight generators passing by the vertex, therefore contained in the surface, (which are the X_0 and Y_0 axes) and the perpendicular to both (which is the Z_0 axis). As generators do not need to form a straight angle, this is a bi-orthogonal system in the general case, being called skew system, for this reason.

Another coordinate system used with surfaces is the Gaussian system based on the lines of maximum curvature. In the case of the hyperbolic paraboloid, these lines of maximum curvature are the parabolas passing through the vertex. Measurements are made, as in the two previous systems, in length dimensions, being the unit of length an arbitrary length chosen for that purpose. In this case, the formulas are also parametric. [Gauss studies also gave origin to a general classification of surfaces, depending on the ratio of the radii of curvature of the lines of maximum curvature on any point. Hypars are said to have negative Gaussian curvature].

The Meyer System To define the surface by the formulas based on parameters α and β mentioned previously, an entirely new system of coordinates was developed[15], having one characteristic similar to the skew system and one similar to the Gaussian system. The Meyer system has two families of straight axes contained in the surface, just as the X_0 and Y_0 axes of the skew system, and measurements are taken on the surface, as in the Gaussian system. However, the two big differences with this system are that the selected "axes" are not the lines with maximum curvature, but, on the contrary, the ones with zero curvature, and that the values of the coordinates do not represent lengths but are adimensional values. Furthermore, these adimensional values are obtained by a fraction where the denominator is a different length for every straight generator.

The system is based on the property that all points located on a straight generator can be associated with a fixed value of a parameter α, defined as the ratio of the distance, measured on a straight generator of the other family, between that straight generator and another straight generator of its own family chosen as origin ($\alpha = 0$) to the distance, measured on the same straight generator, between the origin and another straight generator of its family chosen as unity ($\alpha = 1$). The same is true interchanging families of straight generators with a parameter β. In this way, by the intersection of two straight generators a point on the surface of the hyperbolic paraboloid is bi-univocally defined; *i.e.*, any point P on the surface is associated with just one pair of values of α and β, and this pair of values of α and β is associated only to that point P.

Probably the most important property of these parameters is that they are invariant in any projection parallel to the Z_0 axis whatsoever, allowing the designer to work in the plane that best fits his or her needs, usually on a horizontal plane, *i.e.*, a plan view of the structure. [This property was emphasized by an article[22] and Chapter 2 of a book[4], written by the main author of the article, presenting the method as if it were theirs, with no mention of its origin].

This property is also very useful for finding heights of points or distances of them to given planes, or finding out if given points belong to the surface, or locating intersections between special surfaces. This possibility can be used with advantage when combinations of HP shells are desired[8].

Historically, the use of hypars was usually restricted to cases in which their Z axis was vertical or had a controlled rotation, because otherwise it was quite complicated to find the position of that and the other axes, as well as the coefficients that define the surface and are necessary for structural design. In the pre-computer years, it used to take a long, long time.

With the Meyer parametric method, the time required for the calculations was reduced from about a month to less than half a working day. With modern computers, it is a matter of minutes or even seconds, besides allowing interaction which benefits design capabilities.

Structure of the Package The professional package here presented starts by requesting four ordered points on the surface to be defined, with the condition that they belong to two straight generators in each direction. [The designer should clearly know which are these straight generators, because four points belonging to two straight generators in each direction, can produce three different hypars, as this is the number of ways that, topologically, four points can be connected by lines]. This is all the information that is needed in order to define all the geometric characteristics of the hypar.

With these points, which can be introduced either by their coordinates or by picking them on the screen, the package can immediately provide the following information:

- axes X_0, Y_0 and Z_0 and their unit vectors x_0, y_0 and z_0;
- angle ϕ between Z_0 and the original third coordinate Z, usually vertical;
- warping coefficient k; and
- angle ω between axes X_0 and Y_0.

In a second stage, it is possible to do any of the two following inverse operations:

- given a point on the surface, find out the corresponding values of α and β, or
- given the values of α and β, find the point.

Knowing the point, it is possible to find the angle between the straight generators passing through it, or the local coordinate system u, v and w, defined by those two generators and the normal to both, or the angle between this last axis and the equivalent axis of the original coordinate system.

As far as HP shell structures are concerned, everything we have seen so far is pure geometry, no structural calculations. From this last point of view, various methods have been proposed, the best known of them all being the famous article[3] by Felix Candela, which provides general formulas for membrane stresses, quite adequate considering the capability of this type of structure to support any kind of loads with stresses contained in its surface. As early as 1964, one of the Authors developed in conjunction with Julio Damy a computer program based on this method, and published it in a companion paper[6] to the one in which he presented his parametric method[15]. Many more, by other programmers, followed along the years. In order to use this method of analysis and design, there is only one piece of information missing: components of the external loads according to the three axes of the skew system of coordinates. They can be easily produced, using the professional package.

All this is useful from a design standpoint. From the construction point of view, after calculations have been done according to the last paragraph or any other method, the package can also render quite a few useful things, like total surface area, list of scaffolding and forms, and a materials estimate, like concrete components and reinforcing bars.

Advantages of HP Shells Apart from the fact that, like any other shell, the hypar has the built-in capacity to withstand loads by its shape and not by the amount or strength of the materials used, there are some distinct advantages in using it as the shape for concrete shells:

- being a surface of double curvature, it has the capacity to sustain loads, even concentrated loads, in a membrane state.
- being a ruled surface, its formwork for the casting of concrete can be made up of just straight elements (boards and joists) and not curved ones as in most other shells, with the corresponding reduction of cost in formwork materials and personnel.
- resulting stresses are usually quite low, allowing very thin surfaces (in the order of 4 cm of concrete) and a very small amount of reinforcement (usually only temperature reinforcement in most of the area) almost independently of span, with the corresponding economy of reinforced concrete materials and personnel.
- different cutouts from the surface according to either straight or curved borders, and combinations of them, may create countless different structures, with an almost infinite diversity of esthetic and functional shapes.
- varying the values of the warping coefficient k and the angle ω between the X_0 and Y_0 axis, it is

possible to obtain surfaces with quite different structural properties and visual appearance, great variations being obtained also just by changing the position of the Z_0 axis.
- proper combinations of these characteristics may be so powerful that with just this surface it is possible to adjust a structure to almost any practical situation.

CONCLUSIONS

After a brief introduction of vector analysis, as well as AutoCAD and AutoLISP languages, a presentation is made of a library of functions, written in AutoLISP, prepared to deal with vectors within a graphical environment. This library of macros, called AutoVEC by the Authors, provides a simple but powerful and flexible means to manipulate vectors and other entities defined in a vectorial way. A list of its near seventy commands is presented in Annex. A detailed description of the way each one of them operates is available upon request. Although AutoVEC can be used almost by itself, those who master AutoCAD can benefit most.

Through an algorithm, present at all times, stereoscopic images of figures are available for the asking. This becomes very powerful especially for the pedagogic module but also for architects and engineers using the professional module. Although a few people need some seconds to adjust their sight, this program works very well making it possible to see spatial figures from any point of view. An unexpected by-product of three-dimensional representation is the development of what is called spatial imagination on the part of the users.

The pedagogic module is tailored to the teaching of Descriptive Geometry. It conducts students by the hand from an explanation of Monge's projection method to the representation of such a complex spatial surface as the hyperbolic paraboloid, allowing a thorough understanding of this most interesting discipline. This computer aided teacher for college students functions so well that it had outstanding results also in a test conducted with school children. However, the development of this package made the Authors meditate about the *raison d'être* of Descriptive Geometry as a university course in this modern world of computers.

The professional module is, for the time being, mainly oriented to shell structures whose surface is a hyperbolic paraboloid. This geometric surface is considered in a parametric way that resulted in the development of a coordinate system different from those that are used regularly, with important simplifications. Given only four points of the surface arbitrarily placed in space, the package, besides representing it three-dimensionally, provides all geometrical data needed to work the structural calculations, besides rendering the components of external loads. In the construction phase, it also gives very practical information. The possibility to try several surface configurations and see them immediately is very useful for architects in the decision phase.

The whole package is presented in menu format.

ACKNOWLEDGEMENTS

This project, known as FCF/01/92, was funded by the Central Research Fund (*FCI*), administered by the Department for the Development of Research (*DDI*) of the Multidisciplinary Center for Technological Research (*CEMIT*) of the National University of Asunción.

Physical premises and computer facilities were provided by the Department of Applied Engineering of the School of Engineering of the same University, whose kind members went "beyond the call of duty" day after day. The authors heartily thank them.

REFERENCES

1.- Candela, F. 'General Formulas for Membrane Stresses in Hyperbolic Paraboloid Shells' *ACI Journal*, Vol. 57, No. 4, Detroit, USA, 1960.
2.- Christiansen, J. (Ed). *Hyperbolic Paraboloid Shells - State of the Art*, ACI Special Publication SP-110, Detroit, USA, 1988.
3.- Damy, J.E. 'Computación Electrónica de los Esfuerzos de Membrana en Paraboloides Hiperbólicos' *Revista IMCYC*, Vol. 2, No. 9, México, México, 1964.
4.- Faber, C. *Candela - The Shell Builder*, Reinhold, New York, USA, 1963.
5.- Feliciángeli, H. 'Imágenes Estereoscópicas producidas por Computadora' *Tercer Panel de Informática*, Universidad Católica Nuestra Señora de la Asunción, Asunción, Paraguay, 1984.
6.- Gordon, V.O. and Sementsov-Oguiyevski, M.A. *A Course in Descriptive Geometry*, Mir Publishers, Moscow, Russia, 1980.

7.- López Fernández, J. and Tajadura Zapirain, J.A. *AutoCAD Avanzado - Versión 11*, McGraw Hill, Madrid, España, 1992.
8.- Meyer, L.F. 'Computer-Aided Design of HP Shells' *American Concrete Institute Annual Convention*, Boston, Massachusetts, USA, 1991.
9.- Meyer, L.F. 'El Restaurant del Lago en el Nuevo Chapultepec' *Proceedings of the Jornadas Sudamericanas de Ingeniería Estructural*, Caracas, Venezuela, 1967.
10.- Meyer, L.F. 'Una Forma Paramétrica de definir el Paraboloide Hiperbólico' *Revista IMCYC*, Vol. 2, No. 8, México, México, 1964.
11.- Monge, G. *Geometrie Descriptive*, Paris, France, 1799.
12.- Pál, I. *Geometría Descriptiva con Proyecciones Estereoscópicas*, Ediciones Aguilar, Madrid, España, 1965.
13.- Principe Júnior, A.R. *Noçôes de Geometria Descritiva*, Vol. 1 and 2, Nobel, Sâo Paulo, Brasil, 1986.
14.- Pritulenko, P.V. *Plane Figures and Sections: How to Construct them given Specific Conditions*, Mir Publishers, Moscow, Russia, 1980.
15.- Santaló, L.A. *Vectores y Tensores con sus Aplicaciones*, EUDEBA, Buenos Aires, Argentina, 1981.
16.- Scordelis, A.C., Ramirez, H.D. and Ngo, D. 'Membrane Stresses in Hyperbolic Paraboloid Shells having an Arbitrary Quadrilateral Shape in Plan' *ACI Journal*, Vol. 67, No. 1, Detroit, USA, 1970.
17.- Tajadura Zapirain, J.A. and López Fernández, J. *AutoLISP - Versión 11*, McGraw Hill, Madrid, España, 1991.
18.- Tedesco, A. 'Shell at Denver - Hyperbolic Paraboloidal Structure of Wide Span' *ACI Journal*, Vol. 32, No. 4, Detroit, USA, 1960.
19.- Thomas, R.M. *Encyclopedia AutoCAD - Release 11*, SYBEX, San Francisco, USA, 1991.

ANNEX: List of AutoVEC Macros

I. Operations with Vectors

1. DFVC (P1 P2)	DEFINES A VECTOR
2. MDVC (V)	MODULUS OF A VECTOR
3. VRSR (V)	UNIT VECTOR IN ONE DIRECTION
4. OPVC (V1 OP V2)	ARITHMETIC OPERATION BETWEEN VECTORS
5. ESCV (LMB V)	PRODUCT OF A SCALAR AND A VECTOR
6. PESC (V1 V2)	SCALAR PRODUCT OF VECTORS
7. PVCT (V1 V2)	VECTOR PRODUCT
8. PMXT (V1 V2 V3)	MIXED PRODUCT OF VECTORS

II. Relationship between Points and Straight Lines

1. DFRC1 (P V)	DEFINITION OF A STRAIGHT LINE GIVEN A POINT AND A DIRECTION
2. DFRC2 (P1 P2)	DEFINITION OF A STRAIGHT LINE GIVEN TWO POINTS
3. PRMR (R)	PARAMETERS DEFINING A STRAIGHT LINE (A PLANE)
4. PRC1 (R LMB)	POINT ON A STRAIGHT LINE (IN PARAMETRIC FORM)
5. PRC2 (R X)	POINT ON A STRAIGHT LINE (IN ANALYTIC FORM)
6. DPRC (P R)	DISTANCE FROM A POINT TO A STRAIGHT LINE
7. DSTR (R1 R2)	DISTANCE BETWEEN TWO STRAIGHT LINES

III. Relationship between Points, Straight Lines and Planes

1. DFPL (P V)	DEFINITION OF A PLANE GIVEN ITS NORMAL AND A POINT
2. DFPL2 (P R)	DEFINITION OF A PLANE GIVEN A POINT AND A STRAIGHT LINE
3. DFPL3 (P1 P2 P3)	DEFINITION OF A PLANE GIVEN THREE POINTS
4. DPPL (P PL)	DISTANCE FROM A POINT TO A PLANE
5. PPLN (PL X Y)	POINT ON A PLANE
6. INRP (R P)	INTERSECTION OF A STRAIGHT LINE AND A PLANE
7. INPL (PL1 PL2)	INTERSECTION OF TWO PLANES

IV. Angular Relations

1. ANGV (V1 V2)	ANGLE BETWEEN VECTORS
2. RDGR (FI)	CONVERSION OF RADIANS TO SEXAGESIMAL DEGREES
3. GRRD (GR)	CONVERSION OF SEXAGESIMAL DEGREES TO RADIANS
4. ANGR (R1 R2)	ANGLE BETWEEN TWO STRAIGHT LINES (PLANES)
5. ANRP (R1 PL)	ANGLE BETWEEN A STRAIGHT LINE AND A PLANE

V. Projections of Points

1. XRCN (P R)	ORTHOGONAL PROJECTION OF A POINT ON A STRAIGHT LINE
2. XPLN (P PL)	ORTHOGONAL PROJECTION OF A POINT ON A PLANE
3. XRCG (P R PL)	GENERAL PROJECTION OF A POINT ON A STRAIGHT LINE
4. XPLG (P PL R)	GENERAL PROJECTION OF A POINT ON A PLANE
5. XPPL (CP P PL)	CENTRAL PROJECTION OF A POINT ON A PLANE
6. XFMPL (C FPT PL)	CENTRAL PROJECTION OF A FAMILY OF POINTS ON A PLANE
7. XFMST (VS FPT)	STEREOGRAPHIC PROJECTION OF A FAMILY OF POINTS

VI. Data Generation

1. FMPT ( )	MANUAL GENERATION OF A FAMILY OF POINTS
2. FMPT1 ( )	DEFINITION OF A FAMILY OF POINTS
3. SUBFM ( )	DEFINITION OF A SUB-FAMILY OF POINTS
4. NFM (FPT SQ)	AUTOMATIC GENERATION OF A FAMILY OF POINTS
5. NCFM (FPT)	NUCLEUS OF A FAMILY OF POINTS
6. MPRC1 (R LMB N)	GENERATION OF POINTS ON A STRAIGHT LINE (PARAMETRICALLY)
7. RTPE (P1 EJ FI)	ROTATION (IN SEXAGESIMAL DEGREES) OF A POINT AROUND AN AXIS
8. RTP1 (P1 EJ FI)	ROTATION (IN RADIANS) OF A POINT AROUND AN AXIS
9. PCIRC (P EJ FI N)	GENERATION OF POINTS ON A CIRCUMFERENCE

VII. Operations between Points, Straight Lines and Families of Points

1. INFM (P FPT)	BELONGING OF A POINT TO A FAMILY
2. OPFM (FP1 OP FP2)	OPERATION BETWEEN FAMILIES OF POINTS
3. HMTC (C FC FPT)	HOMOTECY OF A FAMILY OF POINTS
4. HOMTC ( )	HOMOTECY OF A FAMILY OF POINTS
5. TRNS (VC FPT)	TRANSLATION OF A FAMILY OF POINTS
6. TRANS ( )	TRANSLATION OF A FAMILY OF POINTS
7. RTFM (FPT EJ FI)	ROTATION (IN SEXAGESIMAL DEGREES) OF A FAMILY OF POINTS AROUND AN AXIS
8. RTPP (FPT EJ FI)	ROTATION (IN RADIANS) OF A FAMILY OF POINTS AROUND AN AXIS
9. ROTFM ( )	ROTATION OF A FAMILY OF POINTS

VIII. Family of Straight Lines

1.	HZRC (FPT CH)	FAMILY OF CENTRAL STRAIGHT LINES
2.	HZRP (FPT VC)	FAMILY OF PARALLEL STRAIGHT LINES
3.	FMRC (FP1 FP2)	FAMILY OF NON-PARALLEL NON-CENTRAL STRAIGHT LINES
4.	IHRC (FPT CH PL)	INTERSECTION OF A FAMILY OF CENTRAL STRAIGHT LINES WITH A PLANE
5.	IHRP (FPT VC PL)	INTERSECTION OF A FAMILY OF PARALLEL STRAIGHT LINES WITH A PLANE
6.	IFMR (FP1 FP2 PL)	INTERSECTION OF A FAMILY OF NON-PARALLEL NON-CENTRAL STRAIGHT LINES WITH A PLANE

IX. Plotting of Straight Line and Polygonals

1.	SGMN (P1 P2)	PLOTTING OF A SEGMENT OF A STRAIGHT LINE GIVEN TWO POINTS
2.	SGM1 (P VC)	PLOTTING OF A SEGMENT OF A STRAIGHT LINE GIVEN A POINT AND A VECTOR
3.	POLGA (FPT)	OPEN POLYGONAL
4.	SBPOL (FPT SQ)	OPEN SUB-POLYGONAL
5.	POLGC (FPT)	CLOSED POLYGONAL

AUTHORS' INDEX

BIN TRAVERLER FORM

Cut By: Laura Rivas #11 **Qty** 15 **Date** 10.19.20

Scanned By:____________ **Qty**______ **Date**________

Scanned Batch ID's

__________ __________ __________

Notes / Exceptions

____